EXPLORING THE COSMOS

I leave to children the long, long days to be merry in, in a thousand ways, and the Night and the Moon and the train of the Milky Way to wonder at; and I give to each child the right to choose a star that shall be his, and I direct that the child's father shall tell him the name of it, in order that the child shall always remember the name of that star after he has learned and forgotten his astronomy.

Excerpt from last will and testament by unknown author

EXPLORING THE COSMOS

Third Edition

LOUIS BERMAN
University of San Francisco

J. C. EVANS
George Mason University

Little, Brown and Company
Boston   Toronto

Library of Congress Catalog Card No. 79-90887

ISBN 0-316-091766

9 8 7 6 5 4 3

HAL

Published simultaneously in Canada by Little, Brown & Company (Canada) Limited

Printed in the United States of America

ACKNOWLEDGMENTS

We gratefully acknowledge permission to use the following material.

Page ii: background photograph from Hale Observatories.

Chapter 1

Page 1: The Bettman Archive; background photograph by permission of Harvard College Observatory, Cambridge, Mass. *Fig. 1.1:* left and right, NASA. *Fig. 1.2:* Hale Observatories. *Fig. 1.3:* photograph from Hale Observatories. *Fig. 1.4:* left, Lick Observatory; right, Hale Observatories. *Fig. 1.5:* Hale Observatories. *Fig. 1.6:* Department of Environment, England. *Fig. 1.7:* left, The Bettman Archive; right, courtesy of The American Museum of Natural History. *Page 9:* from Robinson Jeffers, "Margrave" in *Thurso's Landing,* pp. 135–147. Copyright, 1932 by Robinson Jeffers. Reprinted by permission of Random House, Inc. *Fig. 1.8:* left, courtesy Hayden Planetarium; right, Jay M. Pasachoff.

Chapter 2

Page 13: Cambridge University Library. *Fig. 2.5:* Lick Observatory. *Fig. 2.6:* photographs from Lick Observatory. *Page 26:* Culver Pictures. *Page 29:* Brown Brothers.

Chapter 3

Page 33: Cambridge University Library; background photograph by permission, Harvard College Observatory, Cambridge, Mass. *Page 36:* courtesy Yerkes Observatory. *Page 42:* The Bettmann Archive. *Fig. 3.7:* courtesy Yerkes Observatory. *Fig. 3.8:* Hale Observatories.

Chapter 4

Page 51: American Institute of Physics Library, Margaret Russell Edmondson Collection. *Fig. 4.1:* The Bettman Archive. *Fig. 4.5:* from Paul G. Hewitt, *Conceptual Physics,* 2nd ed., p. 455. Copyright © 1974, 1971 by Little, Brown and Company (Inc.). *Fig. 4.7:* MIT Department of Physics, Professor A. P. French and Douglas Ely. *Fig. 4.12:* Lick Observatory. *Fig. 4.20:* Hale Observatories.

Chapter 5

Page 76: Mount Wilson and Palomar Observatories. *Fig. 5.2:* University of San Francisco, Department of Physics. *Fig. 5.3:* Leiden Observatory. *Fig. 5.4:* top three photographs, Sacramento Peak, Association of Universities for Research in Astronomy; bottom photograph, The Kitt Peak National Observatory. *Fig. 5.6:* The Kitt Peak National Observatory. *Fig. 5.7:* top, from left to right, Science Museum, London; Hale Observatories; Hale Observatories; middle, from left to right, The Kitt Peak National Observatory; Lick Observatory; Gary Ladd; bottom left, courtesy of Arecibo Observatory; bottom right, National Radio Astronomy Observatory. *Page 85:* Culver Pictures. *Fig. 5.16:* NASA.

Chapter 6

Page 98: NASA. *Fig. 6.2:* prepared from a *National Geographic* map. *Page 121:* from Christopher Fry, *The Lady's Not for Burning* (New York: Oxford University Press, 1950). Reprinted by permission. *Page 123:* from Christopher Morley, "The Hubbub of the Universe" in *Translations from the Chinese* (New York: Doubleday, Page & Company, 1927). Reprinted by permission of Harper & Row. *Fig. 6.3:* redrawn from Peter J. Wyllie, "The Earth's Mantle," *Scientific*

(Credits continued on page 477.)

Preface

Our goal in preparing this new edition of *Exploring the Cosmos* has been to preserve and enhance aspects of the book with which our users have been most pleased while making changes that bring it up to date, increase its teachability, and further interest the student. The third edition reflects our sense that students in introductory astronomy courses are served best by a text that offers a solid background in a broad number of important areas, leaving the instructor free to select particular topics for special emphasis in lectures. Our book is as comprehensive as we feel an introductory level book should be. It contains a balance of topics that presents an overview of the major fields of astronomy and their relation to other sciences. As in the previous edition, organization of the text is traditional, beginning with the history of astronomy and its basic concepts and tools, followed by a discussion of the solar system, and ending with a treatment of stellar and galactic astronomy. Despite the many different organizational approaches that others have tried recently, we believe that our organization is the most natural approach to the teaching and learning of astronomy.

Though the overall organization of the third edition is similar to that of the second edition, we have made a number of changes to increase the book's flexibility and its interest to students, and to introduce more up-to-date views. The book now divides easily into four groupings. Chapters One through Five cover the history, conceptual foundations, and tools of astronomy. Included here are an overview of the contents, structure, and scale of the universe, earth-sky relationships, and the concepts of motion and gravity. Relativity is introduced in Chapter Three of this edition and referred to throughout the text as needed, allowing us to show that relativistic considerations are an integral part of our understanding of celestial phenomena.

As before, Chapters Six through Ten are devoted to the solar system, but we have made changes here to emphasize the results from space astronomy and to permit us to undertake a more comparative treatment of the planets. This section also includes a new chapter on life in the solar system, an addition we felt would be particularly stimulating to the students at this point in the course.

The third section, Chapters Eleven through Fifteen, contains discussion of the nature, variety, and evolution of stars in which we intersperse observational material with theoretical considerations. The chapter on the death of stars brings together a discussion of compact stars—white dwarfs, neutron stars, and black holes—and their relation to supernovae and pulsars.

The last section, Chapters Sixteen through Nineteen, covers galaxies, cosmology, and exobiology. Unlike the preceding edition, Chapters Sixteen and Seventeen cover all the material on galaxies, beginning with our Galaxy and including normal and active galaxies, quasi-stellar objects, and galaxies that emit X-ray, ultraviolet, infrared, and radio radiation in excessive amounts.

In addition to organizational changes, this edition includes an updating of all information, in both the text and the illustrations. For example, coverage of the planets includes the results of the *Voyager 1* and *2* missions to Jupiter and the journey of *Pioneer 11* to Saturn, with new illustrations and a special photo essay that contains photos from *Voyager*. The stellar and galactic sections include a discussion of mass loss, treatment of the evolution of binary systems, new information on interstellar molecules, and recent work on the extent and variety of active galaxies.

We have also made important changes in our treatment of mathematics. We have found that students in this course vary in their ability to feel comfortable with mathematical applications of basic concepts. Since some students benefit greatly from mathematical applications and others do not, we have placed the applications in boxes as much as possible, to be used as needed to supplement the text. In order to present a more qualitative view of astronomy, we have also put quantitative facts in tabular form and in figure captions wherever possible.

Finally, we have added an appendix on coordinate and time systems to this edition and also short biographies on a number of significant astronomers, from Ptolemy to Fred Hoyle. The biographies not only serve to acknowledge the contributions these individuals have made to astronomy, but they also emphasize the human dimension of science. To quote from Jacob Bronowski, "Science is a very human form of knowledge ... there is no God's eye view of nature ... only a man's eye view."

In preparing the third edition of *Exploring the Cosmos,* we have been encouraged and helped by many users of the second edition who have generously offered suggestions and recommendations. We would like to acknowledge our great indebtedness to them, as well as to a number of colleagues and teachers who have reviewed all or parts of the manuscript. Our special thanks go to:

Dr. Raymond White, University of Arizona
Dr. Tobias Owen, S.U.N.Y. at Stony Brook
Dr. Thomas Gehrels, University of Arizona
Dr. Richard Teske, University of Michigan
Dr. Richard Sears, University of Michigan
Dr. William Bidelman, Case Western Reserve
Dr. Susan Lamb, University of Missouri, St. Louis
Professor Stephen Lattanzio, Orange Coast College

We would particularly like to thank Bill Bidelman for his close association and his inspiration in the writing of the second and third editions.

Finally, we would like to express our deepest appreciation to a group of hardworking professionals at Little, Brown and Company. They have helped us to take ideas, notes, and random scraps of paper, and turn them into a book. In particular, we would like to recognize Ian Irvine, Science Editor, whose efforts made the third edition a reality. Also our thanks go to Joanne Hale, Editorial Assistant in Science. Janet Welch as book editor, Wayne Ellis and Catherine Dorin as book designers, and Tonia Noell-Roberts as art editor, have used their professional skills to produce a book that is readable and visually exciting, and for this we are deeply grateful.

Contents

1
Introduction to the Cosmos

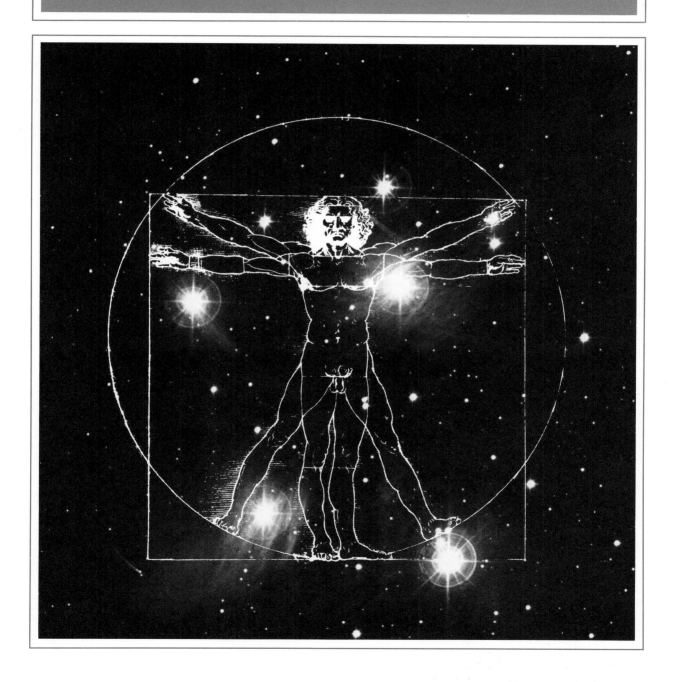

1.1
The Cosmic Design

Our view of the universe today is far different from the view held a thousand or even a hundred years ago. By the word *universe* we mean not only everything that we know exists but also whatever will be discovered in the future; the universe is everything. We begin *Exploring the Cosmos* with a brief overview of the universe as we think we understand it today.

COSMIC OBJECTS:
FROM THE EARTH TO THE GALAXIES

Only within the last four hundred years have we cast aside an earth-centered view of the cosmos. The earth, we have discovered, is not the center of the universe. It is only one of four rocky types of planets (see Figure 1.1) circling the sun. Beyond these there are four large planets made up of gases and liquids; and beyond them is the small and mysterious Pluto. Thousands of smaller solid objects—the asteroids and

◀ From the Stars Has Descended Man.

meteoroids—also orbit the sun; if we put all these together, they would make a body only a fraction of the earth's size. Ranging in wide orbits about the sun are countless bodies of frozen gases and dust; we call these bodies comets. Between these various objects is a near void containing only a few atoms and

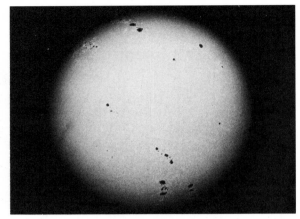

FIGURE 1.2
The sun, photographed in white light under excellent conditions, showing a large number of sunspots. Presumably other stars, if as close to us as the sun, would have a similar appearance.

FIGURE 1.1
Portion of Mercury's sun-illuminated side (left) and the giant planet Jupiter (right). Mercury is heavily cratered like the moon, with craters ranging up to nearly 200 kilometers in diameter. Like the other rocky type or terrestrial planets—Earth, Mars, and Venus—Mercury probably has a metallic iron-rich core and an outer silicate mantle. The banded appearance of Jupiter's atmosphere results from currents drawn out by the planet's rapid rotation. Unlike the terrestrial planets, the large outer planets, or Jovian planets, are composed mainly of hydrogen and helium; they may have a small rocky core.

molecules and some tiny solid particles. This is the picture of our solar system that astronomers have pieced together.

The central object of the solar system is the sun, a sphere of glowing gases so large it could hold 1.3 million earths (see Figure 1.2). Yet the sun is only one rather ordinary star among 200 billion stars or so in our Galaxy—a system of stars, gas, and dust molded by gravity. From the earth we see our Galaxy as the luminous band of stars familiar to us as the Milky Way, shown in Figure 1.3. From outside the Milky Way we would see that its true shape is a spiral disk, a disk so large that a ray of light takes a hundred thousand years to cross from one side to the other as shown in Figure 1.3a and b. Also, we would see that our sun lies in the disk away from the center of the Galaxy, toward its outer edge. The Galaxy rotates like a giant pinwheel, taking a couple of hundred million years for the sun to complete one revolution. In addition to its billions of individual stars, the Galaxy has bright patches of hot gas called gaseous nebulae, immense clouds of cold dark gas and dust, and clusters of stars like raisins stuck in and around the disk.

Astronomers estimate that billions of galaxies can be photographed with the largest telescopes; our Galaxy is among the larger ones (see Figure 1.4). Within each there are millions to hundreds of billions of stars, of assorted size, brightness, and age, arranged singly or in groups. Galaxies themselves appear to come in a variety of sizes, brightnesses, and forms. Some are spiral-shaped; others have bar shapes or elliptical shapes; and some have no discernible shape at all.

Clusters of galaxies are a frequent pattern throughout the universe. One example is shown in Figure 1.5. These clusters are the largest clearly defined collections of matter that we are able to discern. There is convincing evidence, however, that clusters of clusters of galaxies—superclusters—also exist.

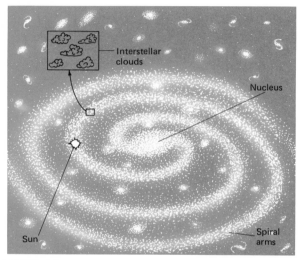

FIGURE 1.3
Artist's conception (above) of the Milky Way Galaxy as seen from north of the galactic plane. Below, a photographic mosaic of the Milky Way from Cassiopeia to Sagittarius. Shown here is a composite mixture of stars, bright discrete patches called gaseous nebulae, and dark clouds of gas and dust.

> Science is a very human form of knowledge. We are always at the brink of the known, we always feel forward for what is to be hoped. Every judgement in science stands on the edge of error, and is personal. Science is a tribute to what we can know although we are fallible. In the end the words were said by Oliver Cromwell: "I beseech you, in the bowels of Christ, think it possible you may be mistaken."
>
> Jacob Bronowski

Furthermore, the universe is expanding—not until this century did astronomers learn this fundamental fact. Galaxies appear to be separating from each other at speeds approaching a large fraction of the speed of light, and all of space is filled with cold radiation coming from everywhere. The most widely accepted explanation for the meaning of the expansion is that about 15 to 20 billion years ago a superhot and superdense fireball exploded, beginning the universe's expansion. Within this primeval fireball the galaxies were eventually formed as it cooled. The cold radiation found throughout the universe, astronomers now believe, is an echo of that initial big bang. How long will the expansion continue? How much is it slowing down? Will the universe eventually cease to expand and begin to contract? There is still much to learn before we have answers to these questions.

In recent years astronomers have discovered objects that stretch our imaginations and challenge our knowledge. Two of the most exotic of these objects are pulsars and black holes. A pulsar (also called a neutron star) is smaller than the earth; it spins at such a fantastic rate that it completes one full turn in a few seconds. A black hole, apparently, is a star that has collapsed in upon itself until it may be no larger than a few city blocks and has a gravitational field so strong that even light cannot escape from it.

A CHANGING, EXPANDING UNIVERSE

The universe not only contains objects that were beyond our comprehension a few decades ago; it also behaves differently from the way astronomers had long thought. Early telescopic observations seemed to reveal a universe that was quiet, orderly, and predictable. It appeared to be populated with unchanging stars and galaxies. Occasionally stars would explode with the brilliance of a million or a hundred million suns, but these objects seemed the rare exception in an otherwise placid universe.

With the advent of radio astronomy in the late 1940s, our picture of the universe began to change dramatically. Evidence mounted that the universe is not static; change, even violent change, is an integral part of it. Certain kinds of galaxies, for example, eject great quantities of matter. Quasars, objects somewhat like galaxies, emit enormous amounts of energy while changing at a relatively rapid pace.

OF TIME AND SPACE: ASTRONOMICAL SCALE

Astronomy not only has shown us that earth is not at the center of things but it has also given us a glimpse of time and distances otherwise beyond our everyday experience. In the chapters that follow we will talk of objects trillions of trillions of miles away. In some cases, light from the object that we now see left billions of years ago, even before the formation of the earth, and is only now arriving. That light tells about the object as it existed in the past, not what it is like today. These times and distances are almost beyond imagining. To help grasp what they mean, we can make some simple comparisons.

Imagine that the solar system could fit on the top of a typical dining room table. Let the size of the solar system (the distance from the sun to Pluto) be about 40 inches in this model. Then the distance from the earth to the sun (which in reality is approximately 100 million miles) would be about equal to 1 inch, and the radius of the sun would be smaller than a period on this page. The nearest star would be a little over 4 miles away from our table, and the radius of the Milky Way would be about 50,000 miles—more than twelve times the actual radius of the earth. If we imagine the solar system to be the size of a table, then the great spiral galaxy in the constellation Andromeda, which is visible with the naked eye on a clear dark night, would be about ten times the actual distance to the moon. Finally, the most distant galaxies in our model would be about sixty times the actual distance to the sun. In other words, if our solar system model would fit on a

FIGURE 1.4
A spiral-shaped galaxy, M81, in Ursa Major (left). The nebulous knots scattered along the spiral arms are regions containing luminous stars and glowing gases similar to the Orion Nebula in our Galaxy. At right, an enlarged photograph of the small elliptical galaxy NGC 205, which is a satellite of the nearest spiral galaxy that is known as the Andromeda Galaxy. M81 is about five times farther away than is NGC 205, which is so far from us that it takes light over 2 million years to reach us traveling at the rate of 300,000 kilometers per second (or 186,000 miles per second).

FIGURE 1.5
A cluster of galaxies in the constellation Hercules, which is almost two hundred times farther away than M81 in Figure 1.4. Photographed with the 200-inch (5.1 meter) Hale reflector.

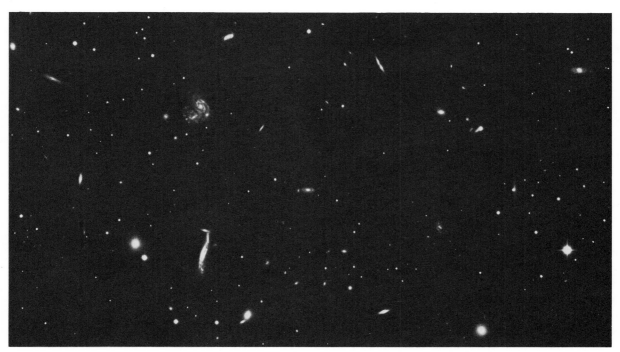

large table, then the model of the universe would be one and a half times the *actual* size of our solar system.

We can compare the sizes of objects in the universe in a similar way. The universe is approximately a million billion billion billion billion (1 followed by forty-two zeros) times larger than the nucleus of an atom.

> Is it not astonishing that nearly all the inhabitants of our Planet, down to our times, have lived and died without knowing where they are, and without having the slightest conception of the marvels of our Universe?
>
> Camille Flammarion

Consider some objects intermediate in size between a nucleus and the universe:

- An atom is approximately 10,000 to 100,000 times larger than its nucleus.

- A human is 10 billion times larger than the atom.

- Humanity's home, the earth, is 10 million times larger than a human.

- The radius of the sun is about 100 times that of the earth.

- The Milky Way Galaxy has a radius 1 trillion (1000 billion) times that of the sun.

- Clusters of galaxies are hundreds of times larger than the individual galaxies within them.

- Finally, the visible universe is on the order of thousands of times larger than a cluster of galaxies.

If you have difficulty comprehending such immense sizes and distances, take heart; it is an overwhelming scale even for astronomers. For convenience astronomers measure distances in units such as light years (defined on page 48) and express large numbers in powers of ten. A hundred, for example, is written as 10^2 and a thousand as 10^3 (that is, $10 \times 10 \times 10$). A hundred times a billion then becomes $10^2 \times 10^9 = 10^{11}$, which is a hundred billion. (For a full explanation, see Appendix 2.)

Another series of comparisons may help to convey a sense of our place in the astronomical scale of time. Scientists estimate that the big bang from which the universe evolved took place some 15 to 20 billion years ago. If we shrink this time span to 1 year, we come up with the following "universal" calendar. The earth would have formed in mid September; the oldest known signs of life would have appeared in October; and the first mammals would not have appeared until December 26. All the events of human prehistory and history—from before the first known stone tools were made millions of years ago until *Voyager 2* beamed photos of Jupiter back to the earth—would occur in the last hour of New Year's Eve.

1.2 Heritage of Modern Astronomy

The picture we have just sketched of the universe is a fairly new one. But for thousands of years people watched and wondered about the heavens and their origins. Both our imagination and the oldest written records tell us that humans must have observed how the sun, moon, planets, and stars came and went in regular cycles across the sky, as we will discuss in Chapter 2. To our ancestors, events in the heavens could evoke both wonder and terror. But people also found uses for nature's regularity: they turned the motions of the heavens into daily, monthly, and yearly timekeepers. The ancients used their knowledge of the heavens to determine when to sow and harvest, to travel to distant places with only the bright stars to navigate by, and to choose locations for religious temples. To assist in these efforts, they invented the sundial, the water clock, and star-sighting devices for marking time and recording astronomic events (see Figure 1.6).

All advanced civilizations in the past—in China, Central America, Mesopotamia—treasured astronomical knowledge not only for its practical value but also for its own sake and for what it told them about the nature of their world. They developed concepts relating to the origin, structure, and evolution of the universe; these topics are the subject of *cosmology*.

ANCIENT COSMOLOGICAL THOUGHT

Out of the great dark void the universe was created. That is the story of the beginning in most ancient cosmologies. Once created, it had to be shaped into an ordered world. Only then could human beings inhabit it. The Bible's version of the beginning of the universe appears in Genesis: "In the beginning, God created the heavens, and the earth. And the earth was without form and void; and the darkness was upon the face of the deep." Six days of labor fashioned the earth as we know it.

Other cultures had their descriptions of the beginning of the world. In Greek mythology Chaos comes first; from it issued Erebus, the darkness under the Earth, and black Night, who conceived Aether and Day. From Earth came the firstborn, which was Heaven, who was destined to harbor Earth's other children; these from their Olympian heights fought among themselves for mastery of the world. The Asiatic Indian account of creation has the same theme: in the beginning all was without form and substance; darkness and space enveloped the undifferentiated waters. Father Sky and Mother Earth conceived the gods, who molded the earth's geography (see Figure 1.7).

To early Egyptians the universe was a large rectangular box in which the Mediterranean area lay at the bottom, with Egypt in the center. Above stretched the vault of the sky, from which hung celestial lamps, the stars (see Figure 1.7). At the four points of the compass were lofty peaks connected by a procession of mountains; the four peaks supported the heavenly dome. Around the mountains ran a great celestial river, whose main branch was the Nile. Each day the sun god Ra, with his attendants, was carried westward in the solar boat along the celestial river. As the day ended, he disappeared behind an eternally shrouded dark region, reappearing in the east on the following day.

Across the Red Sea from Egypt lived two other civilized peoples in ancient Mesopotamia, the Babylonians and the Chaldeans. Their cosmology resembled that of the Egyptians: an enclosed chamber

FIGURE 1.6
Sunrise over the "heelstone" on the day of the summer solstice at Stonehenge, an ancient observatory constructed between approximately 2600 B.C. and 1500 B.C. Its ruins lie on Salisbury Plain, Wiltshire, England.

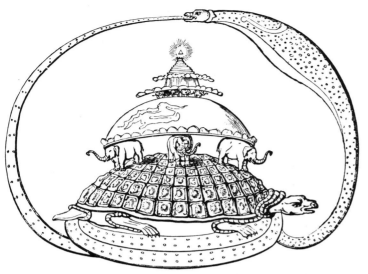

FIGURE 1.7

The ancient Hindu view of the universe. A giant cobra enclosed the universe. The tortoise, symbol of force and creative power, floated on a sea of milk. Upon its back stood four elephants supporting the earth at the east, west, north, and south points of the horizon. Crowning the universe was a triangle, the symbol of creation. At right, the early Egyptian universe. The sun god Ra (upper left) traveled along the celestial river in his boat. The star lamps, suspended from the ceiling of the sky, were lit and extinguished by the gods.

> I believe most simply in the nobility of this great effort to understand nature, and what we can of ourselves, that is science.
>
> J. Robert Oppenheimer

had their part of the world at the center. The earth floated on the surrounding eternal waters; beyond rose a great wall supporting the sky's metal dome.

We can see a common thread in most of these cosmologies: an inclination for particular groups to place themselves at the center of the universe. This egocentric aspect evolved later into a geocentric one—that is, the whole earth was put at the center of the universe, a slightly less narrow point of view. In any event, primitive cosmologies were only idealized sketches against which the activities of nature took place. Little if any aspect of natural phenomena was actually incorporated into the cosmological picture. Not until centuries later were explanations of the details of nature considered to be part of cosmology.

ASTROLOGY

To many ancient peoples the heavens were more than the source of cosmological speculation. To many the heavens held keys to their fate. The celestial gods, they believed, controlled human destiny; *astrology* became the way to understand their mysterious activities and comprehend the course they dictated for human events.

Egyptians and Babylonians indulged in astrology, but the Chaldeans made a fine art of it. Every night in their observatories the Chaldean priest-astrologers, watching the nightly changes in the heavens, performed rites to interpret the gods' will for humankind. Historians say the biblical tower of Babel[1] was a Chaldean religious and astronomical observatory. It was a pyramid constructed of seven layers of stone, one for each principal celestial body—sun, moon, and the five known planets. Atop this pyramid the priests observed the heavenly bodies in motion.

Astrology can claim one constructive act: it spurred the study of astronomy. As astrologers observed the sky wanderers in an attempt to improve their horo-

[1]The Chaldean translation is "Gateway to the Highest God"; in Hebrew *Babel* signifies "confusion."

scopes, they paved the way for the large advances in planetary astronomy made in later centuries. Right up to the seventeenth century some of the leading astronomers were practicing astrologers (including Johannes Kepler, whose contributions we will examine in Chapter 2).

Today the gap between astrological predictions and scientific reality is overwhelming. To think that stars and planets control human affairs by their configurations is a preposterous delusion. As Shakespeare commented in Sonnet 14:

Not from the stars do I my judgement pluck;
And yet methinks I have astronomy
But not to tell of good or evil luck,
Of plagues, of dearths, or season's quality.

The amount of knowledge of the natural world that has accumulated is enormous, and the explanations that we have developed grow in sophistication. This is largely the result of a dramatic expansion of information that began in the sixteenth and seventeenth centuries, when science turned more heavily to observation, experimentation, and analysis. Important to the development of what was then called the new experimental science was its effect on technology. New discoveries in science have at times led to new tools and techniques. But better instruments have also led at times to more precise experiments, which in turn have stimulated developments in scientific explanations. These scientific advances led again to better instruments, the cycle repeating itself again and again.

1.3
Anatomy of Science

Our understanding of the natural world has changed dramatically through the centuries, with every age extending, modifying, and refining the science of the preceding age, and in turn being subjected to the critical evaluation of the age that followed it. But at each stage it has taken feats of intellect, imagination, and courage to arrive at better explanations of physical phenomena. Even when ideas have later been proved wrong or incomplete, they have often been important springboards for new conceptual developments. Many ideas developed long ago seem simple. Yet considering our views of the natural world, the available facts, and the state of technology, these discoveries required perseverance and creativity equal to that of the best minds in our own age.

1.4
Astronomy Today

The word *astronomy* is derived from the Greek words meaning "law and order." Today we define *astronomy* as the science that treats of celestial bodies. Since the time of the early Greeks two and a half millennia ago, astronomers have shifted their emphasis from observing the motion of the heavens to studying the physical nature of the universe. How is the universe expanding? What is the source of the stars' light? How do galaxies form and evolve, or do they evolve? These are among the questions that intrigue astronomers today. Changes in astronomy have altered not only the questions astronomers ask but also the way they work (see Figure 1.8); these changes have spawned concepts, knowledge, techniques, and equipment that have helped mold the modern world.

And the earth is a particle of dust by a sand-grain sun,
    lost in a nameless cove of shores of a continent.
Galaxy on Galaxy, innumerable swirls of unnumerable stars,
    endured as it were forever and humanity
Came into being, its two or three million years are a moment,
    in a moment it will certainly cease out from being
And galaxy on galaxy endure after that as it were forever. . . .

Robinson Jeffers

FIGURE 1.8
Old and modern astronomy compared. The painting at left is *The Astronomer* by D. Owen Stephens. "In Nature's infinite book of secrecy / a little I can read" (Shakespeare, *Antony and Cleopatra,* act I, scene 2). At right, the telescope operation controls in the *International Ultraviolet Explorer* satellite control room at the Goddard Space Flight Center of the National Aeronautics and Space Administration. The astronomer can view results from the satellite on the small monitor to the left and the large screen behind the console. Operating instructions are sent and data received via a digital computer.

Today the universe is a cosmic laboratory in which there exists an enormously greater range of matter and energy with which to test fundamental concepts than can be found in any terrestrial laboratory. Temperatures in the universe vary from a few degrees in the cold reaches of interstellar space to billions of degrees inside stars. Densities can be as low as one or two atoms per cubic centimeter between stars to over a trillion billion times the density of ordinary terrestrial matter inside a neutron star. The energies of some cosmic particles exceed the energies of the molecules in your room by as much as a trillion trillion times.

Astronomy today still involves observation and is still a relatively pure science, devoted to acquiring knowledge about the universe. There are still few opportunities for astronomers to work outside the university in the practical world of technology and commerce. Perhaps partly as a result, astronomers are a smallish group among scientists. In this country one in about two hundred scientists is an astronomer. Compare that ratio with the one in ten who is a physicist and the one in three who specializes in chemistry.

In other ways, though, astronomy has changed drastically in the last thirty years. The telescope is no longer the most expensive scientific instrument, not even the most expensive astronomical instrument. Astronomy like other sciences has grown more complicated and more specialized, with many subdivisions. Some astronomers, *astrophysicists,* specialize in

> The most beautiful thing we can experience is the mysterious. It is the source of all true art and science. He to whom this emotion is a stranger, who can no longer pause to wonder and stand rapt in awe, is as good as dead: his eyes are closed.
>
> Albert Einstein

precise measurements of the radiation emitted by astronomical sources and the theoretical study of their composition, structure, and evolution. *Solar physicists* study the physical characteristics of the sun and its effects on the earth's environment. Other astronomers are specialists studying our Galaxy, or the structure and contents of galaxies and evolution of the universe. And the list of astronomy's subdivisions could go on.

Contributions from other sciences have done much to bring these changes about. A revolution begun in physics after the turn of the century has led to a new understanding of atomic structure and light, which in turn has been used by astronomers to help us understand something about the stars. In the late forties radio astronomy brought in to the profession radio engineers, electronics specialists, and physicists. As a result of such cross-fertilization among the sciences, astronomical knowledge has grown, and astronomy is influenced by and influences the other physical sciences.

Walk into a modern observatory and you will see that it resembles any laboratory in the physical sciences; it is not just a place where an astronomer looks through a telescope. In fact, the astronomer's job is not, as many people think, to spend his or her working hours looking through a telescope. The astronomer may spend weeks or months analyzing and interpreting the results of a few hours of work with the telescope. Much of the astronomer's working day may be spent with computers, analyzing data, or with new, extremely sensitive devices for measuring all kinds of radiation, not just visible light. (In Chapter 5 we will examine some of astronomy's new tools, such as radio, ultraviolet, X-ray, and gamma ray telescopes.) Furthermore, some astronomers rarely even work with telescopes. Instead they use mathematical models to work out theories explaining other people's observations. The theoretician and the observer working together have given us many of our recent advances in astronomy.

To modern people, as to the ancients, the study of the heavens can be a source of wonder, of pleasure, or of philosophical contemplation. What it tells us about the physical structure of the universe can shape our understanding of the world. We have listened briefly for signs of intelligent life transmitted from other worlds and looked for signs of life on other planets in our solar system. Perhaps, someday, somewhere, we will find extraterrestrial life, perhaps even intelligent life. In the meantime, astronomers are learning ever more about the cosmic world.

SUMMARY

Less than four hundred years ago we discovered that the earth is not the center of the universe. It is one of nine planets circling the sun, the center of the solar system. Our sun, in turn, is only one of about 200 billion stars in the Milky Way Galaxy, which itself is only one galaxy among billions in the universe. Though it may appear static and serene from telescopic observations, the universe is actually in constant motion, with evidence now pointing to change, even violent change, as an integral part of it. The scale of the universe is enormous. The range of sizes of known objects—from atoms to clusters of clusters of galaxies—and of the astronomical time scale—from the origin of the universe to the present time—is so vast as to be almost beyond our ability to imagine it.

The study of the heavens has always been important to humankind, both for its practical and its philosophical uses. Early cultures invented systems (or cosmologies) to explain the origins, structure, and evolution of the universe, though without extensive reference to natural phenomena. And many ancient peoples tried to foretell the future through interpretations of astronomical data, a practice called astrology. Modern astronomy, though tracing its origins to the interest ancient peoples took in the heavens, focuses rather on the physical nature of the universe and attempts to answer questions about its structure and evolution. In exploring these questions, astronomers today borrow concepts from other physical sciences to help further our understanding of astronomical phenomena, and use advanced technologies for collecting and analyzing data. But astronomy has also contributed ideas and technological advances to other sciences and to society—from the laws of motion and gravitation to the development of optical equipment.

REVIEW QUESTIONS

1. Discuss the range of sizes and distances for matter in the universe compared to that on the earth.

2. In what way are astronomers justified in claiming that astronomy is the oldest science? Could not mathematics, physics, chemistry, or biology claim priority?

3. In what way does modern astronomy owe a debt to astrology as it was practiced in ancient times?

4. Why are there so few astronomers in comparison with researchers in the other scientific professions?

5. List several practical achievements that have materialized as the result of astronomical research.

6. How has humankind benefited from the exploration of space?

7. Astronomy has also been classed as a cultural science—one with attached aesthetic values. Can you give any reasons why this is so?

8. What is the justification for the description that the universe is a cosmic laboratory?

2
Early Descriptions of the Cosmos

> But who shall dwell in these worlds if they are inhabited? Are we or they Lords of the World? And how are all things made for man?
>
> Johannes Kepler

In the historical development of our conceptual ideas about the universe, an understanding of the common cyclic phenomena of the sky played an important part. Many of these phenomena were not discovered in a literal sense, but have been known since long before our ability to write about them. A good example is the daily rising and setting of the sun. For this phenomenon, what is new and has changed with time is "an explanation" of why it rises and sets. Space does not permit us to sketch the evolution of all the explanations of these various phenomena. Therefore, in the first section, we will discuss most of the common sky phenomena, which were so influential in channeling the developments of astronomy, though the explanations will reflect our present understanding. This will make it much easier to understand the historical events discussed in the last three sections of this chapter.

2.1 Ancient Astronomy: The Sky

CELESTIAL SPHERE

To our ancestors, the sky must have looked like the inside of a huge dome covering the earth as far as they could see. Little wonder that ancient peoples thought of the earth (and themselves) as the center of a sphere, with the stars emblazoned on that sphere. This imaginary sphere is still used and known to astronomers as the *celestial sphere* (see Figure 2.1).

◀ Geocentric model of the universe from *The Cosmographical Glasse, conteinyng the pleasant Principles of Cosmographie, Geographie, Hydrographie, or Navigation,* by William Cunningham, London, 1559.

Watching the sky during the night, early peoples saw the stars rise above the eastern *horizon*—the circle that divides the visible celestial hemisphere from the invisible one—cross the sky, and later set below the western horizon. Observing this daily behavior of the stars was one of the first steps in human understanding of the sky. Ancient peoples apparently thought that the stars rose and set because the celestial sphere actually rotated, carrying the stars from east to west. Today we know that the rising and setting of the stars is caused by the earth's rotation from west to east.

After careful study of the sky, people must then have noticed that the stars appear to move around two points on the celestial sphere: the *north* and *south celestial poles* (see Figure 2.1). To ancient peoples these were the ends of the axis around which the celestial sphere actually rotated. To us they are projections of the earth's axis of rotation, through the north and south celestial poles, onto the imaginary celestial sphere. (Polaris, the *North Star*, now lies within one degree of the north celestial pole and is a relatively bright marker of its position.)

The *zenith* is the point directly overhead for any observer. The imaginary arc on the celestial sphere running from the north point of the horizon through the celestial pole and the zenith to the south point of the horizon is the *celestial meridian*. This line is the dividing line between rising and setting, because the highest position above the horizon that each star reaches in its daily motion is on the celestial meridian. For this reason it is useful for developing a conceptual frame for timekeeping. (Further descriptions of the systems of measurement used to locate objects on the celestial sphere are discussed in Appendix 3; timekeeping is covered in Appendix 4.)

Not all stars rise above and set below the horizon daily. Some stars remain either above or below the horizon, depending on the observer's latitude. As ancient peoples traveled to different latitudes, they must have noticed that a different pattern of stars could be seen. For example, as observers traveled northward, they could see stars near the northern horizon that previously had not been visible. Here was evidence that the earth was a sphere.

CONSTELLATIONS

In the clear skies of the Tigris, Euphrates, and Nile valleys, where the earliest civilizations flourished more than five thousand years ago, watchers of the heavens observed groupings of stars, the *constellations*. They saw in these groupings mythological beings, animals, and monsters who reigned over heaven and earth, and they attached names to the constellations accordingly. It is obvious to anyone who has examined the seemingly disordered jumble of constellation patterns that the ancients possessed a lively imagination in depicting the creatures and objects that populated their skies. The eminent British astronomer John Herschel once complained: "Innumerable snakes twine through long and contorted areas of the heavens where no memory can follow them; bears, lions, and fishes, large and small, confuse all nomenclature."

The names and shapes of our constellations are part of our heritage from the Greeks of antiquity. However, they did not originate in Greek culture but appear to have stemmed from earlier civilizations in Mesopotamia and from other peoples of the Near East. (Some constellation figures are shown in Figure 2.2.) Greek astronomers observed and named forty-eight constellations. Forty more were added, most of them in southern skies, by European mapmakers and astronomers in the seventeenth and eighteenth centuries.

Note that the "catch figure" or *asterism* often associated with a constellation should not be mistaken for the entire constellation. An example is the asterism of the Big Dipper, which is the recognizable figure for the constellation Ursa Major, the Great Bear.

Standing in the midlatitudes of either hemisphere, with practice you could see, as a year goes by, four-fifths of the constellations. Of the eighty-eight constellations, about half of them lie in the Milky Way or near its borders. As you learn the constellations, hearing the name of one will bring to mind both its shape and its place in the sky, just as earthbound place names and their locations become familiar. Keep in mind that the stars of a constellation form an apparent grouping as seen from the earth and are not necessarily in close proximity in space.

NAMING THE STARS

To go with a constellation's name, its brighter stars are designated by lowercase Greek letters, assigned approximately in descending order of brightness. When a letter is added to the constellation's Latin name, the case ending changes to possessive or genitive forms, such as α Orionis. This scheme was invented by Johann Bayer for his 1603 map of the heavens. Over a hundred of the brightest stars also have Greek, Latin, or Arabic names. Brightest of the stars in the constellation Leo is α Leonis, which has its own proper name,

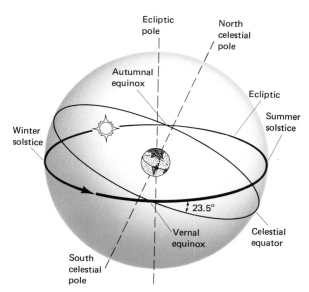

FIGURE 2.1

The celestial sphere. Envisioned by the ancients, the celestial sphere had the earth at the center with the stars emblazoned on the sphere of the sky. The ancients thought the stars rose and set because the celestial sphere rotated, carrying the stars from east to west. All stars appear to move around two points on the celestial sphere, the north and south celestial poles—projections of earth's axis of rotation. The earth's equator projected on the celestial sphere becomes the celestial equator. It intersects the ecliptic, the yearly apparent path of the sun, at the vernal and autumnal equinoxes (21 March and 22 September) at an angle of 23.5°, since that is how much the earth's axis is tilted from the perpendicular to the plane of the ecliptic. The solstices (22 December and 21 June) are the points on the ecliptic 90° from the equinoxes. These imaginary circles and points still play a useful role in modern astronomy, as discussed in Appendix 3.

Regulus. Second brightest is β Leonis (Denebola).

But, the Bayer scheme quickly ran out of Greek letters (the limit was twenty-four) for the 5400 naked-eye stars that are dotted over the entire sky. John Flamsteed, first Astronomer Royal of England, cataloged the stars in 1725 and took care of the shortage by simply numbering stars west to east across the constellation. A particular star in Cygnus is identified as 61 Cygni.

Fainter stars may be identified by their number in any one of several star catalogs. For example, Lalande

21185 is entry 21,185 in Lalande's catalog, which has numbers for the brighter stars as well. There are still other systems for naming the stars. Enough different designations exist today to confuse the astronomer.

THE SUN AND THE SEASONS

The different seasons are caused by the tilt of the earth's axis of rotation and by the earth's revolution about the sun once a year (see Figure 2.3). Relative to the stars, the sun appears to move eastward about one degree each day. Because of this, a particular star rises about four minutes earlier each night than the night before. And, at the end of one month, the star rises approximately two hours earlier than the previous month. By the end of one year, the nightly change adds up to twenty-four hours, and the annual cycle of the heavens begins again.

Over the centuries, priest-astronomers tracked the sun's yearly path across the sky, a path called the *ecliptic* (see Figure 2.1). The earth's geographic equator projected onto the celestial sphere is called the *celestial equator*. The celestial equator intersects the ecliptic at two points: the *vernal equinox*, which the sun reaches about 21 March, and the *autumnal equinox*, which the sun reaches about 22 September. The priest-astronomers found that the moon and the planets moved almost entirely along a narrow band of sky, the *zodiac*, which was 16° wide and centered on the ecliptic. They divided the zodiac into twelve constellational divisions or *signs*, through which the sun passed in successive months (see Figure 2.2).

The earth's axis of rotation is tilted 23.5°, so that in the Northern Hemisphere we're inclined away from the sun in December and toward it in June (see Figure 2.3). Consequently, the amount of sunlight falling on

FIGURE 2.2

Signs of the zodiac and the constellations, with the earth at the center of the celestial sphere. Viewed from the earth, the moon and the planets move almost entirely along the band of sky known as the zodiac. Ancient priest-astronomers divided the zodiac into twelve constellational signs, but today these signs and the position of the constellations are out of alignment because of the earth's precession (see Figure 2.4).

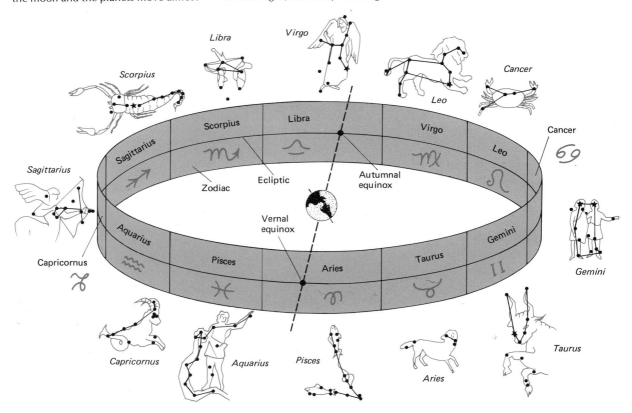

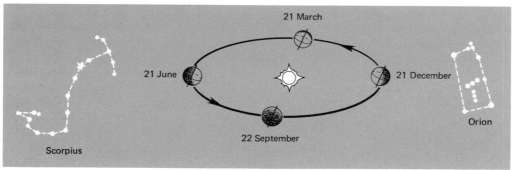

FIGURE 2.3

Seasonal change in the appearance of the sky. These changes provided early peoples with a yearly calendar of events for human activities. Today we know the change is caused by an inclined earth that orbits the sun. Shown are the earth's inclinations at the times of vernal equinox, 21 March; summer solstice, 21 June; autumnal equinox, 22 September; and winter solstice, 22 December. Orion and Scorpius are winter and summer constellations respectively.

the surface of either hemisphere varies depending upon whether the hemisphere is inclined toward or away from the sun.

In the Northern Hemisphere spring begins on or about 21 March, when the sun crosses the celestial equator from south to north at the vernal (or spring) equinox. On this day all places on earth experience twelve hours of daylight and twelve hours of darkness. Summer starts on or about 21 June, when the sun reaches its maximum distance of 23.5° north of the celestial equator at the *summer solstice*. During our summer the hours of daylight are longest and the sun is highest in the sky. Autumn begins on or about 22 September, when the sun crosses the celestial equator from north to south at the autumnal equinox and the days and nights are again equal over the earth. Winter begins on or about 22 December, when the sun is farthest south of the celestial equator by 23.5° at the *winter solstice*. Our periods of daylight are now the shortest and the sun is lowest in the sky. In the Southern Hemisphere the seasons are reversed; for example, Christmas there occurs during the warm summer months.

The shortest seasons in the Northern Hemisphere are autumn and winter. This is because the earth's orbital motion is slightly faster during this half of the year when it is closer to the sun, as contained in Kepler's second law (see page 29). In the Southern Hemisphere the longest season is winter. The situation between the two hemispheres will be reversed in about thirteen thousand years because of precession (see Figure 2.4 on page 18).

> Finally we shall place the Sun himself at the center of the Universe. All this is suggested by the systematic procession of events and the harmony of the whole Universe, if we only face the facts, as they say, with both eyes open.
>
> Copernicus

PHASES OF THE MOON

The rising and setting of the sun, and the period of the phases of the moon were important cyclic events for ancient man. Both of the repetitive cycles of the sun and the moon were intricately involved with man's concept of time. Our present day understanding of the reasons for the phases of the moon predates Aristotle, who was aware that the moon "shines" by reflecting sunlight.

The parallel rays of the distant sun always illuminate one-half the moon's surface as well as one hemisphere of the earth; we see the rays of the sun reflected off the moon. When the moon is between the earth and the sun—the time of a *new moon*—its dark side faces us (see Figure 2.5) and we do not see it at all. Within a few days, because the moon moves eastward relative to the sun, a thin crescent appears low in the western sky after sunset, setting shortly after the sun sets. In the next few days the waxing (growing) crescent appears higher in the sky after sunset and therefore sets later on consecutive nights. One week after new moon, the moon is at *first quarter* and will be on the observer's celestial meridian at sunset; it will set about six hours after the sun. In the following week the gibbous moon (more than half a crescent, but not yet full) waxes toward full as it continues its easterly movement around the earth. Two weeks after new moon, the

PRECESSION OF THE EQUINOXES

In the second century B.C. the Greek astronomer Hipparchus compared the positions of the principal stars in the zodiac with those astronomers had noted over a hundred years earlier. He found that the vernal equinox had shifted westward about 2° along the ecliptic. What he was observing is called the *precession of the equinoxes.*

The earth is not a perfect sphere. It tends to bulge out in the equatorial region. Both the sun and the moon try to pull the earth's equatorial bulge into the terrestrial and lunar orbital planes. Acting like a spinning top, the rotating earth resists this pull. The effect of the sun's and the moon's attraction, and the earth's resistance causes the earth's axis

of rotation to move slowly westward around the pole of the ecliptic (see Figure 2.4). Because of the precession of the axis of rotation, the points of intersection between the celestial equator and the ecliptic shift westward along the ecliptic at a rate of about 50 seconds of arc per year. This causes the equinoxes, which are the points of intersection, to precess completely around the ecliptic in 26,000 years.

Two thousand years ago the vernal equinox (see Figure 2.2) lay in the constellation of Aries. Three thousand years prior to that the vernal equinox was located in the constellation of Taurus. Today the vernal equinox occupies a position in the constellation of Pisces.

The "Age of Aquarius" dawns when the vernal equinox moves into the constellation of Aquarius, about a thousand years from now.

The axis of the earth retains its tilt of 23.5° throughout the cycle of precession. Today the axis is directed toward a point on the sky less than one degree from Polaris. In ancient Egypt the axis pointed in the direction of the star Thuban (α Draconis), which was then only a few degrees from the pole around which the heavens rotated. About A.D. 14,000 the very bright star Vega (α Lyrae) will be "the North Star" and mark the approximate position of the north celestial pole for our descendants.

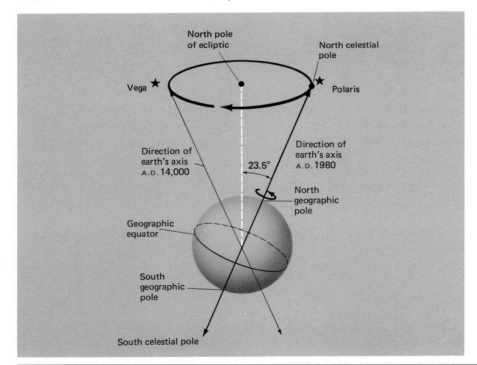

FIGURE 2.4
Precession. Resisting the pull from the sun and the moon, the earth's axis of rotation describes a cone about the perpendicular to the plane of the ecliptic in a period of 26,000 years.

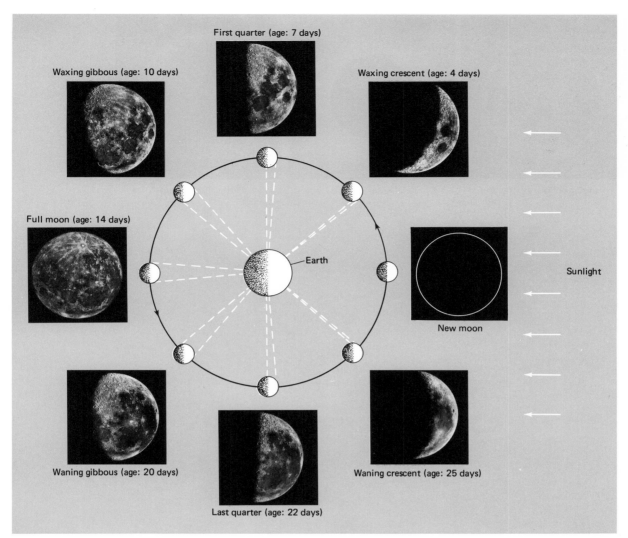

FIGURE 2.5
Lunar phases. The portion between the dashed lines represents the part of the moon we can see from the earth. The moon's age is counted from the time of the new moon. Since the moon keeps basically the same side toward earth, brightening and darkening during the cycle of phases occurs on the face toward earth.

moon is *full* and it lies on the opposite side of the earth from the sun. We see it rise at approximately 6 P.M. and set at about 6 A.M. One week later we see the moon at *last quarter* rise at about midnight and set at about noon. Finally, we see the waning crescent moon rise shortly before sunup as it is about to overtake the sun one month from the start of new moon.

LUNAR AND SOLAR ECLIPSES

When the moon comes directly between the earth and the sun (see position 1 of Figure 2.6), the moon's shadow, or part of it, falls on the earth and the sun is eclipsed. The moon's orbit is inclined to the earth's orbital plane (the ecliptic plane) by about 5°, so a *solar eclipse* is possible only when the moon is near a new phase and, in addition, is at or near one of the two points in its orbit known as *nodes* (where its orbit intersects the plane of the ecliptic). This lineup occurs at least twice each year; at most and rarely, five times a year. The eclipse may be partial, total, or annular (ringlike). The *umbra* or shadow proper is the darker-toned portion; the *penumbra* or partial shadow

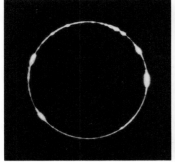
Annular eclipse of the sun.

Total solar eclipse.

Partial lunar eclipse.

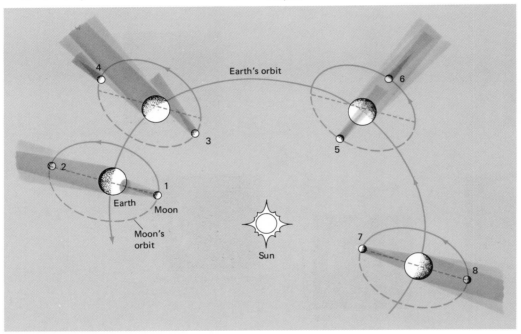

FIGURE 2.6

Solar and lunar eclipses. The lines with short dashes indicate the line of the nodes—where the moon's orbit crosses the plane of the earth's orbit. The lines with longer dashes show that portion of the moon's orbit that lies below the earth's orbital plane. The diagram shows eight positions of the earth and the moon relative to the sun as seen from north of the plane of the ecliptic:

1. Total eclipse of the sun: the new moon is at or near the line of nodes.

2. Total eclipse of the moon: the full moon is at or near the line of nodes.

3. Partial eclipse of the sun: the moon has moved far enough from the line of nodes so that the earth intercepts only the penumbra portion of its shadow.

4. Partial eclipse of the moon: the full moon intercepts part of the penumbra of the earth's shadow; this penumbral lunar eclipse is not visible to the naked eye.

5. No eclipse: the new moon is too far removed from the line of the nodes; its shadow falls below the earth.

6. No eclipse: the full moon is too far removed from the line of nodes; the moon passes above the earth's shadow.

7. Annular eclipse of the sun: the new moon's umbral shadow is too short to strike the earth even though the moon is at or near the line of nodes; however, a partial solar eclipse is visible outside the extended umbral shadow.

8. Total eclipse of the moon: the full moon is at or near the line of nodes.

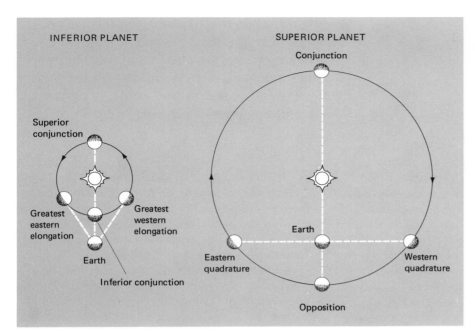

FIGURE 2.7
Configurations of an inferior
plane and a superior planet.
The planet's motions are shown
as they are seen from the
earth; viewed from the sun,
the planets move in the same
direction—eastward. Relative
to the earth-sun line, the
angular distance of the planet
from the sun is called its
elongation. For both the inferior and the superior planet,
there are four values of elongation that carry special names;
these are shown beside the
planet.

is the lighter-toned portion. If you stand in the umbra of the moon's shadow, you see a total eclipse of the sun; if you stand in the penumbra of the moon's shadow, you see a partial eclipse. The penumbra cast on the earth covers a much larger area than the umbra, making a partial solar eclipse visible to many more observers than is a total eclipse.

Because the average length of the moon's tapering shadow is not quite equal to the moon's mean distance from the earth, a small difference separates the total eclipse from the annular eclipse of the sun. An annular eclipse takes place when the new moon is at or near the point farthest from the earth (apogee) and when its umbral shadow is too short to reach the earth (see position 7 in Figure 2.6). Under these conditions the slightly smaller, dark disk of the moon is surrounded by a brilliant ring of the still-exposed sun. Under the most favorable conditions the width of the moon's shadow is about 270 kilometers across. Totality then lasts longest (maximum length is about 7½ minutes) in the equatorial zone, where the velocity of the moon's eastward-traveling shadow in relation to earth's eastward rotation is smallest. The last such eclipse took place on 30 June, 1973 over the northern part of South America, the eastern part of the Atlantic Ocean, and northern Africa. The next seven-minute total solar eclipse will occur on 11 July, 1991 and will be visible in Hawaii, Central America, and Brazil.

Usually if an eclipse of the sun occurs, an eclipse of the moon precedes or follows it by two weeks. Earth, moon, and sun then are sufficiently in line for the earth's shadow to fall on the full moon. The earth has nearly four times the moon's diameter, so its shadow is about four times wider at the base and four times longer than the moon's shadow. Lunar eclipses may be partial or total (positions 2, 4, and 8 in Figure 2.6). All inhabitants on the dark side of the earth may see a lunar eclipse at the same time.

A year may bring as many as three lunar eclipses—or none at all. More often we will have two eclipses of the sun and two of the moon in each calendar year. Centuries of observing eclipses taught the ancients that eclipses recur at regular intervals. After eighteen years and ten or eleven days, an interval named the *saros*, the circumstances of an eclipse are repeated approximately.

CONFIGURATIONS OF THE PLANETS

Ancient astronomers devised systems to describe the configurations of the planets. These earlier descriptions are the basis for our present-day definitions of the configurations (see Figure 2.7). Between the earth and the sun revolve Mercury and Venus, the two *inferior planets*. Because their orbital periods are shorter than that of the slower-moving earth, they overtake and pass it. From the earth, relative to the earth-sun line, Mercury and Venus appear to move counterclockwise around the sun while swinging from one side of it to the other. The angular distance (in degrees) either planet appears east or west of the sun is its *elongation*. From its position closest to earth—called *inferior conjunction*—when it is in line with the sun, the planet appears to move rapidly west of the sun as its phase changes from new to crescent (see Figure 2.7). When an inferior planet reaches its greatest angular distance west of the sun at *maximum western elongation*, it is conspicuous as a morning star; its phase is quarter. Thereafter, the inferior planet appears to reverse its course and move toward the sun and *superior conjunction*. It is now farthest from the earth and its phase is full. Past superior conjunction the inferior planet swings east on its way toward *maximum eastern elongation*, when it becomes the evening star in a quarter phase. It moves back toward inferior conjunction to complete the circuit and the cycle of moonlike phases.

Planets with orbits outside earth's are called *superior planets* (Figure 2.7). Their orbital periods are longer than that of the earth, so their motion relative to the earth-sun line appears to be clockwise around the sun. When a superior planet is nearest to us and also brightest, it is in *opposition*—opposite the sun in the sky and visible throughout the night. As the superior planet moves 90° east of the sun, it passes through *eastern quadrature*, when it rises at noon and is an evening star. Continuing on, the superior planet moves westward. It reaches *conjunction* when it is farthest from the earth, at which time it rises and sets with the sun. Next the planet passes through *western quadrature*, 90° west of the sun. Rising at midnight, it is a morning star. Finally, the superior planet returns to opposition, its cycle of configurations complete. As seen from the earth, superior planets do not exhibit a cycle of phases.

The length of time for one orbit of a planet around the sun is known as its *sidereal period*. It is the time taken to complete a 360° circuit around the sky relative to the stars. From the earth we actually observe the *synodic period*—the time it takes a planet to return to a particular alignment with the sun as viewed from the earth (such as from opposition to opposition). The synodic and sidereal periods differ because the earth is advancing in its own orbit as a given planet revolves around the sun.

Now that we have surveyed some of the common earth-sky relationships, and the regularity of the heavens, we will discuss the historical development of astronomy leading up to the modern dynamics of planetary motion.

2.2
Greek Cosmology:
The Geocentric System

THE EARTH TAKES SHAPE

The Greeks inherited much of their astronomy (and astrology) from the Egyptians and the Babylonians. After the seventh century B.C. astronomy shed much of its astrological influences, and it began to form the science we know today.

Some early natural philosopher's vision of the earth was that of a flat disk afloat in water or riding in midair. Later it seemed impossible to doubt that the earth was spherical—how else, argued Aristotle, could it project a circular shadow when it eclipsed the moon? Eratosthenes (273–193 B.C.), geographer and librarian of the museum at Alexandria in Egypt—convinced that the earth was a sphere and that the sun was far enough from the earth for its rays to be parallel when reaching the earth—was able to measure its circumference, as shown in Figure 2.8. He chose observing stations at Alexandria and at Syene to the south (where the Aswan Dam is now located on the Nile River). He decided to do the experiment at the summer solstice, at local noon time, which comes at the same moment at both sites because they are very nearly on the same meridian of longitude. Why did he choose this day? It was a matter of convenience, because the sun was directly overhead at local noon at Syene. Legend says he looked down a well and observed that no shadow was cast on its sides, confirming that the midday sun was at its zenith.

At noon an observer in Syene observed that the sun was directly overhead, while an observer in Alexandria found the sun to be 7° south of the zenith. Measurers paced off the distance between the two cities as about 4,900 stadia (1 stadium ≅ 0.16 kilometers). Because a straight line cuts two parallel lines at equal angles, the angle at the center of the earth is equal to the zenith angle, 7°. Working a simple proportion, he could find the earth's circumference: C:4,900 stadia = 360°:7°, or C = 252,000 stadia, or about 40,320 kilometers. Purely by chance Eratosthenes came extremely close to today's mean value of 40,030 kilometers.

THE GEOCENTRIC SYSTEM DEVELOPS

Even though the spherical shape of the earth was, in general, accepted by the early natural philosophers, they still thought the earth was an immovable globe, surrounded by the fixed sphere of the heavens. Certainly the most immediately obvious and simplest model was one in which the earth, apparently stationary and solid, was located at the center of the universe. Between the earth and the heavens was a second sphere, a rotating one that would account for the stars' daily rising and setting.

However, even with this model a problem remained. How was one to explain the other movements in the universe—those of the sun, moon, and five planets,[1] all of which seemed to wander among the stars? Never changing their course, the sun and moon move eastward relative to the celestial sphere, with the planets generally following the same path from west to east. The planets confuse matters, though, because sometimes they temporarily reverse their direction of motion, which is called *retrograde*

[1]The word *planet* is of Greek origin and signified "wandering star," in contrast with the fixed stars.

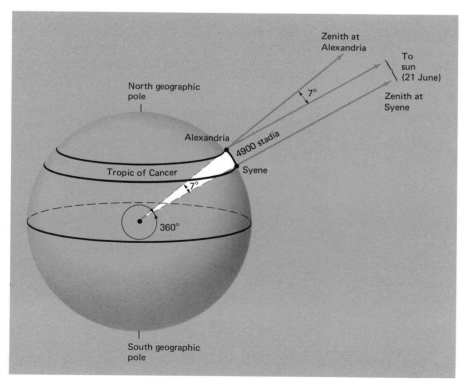

FIGURE 2.8
Eratosthenes' determination of the earth's circumference. Observations of the sun's position, together with the measured distance between Alexandria and Syene, enabled Eratosthenes to calculate earth's circumference as described in the text.

motion—and execute a closed or open loop (⌒ or S), then continue traveling eastward.

Thus explaining the universe with a logical system was a challenge for any thinker, and it absorbed the most eminent natural philosophers of ancient Greece. For centuries various schools of philosophy proposed, debated over, and elaborated on several geocentric (earth-centered) theories.

The Greeks, with their taste for balance and symmetry, reasoned that nature arrayed her celestial bodies in the perfect geometric figure, the sphere, and combined them with flawless circular movement. Their philosophers, beginning with Plato (427?–347 B.C.) or earlier, thought that planetary movements were accounted for by combinations of uniform circular motions with the earth at the center.

Eudoxus (408–355 B.C.) suggested twenty-seven concentric transparent shells rotating uniformly around the earth. The stars had one shell, the sun and moon three each, and the five planets, four shells apiece. Properly aligned and with appropriate rates of rotation, the shells roughly approximated the observed phenomena. The exterior shell in all sets produced the daily westward motion of the sky along with the stars. The prime bodies (sun, moon, planets) required their own shells with poles that could rotate independently of the sky's pole. The inner shells rotated at different rates, some in opposite directions. These produced the daily, nonuniform, easterly motion of sun and moon and the planets' loops. To bring the planets' motion closer to reality, Aristotle (384–322 B.C.) added twenty-nine more shells. However, even his scheme did not explain changes in the planets' brightnesses or reproduce precisely the variations in the motions of the sun, moon, and planets over very long periods of time.

PTOLEMY'S GEOCENTRIC MODEL

Hellenistic culture spread throughout the eastern Mediterranean world and centered in Alexandria after about 300 B.C. There a new center for science was started. Alexandrian astronomers set themselves to removing discrepancies between the geocentric theory and the motions they observed. Thus geocentric theory continued to dominate philosophical thought—except for a brief departure in the third century B.C.—until the sixteenth century.

During the third century B.C. a *heliocentric* (sun-centered) scheme was proposed by Aristarchus (320?–?250 B.C.). He thought it natural to put the largest and only self-luminous body, the sun, at the center of the system. According to Aristarchus, the heavens moved each day because the earth spun on its axis. Annual changes in the sky and the planets' irregular motions could be explained by assuming that they and the earth revolved about the sun.

Aristarchus's contemporaries objected. Why suggest that the earth moves when no one has evidence for that movement? If the earth did rotate, unattached objects would fly off. Besides, if the earth moved around the sun, the motion would show up in an apparent shift in position of the nearby stars as seen against more distant ones, and no such shift was ever observed. (We know today that, because of the earth's orbital motion, these stars do appear to change place. But they are so far away from us that the apparent angular displacement, or *parallactic shift*, is extremely small, even for the closest stars, and can be measured only with a large telescope.)

Alexandrian astronomers, therefore, rejected the heliocentric theory and looked for some way of improving the geocentric picture. To work, the theory would have to represent more accurately the many small cyclic changes and the large general motions of the world system.

They found the system they wanted in the work of several astronomers, especially Hipparchus (190?–?120 B.C.), Appollonius (261?–?190 B.C.), and Ptolemy (A.D. 100?–?170). This system had a number of combinations of circles and off-center motions. The final version, which retained the circular motions and uniform rates from Aristotle's system, was published in the *Great Syntaxis of Astronomy,* an astronomical encyclopedia compiled by the last of the great Alexandrian astronomers, Claudius Ptolemy.

The Ptolemaic system had each planet (as in Figure 2.9a) moving uniformly around a small circle called an *epicycle.* The epicyclic center in turn revolved uniformly around the earth on the circumference of a large circle called the *deferent.* Properly combining rates in the epicycle and deferent, the planetary motions could be made mostly direct and occasionally retrograde. The sun and moon had no epicycles. They completed their turns around the earth in 365.25 days and 27.3 days, respectively. Each of the two inner planets, Mercury and Venus, moved on their individual epicycles in their particular period of phases. The center of their epicycle, always located on the earth-sun line, described a complete revolu-

FIGURE 2.9
Simplified schematic of the Ptolemaic system. Each planet in part a moves along a small circle called an epicycle, which in turn revolves on the circumference of a large circle called a deferent. On the side of the epicycle farthest from the earth, the planet appears to move in the same direction as the deferent—eastward, or direct. As it ap-proaches the half of the epicycle closest to the earth, the planet appears to slow down, temporarily halt, then quickly reverse its direction and halt once more before resuming its direct motion, when it returns to the other half of the epicycle. The planet has apparently executed a complete loop, in accordance with the observations shown in part b.

Eccentric motion, shown in part c, is a refinement of epicyclic motion. In this model the earth is not located at the center of the deferent. A second refinement is to allow the center of the epicycle to move uniformly about a point called the equant.

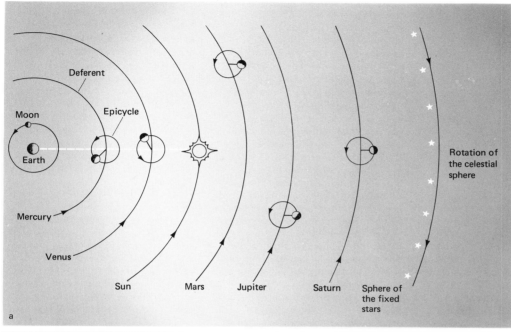

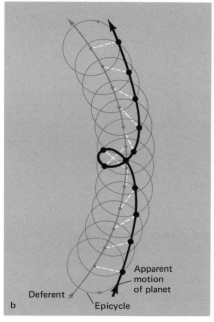

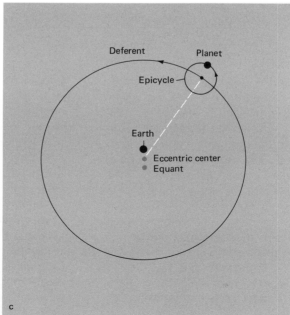

Ptolemy was the last, and perhaps the greatest, of the astronomers of antiquity. He lived most of his life in Alexandria, Egypt, which at the time was the cultural capital of the civilized world. Here scholars from many lands gathered to browse through its magnificent library. Ptolemy was both an observer and theoretician. He was the first to point out the effect of atmospheric refraction on the position of a heavenly body. And he asserted that natural phenomena require the simplest hypothesis that will coordinate the facts.

To Ptolemy must be given credit for collecting and handing down for posterity the scattered observations and principal discoveries of his Greek predecessors. These were incorporated in his famous work *Mathematical Syntaxis.* Translated later into Arabic about A.D. 827, it became known as the *Almagest* ("Greatest"). The *Almagest* is a monumental encyclopedia consisting of thirteen books. The work includes Ptolemy's refinement of the geocentric epicyclic theory into its most elegant and final form; the Hipparchus (190?–120 B.C.) catalog of some one thousand stars, plus some additions by Ptolemy; mathematical definitions and formulas; and tables by which the positions of the sun, moon, and planets can be predicted, derived from epicyclic theory. The predictions proved satisfactory within the limits of the naked-eye observations of the planetary motions. (It should be emphasized that the Ptolemaic model was a geometric representation of the planetary motions and not a physical model.) The *Almagest* became known in Western Europe through a Latin translation from the Arabic in 1175. So highly prized was the *Almagest* that it served as the standard treatise of astronomy for some fourteen centuries prior to the time of Copernicus.

The veracity of Ptolemy's observational data has been questioned by Robert R. Newton in his book *The Crime of Claudius Ptolemy* (Baltimore: Johns Hopkins University Press, 1977). He stands accused of being "the most successful fraud in history" and is charged with faking the observations of his predecessors to support his theory of planetary motion. The figures that Ptolemy quoted, according to historians of science, were the ones that agreed best with theory, out of a larger body of observations. Their consensus, however, is that the case against Ptolemy is not proven.

Ptolemy also wrote a four-volume textbook on astrology known as the *Tetrabiblos,* which deals with various astrological influences and the casting of horoscopes. It still serves as the main reference for modern astrologers. Other works by Ptolemy include a compilation of geographical maps and a discussion of optics, including light, color, reflection, and refraction.

tion around the earth in one year. Mercury and Venus, therefore, swung from one side of the sun to the other as "morning" and "evening" stars. Each of the three outer planets—Mars, Jupiter, and Saturn—moved once around their individual epicycles in their synodic periods. The period of the motion of the center of the epicycle on the deferent was thought to correspond to the sidereal period of the planet.

Refinements were made to better account for variations among the planets and the nonuniform motions of sun and moon, which were off-center motions. These devices worked in the following way. The earth (see Figure 2.9c) was set off from the center of the deferent so that a planet's *eccentric motion,* as observed from the earth, was nonuniform and varied in distance from the earth. The second refinement was to allow the center of the epicycle to move uniformly about another point, the *equant,* rather than the deferent's center. However, the center of the epicycle was no longer in strictly uniform circular motion; its motion was only uniform as seen from the equant and it was only circular as seen from the center of the deferent (the earth is located at neither the center of the deferent or the equant).

Ptolemy and astronomers before him made the geocentric system a mathematical representation of the planets' movements, not a physical model. Today's astronomy is descended from ancient Greek astronomical thought, through Arab culture, from which it spread across medieval Europe and eventually everywhere.

2.3
The Copernican Revolution: Heliocentric Cosmology

DECLINE AND REVIVAL OF ASTRONOMY

The golden era of Greek astronomy was ending by the fourth century A.D. Conspiring with Roman conquerors in that year, Bishop Theophilus partially destroyed the library at Alexandria, under the delusion that the heathen collection was a menace to Christianity's growth. Early Christian theologians had boundless religious zeal but had little interest in the conceptual structure of science and not much use for pagan Greek methods of philosophical inquiry. The heavenly sphere, they insisted, was not a sphere at all but a vault whose stars were moved by the angels.

The Romans were not strongly interested in astronomy but readily took to astrology. Cicero, the eloquent Roman orator (106–43 B.C.), firmly believed in the geocentric system. The universe, he proclaimed, consisted of nine moving globes holding the stars, planets, sun, and moon, all centered on a fixed and motionless earth. The philosopher Seneca (4 B.C.?–A.D. 65) questioned whether the earth stood still or rotated. As a few others had done, he suggested that celestial bodies rose and set because people themselves rise and set as the earth rotates.

The Arab peoples, new masters of the Byzantine Empire in the seventh century, resurrected the science of astronomy. We know of Ptolemaic cosmology because of the Arabic translation, *The Almagest*, of Ptolemy's original work, which was lost. Diligent observers and skillful calculators, the Arab peoples essentially contributed no new astronomical ideas, preferring instead to elaborate on the Ptolemaic notion of the universe. Under the pressure of Moslem religious zeal, Greek natural philosophy fared somewhat better than it did under Christian domination, particularly in mathematics.

Halfway through the thirteenth century, knowledge of astronomy had spread throughout Europe as Greek manuscripts were translated into Latin in the newly founded European universities. For example, under Spanish King Alfonso X (1221–1284), who was keenly interested in astronomy, astronomers compiled an encyclopedia of astronomical knowledge. The Renaissance blossomed in the next two centuries, ending the dominance of ecclesiastical concerns in medieval thought and beginning the development of a broader range of intellectual considerations, including cosmology. Renaissance scientists were creative in picturing the physical world, because they were not dominated by past dogmas. They prepared the way for a deep change in scientific thought and viewpoint.

THE COPERNICAN SYSTEM

Cultural ideas, including astronomy, proliferated after 1430 A.D., the year the printing press was invented. Although the Ptolemaic system had been immensely successful in describing general aspects of planetary motion for over thirteen centuries, there had arisen by this time some easily recognizable discrepancies in the observed and predicted positions of some of the planets. About the time the New World was being discovered, Nicolaus Copernicus (1473–1543), a Polish canon of ecclesiastic law and astronomer, began wondering if any arrangement of the planetary system might be simpler, more reasonable, and more aesthetically pleasing than the Ptolemaic one. He resurrected Aristarchus's heliocentric idea and built a new cosmology based on it. After nearly four decades of study, Copernicus's monumental book *On the Revolutions of the Heavenly Orbs* was finally published in the year of his death, 1543. Dedicating the work to Pope Paul III, he died without seeing his theory accepted, save by a few friends to whom he had given his manuscript years before its publication.

To soften the impact of this revolutionary theory on the church, the theologian in charge of the book's printing, Andreas Osiander, anonymously inserted this apologetic note to the reader: "It is consequently permissible to imagine causes (of the planetary motions) arbitrarily under the sole condition that they should represent geometrically the state of the heavens, and it is not necessary that such hypotheses should be true or even probable."

But this apologia did not help. "Contrary to scriptural revelations," European religious leaders said of heliocentric cosmology. Humanity was threatened; the earth was no longer the center of the universe. Martin Luther attacked Copernicus as "the fool who would overthrow the whole science of astronomy." Publicly advocating ideas contrary to the church's established doctrine was, of course, punishable as heresy. Outspoken philosopher Giordano Bruno

(1548?–1600) committed such a crime. He lectured on the Copernican theory, saying the universe was infinite and contained stars like the sun with orbiting planets that might be inhabited. Charged with heresy and apostasy by the Inquisition in 1592, he would not recant to save himself, was imprisoned, tortured, and in 1600, was burned at the stake.

Because Copernicus believed still in the Greek idea that heavenly bodies must move in perfect circles, he had to explain the deviations from uniform motion. To clear those up he injected a number of epicycles and

FIGURE 2.10
The heliocentric model, showing retrograde motion of a planet. The diagram shows ten positions of the earth and a planet as they orbit the sun and the corresponding apparent positions of the planet in the sky. Because the earth is moving more swiftly it overtakes the planet. As the earth passes, the planet appears to backtrack (between points 4 and 6) before it seems to resume its forward motion.

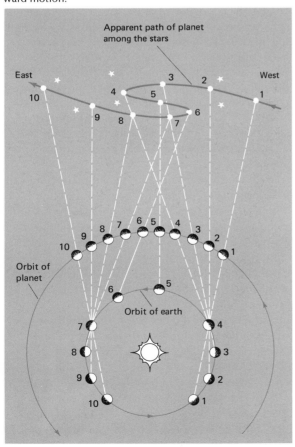

other mathematical structures. His system, then, was not much more accurate than Ptolemy's but the Copernican system was a tremendous leap for astronomy. Figure 2.10 shows that the heliocentric model was as capable of explaining retrograde motion as was the geocentric model. Thus it included all the observed motions that were included in the geocentric model. This change led, in the next century, to the realization that celestial physics was not a supernatural matter but only an extension of terrestrial physics; Isaac Newton was later to make that clear.

2.4
Copernican Successors: The Rise of Modern Astronomy

BRAHE'S CONTRIBUTIONS
Appearing at an opportune time, the right man for the next advance in astronomy was Danish nobleman-astronomer Tycho Brahe (1546–1601). With financial help from King Frederick II he constructed in 1582 a superbly equipped observatory on the Island of Hveen, about 32 kilometers (20 miles) northeast of Copenhagen. There, with the most accurate pretelescopic observing instruments ever designed, Brahe determined positions with a precision of one minute of arc, far surpassing any previous measurements.

Brahe observed the sun, moon, and stars regularly instead of haphazardly as others had in the past. An uninterrupted record of their movements over many years was thus available for study and analysis. Brahe had reservations about adopting the entire heliocentric theory. He accepted the idea that the five planets revolved around the sun but not the idea that the heavy and sluggish earth moved. Earth's motion would be felt, he argued—and besides, a moving earth was contrary to scriptural belief. Neither could he detect the earth's orbital motion by parallactic shifts in the positions of the brighter stars. Consequently, Brahe's cosmological system was a compromise: the planets orbited the sun while the sun and moon, in turn, orbited a fixed earth.

KEPLER AND LAWS OF PLANETARY MOTION
In the years just prior to 1600 the Renaissance and Reformation were coming to an end. Copernicus's works were read by a few astronomers who recognized the computational advantages of the Coper-

NICOLAUS COPERNICUS (1473–1543)

The 500th anniversary of Copernicus's birth was celebrated throughout the world in 1973. Various governments, including ours, issued commemorative stamps in his honor; the last of the two orbiting astronomical observatories, launched in 1972 and still in orbit, was named *OAO-Copernicus*; and histo-rians of science met to eulogize the accomplishments of the man who revolutionized astronomy with his planetary heliocentric system.

After the death of his father when Copernicus was ten years old, an uncle, who was also a bishop, raised him and saw to it that he had an excellent education. Copernicus studied mathematics, philosophy, astronomy, and astrology at the University of Cracow; he studied law and medicine at the universities of Bologna and Padua. When he returned to Poland, he lived for a while in his uncle's castle. There he spent time as a physician, engaged in diplomatic activities, and undertook various administrative duties. After he was elected a canon through his uncle's influence, he had sufficient income to devote more of his time to astronomy, his first love.

Beginning in 1512, Copernicus set himself the task of examining critically the various systems of the world that had been proposed in the past. After several decades of study he became dissatisfied with the complexity and improbability he found in the Ptolemaic system. Placing the sun at the center of the solar system simplified matters. "... In the center of everything the sun must reside; ... there is the place which awaits him where he can give light to all the planets." So wrote Copernicus. In his development of the heliocentric system he retained the notion of uniform circular planetary motion. He was thereby compelled to introduce a number of epicycles and eccentrics in order to account for the variable movements of the planets.

In 1530 Copernicus circulated a summary of his ideas among his friends. Knowledge of it spread to others. Eventually the fruits of his full labor appeared in print in 1543. To a cardinal friend who had inquired about his theory, Copernicus wrote: "Although I know the thoughts of a philosopher do not depend on the judgment of the many, yet when I considered how absurd my doctrine would appear, I long hesitated whether I should publish my book." His book is divided into six volumes. Included are discussions of the heliocentric concept, the geometry of the spheres of the heavens, and the earth's motions, including precession, lunar theory, and the planetary motions.

nican system but were not willing to take seriously its philosophical and physical implications. However, a devoted Copernican, Johannes Kepler (1571–1630), the German assistant and successor to Tycho Brahe, was to change its acceptance. Brahe's observations of the celestial bodies bore fruit, when Kepler pulled from Brahe's records his history-making discoveries. The question that concerned Kepler was: what was the clockwork that governed the celestial machinery? After seventeen years of labor, during which he rejected many ideas because they did not fit Brahe's observations, Kepler explained in two books, published in 1609 and in 1619, how the planets moved. His science was uncompromising and conformed to today's standards: a theoretical model must satisfy all observational facts or fail. "By the study of the orbit of Mars," he said, "we must either arrive at the secrets of astronomy or forever remain in ignorance of them."

Kepler solved the problem on which so many astronomers had labored for centuries. The orbits of the planets are *ellipses*, not circles, and their varying motions are due to their varying distances from the sun. These are his first two laws, briefly stated (see Figure 2.11). The third states the relationship between the planets' orbital periods and their distance from the sun.

KEPLER'S FIRST LAW (LAW OF ELLIPTIC ORBITS): *Each planet moves in an elliptic orbit around the sun with the sun occupying one of the two foci of the ellipse.*

KEPLER'S SECOND LAW (LAW OF AREAS): *The imaginary line connecting any planet to the sun sweeps over equal areas of the ellipse in equal intervals of time.*

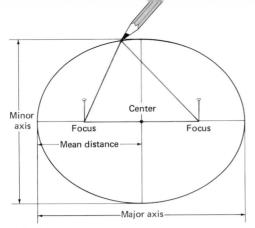

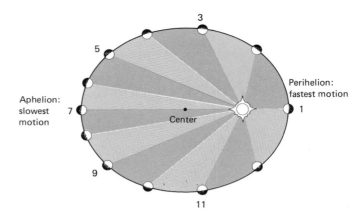

FIGURE 2.11

Kepler's first two laws. Kepler said that the planets move in elliptical orbits. (To draw an ellipse, loop a string taut around two tacks—the foci—and a pencil.) An elliptical orbit is shown with the focus (sun) off center. The mean distance of the planet is equal to one-half the length of the major axis, whose end points are 1 and 7. Because the areas of all sectors, gray and colored, are equal, the planet passes through the numbered positions in equal intervals of time in accordance with Kepler's law of areas. The planet moves fastest when near the sun and slowest when farthest away.

KEPLER'S THIRD LAW (HARMONIC LAW): *The square of any planet's period of orbital revolution is proportional to the cube of its mean distance from the sun.*

Kepler's laws are universal. They apply to any two bodies gravitationally bound to each other, whether in the solar system or elsewhere in the universe. Kepler buried forever the notion that the planets moved in perfectly circular orbits because nature decreed that the heavenly bodies must show perfection in their movements. Of course, bodies may move in circular orbits when the two focal points coincide with the center of a circle, which is the limiting form of the ellipse. Kepler did not know why the planets moved by these empirical relationships that he had established from Brahe's observations. He vaguely sensed that bodies have a natural magnetic affinity for each other and guessed that the sun has an attractive force. It remained for Newton, half a century later, to formulate a unified theory of motion, which includes planetary motion.

The substitution of a kinematic planetary model for a purely geometric one was Kepler's primary achievement. He prepared the way for the modern theory of force and the mathematical analysis of motion in terms of forces. A chief proponent of this new idea was the Italian physicist-astronomer Galileo Galilei (1564–1642). From his experiments with bodies in motion and the forces controlling them came the foundations of modern mechanics.

GALILEO AND HIS TELESCOPE

Mechanics was one of his accomplishments, but Galileo also revolutionized astronomy in 1609 with an optic tube, or telescope, of his own construction. As the first telescopic explorer of the heavens, he established his place in history through his discoveries: Jupiter's four large satellites, craters and mountains on the moon, the phases of Venus, individual stars in the Milky Way. Kepler had been the first to demonstrate that the heliocentric system was valid, and Galileo gave the Copernican theory observational support. For example, Galileo observed that, on a smaller scale, Jupiter's satellites moving around the planet were analogous to the planets orbiting the sun. They were obviously heavenly bodies not in orbit about the earth, and here also was evidence disputing Aristotle's contention that a moving earth would leave the moon behind. Jupiter retains its satellites; logically, then, the earth can move around the sun without losing its satellite.

Theological hostility loomed over Galileo for supporting Copernican cosmology. Pope Paul V instructed his emissary Cardinal Bellarmine to warn Galileo against teaching or upholding Copernican doctrine, and from the Holy Office in February 1616 came this stern decree:

The following propositions are to be censured: (1) that the Sun is at the center of the world and the universe. . . . Unanimously, this proposition has been declared stupid and absurd as a philosophy, and formally heretic because it contradicts in express manner sentences in the Holy Scripture. . . . (2) that the Earth is not the center of the world and motionless, but changes its place entirely according to its diurnal movement. Unanimously, this proposition is declared false as a philosophy. . . .

But liberal Pope Urban VIII took office, and Galileo obtained permission to teach both the Ptolemaic and Copernican systems; he was, though, to present the latter as an unproven alternative. Encouraged, Galileo began work on his masterful astronomical commentary, which passed censorship and was published in

1632: *The Dialogues of Galileo Galilei on the Two Principal Systems of the World: The Ptolemaic and Copernican.* Powerful enemies soon convinced the Pope that Galileo had cast the Ptolemaic system in an unfavorable light. The book was officially banned. The great scientist was publicly humiliated before a papal tribunal and forced to recant his heretical views. In his declining years he was placed under house detention, lasting until his death in 1642. The official condemnation of Galileo was not lifted until 1965.

GEOMETRICAL AND MATHEMATICAL FORM OF KEPLER'S LAWS

The ellipse, a family of mathematical curves, is important in discussing the orbits of bodies about each other (other families of curves will be discussed in Chapter 3). Roughly speaking, an ellipse is a circle with the opposite ends of a diameter pulled outward, distorting it into an oval-shaped figure. The long axis of the ellipse (see Figure 2.11) is called the *major axis*, and perpendicular to it through the center of the figure is the *minor axis*. There are two points on the major axis, called the *foci* (the singular form is *focus*), about which the figure is roughly symmetric. The sum of the distances from each of the foci to every point of the ellipse is a constant. This immediately suggests a means of drawing an ellipse: loop a piece of string around two tacks (the foci) and a pencil as shown in Figure 2.11. In a planet's orbit the sun occupies one focus; the other one is empty. The further the foci are from each other, the more elongated the ellipse; the closer together they are, the more nearly circular the ellipse.

The mean distance of the planet from the sun is the average of the distance between the point of closest approach, called *perihelion*, which is located at one end of the major axis (point 1 in Figure 2.11), and the most distant point of the orbit, called *aphelion*, which is located at the other end of the major axis (point 7 in Figure 2.11). The average is one-half the length of the major axis or the semimajor axis, as shown in the figure. The alternate gray and colored sectors in the figure are of equal area. Therefore, according to Kepler's second law, the planet passes through the numbered positions in equal intervals of time.

In mathematical shorthand we write Kepler's third law as (see Appendix 2 on measurements and computations)

$$P^2 = Ka^3$$

where P is the sidereal period of revolution, a is the mean distance of the planet from the sun, and K is a constant of proportionality depending upon the units for P and a. Even though Kepler did not know the numerical value of the constant K, he could eliminate the constant by assuming that K was the same for each planet. He then considered the ratio of the equation for any planet to that for the earth:

$$\frac{P^2}{P_\oplus{}^2} = \frac{a^3}{a_\oplus{}^3}$$

where the subscript $\oplus$ stands for the earth. If we express the planet's sidereal period in years and its mean distance in units of the mean earth-sun distance (known as the *astronomical unit* and now known to be equal to 149,597,871 kilometers), then $P_\oplus = 1$ and $a_\oplus = 1$, and Kepler's third law becomes

$$P^2 = a^3$$

Example: Let us calculate the mean distance of Jupiter from the sun given that its sidereal period is about 11.8 years. From Kepler's third law we obtain an expression for a

$$a = \sqrt[3]{P^2} = \sqrt[3]{(11.8)^2} = 5.2 \text{ astronomical units.}$$

SUMMARY

The astronomy of the ancient world was primarily to observe and record sky phenomena, such as the daily motion of the stars, the motion of the sun, the seasons, phases of the moon, lunar and solar eclipses, and the motions and configurations of the planets. For these common and repetitive occurrences in the sky, there are no historical dates on which human beings first recognize that they were happening. Much of the motivation for watching these events was that they marked the passage of time. But, they were also aspects of the earth's relationship to the universe and as such stimulated the ancients' thinking about the nature of their world (cosmological thought).

The most eminent Greek philosophers strained for a system that would explain celestial movements. For centuries geocentric theories were disputed over. The last great Greek philosopher, Claudius Ptolemy, gave the earth-centered universe its most elegant form about A.D. 140. For fifteen centuries more it was the supreme astronomical theory, buttressed by the weight of Aristotelian and theological authority. Copernicus was its first strong challenger, proposing a heliocentric model of the universe. His successors, Kepler and Galileo, confirmed the idea in the next century. Kepler found that the orbits of the planets are ellipses and that the orbital speed and period of any planet depend on its distance from the sun. Kepler demonstrated that the heliocentric concept of Copernicus was valid; Galileo, using a telescope to explore the heavens, found evidence to support the theory.

REVIEW QUESTIONS

1. How did the ancients differentiate between a planet and a star?

2. What is the zodiac?

3. How did the Ptolemaic system account for the retrograde motions of the planets and the non-uniform motions of the sun and moon? How does the Copernican system account for these motions?

4. What observational evidence did Copernicus have that his system was the correct interpretation of planetary motions?

5. Why does the Copernican revolution represent a turning point in the history of science?

6. Why was Tycho Brahe reluctant to fully accept the Copernican system?

7. State in your own words, with the help of a diagram if necessary, Kepler's three laws of planetary motion.

8. How was Kepler able to determine the correct relative distances of the planets from the sun?

9. What were Galileo's contributions to astronomy?

10. Why are the signs of the zodiac out of step with the constellations of the same name?

11. Why are there different constellations associated with the four seasons?

12. Consulting the winter sky map in Appendix 3, about what time of night would you see the constellation of Orion due south in mid February?

13. Facing north at 8 P.M. on 5 November, where would you find Ursa Major in the sky relative to the horizon?

14. Why is the brightest star in the sky, Sirius, not visible in the summer?

CONCEPTS OF MOTION

Modern astronomy needed one last, vital foundation stone. Isaac Newton (1642–1727) supplied it when he developed his concept of gravity and the physical laws relating to matter, force, and motion. These Newton published in 1687 in his *Principia*, whose complete title, translated, is *The Mathematical Principles of Natural Philosophy*. The concept of gravity, it became clear, was an abundant source of solutions to old problems. With this one theory Newton accounted for the rise and fall of tides, described more accurately the motions of the moon and the planets, and explained why the earth is flattened at the poles. His theory also explained the precession of the equinoxes (see page 18) and opened the door for the discovery of the planet Neptune.

To prepare for the study of Newton's laws, and for ideas that will come up in later chapters, let us consider some fundamental concepts of motion. The first term we will define is sometimes used as if its meaning were obvious. *Force* is the concept, and we can visualize it as a push or a pull that makes a body change its *state of motion*. We should note that rest (or no apparent motion) is one possibility for a state of motion. The most familiar forces to us are *mechanical forces;* these are exerted by one body in contact with another, such as a batter hitting a baseball. *Fields of force,* which are exerted without contact between bodies, comprise the other class of forces. There are four types of fields of force (see page 73). These are the *strong and the weak nuclear forces,* which act on the subatomic scale; *electromagnetic forces,* which can be either attractive or repulsive; and the *gravitational force,* which exerts a pull or attraction but never a push or repulsion. Empty space between matter is no barrier to the fields of force, as everyday experience has taught us in the case of the gravitational force.

◀ The title page of Newton's famous publication *The Mathematical Principles of Natural Philosophy*. The noted diarist, Samuel Pepys, was president of the Royal Society at the time.

Every material object possesses a property called *inertia*, the resistance it offers to any change in its state of motion. The more matter a body has, the greater is its inertia. *Mass* is a measure of the amount of matter a body contains; therefore, mass measures the body's inertia. Massive bodies are more resistant to a change in their state of motion than are less massive bodies, as we know from the common experience of moving rocks. A material body has the same mass regardless of where in the universe it is located, but its weight depends on its position relative to various attracting masses. An object's *weight* is a measure of the gravitational force that an attracting body exerts on the object. For example, a person weighing 90 kilograms on the earth's surface would weigh 15 kilograms on the moon's surface because the moon's gravitational pull is one-sixth that of the earth's. However, the person's mass is the same whether on the moon or on the earth.

An important concept related to mass is an object's compactness or *density*. Density is defined as the amount of matter (mass or weight) in each unit of volume. Water has a density of 1 gram per cubic centimeter, while that of lead is 11.3 grams per cubic centimeter. If matter is distributed unevenly throughout the body, as in the earth, then the mass divided by the volume gives a *mean density*.

To fully describe the state of motion of a body, we must have definitions for the concepts of distance, time, velocity, and acceleration.

Distance is a familiar idea, but for our use we need the notions of the origin and the direction in which it is being measured. We therefore construct a *reference frame*, with an origin from which distances can be measured relative to some reference direction. As an example, let's consider the distances to the planets. To measure them, we must first select a reference frame, such as the sun as the origin; and as our reference direction we might choose the direction toward a given star lying in the ecliptic plane. This is only one of several possible reference frames that could be used to measure the distance to the planets.

Equally familiar from our everyday experiences is the idea of *time*. However, our intuitive concept of time is based upon changing patterns and events in our lives, which is quite different from our intuitive concept of distance—the separation between tables and chairs, for example. Time can be measured from an arbitrarily chosen origin such as a historical event. But, unlike distance, time moves only forward and never backward in common experience. Newton and most scientists after him believed in an absolute time and an absolute space that are unchanging qualities of the universe. It was not until the late 1800s that Ernst Mach (1838–1916), an Austrian physicist-philosopher, began questioning the concept of an absolute time. Few took Mach's criticism seriously in his lifetime. Later Albert Einstein (1879–1955), a German physicist, would be influenced by Mach's thinking in developing his theory of relativity. However, in the period between Newton and Einstein, the concepts of absolute time and space were the pervasive influences in the development of science.

For a moving body the distance traversed divided by the elapsed time is the *speed* of the body. If we include the direction of motion along with the speed of motion, then we have the *velocity* of the body. It is important to remember that a change in the state of motion or velocity of a body, which is known as *acceleration*, can be due to a change in the speed or the direction of motion or both. Thus acceleration is measured as the rate of increase or decrease of a body's speed or as the rate at which its direction of motion changes. Since distance depends upon a frame of reference, then so also will velocity and acceleration. An airplane's velocity, for example, is usually measured relative to the surface of the earth, say, 600 miles per hour. However, we could measure it relative to the sun, in which case the earth's rotational and orbital velocities would have to be added to that of the airplane. But any fixed point on the surface of the earth is continuously changing velocity—or accelerating—due to the spin of the earth on its axis of rotation and to the revolution of the earth about the sun. Hence what we really want to talk about in the case of the orbital motion of planets is the velocity at one single moment of time, since an instant later the velocity is different due to acceleration. This concept is called *instantaneous velocity*.

As you might guess, there is more to defining the motion of a body than simply declaring a value for its velocity. For example, in the collision of two billiard balls moving with different velocities, it may appear that they simply exchange their states of motion, so that the total quantity of motion is conserved and just redistributed between them. But for two billiard balls with different masses moving with the same speed, the more massive one is able to transfer a greater quantity of motion in a collision than the small-mass ball. Thus the concept of quantity of motion, or

momentum, involves both the velocity and the mass of the body. To find momentum, we multiply the body's mass by its velocity. It is not difficult to visualize philosophically the possibility that the total quantity of motion or momentum in the universe is a constant—that is, the total momentum of the universe is conserved. However, the interaction of various bodies in the universe with each other redistributes the momentum of individual bodies. This concept of *conservation of momentum* is due to the French philosopher René Descartes (1596–1650).

> I do not know what I may appear to the world, but to myself I seem to have been only like a boy playing on the seashore and diverting myself in now and then finding a smoother pebble or a prettier shell than ordinary, whilst the great ocean of truth lay all undiscovered before me.
>
> Isaac Newton

NEWTON'S LAWS OF MOTION

We know how to measure how much a body moves, but can we predict how it will do so? In his book *Principia*, Newton formulated three laws that describe and predict how things move. The first law is about uniform motion.

NEWTON'S FIRST LAW: *A body remains at rest or moves along a straight line with constant velocity as long as no external force acts upon it.*

If a body is in motion in a straight line, it continues to move along that line without changing its speed or direction as long as no force acts on the body. Also, its momentum remains the same.

Newton's second law deals with nonuniform motion, or velocity changes, and why it occurs.

ISAAC NEWTON (1642–1727)

At his birth on Christmas day (1642) in Woolsthorpe, Lincolnshire, Newton was so tiny and frail that he was not expected to live. Despite his boyhood frailty, Newton lived to the age of eighty-five. As a delicate child he was a loner, more interested in reading, solving mathematical problems, and mechanical tinkering than taking part in the usual boyish activities. Up to the time Newton entered Cambridge University in 1661, however, there was no inkling of his mental prowess. His shyness kept him from making friends easily and he did not mix with his more boisterous fellow students. At the uni-

versity he took courses in Latin, Greek, Hebrew, logic, geometry, and trigonometry, and he attended lectures in astronomy, natural philosophy, and optics. His leisure time was spent reading works by Kepler and by Descartes, the inventor of analytic geometry, and filling his notebooks with remarks on the refraction of light, the grinding of lenses, and the extraction of roots of algebraic equations.

Newton received his BA degree—without any great distinction—and then returned to his home because of the Great Plague that was sweeping Europe. Practically all his time in his early twenties was spent at his home in Woolsthorpe. These were the most profitable years of his life. During this productive period he discovered the expansion of the general binomial $(a + b)^n$; he invented the fluxions (differential calculus); he demonstrated the composite nature of white light with prisms; he discovered the law of gravitation; and he laid the foundations of celestial mechanics.

In 1668 Newton constructed the world's first reflecting telescope. It had an aperture of 1 inch and a tube length of 6 inches, which led Newton to say of it: "This small instrument, though in itself contemptible, may yet be looked upon as the epitome of what may be done this way." Not satisfied with his first effort, he completed an improved

and somewhat larger reflector with an aperture of nearly 2 inches. This one, together with a drawing, is owned by the Royal Society (see page 82).

The publication of his *Principia* (1687), embodying his mathematical principles and his ideas on gravitation and the system of the world, marked the near close of Newton's creative career in science. The *Principia* represents the thought and study of more than twenty years, and it ranks in importance with Ptolemy's *Almagest* and Copernicus's *De Revolutionibus*. His treatise on *Opticks* appeared in 1704, but most of it was written many years earlier.

Many tributes followed Newton's death in 1727. One in particular that stands out was made by the great French mathematical astronomer Lagrange, who said: "Newton was the greatest genius who ever lived, and the most fortunate, for we cannot find more than once a system of the world to establish." Newton himself acknowledged: "If I have seen further than other men, it is because I have stood upon the shoulders of giants." In poet Alexander Pope's *Epitaph for Newton* are these lines:

Nature and Nature's laws lay hid in night;
God said, "Let Newton be!" and all was light.

NEWTON'S SECOND LAW: *A body acted upon by a force will accelerate in the direction of the applied force. The greater the force, the greater the acceleration will be.*

Acceleration is proportional to the magnitude of the acting force—as one increases, the other does, too—and inversely proportional to the body's mass—as one increases, the other decreases. If a moving body is subjected momentarily to an external force, it will momentarily accelerate in the direction given it by

the applied force. Its velocity changes to a new value, but its new velocity is not necessarily in the same direction as the force. But if the force is applied continuously, then a continuous change of velocity, or acceleration, takes place. For example, as a planet orbits the sun, there is a continuous change of speed and direction, or a continuous acceleration. The change must be due to some continuously applied external force. In this case, we know that that force is the gravitational attraction of the sun.

Newton's third law deals with the relation between forces.

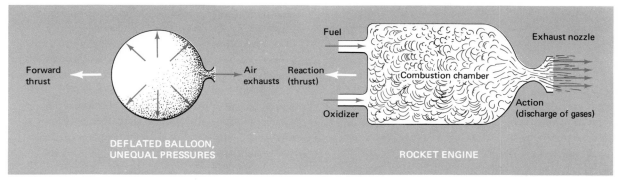

FIGURE 3.1
The principle of action and reaction in a
balloon and in a rocket engine.

NEWTON'S THIRD LAW: *A body subjected to a force
reacts with an equal counterforce to the applied
force; that is, action and reaction are equal and
oppositely directed.*

Two forces are involved here: action and reaction.
One force never acts alone. We can see equal and
opposite reactive forces every day. A bird taking off
from an overhead power line produces a reaction
that moves the line backward. Water forced out of
a lawn sprinkler produces a backward reaction that
rotates the sprinkler.

ROCKETS: AN APPLICATION OF
NEWTON'S THIRD LAW

From space our knowledge about the solar system
has expanded incredibly, as we will discuss in later
chapters. Yet no escape from earth to find this knowledge would have been possible without an old idea
that is nevertheless part of our new technology: the
rocket.

A rocket engine operates on the principle of action
and reaction, as expressed in Newton's third law of
motion. If you hold the neck of an inflated balloon
(see Figure 3.1) to keep the compressed air from escaping, nothing happens—the internal gas pressure is
equally balanced in all directions. But if you release
the neck, the air rushes out (action), causing the
deflated balloon to move in the opposite direction
(reaction).

In a liquid propellant rocket engine, a mixture of
fuel and oxidizer (such as liquid hydrogen and liquid

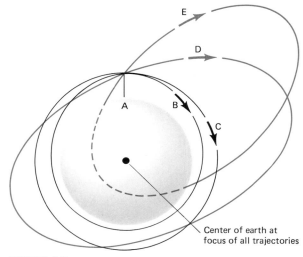

FIGURE 3.2
Rocket trajectories: A, launch point; B,
elliptic orbit (injection approximately
horizontal with a velocity of slightly
under 8 km/s); C, circular orbit (injection horizontal to the earth's surface
at 8 km/c); D, elliptic orbit (injection angle greater than 0° with a
velocity somewhat over 8 km/s but less
than the escape velocity of 11 km/s); E,
intercept orbit (injection angle too
high).

oxygen respectively) is pumped into the rocket's combustion chamber. There it explodes in a hot blast of
gases that are directed outward at high speed through
the exhaust nozzle. As stated in Newton's third law,
the force of the chemical reaction produces a thrust
that accelerates the rocket forward. As the burning
continues, the rocket continues to accelerate until it
reaches its maximum velocity at burnout. The type of
orbit achieved is determined by the final injection
velocity and the angle the injection makes with the
horizontal at the earth's surface (see Figure 3.2).

3.2
Gravity

NEWTON DERIVES THE LAW OF GRAVITATION

The recognition that the earth exerts a pull or tug upon an object which is similar to the tug an individual might exert on that same object was not original with Newton. However, it was Newton who recognized that the pull of the earth could extend all the way to infinity. This was a universal phenomenon to which all material bodies were subject.

Without bogging ourselves down in complications, we can follow Newton's path of reasoning to establish the law of gravitation. Taking Kepler's first law (Section 2.4) and his own second law of motion, Newton made this inference: If a planet moves in a curved path—in this case an ellipse—a force must be acting on it. Analyzing Kepler's second law mathematically, Newton showed that the force acting on a planet must be a *central force*, the sun. Only one kind of force, Newton deduced, would satisfy Kepler's requirement that the sun must be at the focus of the ellipse and still be consistent with Kepler's third law relating the planets' periods to their distances from the sun. The force between the planets and the sun must be an *inverse-square force*. That is, the force must weaken as the square of the distance between a planet and the sun increases. Several contemporaries of Newton

had suspected this relationship, but they could not prove it.

Newton then took his third law of motion and assembled his results in one comprehensive statement, the law of gravitation. He showed that it was universal by applying it to a falling apple and the earth, to the moon's motion around the earth, and to the planets revolving around the sun. He even imagined gravitation at work beyond the solar system, a thought that was verified later.

NEWTON'S LAW OF GRAVITATION: *Objects in the universe attract each other with a force that varies directly as the product of their masses and inversely as the square of their distances from each other.*

Newton proved that spherical bodies act as if their *gravitational mass* is concentrated at their centers. This simplifies their mathematical treatment because the distance between their centers is ordinarily used in calculating their mutual gravitational attractions. Figure 3.3 shows examples of how the force of gravity varies under different circumstances.

Newton tested his law of gravitation by comparing the acceleration of a falling body such as an apple at the earth's surface, which is 980 centimeters per second squared, with the moon's acceleration at a distance of 60 earth radii from the earth's center. Gravitational force decreases as the square of the distance;

MATHEMATICAL FORM OF NEWTON'S THIRD LAW

In mathematical form Newton's third law can be written as

$$F_1 = F_2$$

where F_1 is the force acting on body 1 and F_2 is the reactive force acting on body 2. From a combination of Newton's second and third laws, it follows that

$$m_1a_1 = m_2a_2$$

where the subscripts refer to body 1 and body 2.

Example: The force of attraction the earth has for you (F_1), standing on its surface, is *equal* and *oppositely directed* to the force of attraction you have for the earth (F_2). Both the earth and you are accelerated. But because the earth's mass is very large, the acceleration it gets is infinitesimal compared with that you get, since your body's mass is very small. That is why we can see the acceleration of a body falling at the earth's surface (980 centimeters per second squared) but not the acceleration of the earth, which is about 1.5×10^{-21} centimeter per second squared for a 90-kilogram person. Only when two masses are more nearly the same can we easily observe the accelerations of both bodies.

therefore, the moon's acceleration should be $1/60^2$ or 1/3600 that of the apple's acceleration, or about 0.3 centimeter per second squared. From knowledge of the moon's orbit, Newton calculated its acceleration; the result agreed well with that calculated from his law of gravity.

PLANETS, ORBITS, AND NEWTON

Newton used his laws of motion and gravitation to show that Kepler's third law was only an approximation to the actual relation. The actual form of the law is much more useful, for it allows us to determine the masses of other celestial bodies (see pp. 259–265).

Newton also showed that the orbit of a body revolving around a central force always matches in shape one of the class of curves called *conic sections*, illustrated in Figure 3.4.

These curves are called conic sections because they are formed when we pass a plane (like a knife blade) through a cone at different angles. For the *ellipse*, of which a planetary orbit is one example, the cutting plane intersects opposite sides of the cone's slant edge. For a *circle* the plane cuts the cone at right angles to the vertical axis. The other two conic sections are open at one end. The *parabola* is formed when the plane passes through the cone parallel to its slant

THE LAW OF GRAVITATION AND THE MASS OF THE EARTH

If m_1 stands for the mass of one body, m_2 for the mass of a second body, d for the distance between their centers, F for the mutual force of gravity between them, and G for the constant of gravitation, the law of gravitation can be stated mathematically as

$$F = \frac{Gm_1m_2}{d^2}$$

The constant of gravitation G was first measured in the eighteenth century by Henry Cavendish (1731–1810). One method for measuring it makes use of a torsion balance, a device consisting of a lightweight rod that is suspended at the center by a fine quartz fiber and carries a small sphere at each end. If a body of large mass is brought close to one of these spheres, the fiber will twist as a result of the gravitational attraction between the small sphere and the large body. From the known masses of the two bodies m_1 and m_2, their separation d, and the amount of twist, which depends on the gravitational force F, we can calculate the constant G. Its numerical value in the metric system is 0.0000000667 (6.67×10^{-8}) centimeter cubed per gram second squared.

Galileo is said to have demonstrated that balls of different weights dropped from the Leaning Tower of Pisa fell with a constant acceleration. Why? From the second law of motion we know that a body of mass m, subjected to the earth's gravitational force of attraction F, undergoes an acceleration at the earth's surface of $g = F/m$. From the law of gravitation this force is $F = GmM_\oplus/R_\oplus$, where $M_\oplus$ is the mass of the earth and $R_\oplus$ is the separation between the centers of the two bodies, or the earth's radius. Assuming that we know G, we have

$$mg = \frac{GmM_\oplus}{R_\oplus{}^2} \quad \text{or} \quad g = \frac{GM_\oplus}{R_\oplus{}^2}$$

where the mass of the attracted body has canceled out. Thus the acceleration of the attracted body does not depend on its own mass; it depends on the mass of the attracting body, which in this case is the earth.

Example: The observed acceleration g at earth's surface is 980 centimeters per second squared. From the known values of the gravitational constant G and the earth's radius (6.38×10^8 centimeters), we can find the mass of the earth by rearranging the equation above to obtain

$$M_\oplus = \frac{gR_\oplus{}^2}{G} = \frac{(980 \text{ cm/s}^2)(6.38 \times 10^8 \text{ cm})^2}{(6.67 \times 10^{-8} \text{ cm}^3/\text{g} \cdot \text{s}^2)} = 5.98 \times 10^{27} \text{ g}$$

and from this the earth's mean density is

$$\rho_\oplus = \frac{M_\oplus}{(4/3)\pi R_\oplus{}^3} = 5.5 \text{ g/cm}^3$$

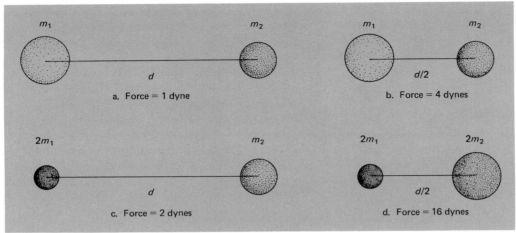

a. Force = 1 dyne

b. Force = 4 dynes

c. Force = 2 dynes

d. Force = 16 dynes

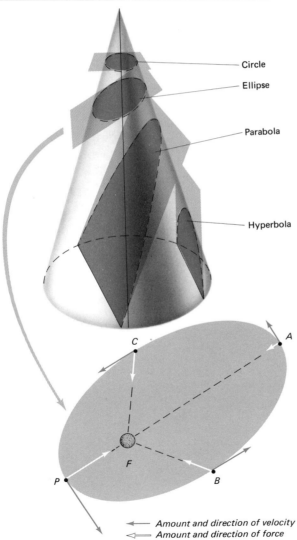

Circle

Ellipse

Parabola

Hyperbola

⟵ Amount and direction of velocity
⟸ Amount and direction of force

FIGURE 3.3

Comparison of gravitational forces. The greater the mass of the body, the darker the shading in the drawing. If the gravitational force between the two masses m_1 and m_2 at distance d (part a) is equivalent to 1 gram centimeter per second squared, or 1 dyne, then at half the original distance but with the same masses the force is quadrupled (part b). If we replace m_1 with a smaller but denser body (part c), having doubled the mass, we double the original force. Doubling both masses but halving the original distance (part d) results in a gravitational force 16 times as strong as the original force. Note that the mass of the body is the important element, not the size. The relationship can be expressed as $F \propto m_1 \cdot m_2/d^2$.

FIGURE 3.4

Permissible orbits under the law of gravitation. The orbital eccentricity e, which is a measure of how stretched out the orbit is, for each shape is as follows: circle, $e = 0$; ellipse, $0 < e < 1$; parabola, $e = 1$; hyperbola, $e > 1$. The factors involved in shaping these orbits are clarified by the lower drawing, which takes the case of the ellipse. At A the body is moving at right angles to the direction of the attracting force F, which tends to pull it toward C, causing it to change direction toward C and to speed up. Closer in at C the tangential velocity increases because of the stronger force. As the body approaches closer to F, the increasingly stronger attractive force causes the body to curve around it, attaining maximum velocity at P. As the body swings outward past P, its motion is retarded by the attractive force; after passing B the body decelerates until it finally reaches A; and the action repeats.

edge. The *hyperbola* is formed when the cone is intersected at an angle which is between that for the parabola and parallel to the vertical axis (as shown in Figure 3.4). In a parabolic or hyperbolic orbit the body will pass by the attractive central force only once, receding toward infinity, never to return. The different members of the sun's family move around the sun in closed paths, either ellipses or something approaching a circle. An object approaching the sun from outside the solar system would, when attracted by the sun, travel around it in a parabolic or hyperbolic orbit. If its motion is significantly influenced by the gravitational attraction of a planet during a near encounter with a planet such as Jupiter, it might be forced into an elliptical orbit around the sun, in which case we say it has been gravitationally captured.

3.3
Relativity

EINSTEIN MODIFIES NEWTON'S LAW

The Newtonian theory of gravitation has been very successful in explaining most gravitational problems involving the material objects of the macroscopic world. However, in another world, the submicroscopic universe of the atom, nuclear and electromagnetic forces dominate the very weak gravitational force, and they determine form and behavior in this realm. Also, for the movement of bodies at high velocities or in the presence of very strong gravitational fields, such as that of a star, Newton's theory of gravitation does not adequately describe what happens. Relativity has changed cosmology profoundly, helping to explain cosmic reality, as we shall discuss in Chapter 18.

The special theory of relativity of 1905 was not originally designed as a refinement of Newtonian gravitational theory but was addressed by Albert Einstein (1879–1955) to some aspects of electromagnetic radiation, that is, light. *Special relativity theory* can deal only with physical phenomena in which gravitational forces are not involved. By 1916 Einstein had worked out another, more comprehensive theory, which was an alternative to the gravitational theory of Newton. He incorporated in his *general theory of relativity* a description of nonuniform or accelerated motion between observers. The theory of relativity has important implications for our ideas of motion, mass, and the geometry of space and time.

NEWTON'S MODIFICATION OF KEPLER'S THIRD LAW

Recall Kepler's third law relating a planet's period of revolution to its mean distance from the sun:

$$P^2 = Ka^3$$

Newton worked out a more precise version of Kepler's law, which is expressed mathematically as

$$P^2 = \frac{4\pi^2 a^3}{G(M_\odot + M_p)}$$

where $K = 4\pi^2/[G(M_\odot + M_p)]$, $M_\odot$ is the mass of the sun, M_p is the mass of the planet, P is the orbital period, and a is its mean distance from the sun. Notice that K is not quite a true constant as Kepler had supposed, because the mass of each planet, though a small fraction of the solar mass, is not truly zero.

Example: Let us determine the mass of the sun. Applying Newton's modification of Kepler's third law to the earth and the sun, we have

$$M_\odot + M_\oplus = \frac{4\pi^2 a^3}{G P_\oplus^2} = \frac{(39.5)(1.496 \times 10^{13} \text{ cm})^3}{(6.67 \times 10^{-8} \text{ cm}^3/\text{g} \cdot \text{s}^2)(3.156 \times 10^7 \text{ s})^2} = 1.99 \times 10^{33} \text{ g}$$

where $M_\odot$ is the mass of the sun; $M_\oplus$ is the mass of the earth; a is the earth's mean distance from the sun, 1.496×10^{13} centimeters; G is the gravitational constant, 6.67×10^{-8} centimeters per gram second squared; and $P_\oplus$ is earth's period of revolution around the sun, 3.156×10^7 seconds. Earth's mass (6×10^{27} or 0.000006×10^{33} grams) is insignificant here and may be neglected. Thus the mass of the sun is 1.99×10^{33} grams.

ALBERT EINSTEIN (1879–1955)

In 1979 many countries celebrated the 100th anniversary of Einstein's birth. To mark the occasion, a large bronze statue of Einstein was sculpted and placed on the grounds of the National Academy of Sciences. The second *High Energy Astronomy Observatory (HEAO-2)*, launched November 1978, was renamed the *Einstein Observatory.*

As a child Einstein developed slowly. During his early years he showed no special aptitude in elementary and secondary school, whose rigid methods of instruction he disliked. He was fascinated by mathematics and science, subjects that he studied on his own. He became a high school dropout, and he left school to join his family in Milan.

Two years later he was able to enroll at the Swiss Federal Institute of Technology in Zurich after making up a number of subject deficiencies. At the institute the academic fare did not suit him either, so he continued to follow his own inclinations. He managed, however, to pass the required examinations for his degree by cramming from the excellent lecture notes supplied by a close friend. In the two years following his graduation in 1900 he subsisted on odd teaching jobs. In 1902 he secured a position as patent examiner at the Swiss patent office in Bern.

During the next seven years he worked at the patent office, and without any academic connections, he found time to publish, at the age of 26, three trailblazing papers. One dealt with the random thermal motions of molecules in colloidal solutions, called the Brownian movement. In 1827 Robert Brown, an English botanist, had observed through a microscope the zigzag paths of tiny pollen grain particles in a drop of water being buffeted by the much smaller, invisible atoms and molecules in the fluid. Einstein worked out the correct mathematical expression for this action on the basis of the random thermal motions of the atomic and molecular constituents, and he did this at about the time the structure of

the atom was first being probed. His second paper reinforced the quantum theory of light developed by Max Planck in 1900. Einstein established the photon nature of light in accounting for the recently discovered photoelectric phenomenon. For this contribution Einstein was awarded the Nobel Prize in physics in 1921. His third and most famous paper dealt with the special theory of relativity.

Einstein's last years were spent mostly in a vain search for a unified field theory, seeking a universal force that would link gravitation with the electromagnetic and subatomic forces. Others who have tried since have not been successful either. Einstein was filled with reverence for the works of nature and said: "The most incomprehensible thing about the world is that it is comprehensible." He thought of himself more as a philosopher than as a scientist. In a way he followed the Greek philosophers, who tried to account for natural phenomena by logical deductions instead of experimentation. He succeeded where the ancients failed because he could draw on the insight of his predecessors and the powerful analytical tools of mathematics developed in the two thousand years since Plato and Aristotle, combining them with his unerring cosmic perception.

What is *relative* about relativity? Let us consider an example. If we look out the window of our train, watching another train apparently moving on the next track, who is to say which train is moving, which is not, or whether both are moving? And so it is on a larger scale—we cannot ascertain anything concerning our possible absolute motion in the universe. Einstein brought the three dimensions of space together with that of time and taught us that when we measure length and time, we find no absolute results—the answers are relative, depending on the observer. Two people who are moving relative to each other see the same events going on at different places and different times; each experiences the event in his own respective coordinate system, or frame of reference. We

need relativity in science—especially in astronomy—because we cannot drop anchor in the universe and passively watch what happens around us. Each of us is both observer and participant, and the first truth we notice in the relativistic universe around us is that *all motion is relative.*

In explaining his general theory Einstein was the first to point out a feature that was generally accepted but never properly understood: the inertial mass of every object has the same value as its gravitational mass. He argued that this equality must mean that "the same quality of a body manifests itself according to circumstances as 'inertia' or as 'weight.'" To the extent that we can measure it, about one part in 10^{12}, inertial mass and gravitational mass are equal.

Newtonian space has three numbers, the spatial coordinates that describe where an object is: forward-backward, right-left, and up-down. Location has nothing to do with time. According to Newton, "Absolute space, in its own nature, without relation to anything external, remains always similar and immovable. . . . Absolute, true and mathematical time, of itself, and from its own nature, flows equably without relation to anything external. . . ."[1]

Einstein's theory connects space with time. In his gravitational theory he described the geometry of four-dimensional space-time. The space-time continuum shows deformities—that is, *space curvature*—in the vicinity of material bodies. The more massive the body, the stronger is this curvature. From Newton's viewpoint, an object moves in a curved path in response to the gravitational force of an attracting mass. For Einstein an object moves in a curved path as a natural consequence of the space curvature produced by another mass. If no mass is there, the space-time geometry is flat, showing no warping, and a body moves uniformly in a straight line. In the Newtonian world it moves that way because no force is acting upon it.

SPECIAL THEORY OF RELATIVITY

In the nineteenth century, space was imagined to be filled with a stationary invisible medium, *ether*, which carried electromagnetic waves (as air transports sound waves). If light must move through such a medium, its speed should differ depending on its direction relative to an observer. Thus the earth's true (absolute) velocity could be measured against this stationary medium by timing the speed of light in various directions.

American physicists Albert Michelson (1852–1931) and Edward Morley (1838–1923) sought to detect a difference in the speed of light beams propagated over the same distance, one parallel to the earth's motion around the sun and one simultaneously transmitted perpendicular to it.

The time light would spend moving across the earth's path, back and forth, was calculated to be $\sqrt{1 - (v^2/c^2)}$ shorter than the time it would take light to move *parallel* to the earth's path in the same direction and then in the opposite direction; here v

[1] From Newton's *Mathematical Principles of Natural Philosophy and His System of the World*, translated by Andrew Motte in 1729, revised and annotated by Florian Cajori (Berkeley: University of California Press, 1947), vol. 2, p. 6.

is the absolute velocity of the earth's orbital motion and c is the velocity of light. It was reasoned that the minute difference in the relative speeds of the light beams would lead to an evaluation of the earth's absolute motion in space.

But no matter in which direction measurements were made or at what time of year the experiment was performed, the result never changed: there was no measurable difference in time between the light beams. We cannot measure earth's absolute velocity in space.

The Michelson-Morley experiments are consistent with Einstein's special theory. Einstein rejected the notion of the ether, and to develop the mathematical formulation of his theory, he laid down these two postulates.

FIRST POSTULATE: *The laws of physics are the same for all observers in inertial frames of reference.*

SECOND POSTULATE: *The velocity of light (299,792 kilometers per second) is the same for all observers in space.*

Two observers moving uniformly relative to each other (not accelerating) are said to occupy *inertial frames of reference*. Neither frame of reference is preferred by the laws of physics; there is no way to distinguish one from another. They have equal physical status. Neither observer will be able to see an absolute motion of his own system by any experiments conducted in that system. For example, a person occupying the inside cabin of a ship cruising

> You imagine that I look back on my life's work with calm and satisfaction. But . . . there is not a single concept of which I am convinced that it will stand firm, and I feel uncertain whether I am in general on the right track . . . I don't want to be right—I only want to know whether I am right.
>
> Albert Einstein, when he was 70 years old

at uniform speed in calm seas could not conduct any kind of experiment that would indicate the ship to be in motion or tied to a dock. (But let the ocean become stormy—then the deviation from uniform motion is painfully evident.)

Einstein's second postulate is obviously consistent with the Michelson-Morley experiment: The speed of the light beam in the experiments did not depend on its direction; it did not depend on the light source; it was not affected by the earth's motion. In other words, the speed of light is independent of the relative velocity of the source and the observer. The concept of the ether was no longer needed and so was discarded.

On the basis of his two postulates, Einstein derived three important formulas for length, mass, and time in which the factor $\sqrt{1 - (v^2/c^2)}$ called the *Lorentz contraction factor*[2] plays a crucial role. The relativistic length, mass, and time interval measured by an observer A who sees the object in motion, is related by the Lorentz contraction factor to the objects' length, mass, and time interval, measured by an observer B who sees the object at rest. The effect of the Lorentz factor on length, mass, and time is shown graphically in Figure 3.5. These formulas are unimportant for all ordinary speed ($v/c \approx 0$) but must be reckoned with at speeds comparable to the velocity of light (see page 45).

What do the formulas tell us? The observer A, who sees the object moving, will measure a shorter length for the object in the direction in which the object is moving than will observer B, to whom the object looks stationary. For example, observer A would notice that B's rocket ship has contracted in length while passing by him compared with his own rocket ship which is at rest in the laboratory. Observer A will also measure a greater mass for B's moving rocket ship than will observer B; and observer A will measure a longer time between two events taking place on B's rocket ship than will observer B for the same two events. The latter effect is called *time dilation*, or spreading out of time. The faster observer B travels relative to the speed of light, the slower his clock appears to run as observer A sees it (observer A's own clock is seen by observer B also to be running slow). Neither observer is aware of this effect on his own clock. Thus mea-

[2] Named after the Dutch physicist who derived it in 1904 from his mathematical analysis of electromagnetism.

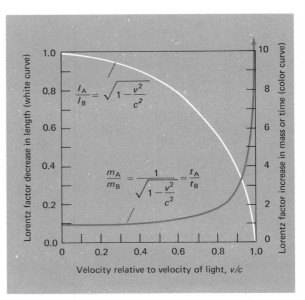

FIGURE 3.5
Change in length, mass, and time with velocity.

surements of length, mass, and time vary with the frame of reference.

So contrary to ordinary notions are the concepts of relativity theory that it sounds farfetched, seems to violate common sense, and appears to be too abstract to be of any consequence. This is far from the case, for it has amalgamated old ideas of space and time into a unified arrangement that has led the way to new and unsuspected revelations that do make it possible to test its consequences. At the appropriate points in the remaining chapters, we have inserted these tests of relativity theory as supplemental reading.

GENERAL THEORY OF RELATIVITY

Einstein advanced his theory of relativity greatly in 1916 by making it apply to observers moving non-uniformly relative to each other. Nature's fundamental laws, he reasoned, remain invariant throughout the universe in all frames of reference, whether the observers are accelerated or not.

In his second law of motion Newton showed that the force it takes to accelerate a body is proportional to its inertial mass. (*Inertia* is the resistance a body offers to an applied force.) He was also aware that the gravitational force on a body is proportional to its gravitational mass. Otherwise, bodies of different masses would not fall to the ground at the same rate—that is, with constant acceleration, as we know they do (see page 39). In 1889 Hungarian physicist Baron von Eötvös first proved experimentally and very precisely that *inertial mass and gravitational mass are equivalent,* an equality that had long been taken for granted. Modern experiments confirm that the inertial mass and the gravitational mass are the same to about one part in 10^{12}.

Einstein argued that it is impossible to distinguish between the effects of an inertial or a gravitational force on accelerated motion. He worked the idea into this principle.

PRINCIPLE OF EQUIVALENCE: *A gravitational force can be replaced by an inertial force that is due to accelerated motion without any change in the physical activity.*

Imagine a Newtonian observer in a rocket ship constantly accelerating at 1 g (see Figure 3.6). At 1 g of acceleration the observer feels at home—that is, as he would on the earth. A ball is in one hand (tennis ball or billiard ball, it does not matter). When the ship is in position 1, the occupant releases the ball. It continues to move upward with the velocity the ship had at the moment of release (the successive positions along the broken line in Figure 3.6). If the ship were moving upward at *constant velocity,* the

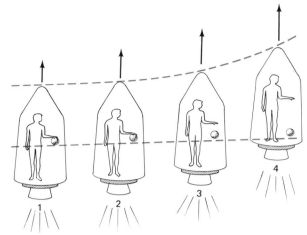

FIGURE 3.6
Experiment in an accelerating rocket ship illustrating the Principle of Equivalence. To the observer in the rocket ship, accelerated motion mimics the action of gravity.

ball would remain suspended in the same place because ship and ball move the same amount. But the ship is accelerating, so that the floor moves upward faster than the ball, colliding with the ball in position 4. The occupant imprisoned in the spaceship attributes this to the force of gravity.

From a vantage point outside the ship, however, an Einsteinian observer has been watching what is going on inside. The observer's explanation of the sequence

RELATIVE VELOCITIES

An aspect of relativistic and Newtonian motions that makes an informative comparison is velocity. If v_A and v_B are the velocities in space relative to some fixed system of coordinates of two observers A and B, then their relative velocity, v_{AB}, is, according to Einstein,

$$v_{AB} = \frac{v_A + v_B}{1 + (v_A v_B / c^2)}$$

where c is the velocity of electromagnetic waves (about 300,000 kilometers per second). But if A's and B's velocities are insignificant compared with the velocity of light ($v_A/c \approx 0$ and $v_B/c \approx 0$), Einstein's formula reduces to the Newtonian expression:

$$v_{AB} = v_A + v_B$$

Example: If two light beams are transmitted simultaneously in opposite directions from the same point, they will recede from each other at the relative velocity of light ($c = 300,000$ kilometers per second) and not at 600,000 kilometers per second (or 2c). We verify this answer by substituting $v_A = v_B + c$ in Einstein's formula, and we get $v_{AB} = c$.

of events is simple: All those actions are explained by the ship's accelerated motion. Who needs the force of gravity? Einstein pointed out that each observer has a right to his or her own description of events. We can replace the force of gravity by a fictitious force—an inertial force caused by an accelerated motion. An inertial force is not a real force but an effect of the nonuniform motion of the observer's frame of reference. You can experience fictitious force from an accelerated motion by standing on the floor of a merry-go-round. You feel a force that tends to move you toward the rim; it is called the centrifugal force.

To the relativist, spatial curvature is dictated by the presence of material bodies. If no mass is there, the curvature of nearby space is zero, and a body in this kind of flat space moves uniformly in a straight line. In the *warped space geometry* that surrounds a large mass, less massive objects move along curved paths. A planet's elliptic motion is an example: the planet moves in a curved path in the warped space surrounding the sun.

The force we call gravitation, then, is nothing more than natural behavior by bodies moving within the geometrical framework of the space-time world. The Newtonian would say the body moves according to action from a distance dictated by a force called the law of gravitation. The Einsteinian says a body moves naturally in response to the local structure of curved space-time. Newtonian theory describes physical events adequately in the vast majority of instances, permitting the use of simpler mathematical methods.

3.4 Cosmological Advances After Newton

THE UNITY OF THE WORLD
In the decades following Newton, astronomers expanded Newtonian theory and developed some very powerful techniques for analyzing planetary and stellar motions. They were able to predict the complex interactions between bodies in the solar system and to compare these results with observations, revealing a unity between the motions of common experience and the cosmic world. Between these and subsequent developments in the technology of the electronic computer, astronomers have been able to calculate the paths of planetary probes, which have launched us into a new age of exploration.

During the last half of the eighteenth century and into the next century, astronomers gathered data on all sorts of astronomical phenomena. Foremost of the observational astronomers of the eighteenth century was William Herschel (1738–1822). He started his career as a professional musician—organist, conductor, and composer. A textbook on astronomy so captivated him, at age thirty-five, that he began observations of the heavens with a telescope of his own making. Eventually he built a reflector of 0.5-meter diameter and another more than twice as big (see Figure 3.7). With these and his skill as an observer, Herschel made many remarkable discoveries. Recognizing his discovery of the planet Uranus on 13 March 1781, King George III knighted Herschel and gave him an annual stipend of 200 pounds sterling. He spent the rest of his life with his sister Caroline and son John, both of whom assisted him in prodigious astronomical research. Herschel surveyed the Milky Way extensively with telescopes of different sizes. He cataloged about eight hundred double stars, finding that many of their components were orbiting each other. Thus he had found the first positive evidence that the law of gravitation extends far beyond the solar system. And from his analysis of the motions of nearby stars, Herschel correctly deduced that the sun's motion relative to the nearby stars of the Galaxy is taking it toward the constellation Hercules.

COSMOLOGICAL SPECULATIONS
Whether the universe "have his boundes or bee in deed infinite and without boundes" was the profound question English astronomer Thomas Digges (c.1546–1595) asked himself, as Giordano Bruno had. It appeared in Digges's 1576 book *Perfit Description of the Celestial Orbes*. A few decades later Galileo too wrote of an infinite and unbounded universe in his

It is difficult to say what is impossible, for the dream of yesterday is the hope of today and the reality of tomorrow.

Robert H. Goddard

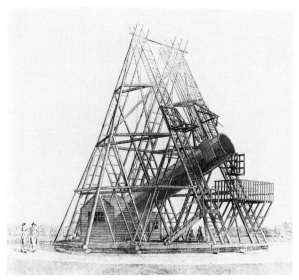

FIGURE 3.7
William Herschel's reflector. This reflector was 12 meters long, with a diameter of 1.2 meters. The tube could be swung up and down and rotated horizontally on a circular rail. The observer sat in a carriage at the top of the tube while an assistant on the ground helped to manipulate the telescope and to record the observations. Because this reflector was unwieldy and dangerous to use, Herschel carried out most of his work with his 6.1-meter-long reflector.

Dialogues. He had seen with his telescope that the Milky Way consisted of myriads of stars.

What is the Milky Way? Swedish philosopher Emanuel Swedenborg (1688–1772) speculated in 1734 that the stars formed one vast collection of which the solar system was but one constituent. Thomas Wright (1711–1786) of England theorized in his 1750 work, *An Original Theory of the Universe*, that the Milky Way seems to be a bright band of stars because the sun lies inside a flattened slab of stars. He also suggested that other milky ways are in the universe. Immanuel Kant (1724–1804), the noted German philosopher, went beyond Wright's idea, suggesting in 1755 that the small oval nebulous objects seen with telescopes were other milky way systems or "island universes"; the phrase captured popular fancy a century and a half later. A modern photograph of Kant's island universes is shown in Figure 3.8.

Ignoring these speculations, we can ask what observational evidence people had about the Milky Way's structure. In 1785 Herschel gave the first quantitative proof that the Milky Way was a stellar structure shaped like a flat disk, a grindstone. He obtained his evidence by sampling stars in selected areas of the sky. A cross-sectional view of his Milky Way system appears in Figure 3.9.

Herschel and others had no accurate knowledge of stars' brightnesses or distances. No one suspected that starlight might be dimmed by obscuring material between the stars. The faintest stars then observable in his telescope were about two hundred times fainter than the faintest stars that could be seen with the naked eye. Despite all these handicaps, Herschel delineated the Milky Way system's general structure and extent—a major work. Later studies substantiated our stellar system's disklike shape.

PREPARATION FOR MODERN COSMOLOGY

William Herschel and his son John (1792–1871), who carried his father's surveys to the southern skies, found over five thousand nebulous objects, cataloged in 1864 by John. Shortly thereafter the third Earl of Rosse examined many of these objects with a 1.8-meter reflecting telescope on his Irish estate. Several of the larger nebulosities appeared to consist of gaseous clouds, often with filamentary or ring-shaped forms; still others were resolvable into clusters of stars.

As more nebulosities were found, the conviction grew that they belonged to one grand system, the Milky Way. If these objects were outside the Milky Way system, they could logically be expected to be spread more or less uniformly in all directions. But the nebulae were not uniformly distributed over the sky; their number increased in either direction away from the plane of the Milky Way. No one knew that this effect was illusionary: light was simply dimmed by the cosmic haze in our Galaxy's plane. The nebulae's apparent distribution and the mixed appearance of stars and gas in the same nebula seemed to outweigh the possibility that some nebulae might be unresolved stellar systems, separate from the Milky Way. The true nature of the Milky Way, the nebulae, and what they had to do with each other and the Milky Way remained uncertain.

After William Herschel's pioneering work, the next comprehensive study of the stars' distribution was done by Dutch astronomer Jacobus Kapteyn (1851–1922) toward the end of the nineteenth century. Using photographic star counts, mean stellar dis-

FIGURE 3.8
A cluster of distant galaxies in Corona Borealis, photographed with the 5.1-meter Hale reflector. The small, fuzzy, lens-shaped objects are the galaxies.

FIGURE 3.9
Herschel's grindstone model of the Milky Way. The sun is the star at the approximate center. The cleavage at the right end is caused by the obscuring Great Rift, which divides the Milky Way into two branches between Cygnus and Centaurus.

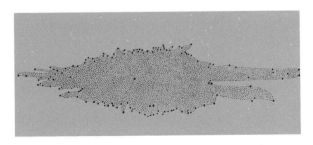

tances, and stellar brightnesses, he developed statistical methods that are still useful today. After long and laborious research and several preliminary versions, in 1922 he published his final statement on the shape, structure, and dimensions of the stellar system. Known as the Kapteyn universe, his was a sun-centered model of the Galaxy, 50,000 light years in diameter and 6,000 light years thick. A *light year* is the distance light travels, at the rate of approximately 300,000 kilometers per second, in one year, or about 9.5×10^{12} kilometers.

Five years before Kapteyn published his version of the stellar system, the American astronomer Harlow Shapley (1885–1972) had arrived at a markedly different conclusion about the sun's location and dimensions. Studying photographs taken with the 1.5-meter Mount Wilson reflector, he found 69 globular star clusters that formed a spheroidal system centered on our flattened stellar system (see Figure 3.10). The overall system, Shapley said, is 300,000 light years

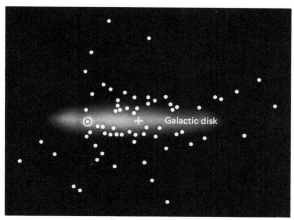

FIGURE 3.10
Shapley's globular cluster distribution, proposed in 1917. The globular clusters are represented as white dots, the sun is shown as a circle with a central dot, and the cross marks the center of the Milky Way. Shapley pictured the galactic disk as highly mottled, so that the large concentration of stars around the sun could be the Kapteyn universe, engulfed in our Galaxy.

Galactic disk

across and 30,000 light years thick, and the sun is 65,000 light years away from the center of the system. Shapley concluded that the spiral nebulae would be engulfed by our giant Galaxy and therefore were not island universes as some astronomers believed.

Astronomers accepted Shapley's finding that the sun was not the center of the Galaxy, but they were disturbed by the enormous scale he claimed for our Galaxy. He was challenged by Lick Observatory astronomer Heber Curtis, who insisted that the Milky Way's size had been considerably overestimated. In a great debate before the National Academy of Science on 26 April 1920, Shapley advocated the one-galaxy idea and Curtis the multigalaxy or island universe concept. Their debate was inconclusive. Better observational data were needed, and they were not long in coming.

Distances to the spiral nebulae were derived in 1923 by Edwin Hubble (1889–1953), a Mount Wilson astronomer, by analyzing photographs of the Great Nebula in Andromeda and another great spiral in Triangulum (see page 346). Using the new 2.5-meter reflector, he was able to obtain photographs showing greater detail. They showed that the outer portions of these nebulae could be resolved into swarms of stars. Hubble derived a distance of 900,000 light years for the two spirals (an underestimate of about 100 percent) by a technique described in Chapter 12. Doubt was gone that the spiral nebulae were the island universes Kant imagined long ago. Today we are aware of billions upon billions of galaxies, often in giant clusters, spread over the vast reaches of space.

> Nothing puzzles me more than time and space; and yet nothing troubles me less.
>
> Charles Lamb

SUMMARY

Newton's discovery of the universal law of gravitation was the crowning achievement of the Copernican revolution, providing a framework for understanding most planetary motions and gravitational interactions. Newton's ideas of gravitation, however, do not adequately describe the movement of bodies at high velocities or in the presence of strong gravitational fields. The theory of relativity, proposed by Einstein early in this century, refined Newton's views to account more completely for these phenomena. While Newton's theory describes bodies reacting with each other at a distance, relativity theory states that bodies moving in space are subject only to the local structure or curvature of four-dimensional space-time and not to any attractive force.

Newton's theories nevertheless gave astronomers of succeeding centuries powerful conceptual tools for analyzing planetary and stellar motions. The work of William Herschel, Harlow Shapley, Edwin Hubble, and others has elucidated both the shape of our own Galaxy and its place among billions of other galaxies in an expanding universe.

REVIEW QUESTIONS

1. What is meant by a field of force? Give examples of such fields.

2. What is meant by the inertia of a body? How is it related to its mass?

3. What is the difference between the weight of an object and its mass?

4. Describe the following concepts applied to a moving body: velocity, acceleration, momentum.

5. How does a body originally moving at constant speed in a certain direction act when subjected to an applied force?

6. Explain what is meant by this statement: To every action there is an equal and opposite reaction.

7. If the effect of gravity is not noticeable between persons, why is it so obvious in the planetary motions or in the motion of the moon around the earth?

8. Describe the types of orbits that gravitation imposes on bodies moving around a central attractive force.

9. How does one explain the negative result of the Michelson-Morley experiment?

10. What are the two postulates of the special theory of relativity?

11. What is meant by time dilation?

12. What is the difference between an inertial frame of reference and a noninertial frame of reference?

13. How would you describe the difference between Newtonian space and Einsteinian space-time?

14. State the principle of equivalence in your own words. Illustrate it by means of an example.

15. What is the difference between the way gravitation is viewed by Newton and the way it is viewed by Einstein?

16. How did William Herschel come to the conclusion that the Milky Way system had the shape of a grindstone?

17. How did Shapley discover that the sun was off center in the Milky Way?

18. How was the issue settled that the spiral nebulae were other Milky Way systems? By whom?

4
Celestial Radiation and the Atom

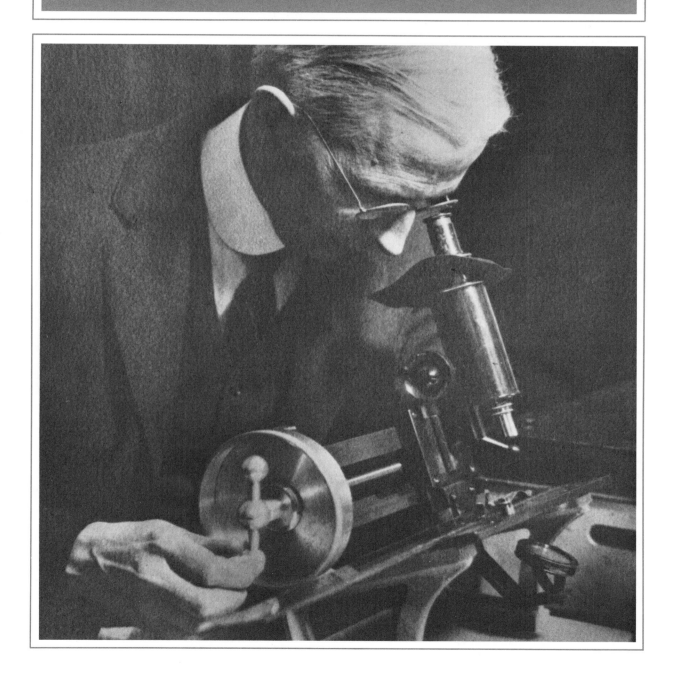

Until about a century ago astronomers were almost exclusively concerned with the positions and motions of celestial bodies. Of the physical makeup or nature of these bodies they knew little or nothing and were really not able to find out. Yet today this concern, which is the field of *astrophysics,* is a prime one in astronomy. This change in orientation is primarily the result of two events in the growing knowledge of the nature of light.

The first event was the development of the *spectroscope,* a device capable of breaking down the white light from a distant source into its component colors. In his *Opticks* Newton described how one saw a rainbow of colors when sunlight was passed through a prism. However, it was William Wollastron (1766–1828) in England and Joseph Fraunhofer (1787–1826) in Bavaria who were primarily responsible for developing the spectroscope (see Figure 4.1).

The second development was the recognition that each different chemical element emits a specific set of colors that is peculiar to it only, much like the fingerprints of an individual. As early as the 1830s this fact was suggested in connection with the presence, identity, and abundance of materials in ores. However, the real beginning of the field of spectroscopy was made in the last half of the nineteenth century by chemist Robert Bunsen (1811–1899) and physicist Gustav Kirchhoff (1824–1887) at the University of Heidelberg.

From his experimental work in the laboratory Kirchhoff was able to formulate three empirical laws of spectroscopic analysis. These laws described the physical conditions under which matter will produce light that has a different spectrum of colors. However, the most important astronomical application was the potential to determine the chemical composition of the sun and stars. By 1864 the English astronomer Sir William Huggins (1824–1910) had identified nine elements in the bright star Aldebaran in the constellation of Taurus. Sir Norman Lockyer (1836–1920) in 1868 detected the presence of an element in the solar spectrum that was unknown on the earth. It was later found in natural gas and still carries its solar name—helium. In the years following, more elements were identified in stars, along with the discovery that the pattern of colors in the white light coming from stars is not the same for

◄ Henry Norris Russell analyzes a spectrogram of a star with a measuring engine.

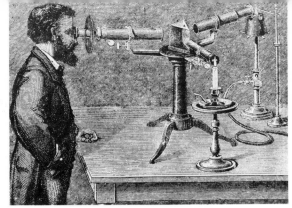

FIGURE 4.1
Spectroscopic analysis. A woodcut from about 1887 of a scientist looking through the eyepiece of a prism spectroscope that is analyzing the light from both a candle (foreground) and a gas flame (background).

all stars. However, similarities were recognized in the spectra of stars so that the stars could be arranged into broad spectral classes.

By the beginning of this century an important relationship had been established between the researcher in the laboratory and the astronomer in the observatory. New means of viewing the universe were being prepared, which have fundamentally altered our concept of the universe. The broadening of the spectrum over which we see radiation coming from different parts of the universe and the ability to move above the earth's obscuring atmosphere are significant elements of today's astronomy.

4.1 Electromagnetic Radiation

ENERGY

Astronomers have learned most of what we know about the stars and the galaxies by analyzing the light coming from them. Light is a form of energy. In science no concept is more important than energy and its conservation principle. *Energy* is a measure of the capacity of a physical system to perform work when the system is changed in such a way that we can describe it before and after the change. Imagine a stream turning a waterwheel as it flows over a dam. The stream performs work when it turns the waterwheel. The energy of the stream is a measure of the capacity of the stream to perform useful work.

Energy does not in general have such properties as size, shape, or color. However, it can be measured, and one unit used by astronomers is the *erg,* the amount of energy that is needed to accelerate a mass of 1 gram at a rate of 1 centimeter per second squared, as it moves 1 centimeter. A housefly flapping its wings a couple of times on a cool morning expends about 1 erg of energy. Thus you can see that an erg is a very small quantity of energy by our everyday standards.

The most important of all physical laws is the law of conservation of energy.

LAW OF CONSERVATION OF ENERGY: *Energy may neither be created nor destroyed, but only transformed from one form to another.*

For example, one form of energy is *kinetic energy,* and it is the energy a body has because of its state of motion. Another form is *potential energy,* and it is the energy a body has because of its position in a field of force. A stone at the top of a hill, for example, can be said to have energy by virtue of its position in the earth's gravitational field. If it is pushed, the stone will roll down the hill, converting potential energy to kinetic energy.

Energy may also be transported from one place to another. Waves are one way of transporting energy. Imagine two people several feet apart holding the ends of a rope; then one jiggles the rope; a wave will travel from one end to the other. No particles are moving from one end of the rope to the other, but a disturbance is traveling along it. Drop a stone into a pond and you also get a wave. A *wave* is a disturbance that is capable of transferring energy from one place to another. Let us see how the wave has been used as one means of describing the physical nature of light.

LIGHT AS WAVES

Three-dimensional light waves are somewhat easier to visualize if we compare them with waves on the surface of a pond. As a way of describing the water wave, we can measure its *wavelength,* which is the distance between the successive crests, or we can measure its *frequency,* which is the number of complete cycles of the disturbance passing a fixed point per second. The speed of a water wave is the distance it travels per unit of time; this is just the length of each wave, its wavelength, multiplied by the number of waves passing a point per unit of time, its frequency.

Thus the *speed* of the wave can be found from the product of its wavelength and frequency. Finally, the last of the quantities used to describe a wave is its *amplitude.* This is the greatest height the crest reaches or the greatest depth to which a trough falls. The greater the amplitude, the more energy the wave transports from one point to another. It is important to remember that the water itself does not move forward, only the disturbance.

In 1862 the Scottish physicist James Clerk Maxwell (1831–1879) showed that light is energy carried in the form of a traveling wave composed of electric and magnetic fields. The electric and magnetic fields vary in intensity, and they are at right angles to each other and the direction in which the wave is propagating (see Figure 4.2). The electric and magnetic fields continually interact, forming an *electromagnetic wave.* These fields maintain themselves and continue to propagate until the energy of the wave is converted to some other form of energy. At that point the electromagnetic wave ceases to exist.

Maxwell's proposal that light is an electromagnetic wave, as we will soon see, was not the last word in explaining the physical nature of light. But visualizing light as waves spreading out from a radiating source helps us to understand many aspects of light. The *speed of light* measured in empty space is 299,792 kilometers per second (or about 3×10^5 kilometers

FIGURE 4.2
Propagation of electromagnetic waves. We do not actually "see" these waves or know exactly what they look like. This is a schematic representation, depicting electromagnetic radiation as oscillations or variations of electric and magnetic fields at right angles to each other. As shown, the electric field intensity varies in the up-and-down direction and the magnetic field intensity varies in the back-and-forth direction. The electromagnetic wave is advancing in the direction shown.

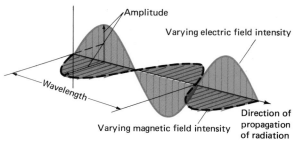

per second in round figures, or 186,300 miles per second). This appears to be the speed limit for all energy transported in the universe. As we discussed in the last chapter, the speed of light is a fundamental constant of nature and apparently has the same value throughout the universe.

VELOCITY OF A WAVE

Consider a wave moving across the surface of a body of water between two points 100 centimeters apart. In part a of Figure 4.3 each wave moves the 100 centimeters in 5 seconds at a constant velocity v, and 5 complete waves pass a fixed point in 5 seconds. The distance between the crests, the wavelength λ (λ is the Greek lowercase letter lambda), is $100/5 = 20$ centimeters. The number of waves passing a point in 1 second, the frequency f, is 1 wave per second. In part b of the figure each wave still moves the 100 centimeters in 5 seconds, but now 10 waves go by a fixed point in 5 seconds. Here the wavelength $\lambda = 100/10 = 10$ centimeters, and the frequency $f = 2$ waves per second. In both examples the velocity of the waves is the same, 20 centimeters per second. Here is a summary of the results:

$$\underline{\quad \lambda \quad} \cdot \underline{\quad f \quad} = \underline{\quad v \quad}$$

Example a: 20 cm $\cdot$ 1/s = 20 cm/s

Example b: 10 cm $\cdot$ 2/s = 20 cm/s

(See Appendix 1 for abbreviations of units.) For light we express this important relationship as follows:

$$\lambda \cdot f = c, \quad \text{or} \quad \lambda = c/f, \quad \text{or} \quad f = c/\lambda$$

where λ is the wavelength, f is the frequency, and c is the velocity of light.

FIGURE 4.3
Wave forms of different lengths and frequencies moving at the same speed across the surface of water. Imagine the star represents the point at which, say, a falling stone enters the water. The kinetic energy of the stone is transformed into the surface wave, which in turn transports the energy across the surface. The solid color lines represent the crests of the wave and the dashed lines represent the troughs.

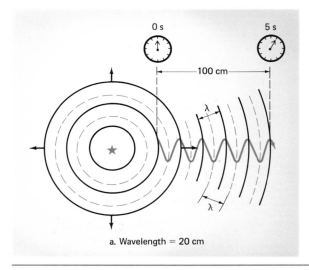

a. Wavelength = 20 cm

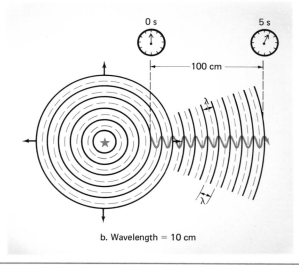

b. Wavelength = 10 cm

ELECTROMAGNETIC SPECTRUM

Our eyes are sensitive to only a very limited portion of the entire range of wavelengths for electromagnetic radiation. If we order this range of wavelengths from the shortest on the left to the longest on the right, we produce an array of wavelengths called the *electromagnetic spectrum,* which is shown in Figure 4.4. Toward the short wavelength end we find that portion to which our eyes are sensitive, called the *visible spectrum.* The physiological response of the eye to the various wavelengths of the visible spectrum is color. Short wavelengths in the visible spectrum are violet, with progressively longer wavelengths producing the response we identify as the range of hues from blue, green, yellow, orange, to red in the color spectrum. Visible light is electromagnetic radiation with wavelengths between approximately 35×10^{-6} and 70×10^{-6} centimeter. These wavelengths correspond to frequencies between 8.5×10^{14} to 4.3×10^{14} hertz. One hertz equals one cycle or oscillation of the wave per second. The lowest frequencies of visible light appear red; the highest frequencies appear violet; and between these are the rest of the color spectrum.

All types of electromagnetic radiation show properties of being a wave; all propagate in the same way with the same speed in empty space; and all transport energy. For convenience, however, we divide the rest of the electromagnetic spectrum into regions according to the radiation's wavelength, such as the ultraviolet or the infrared and so on. We recognize these

> Up to the twentieth century, "reality" was everything humans could touch, smell, see, and hear. Since the initial publication of the chart of the electromagnetic spectrum . . . humans have learned that what they can touch, smell, see, and hear is less than one-millionth of reality. Ninety-nine percent of all that is going to affect our tomorrows is being developed by humans using instruments and working in ranges of reality that are nonhumanly sensible.
>
> R. Buckminster Fuller

different regions not so much because of intrinsic differences in the radiation but because we have different ways of detecting radiation depending on its wavelength. (These detection methods are discussed in Chapter 5.) Gamma rays, X rays, and ultraviolet radiation are the wavelength regions shorter than visible light (see Figure 4.4); most of this radiation coming from outer space is filtered out high above the earth's surface by our atmosphere (see Figure 4.5). Infrared radiation is the first wavelength region beyond visible light and it is also partially absorbed by

FIGURE 4.4

The electromagnetic spectrum, showing the various spectral or wavelength regions. The spectral regions are not divisions imposed by nature. They are based on the means by which we detect radiation of different wavelengths.

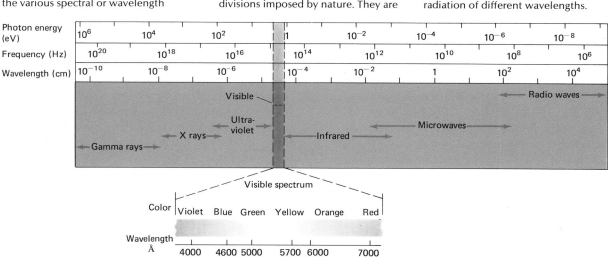

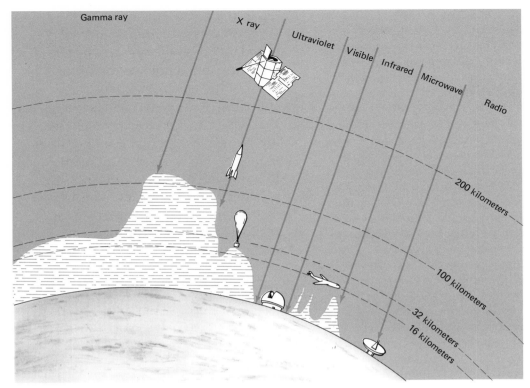

FIGURE 4.5
Obstruction and detection of electromagnetic radiation. The dashed areas represent atmospheric regions in which celestial radiation is obstructed. X rays and gamma rays are filtered out by molecules and atoms of oxygen and nitrogen at the higher levels; ultraviolet radiation, by the ozone layer between 20 and 35 kilometers above the earth's surface. Molecular oxygen, carbon dioxide, and water vapor are the principal absorbers up to 16 kilometers of infrared radiation. Before the space age most of our knowledge of the heavenly bodies came to us through the narrow optical window in the visible region and the broad window in the radio region of the electromagnetic spectrum.

the earth's atmosphere. The next wavelength regions of longer waves are the microwave and radio regions, for which there is a broad electromagnetic window (Figure 4.5) through the earth's atmosphere. Today gamma ray and radio astronomers can measure, respectively, celestial radiation at wavelengths one million times shorter and 100 million times longer than the visible wavelengths.

Various ways of stating wavelengths are more convenient for describing different regions of the electromagnetic spectrum. For the visible spectrum angstroms are convenient. An *angstrom* is 1×10^{-8} centimeter. Visible radiation lies approximately between 3500 angstroms (the violet end of the spectrum) and 7000 angstroms (the red end). X-ray astronomers also use angstroms, but infrared astronomers prefer to measure wavelength in microns (1 micron $= 10^4$ angstroms $= 10^{-4}$ centimeter). For all radiation, however, astronomers use the hertz as the unit for measuring frequency.

WAVE PROPERTIES OF LIGHT
Light traveling through empty space moves in a straight line. However, in our experiences with light we are concerned not with light in empty space but with light coming through various media—light partially absorbed by the atmosphere, scattered by dust, transmitted through a window or a telescope. In these circumstances the speed of light may be slowed and the direction of the light wave may be changed. These changes are best understood through the wave properties of light.

Several properties illustrate the wave characteristics of light. One is *reflection*, which occurs when light

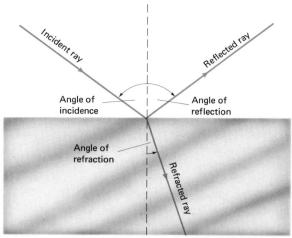

FIGURE 4.6
Reflection and refraction. The angle of incidence equals the angle of reflection. Also, the reflected ray is always in the plane formed by the incident ray and the perpendicular to the boundary (shown as a dashed line). Here the angle of refraction is less than the angle of incidence and also lies in the same plane as the incident ray and the perpendicular.

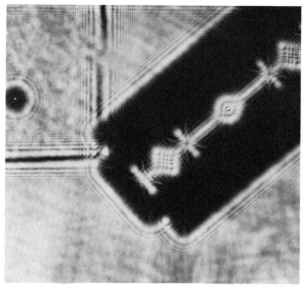

FIGURE 4.7
Shadow-diffraction pattern of a razor blade and ball bearing on a microscope slide. Diffraction bands are seen both inside the hole in the razor blade and around the outer edges of the blade, slide, and ball bearing.

strikes (is incident upon) the boundary between two media of different materials, such as glass and air. When a light ray moving in air reaches the boundary, part of it may be reflected, as shown in Figure 4.6. The reflected ray lies in the plane of the incident ray and the perpendicular to the boundary. The ordinary mirror or looking glass illustrates reflection.

Also, part of the incident ray may be transmitted through the glass rather than being reflected. The transmitted ray does not, however, continue along the same straight line; it is bent toward the perpendicular shown as the dashed line in Figure 4.6. This change in direction is called *refraction*. If the medium into which the ray moves is more dense than that from which it comes, the angle of refraction will be less than the angle of incidence. If its density is less, then the angle of refraction is greater. A good example of refraction is shown by a spoon sticking out of a glass of water: the handle looks bent at the point where the spoon enters the water.

Light shows another wave property, *diffraction*, which is the spreading out of light past the edges of an opaque body (see Figure 4.7). Instead of being propagated in a straight line, light like sound waves, bends around corners. The spread is greater for longer wavelengths. Because light's wavelength is very small, we do not normally observe diffraction in the everyday world. We can see diffraction, though, in the laboratory. Optical instruments, such as telescopes and microscopes, depend upon these wave properties of light for their operation.

Nearly all natural light sources, such as stars, emit electromagnetic waves composed of many wavelengths. How do waves of different wavelengths add to produce a composite wave? If waves of the same wavelength from two sources are superimposed so their crests and troughs coincide, then they are said to be *in phase* with each other, and their amplitudes add to produce a sum greater than the amplitudes of the individual waves; the light is said to *interfere constructively*. If the crests of one set of waves fall on the troughs of the other, then they are said to be *out of phase* with each other, and their amplitudes cancel each other; the light is said to *interfere destructively*. Interference is peculiar to all types of waves; in fact, its occurrence was strong evidence that light is a wave phenomenon. Waves of one or many different wavelengths may interfere constructively or destructively. Such waves are called composite waves, or white light, since that is the physiological response they evoke. If we can add waves together, then we must also be able to separate a composite wave into its constituent wavelengths. Indeed we can, as we will describe in Section 4.2.

BRIGHTNESS OF ELECTROMAGNETIC WAVES

The surface area covered by an expanding sphere of light (or a portion of it) increases as the square of the radius of the sphere, that is, as the square of the distance from the light source. Since the total amount of energy leaving the light source in all directions is the same at any distance, the amount of radiation passing through each unit of area of the expanding sphere must diminish with the square of the distance (see Figure 4.8). For example, suppose at 1 meter from a light source, the apparent brightness of the radiation over 1 square meter of surface area is 1 unit. At twice the distance, each square meter will receive one-fourth of a unit of illumination; at three times the distance, one-ninth of a unit; at four times, one-sixteenth of a unit; and so on. This relationship between apparent brightness and distance is known as the inverse-square law of light.

INVERSE-SQUARE LAW OF LIGHT: *The apparent brightness* b *varies inversely as the square of the distance* d *from the light source; that is,* $b \propto 1/d^2$.

This law is applied in many kinds of astronomical work, as we will see in Chapters 12 and 16.

DOPPLER EFFECT

If an observer is moving relative to a source of light, or the source is moving relative to him, then the observer will see a change in the wavelength of the light. Suppose a stationary light source is radiating concentric waves of one wavelength in all directions. Then observers in any direction, if stationary, would see successive crests of the wave passing them at the same rate at which they were emitted by the source. If, on the other hand, the light source begins to move at uniform speed toward the right (as in Figure 4.9), the two observers O and P along the line of motion would see crests passing them at rates different from that with which they were emitted. Observers Q and R, located at right angles to the moving source, would detect no change in the rate for crests passing them. Observers elsewhere would notice some change, the amount depending on the angle between their radial direction to the source and the line of motion. This phenomenon is known as the *Doppler effect*, named for Christian Doppler (1803–1853), the Austrian physicist who first explained it.

Let us consider observers O and P in more detail. Wave 1 was produced when the light source was at position 1; wave 2 when it was at position 2, and so on. Because of the greater distance the wave travels in reaching observer O, each successive wave crest passes him at a slower rate (lower frequency) than when the source was stationary. Because the waves travel a shorter distance to reach P, the successive crests pass at a faster rate (higher frequency). The wavelength is shifted toward longer wavelengths as the source recedes from O and toward shorter

FIGURE 4.8
Inverse-square law of brightness. Light from a radiating source spreads out as the square of the distance from the source. The amount of illumination reaching surfaces that are equal in area but at different distances from the source decreases with the square of the distance from the source.

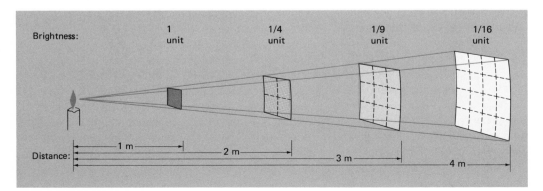

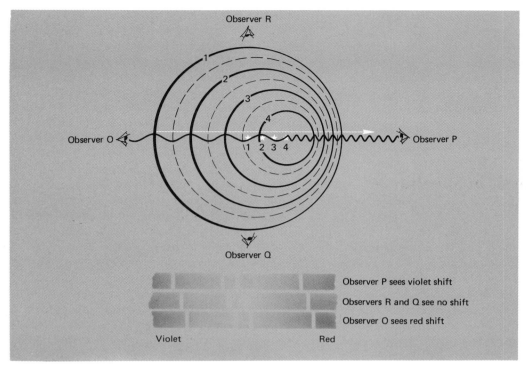

Observer R

Observer O

Observer P

Observer Q

Observer P sees violet shift

Observers R and Q see no shift

Observer O sees red shift

Violet

Red

FIGURE 4.9
Doppler line shifts. In the drawing, as
the light source recedes (points 1–4)
from observer *O*, he sees longer waves
and fewer of them per second passing
by, and the light's spectral lines show a
Doppler shift toward the red. Observer

P sees shorter waves rushing by at a
higher rate, and the spectral lines show
a Doppler shift toward the violet. Observers *Q* and *R* see no change in
wavelength or frequency—and no Doppler shift.

wavelengths as the source approaches *P*. An example
of the Doppler effect familiar to all of us is the change
in frequency of sound waves in the rising and falling
pitch of a train whistle as the train approaches and
then moves away. It is immaterial whether the light
source is in motion, or the observer, or both: the size
of the Doppler effect seen for light waves depends
only on the net relative motion along the line of
sight between the light source and the observer.

The amount of the wavelength shift due to the
Doppler effect is directly proportional to the velocity
of approach or recession, as long as the relative velocity is well below the velocity of light. (Later in this
book we will discuss the relationship when the relative velocity is a substantial fraction of the velocity of
light.) The constant of proportionality is the ratio of
undisplaced wavelength to the velocity of light. This
means that the wavelength shift is greater the longer
the wavelength of the radiation. As an example, if we
are approaching two stationary radiation sources, with

When we step through the gateway
of the atom, we are in a world which
our senses cannot experience. There
is a new architecture there, a way
that things are put together which
we cannot know; we only try to
picture it by analogy, a new act of
imagination. The architectural images
come from the concrete world of our
senses, because that is the only world
that words describe. But all our ways
of picturing the invisible are
metaphors, likenesses that we snatch
from the larger world of eye and ear
and touch.

Jacob Bronowski

one emitting electromagnetic radiation twice the wavelength of the other, then we would expect twice the wavelength shift from that one. At the bottom of this page, we present the mathematical form of the Doppler effect. Because of the continual movement of all bodies in the cosmos, the Doppler effect is an important tool for the astronomer.

4.2
Analysis of Spectra

SPECTROSCOPY

Electromagnetic waves of just one wavelength are known as *monochromatic radiation*. Just as sound waves of various wavelengths are transported simultaneously through the same region of air, electromagnetic waves of different wavelengths can move through the same point in space and superimpose to form composite waves or white light. Stars, for example, are white-light sources, though the color of the composite light from the stars may be white, red, yellow, or blue (see Table 12.5). As stated earlier,

white or composite light can then be *dispersed,* or separated, into its component colors or wavelengths. Let us briefly explain how we can accomplish this by using a triangular piece of glass called a *prism*.

In the refraction of a ray of light, the angle through which light is refracted depends on wavelength; the angle of refraction is greater for shorter wavelengths than for longer ones. Consider white light passing through the slit of a narrow diaphragm and then through a glass prism, as in Figure 4.10. The light disperses into its component wavelengths with short-wavelength light refracted by larger angles than long-wavelength light. Thus waves of different wavelengths disperse in different directions. The result is a rainbow-colored sequence of images of the slit containing an image for each wavelength present in the white light.

Another means of dispersing white light to produce a spectrum is the *diffraction grating*. Unlike the ordinary glass prism, which is transparent only to visible and infrared radiation, the grating is useful over a broad spectrum, from X-ray to infrared wavelengths. In its simplest form the diffraction grating is a plate containing a very large number of very narrow parallel slits, uniformly spaced at distances that are only a

---

MEASURING THE DOPPLER EFFECT

The Doppler displacement can best be measured if we know the wavelength when there is no relative motion between the source and the observer. Let $\Delta\lambda$ (Greek letters delta lambda, where delta means "difference") be the shift in wavelength equal to the difference between the measured wavelength λ_m and the wavelength λ when there is no relative motion, or $\Delta\lambda = \lambda_m - \lambda$. Then the formula relating the Doppler shift for light to the relative velocity is

$$\frac{\Delta\lambda}{\lambda} = \frac{v}{c}$$

where c is the velocity of light (approximately 300,000 kilometers per second) and v is the relative line-of-sight velocity between the observer and the light source, or, as astronomers call it, the *radial velocity*. The wavelength shift is considered positive if the wavelength shifts to the red (distance increasing) and negative if it shifts to the blue (distance decreasing).

Example: The measured red shift of an absorption line at 5000 angstroms in the spectrum of a star was found to be +0.5 angstroms. From the Doppler formula we find the radial velocity of the star relative to the earth to be

$$v = \frac{\Delta\lambda}{\lambda} c = \frac{(+0.5 \text{ Å})(3 \times 10^5 \text{ km/s})}{(5 \times 10^3 \text{ Å})} = +30 \text{ km/s}$$

where the plus sign denotes a velocity of recession. If the algebraic sign of the velocity had been negative (i.e., the wavelength shift had been negative), then the star would have been approaching the earth.

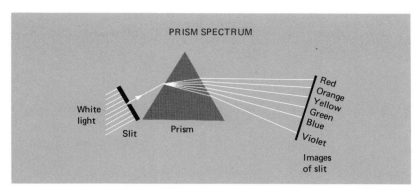

PRISM SPECTRUM

White
light

Slit

Prism

Red
Orange
Yellow
Green
Blue

Violet

Images
of slit

FIGURE 4.10
Two methods of producing a spectrum.
The prism produces a single spectrum
whose dispersion (spreading out of
colors) is greater at the blue end than at
the red end. The grating produces a
bright central image flanked on either
side by multiple number of fainter,
higher-order spectra.

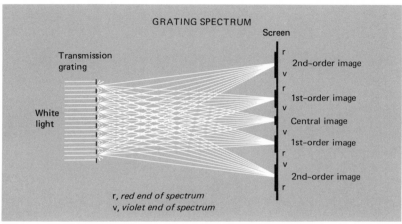

GRATING SPECTRUM

Screen

Transmission
grating

White
light

r
v   2nd-order image

r
v   1st-order image

Central image

v
r   1st-order image

v
r   2nd-order image

r, *red end of spectrum*
v, *violet end of spectrum*

few times the wavelength of light. (By *large number* we mean many thousands of slits per centimeter.) The spectrum is viewed in the direction of the light source, as in Figure 4.10. Since the amount of the bending or diffraction of electromagnetic waves at each slit depends on its wavelength, composite light is dispersed into its component colors. (For more details on these dispersing devices as they are used in an instrument called a spectroscope, see Section 5.2.)

LAWS OF SPECTRUM ANALYSIS

When we analyze light from various astronomical sources, we do not always find a continuous rainbow-colored sequence of wavelengths. Spectra can be classified and interpreted according to laws formulated by the German chemist Gustav Kirchhoff more than a century ago. The three basic types of spectra and the physical conditions under which they are formed are given by Kirchhoff's laws and are illustrated in Figure 4.11.

FIRST LAW—CONTINUOUS SPECTRUM: *The spectrum of a radiating solid, liquid, or highly pressurized gas is an uninterrupted sequence of wavelengths known as a continuous spectrum.*

SECOND LAW—EMISSION OR BRIGHT-LINE SPECTRUM: *The spectrum of a radiating rarefied gas is a set of isolated or discrete wavelengths whose appearance is a series of bright-colored lines that form a pattern characteristic of the chemical composition of the gas.*

THIRD LAW—ABSORPTION OR DARK-LINE SPECTRUM: *Light from a radiating source producing a continuous spectrum will, if it passes through a cooler gas, have certain specific wavelengths characteristic of the cooler gas removed from the spectrum. The spectrum appears continuous except where it is crossed by dark lines, which indicates that these wavelengths have been removed.*

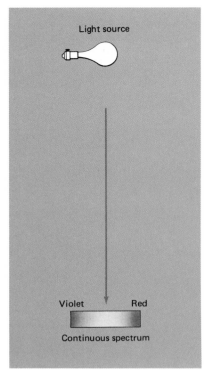

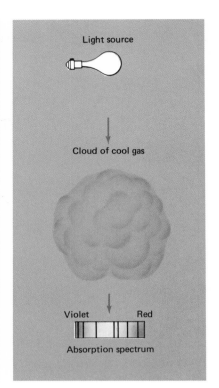

FIGURE 4.11
Kirchhoff's laws of spectrum analysis. The incandescent lamp produces a continuous spectrum (first law); the glowing rarefied gas produces a bright-line or emission spectrum (second law); the rarefied gas (either luminous and cooler than the continuous source or nonluminous) lying in the light path of the continuous source results in a dark-line or absorption spectrum.

Illustrations of the different types of spectra are shown in Plate 1 and Figure 4.11. There are many common examples of light sources whose spectrum is one of the three basic types. As an example, the spectrum of the lighted filament of an electric light bulb is a continuous spectrum. The spectrum of a neon street sign is one example of an emission spectrum. The spectrum of a gas composed of molecules, which consist of two or more atoms, is actually many sets of very closely spaced spectral lines known as emission bands. And, as a final example of the three basic types of spectra, the spectrum of the sun and most stars is an absorption spectrum. We will see additional examples of each type of spectrum at many points in the chapters throughout this text. The important point we ought to remember here is that the type of spectrum of the different light sources tells us something about the conditions in and around that source.

IDENTIFYING THE ELEMENTS

An astronomical light source, such as a star or a gaseous nebula, contains a mixture of chemical species, each either emitting or absorbing its own set of spectral lines. With the aid of the laboratory spectral analysis of the different chemical elements, astronomers can identify individual elements in the light source from the measured wavelengths of its spectral lines, regardless of whether they are emission or absorption lines.

This identification is done in the following way. Light from a celestial body is collected by a telescope and then passed through a spectrograph in order to disperse the white light from the light source and form its spectrum. The photographic plate on which the spectrum is recorded is called a *spectrogram*. As a standard against which wavelengths in the astronomical spectrum can be measured, an emission spectrum of a gas, such as neon or vaporized iron or titanium, is placed above and below the astronomical spectrum. The mechanism for placing the laboratory spectrum on the astronomical spectrogram is a part of the telescope and spectrograph. With these comparison lines, whose laboratory wavelengths are known, the astronomer can determine the unknown wavelengths of the astronomical object's spectral lines. An example of a star's absorption spectrum is shown in Figure 4.12; the absorption spectrum is gray with black absorption lines and the comparison emission spectrum of neon is white spectral lines on a black background.

Kirchhoff's laws of spectrum analysis tell us about the general physical conditions of the light source. And if the spectrum of the light source contains absorption or emission lines, then we can measure their wavelengths and identify the chemical elements that are present. But can more detailed information about the light source be found? Suppose we want to know the temperature of the light source. Can this be done? Yes it can, for special types of light sources known as ideal radiators or blackbodies. In Chapter 12 we will outline the basis for learning more detailed information about stars, such as density, rotation, and so forth.

BLACKBODY RADIATION

All objects radiate and absorb some form of electromagnetic radiation; the wavelength region and the amount of energy generally depend on the body's temperature and physical state. From laboratory experiments and from theory, physicists in the nineteenth century analyzed how various bodies emit and absorb radiation as a function of temperature and wavelength. From this work they developed an idealized radiator called a *blackbody*. This imaginary body can absorb all the radiant energy that reaches its surface and reradiate it with 100 percent efficiency. Real matter is generally less than 100 percent efficient when it radiates.

For our purposes the most important feature of the blackbody is the way in which emitted radiant energy is spread in wavelength, or the *spectral energy distribution*. Scientists have found that the distribution of energy depends only on the blackbody's *temperature* and not on its chemical composition. Figure 4.13 illustrates this point; note how the amount of radiant energy emitted by a blackbody varies with wavelength in a very recognizable way, even for different temperatures. The emission of radiant energy (or the brightness at each wavelength) covers a continuous range of

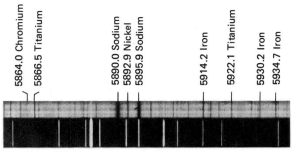

FIGURE 4.12
A portion of the yellow part of the spectrum of an orange-colored giant star. A number of absorption lines are marked, giving the wavelength in angstroms and the elements responsible for them. Below the star's spectrum is a comparison spectrum of neon used to establish the scale of wavelengths.

FIGURE 4.13
Blackbody energy curves according to Planck's radiation law. Note that as the temperature increases, the peak energy shifts toward the shorter wavelengths and the energy emitted at every wavelength increases.

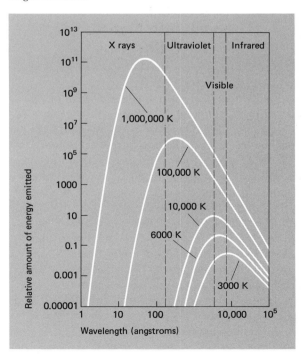

wavelengths, so that the *spectrum of a blackbody is a continuous spectrum;* that is, there are no color bands missing from its spectrum. At room temperature, lampblack (a finely powdered black soot) is very close to being a blackbody because it absorbs almost all of the radiation incident upon it and reflects very little. Fortunately, the radiation emitted by stars tends to be much like that emitted by a blackbody. We will use this fact later in our study of the sun and stars.

In 1900 the German physicist Max Planck (1858–1947) derived a mathematical expression, now called *Planck's law,* that describes the distribution of brightness in the spectrum of a blackbody, which is shown in Figure 4.13. There are two other distinguishing characteristics about the spectrum of blackbody radiation in Figure 4.13. First, the energy emitted by the blackbody is greater at every wavelength as the temperature increases. Thus the total amount of radiant energy emitted increases with increasing temperature, which is known as the *Stefan-Boltzmann law.* Second, the greatest amount of radiation (the peak of the curves in Figure 4.13) is found toward shorter wavelengths (blue end of the spectrum) as the temperature increases. This is known as *Wien's displacement law.*

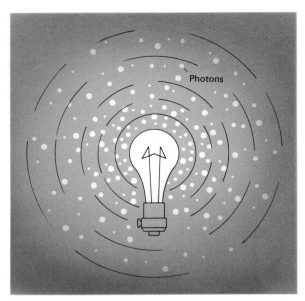

FIGURE 4.14
Photon emission. The incandescent filament of the lightbulb emits photons in all directions in space. Since the source is a continuous source, the photons have a continuous range of energies (i.e., wavelengths).

4.3
The Discrete Nature of Light: Photons

From his theoretical study of the emission of radiation by blackbodies, Planck concluded that they do not emit or absorb energy in a continuous fashion, but they do it discontinuously, in discrete units that later were called *photons.* This means that the energy transported by an electromagnetic wave is not continuously distributed over the wave front; it is located at discrete points in the wave (the photons) moving with the wave front. In 1905 Einstein used Planck's idea of a discrete nature for the emission of light to explain a phenomenon discovered in 1887 known as the photoelectric effect. This effect cannot be understood if light has only a wave nature. Since that time an extensive body of experimental and theoretical evidence has been collected validating the photon concept of light.

What are some of the properties of photons? They

move with the speed of light, have no inertia, are electrically neutral, and are massless. Picture a radiating body as emitting photons of differing discrete energies in all directions (see Figure 4.14). The photons retain their energy while traveling through space. Their arrival rate, or *flux,* at any point in space decreases with the square of their distance from the radiating source. Hence the inverse-square law of light can be understood in terms of numbers of photons (brightness of radiation) as well as in terms of electromagnetic waves.

The energy of each photon is inversely proportional to its wavelength. The shorter the wavelength, the more energetic is the photon; the longer the wavelength, the less energetic is the photon. That is why for example very short wavelength X-ray and gamma ray photons can destroy molecular structures in living tissue while photons of visible light cannot.

Photons may be absorbed by an atom, scattered by particles of matter, or converted into matter by interacting with other photons. They are created inside atoms and in violent collisions between subatomic particles. When they lose their identity, they transfer their energy to some other physical system; and when they are created, they obtain their energy from some other physical system. Their creation and destruction is a classic example of the conservation of energy. The two concepts of light—as a photon and as a wave—are complementary. However, experiments are designed to inquire about either light's wave nature or its corpuscular nature; no experiment will simultaneously yield the discrete and the wave properties of light.

The theory of the discrete nature of light began a conceptual revolution in twentieth-century physics and astrophysics. It was used by the Danish physicist Niels Bohr to formulate a new model for the atom that can be used to understand how light is created and destroyed inside the atom. Before we answer questions about the creation and destruction of photons, we must look at the structure of the atom.

4.4 Atomic Structure

STRUCTURE OF THE ATOM

One of the earliest philosophical speculations dealt with what the material world is made of. Is each substance, such as rock or wood, infinitely divisible, so that its subdivisions always yield the same properties as when whole? Or is there some level of structure below which the subdivisions will show new properties and forms? The Greek philosopher Democritus (460?–?370 B.C.) suggested that the material objects of our experience are actually made up of fundamental units. He called these units *atoms* and he visualized them as indestructible, indivisible, and capable of assuming various forms and shapes. Matter is never born from nothing but springs from new combinations of atoms. The concept of the atom was not widely accepted until the English scientist John Dalton (1766–1844) developed his *atomic theory* about one hundred fifty years ago.

Modern science has shown that the atom, too, can be subdivided into more fundamental units. In 1897 in the work of the English physicist J. J. Thompson (1856–1940), the *electron* was identified as a constituent of the atom. Also in England early in this century, Ernest Rutherford (1871–1937) was able to show that most of the atom is empty space and that nearly all its mass is concentrated in the *nucleus*. The principal constituent of the nucleus is the *proton,* which was identified by Rutherford in about 1919. By 1932

the British physicist James Chadwick (1891–1974) was able to show that a second particle resided in the nucleus, which was called a *neutron.*

The picture of the atom used today is of a nucleus made of positively charged protons and neutrons, which have no charge, with negatively charged electrons in orbits surrounding the nucleus. This conceptual model, with later refinements, accounts satisfactorily for the periodicities in the physical and chemical properties displayed in the *periodic table of the elements* (see Figure 4.15). The chemical identity of each atom of an element is characterized by the number of protons in its nucleus, which determines the element's *atomic number.* The simplest nucleus is that of hydrogen, with atomic number 1. The atomic number of helium is 2, that of lithium is 3, and so on. The proton has a unit of positive electrical charge equal and opposite to the negative charge of the electron; the neutral atom has as many electrons as protons (see Figure 4.16). Atomic nuclei are made up of from one to about 260 protons and neutrons. The atom's mass is concentrated in the nucleus, because the mass of the proton or neutron is 1836 times greater than that of the electron. By convention the masses of atomic particles are measured in atomic mass units; one atomic mass unit is equal to one-twelfth the mass of the most common nucleus of carbon (1 atomic mass unit $= 1.66 \times 10^{-24}$ gram), which contains six protons and six neutrons (see Table 4.1). The proton's mass is 1.00728 atomic mass units and the neutron's mass is 1.00866 atomic mass units.

Although the nucleus of an element contains a fixed number of protons, it may have different numbers of neutrons. These different nuclei of an element are called *isotopes.* Some representative isotopes are listed in Table 4.1. All the elements in the periodic table except for twenty possess two or more isotopes, so that their *atomic weights* depend on the relative proportion of each isotope. Among all elements there are about 300 stable isotopes and 1600 unstable or radioactive isotopes. *Radioactive nuclei* are ones that are undergoing spontaneous changes that make them into the nuclei of elements with smaller atomic numbers. Although there are a large number of unstable isotopes, the majority of all nuclei are stable.

The search for the ultimate indivisible constituents of matter that began before Democritus continues today. Scientists have discovered over a hundred subatomic particles. Is there an even more fundamental

FIGURE 4.15

Periodic table of the chemical elements. The number to the left of the chemical symbol is the atomic number; the number below is the atomic mass.

When the atomic mass is not accurately known, the mass of the most stable isotope is given in parentheses.

1 H hydrogen 1.008																	2 He helium 4.003
3 Li lithium 6.94	4 Be beryllium 9.01											5 B boron 10.81	6 C carbon 12.01	7 N nitrogen 14.01	8 O oxygen 16.00	9 F fluorine 19.00	10 Ne neon 20.18
11 Na sodium 22.99	12 Mg magnesium 24.30											13 Al aluminum 24.30	14 Si silicon 28.09	15 P phosphorus 30.97	16 S sulfur 32.06	17 Cl chlorine 35.45	18 Ar argon 39.95
19 K potassium 39.10	20 Ca calcium 40.08	21 Sc scandium 44.96	22 Ti titanium 47.90	23 V vanadium 50.94	24 Cr chromium 52.00	25 Mn manganese 54.94	26 Fe iron 55.85	27 Co cobalt 58.93	28 Ni nickel 58.71	29 Cu copper 63.55	30 Zn zinc 65.38	31 Ga gallium 69.72	32 Ge germanium 72.59	33 As arsenic 74.92	34 Se selenium 78.96	35 Br bromine 79.90	36 Kr krypton 83.80
37 Rb rubidium 85.47	38 Sr strontium 87.62	39 Y yttrium 88.90	40 Zr zirconium 91.22	41 Nb niobium 92.91	42 Mo molybdenum 95.94	43 Tc technetium 98.91	44 Ru ruthenium 101.07	45 Rh rhodium 102.90	46 Pd palladium 106.4	47 Ag silver 107.87	48 Cd cadmium 112.40	49 In indium 114.82	50 Sn tin 118.69	51 Sb antimony 121.75	52 Te tellurium 127.60	53 I iodine 126.90	54 Xe xenon 131.30
55 Cs cesium 132.90	56 Ba barium 137.34	See below 57-71	72 Hf hafnium 178.49	73 Ta tantalum 180.95	74 W tungsten 183.85	75 Re rhenium 186.2	76 Os osmium 190.2	77 Ir iridium 192.22	78 Pt platinum 195.09	79 Au gold 196.97	80 Hg mercury 200.59	81 Tl thallium 204.37	82 Pb lead 207.19	83 Bi bismuth 208.98	84 Po polonium (209)	85 At astatine (210)	86 Rn radon (222)
87 Fr francium (223)	88 Ra radium 226.02	See below 89-103	104 Rf rutherfordium (259)	105 Ha hahnium (260)	106?												

Lanthanide Series (rare earth elements)

57 La lanthanum 138.91	58 Ce cerium 140.12	59 Pr praseodymium 140.91	60 Nd neodymium 144.24	61 Pm promethium (145)	62 Sm samarium 150.35	63 Eu europium 151.96	64 Gd gadolinium 157.25	65 Tb terbium 158.92	66 Dy dysprosium 162.50	67 Ho holmium 164.93	68 Er erbium 167.25	69 Tm thulium 168.93	70 Yb ytterbium 173.04	71 Lu lutetium 174.97

Actinide Series

89 Ac actinium (227)	90 Th thorium 232.04	91 Pa protactinium 231.04	92 U uranium 238.03	93 Np neptunium 237.05	94 Pu plutonium 239.05	95 Am americium (243)	96 Cm curium (247)	97 Bk berkelium (247)	98 Cf californium (251)	99 Es einsteinium (254)	100 Fm fermium (257)	101 Md mendeleevium (256)	102 No nobelium (254)	103 Lw lawrencium (257)

Legend:
Light Metals
Transition Heavy Metals
Nonmetals
Inert Gases
Lanthanide Series (rare earth elements)
Actinide Series

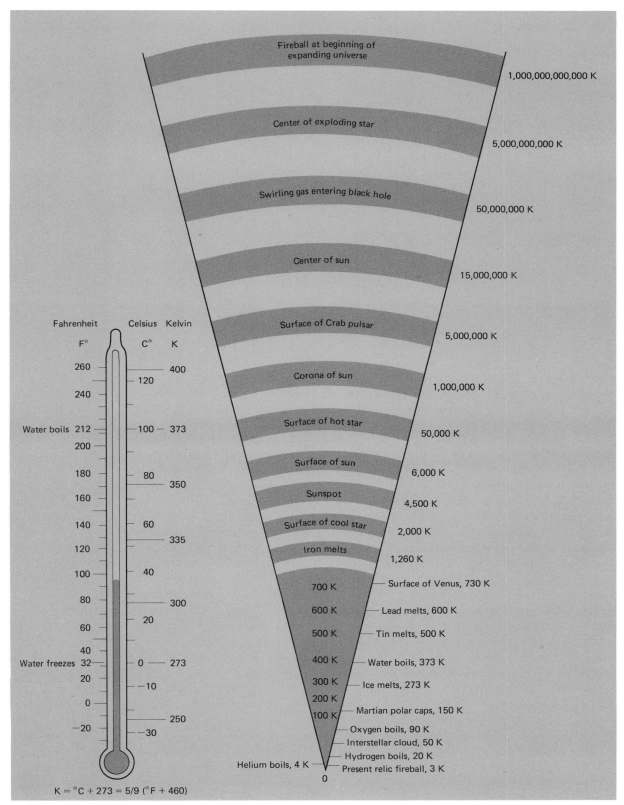

FIGURE 4.16
Different temperature systems and the
temperature range in the universe.

4.4 Atomic Structure   67

TABLE 4.1
Some Representative Isotopes of Several Elements

Element	Atom Number	Atomic Nucleus Symbol[a]	Atomic Weight	Percentage of Abundance[b]	Total Number of Natural Isotopes	Inside Nucleus (Nucleons)		Outside Nucleus
						p[c]	n[c]	Electrons
Hydrogen	1	$^1H_1$	1.0078	99.985	2	1	0	1
Deuterium	1	$^2H_1$	2.0140	0.015		1	1	1
Helium	2	$^4He_2$	4.0026	99.99987	2	2	2	2
Carbon	6	$^{12}C_6$	12.0000	98.89	2	6	6	6
Carbon	6	$^{13}C_6$	13.0034	1.11		6	7	6
Oxygen	8	$^{16}O_8$	15.9949	99.759	3	8	8	8
Aluminum	13	$^{27}Al_{13}$	26.9815	100.00	1	13	14	13
Iron	26	$^{56}Fe_{26}$	55.9349	91.66	4	26	30	26
Uranium	92	$^{235}U_{92}$	235.0439	0.72	3	92	143	92
Uranium	92	$^{238}U_{92}$	238.0508	99.27		92	146	92

[a]The subscript is the number of protons; the superscript is the number of protons and neutrons. The atom is neutral when the number of electrons (which are negative) equals the number of protons (which are positive).
[b]Relative contribution of isotope to total abundance of that element.
[c]Protons and neutrons.

level of structure below these? Evidence grows that indeed many of the subatomic particles may in turn be composed of a few more basic particles, called quarks.

NATURE OF GASES

Now let us consider the behavior of collections of atoms. Matter exists in three states. In a *solid*, atomic particles are bound to permanent positions relative to each other; in a *liquid* the particle bonds are weak and temporary. By contrast, in a *gas* the bonds between atomic particles are lacking and the particles have no permanent positions relative to each other. Most of the matter in the universe is in the form of a gas. Frequently it is in the form of a *plasma*, a high-temperature mixture of electrons stripped from atoms and *ions,* which are atoms from which one or more electrons have been removed. The ion has a net charge: if it has fewer electrons than protons, it is a *positive ion*; if it has more electrons than protons, it is a *negative ion*. Now we need to know something about gases and plasmas.

The particles of a gas can be *molecules* (which consist of two or more atoms), *atoms* themselves, or *ions* and *electrons*. In the molecule the atoms may be the same element, as in the two atoms of the oxygen molecule we breathe, or different elements, such as

the two hydrogen atoms and one oxygen atom in the water molecule. Within a gas, particles dart about rapidly, colliding millions of times each second and changing their direction of motion just as frequently. Each gas particle has a kinetic energy that is proportional to the product of its mass and the square of its speed. After a collision, the speed can be either greater or smaller than before the collision; the kinetic energy of each particle changes in its repeated collisions. However, collectively the gas particles will have some *average kinetic energy*, which changes only when energy is added to the gas or removed from it. Another way of saying this is that the average kinetic energy changes when the gas is heated or cooled. *Temperature* is a measure of the average kinetic energy of the gas particles. The motion of the particles is called *random thermal motion*. It increases as the temperature goes up and it decreases as the temperature goes down. *Absolute zero* is reached when the average kinetic energy is zero. Temperatures in astronomy are usually measured on the absolute *Kelvin* (K) scale. In this system, there are 100 divisions (degrees) between the freezing point (273K) and the boiling point (373K) of water as shown in Figure 4.16.

In our daily lives we often confuse temperature with the effects of heating and cooling. For example, when we heat the air in a vessel, we are increasing the kinetic energy of each particle, which means a higher average kinetic energy or an increase in the temperature. As a result the gas particles move about faster, and they collide more frequently and more violently with their surroundings. If the density of gas particles (the number per unit volume) is quite large, then the hot gas can transfer a great deal of heat to its surroundings. But if the density is quite low, the hot gas may transfer virtually no heat to its surroundings. Thus heating and cooling effects depend on both the temperature of the gas particles and their density. This fact will help us to understand many topics throughout the book.

4.5
Emission and
Absorption of Radiation

THE BOHR ATOM

Early twentieth-century physicists had one bedeviling problem: What makes the atom emit a discrete pattern of spectral lines? By 1913, when the structure of atoms was reasonably well known, Niels Bohr (1885–1962), a Danish physicist, developed a theory for the structure of the simplest atom, hydrogen, whose one electron orbits around a proton. He suggested that the electron can only occupy a selected number of prescribed concentric orbits about the nucleus, rather than having an unlimited and unspecified orbital distance. Also, the electron normally resides in the lowest energy orbit, which is the one closest to the nucleus. The diameter of the first orbit corresponds to the normal size of the hydrogen atom, about 10^{-8} centimeter in diameter (see Figure 4.17).

When the atom absorbs energy, it is *excited*, and the electron appears in one of the outer orbits, which have successively higher energies than the lowest. The electron's change (up or down) from one allowed orbit to another is called a *transition*. An atom in a gas may acquire energy by random thermal collisions with other gas atoms, collisions with subatomic particles such as electrons, or absorption of a photon traveling through the gas. Of all the photons striking the atom, only those possessing an amount of energy equal to the energy difference between a higher energy orbit and the one in which the electron is located will be

FIGURE 4.17
Simplified model of atoms and ions.

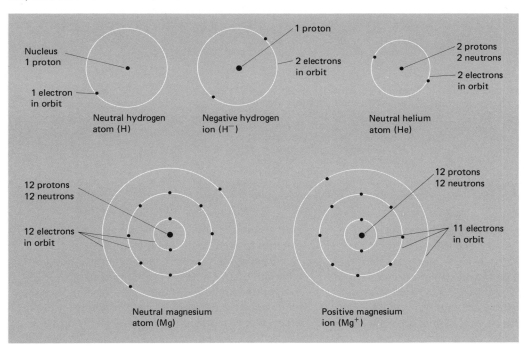

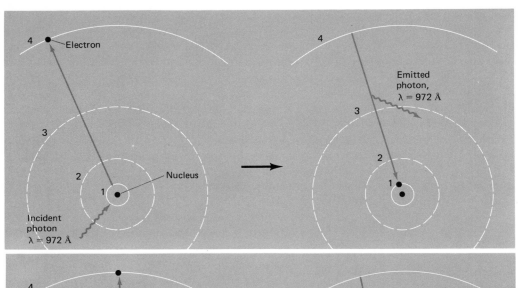

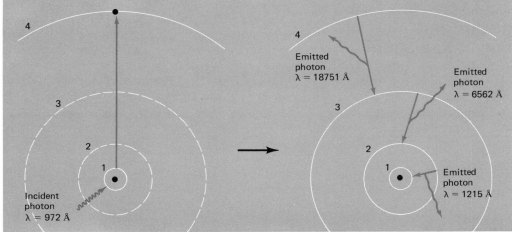

FIGURE 4.18

Two examples of atomic absorption and emission. When the hydrogen atom absorbs a photon (top, at left), it is excited and an electron jumps, for example, from orbit 1 to orbit 4. Within 10^{-8} second the atom radiates a photon of equivalent energy as the electron drops from orbit 4 to orbit 1. Here (bottom), the hydrogen atom absorbs a photon and an electron

jumps from orbit 1 to orbit 4, as in the first example. But this time the energy is quickly released in three steps (at right) as the electron drops from orbit 4 to orbit 3 to orbit 2 to orbit 1, each time giving up a photon of different energy, that is, different wavelengths. The emitted energies of the three photons are equal to the absorbed energy of the incident photon.

absorbed and excite the atom. For example, in the hydrogen atom it takes 10.2 electron volts, or 1.63×10^{-11} erg, of energy to raise the electron from its lowest energy orbit to the next higher energy orbit. Photons with energies below 10.2 electron volts will not be absorbed and excite it. Photons with energies in excess of 10.2 electron volts cannot raise the electron to the second orbit, but they may, if they have the

right amount of energy, excite the atom by lifting the electron to even higher energy orbits.

If a gas atom is excited, then in about a hundred-millionth of a second it will emit the energy in the form of one or more photons. Somewhat like a ball bouncing down a staircase, the electron will drop in succession into one or more lower energy orbits on its way to the lowest energy level, in which it can reside

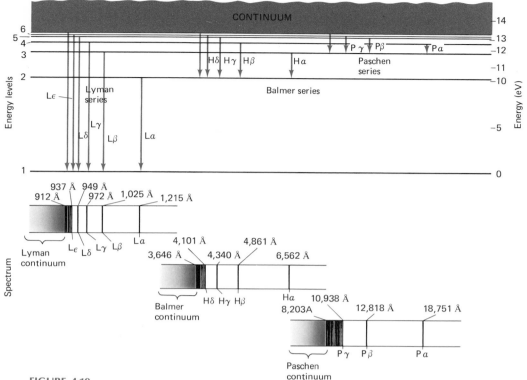

FIGURE 4.19

Energy levels and series in the hydrogen atom. Each series occupies a different portion of the electromagnetic spectrum; each series comes to a limit, known as the series limit. Downward transitions shown here give rise to emission lines, diagrammed below (see also Figure 4.20). Upward transitions give rise to absorption lines.

indefinitely. With each downward transition a photon of electromagnetic radiation is emitted. This photon represents the energy difference between the two orbits associated with the electron's transition. The greater the energy difference, the greater is the amount of energy released in the form of a photon and, consequently, the shorter is the wavelength of the photon. Two examples from the countless electron transitions possible in the hydrogen atom are shown in Figure 4.18. Besides emitting energy spontaneously, an excited atom, before it can emit a photon, may collide with another atom in the gas and transfer energy to it as kinetic energy of motion. In this case, no photon will be emitted.

SPECTRUM OF HYDROGEN

An energy level diagram for hydrogen appears in Figure 4.19. The number of the energy levels corresponds to the number of the electron orbits. The distance between successive electron orbits increases with higher orbit numbers, but the differences in energy between successive orbits grows smaller as the orbit number increases. Compare Figures 4.18 and 4.19, which illustrate this point. Notice in Figure 4.18 that an electron can de-excite in many ways. Suppose a hydrogen atom is excited so that its electron is in the third energy level. Then it may de-excite directly to the ground state, emitting one photon. The photon would have a wavelength of 1026 angstroms, which corresponds to the energy difference between these two orbits in the hydrogen atom. Or it may de-excite to the second energy level, emitting a photon with a wavelength of 6562 angstroms, and then again de-excite to the ground state, emitting a photon with a wavelength of 1216 angstroms. The total energy emitted in both cases is the same, but the wavelengths that result, and hence the spectral lines, differ.

The hydrogen line spectrum in the visible region, known as the *Balmer series*, is prominent in the absorption spectra of most stars. It arises from electron transitions originating on the second energy level of the atom. In the same way, all the possible transitions from the ground level are known as the *Lyman series*, which is in the ultraviolet part of the electromagnetic spectrum. Those transitions from the third level up to higher energy levels are the *Paschen series*, which is in the infrared; and so on for the remaining series, whose lines appear in the far infrared or microwave region of the spectrum. Each series comes to a limit toward shorter wavelengths. The uppermost levels, representing the electron's highest energy orbits, crowd together toward a *series limit*, which represents the point beyond which the proton can no longer bind the electron to it. In this case the electron has been removed from the atom (it has been ionized) and is free to take on any energy.

If the electron is given enough energy, either by collision or absorption of a photon, it can escape the electrical attraction of the nucleus. The atom is then ionized and is a positive ion. The ionized hydrogen atom cannot absorb or reradiate energy in the form of discrete lines until it captures a free electron. It can execute the capture because of the electrical force of attraction between the positively charged nucleus (the proton) and the negatively charged electron. Figure 4.20 shows the Balmer series in the absorption spectrum of an actual star. Note that it converges toward its series limit at approximately 3646 angstroms.

SPECTRUM OF OTHER ELEMENTS

In addition to limits on the size of electron orbits, there is a limit to the number of electrons that may occupy a given orbit. These allowed orbits with a prescribed number of electrons in them are called *electron shells* (see Figure 4.16). In general, as one goes through the periodic table, electrons are added to balance the number of protons in the nucleus by filling the shells from the one closest to the nucleus outward. In hydrogen there is one electron in the innermost shell, which has room for a maximum of 2 electrons. Helium's 2 electrons fill or *close* the shell, so that for the element lithium, the third electron must start a new shell, which is the next innermost. In the second shell there is only room for 8 electrons; in the third, 18 electrons; in the fourth, 32; and so on.

Each element has its own unique set of energy levels. Consequently, the wavelength of the spectral lines originating from electron transitions between various levels is also unique for each element—a clear fingerprint of the element. The amount of energy needed to ionize an atom that has more than one electron varies. For example, to remove the outermost electron from helium takes five times as much energy as to do the same for sodium. Also, for a given element each additional ionization takes more energy to free electrons in the inner orbits, because they are more tightly bound to the nucleus. This multiple ion-

FIGURE 4.20
Balmer series in the spectrum of the star HD 193182, showing the hydrogen lines in absorption converging toward the series limit to the left of H40 at 3646 Å. Beyond that extends the Balmer continuum (a continuous spectrum of diminishing intensity), produced by the removal of the electron from the atoms. Balmer lines between H9 and Hα (the first Balmer line) are also present, although they are not shown here. The comparison spectrum of iron, above and below the star's spectrum, establishes the scale of wavelengths.

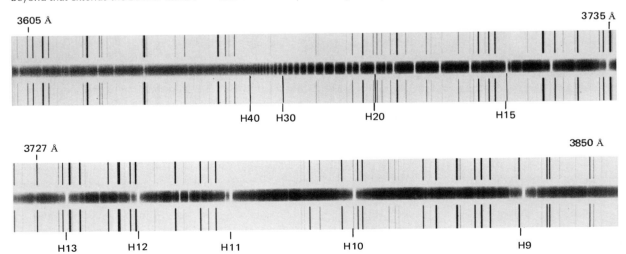

ization in, say, an iron atom brings a corresponding readjustment of the energy levels because of the altered electrical attraction between the positive nucleus of the iron atom and the reduced number of electrons. Redistribution of the energy levels gives different spectral lines with each succeeding ionization of the iron atom. So not only do we see a different set of wavelengths in absorption or emission between different unionized elements, but for the same element the spectrum is different after each ionization.

Using the properties of electromagnetic radiation, the atom's structure, the interaction between matter and energy, and spectrum analysis, astronomers have gained much information about the universe from the radiation it emits. And as we develop an even greater understanding of the nature of radiation—its formation, propagation, interaction with matter, and destruction—we can explore more deeply those dim sources in the outer reaches of the cosmos, almost back to the beginning of time.

QUANTUM THEORY AND THE FORCES OF NATURE

Gravity was the first force in nature to be understood in at least a mathematical sense. Newton's theory shows that, even though separated by enormous distances, pieces of matter can influence each other's state of motion. Although less familiar in many respects, the electric and magnetic forces have been known since ancient times. Like gravity they both weaken as the square of the distance away from their source. In 1873 James Clark Maxwell (1831–1879) showed that a relationship exists among electricity, magnetism, and light—an amazingly unifying step.

In 1924 the French physicist Louis de Broglie (1875–1960) pointed out that, like light, subatomic particles also have a wave nature, as well as a discrete nature. This has been verified experimentally many times. It is now an accepted fact that matter and radiant energy have dual natures in that they show both wave and discrete properties. Taking de Broglie's idea, Erwin Schrödinger (1887–1961), an Austrian physicist, and Werner Heisenberg (1901–1975), a German physicist, independently constructed mathematical theories for atomic structure at about the same time (1925). Their theories were consolidated by Paul Dirac, an English physicist, into the mathematical formulation called *quantum mechanics*, the most rational and logical approach so far for understanding a vast variety of atomic phe-

nomena. In reality there are no discrete electron orbits like those of planets in the solar system. Within the hydrogen atom, for example, are spherical regions surrounding the proton. In these regions the electron is spread into a pattern of standing waves, whose distribution corresponds to a discrete energy state of the atom.

All atomic properties are known to be the consequence of the electrical interaction between the nucleus and the electrons surrounding it. This electromagnetic interaction is responsible for the characteristic structure of each atomic species. These characteristic structures are responsible for the basic forms of matter, from simple rocks and crystals to flowers and even humans. The electromagnetic force between the electron and the nucleus is 10^{39} times stronger than the gravitational force between them; no one has detected, nor is there any prospect of detecting, the effects of gravity in atoms or molecules.

By 1932 it was known that the nucleus was composed of protons and neutrons. This raised the problem of what force holds the nucleus together against the mutual electrical repulsion of the protons for each other. The solution of this question was the discovery of the strong *nuclear force* of attraction. It is about a hundred times more powerful than the electromagnetic

force, but of very short range, and is capable of holding together nuclei with as many as a hundred or so protons.

Finally, a fourth force was discovered around 1935, the *weak nuclear force*, which is about 10^{-5} times as strong as the strong nuclear force, or about a thousandth as strong as the electromagnetic force. This force is responsible for some changes in the nature of the nucleus that occur in radioactive decay. It is also a very short range force. There is recent evidence that suggests that the electromagnetic force, the weak nuclear force, and possibly the strong nuclear force are actually different manifestations of the same force acting differently at different distances between particles. Linking all four forces into one universal expression, the so-called unified field theory, still eludes us.

The reason we are familiar with the gravitational and electromagnetic forces is that they operate on the scale of our experiences. The other two, the strong and weak nuclear forces, are confined to the nuclear scale of existence. The gravitational force increases its intensity with increasing mass, whereas the other forces are independent of mass. In the cosmos, as we will see in later chapters, gravity dominates.

SUMMARY

Astronomers have learned much about celestial bodies by studying the light they emit. Light is a form of electromagnetic energy and has the properties of both waves and particles. Light is reflected, refracted, and diffracted as are other waves. But to understand how light is emitted by sources such as stars, we must utilize the discrete nature of electromagnetic radiation.

Visible light is no more than a glimmer of the electromagnetic radiation emitted by all objects in the universe. The electromagnetic spectrum includes radiation in wavelengths much shorter and much longer than those of visible light, from gamma rays at the short end of the spectrum to radio waves at the long end. Earth's atmosphere keeps most celestial radiation out. The exceptions are the visible region, some parts of the infrared and microwave regions, and the radio region of the spectrum.

Most light from natural sources is composed of many wavelengths and is called white light. By using a prism or diffraction grating, astronomers break up white light into its component wavelengths, forming its spectrum. Light sources produce different kinds of spectra depending on their physical state. There are three basic types of spectra: continuous spectra, bright-line or emission spectra, and dark-line or absorption spectra. Each chemical element produces its own unique pattern of emission or absorption lines, so studying these lines can tell us about the chemical composition of a light source. A continuous spectrum cannot give us this kind of information; but by studying the intensity as it varies in wavelength, we can learn about a light source's temperature, because many light sources, including the stars, come close to acting like blackbodies. A blackbody is an idealized radiator. As temperature increases, it will emit more radiation at every wavelength, and its peak radiation will lie toward the blue end of the spectrum.

In studying blackbodies a new concept of light was developed—it has discrete properties as well as wave properties. These discrete units are known as photons. They move with the speed of light, are electrically neutral, and are massless. Each photon has a discrete amount of energy, inversely proportional to its wavelength.

Electrons in atoms may both absorb and emit photons. When a bound electron acquires the energy of a photon or acquires energy by collision with another particle, it is excited to a higher-energy orbit around the atom's nucleus; then it de-excites, giving up the energy it absorbed and falling to a lower-energy orbit. It gives up that energy by emitting another photon. Each orbit represents a discrete amount of energy, which is different for each chemical element; the photon that is emitted represents the difference between the energy levels of the orbits involved in the transition. These transitions account for either the emission or the absorption lines seen in the spectra of various astronomical bodies.

REVIEW QUESTIONS

1. What are some properties of light that characterize its wave nature?

2. Define the wavelength of light and the frequency of light.

3. What is meant by apparent brightness?

4. Suppose you measured the brightness of a streetlamp that is twice as distant as another identical streetlamp of the same power and found that the more distant lamp's brightness was down by a factor of 75 percent. Would that be normal?

5. List the different regions in the electromagnetic spectrum.

6. What makes a grating superior to a prism for spectrum analysis?

7. How does an astronomer identify the elements present in a star's atmosphere?

8. What kind of spectra do the sun and most stars exhibit?

9. What sort of spectrum do the lights in your classroom show?

10. Is it possible to photograph gamma radiation? X-ray radiation? Infrared radiation?

11. How could you demonstrate that neon signs contain other gases than neon?

12. How does the Doppler effect enable an astronomer to determine the line-of-sight motion of a celestial body?

13. How can a brilliantly glowing body act like a blackbody?

14. Why does a glowing metal object change its color from red to blue as its temperature is increased?

15. If you were told that the normal temperature of the human body is 37°C, would you believe it?

16. How do we know that a hydrogen atom contains only one proton and one electron?

17. Which of the following atoms, neutral or ionized, is possible? (a) 6 protons, 7 neutrons, and 5 electrons; (b) 7 protons, 6 neutrons, and 9 electrons; (c) 7 protons, 8 neutrons, and 7 electrons.

18. Why doesn't mercury, with atomic number 80, when stripped of one electron, change into gold with atomic number 79 (the alchemist's dream)?

19. Why are the energy levels of each atom different?

20. Why will no new natural element be found?

5
Telescopes and Their Accessories

The earth receives electromagnetic radiation of all wavelengths from various places in space. However, only two spectral regions of the electromagnetic spectrum are able to freely penetrate the earth's atmosphere. Most of the electromagnetic spectrum is filtered out by the atmosphere well above the earth's surface, as we saw in Figure 4.4. The two spectral windows in the atmosphere through which we can best observe the universe are called the *optical window,* from about 3,000 angstroms to about 10,000 angstroms, and the *radio window,* which includes the wavelength region from about 1 millimeter to 30 meters. Let us first look at the devices with which we observe objects in the optical window.

5.1 Optical Telescopes

THE FORMATION OF AN IMAGE

In optical astronomy the object with which we work is the *image* of the light source formed by the principal image-forming part of the telescope, which is called an *objective*. The objective of the optical telescope is either a lens or a mirror, as shown in Figure 5.1. Light rays from the light source are *refracted* in passing through a lens and are *reflected* from a mirror. The image is produced where the light rays converge to a position known as the *focus*. The *focal length* of the objective is the distance behind the lens or the distance in front of the mirror to the focus. The image of a star is just a point of light, as shown in Figure 5.1, while that of an extended object such as the moon is inverted, as can also be seen in Figure 5.1. In telescopes using either mirrors or lenses, an eyepiece magnifies the image much as a reading glass magnifies small print. Or a photographic plate may be inserted into the focal plane of the objective instead of the eyepiece, transforming the telescope into a giant camera.

◀ Mount Wilson and Palomar observatories. This moonlight view on Mount Palomar shows the 5.1-meter Hale telescope dome with the shutter open.

PROPERTIES OF AN IMAGE

The image formed by either a lens or a mirror has certain properties that depend upon the diameter or *aperture* of the objective and its focal length. One property is the size of the image. Since the image of a star is a point, size is not an important property for it. However, for an extended object the image size depends upon the angular size of the light source on the sky and also on the focal length of the objective. As an example, the angular size of the moon is about 1/2° on the sky. A telescope with a 100-inch (2.5-meter) focal length produces a lunar image of approximately 1-inch (2.5-centimeter) diameter in its focal plane.

The brightness of an image of a *point source*, such as a star, depends on how much light is intercepted by the objective. Hence its brightness is proportional to the area of the objective or to the square of the aperture. Doubling the aperture increases the area of the objective by four times, concentrating four times as much light into the image.

However, when we photograph an extended object, the surface brightness of the image depends on the amount of radiant energy per unit area of the image. The objective's area (or the square of its aperture) still determines the total amount of energy collected, but the total energy is distributed over the entire image. Thus the larger the image's area, the smaller the energy per unit of area. The image size of an extended object increases in proportion to the focal length, so for a given telescope aperture the surface brightness of the image decreases as the focal length is made longer. It turns out that the brightness of an extended object's image is inversely proportional to the *focal ratio* squared. The focal ratio is the ratio of the focal length (which determines the image size) to the aperture (which determines total energy collected). The smaller the focal ratio, the brighter is the image of an extended object.

How well a telescope discriminates between two adjacent objects or shows fine details is called its *resolving power*. Because of the wave nature of light, the image of a point source produces a *diffraction pattern* (see Figure 5.2); it appears as a bright central spot called a *diffraction disk* surrounded by progressively fainter rings. When the diffraction patterns of two stars that are close together no longer overlap, we can see separate stellar images, as shown in Figure 5.3. The larger the telescope's aperture, the smaller is the diffraction disk of each image. A large aperture there-

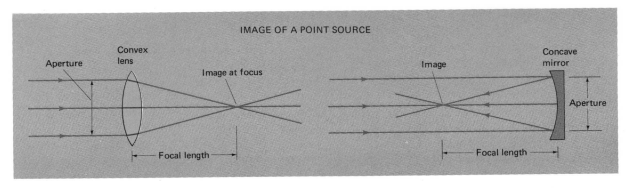

IMAGE OF A POINT SOURCE

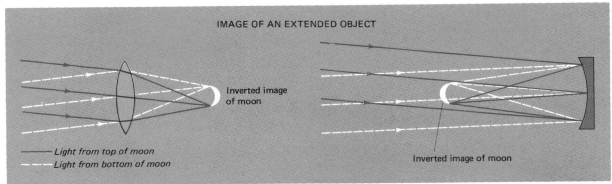

IMAGE OF AN EXTENDED OBJECT

Light from top of moon
Light from bottom of moon

FIGURE 5.1
Comparison of images formed by point sources and extended objects. The extended object's image is inverted because of the way light is refracted by the lens or reflected by the mirror.

FIGURE 5.2
Diffraction pattern of a point source of light. The alternating bright and dark fringes are produced by the reinforcement and cancellation of the light waves with each other. The diffraction disk is the central white image where most of the light is concentrated. This is a highly magnified photograph.

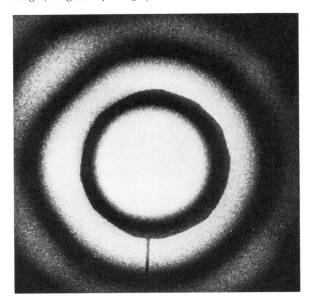

fore improves the resolution of closely adjoining features by making the diffraction effect of adjacent objects overlap less. We define resolving power as the smallest angle between two close objects whose images can just be separated by a telescope. This critical angle is directly proportional to the wavelength of the observed radiation and inversely proportional to the aperture of the objective. To just separate the components of a double star whose angular separation is 1 second of arc (which is very close), we need an 11-centimeter objective (easily attainable in an amateur telescope). To resolve two such close radio sources with a radio telescope that operates on a wavelength of 10 centimeters, the aperture would have to be 21 kilometers, because such waves are approximately 180,000 times longer than light waves, and the resolution decreases as the wavelength increases.

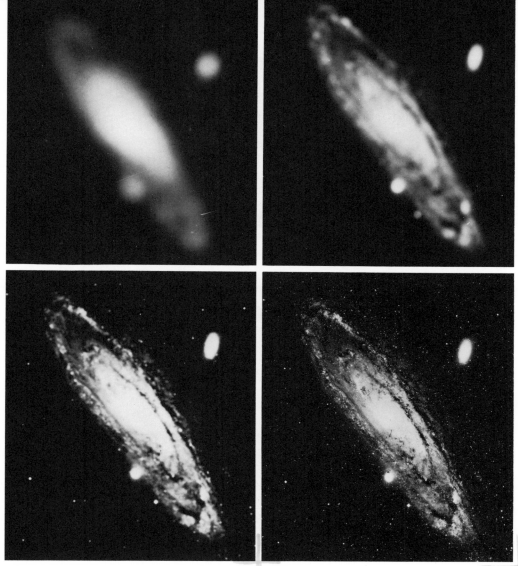

FIGURE 5.3
Resolution of an image. The spiral galaxy M31 as it would appear with resolutions of (top, left) 12 arc minutes; (top, right) 3 arc minutes; and (bottom, left) 1 arc minute; (bottom, right) M31 as seen with a large optical telescope.

VIEWING PROBLEMS

The theoretical resolving power of any optical telescope is never fully realized. The lower layers in the earth's atmosphere are unsteady and turbulent; this turbulence blurs and distorts the star's image and makes it twinkle or *scintillate*. The rapid scintillations break the starlight into many dancing specks of light. In long exposures these merge to form the fuzzy stellar images we see in photographs. When the atmospheric turbulence is low, the stars twinkle or scintillate less and the so-called *seeing* is improved. A planet, though, shines with a steady light, because each point of the

tiny disk visible from the earth twinkles out of step with neighboring points, and we see an average of all the twinkling points.

A technique called *speckle photography*, which can be used with large telescopes, can get around the smearing and wiggling of the image that comes from atmospheric turbulence. By exposing the photographic plate for an extremely short time (less than 1/100 second), we can get each star image to appear as a cluster of sharp specks of different brightness. Then we apply a high-speed light-sensing device and measure the variations in brightness across each

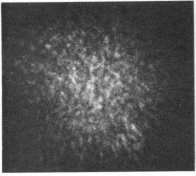
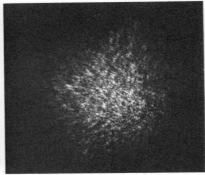
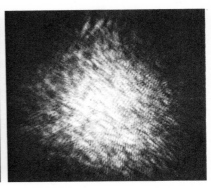

FIGURE 5.4
Speckle images (above) of three stars taken at Kitt Peak National Observatory. The synthesis of such images produces the image of Betelgeuse shown below.

Betelgeuse is seen as an extended image of this enhanced photograph from the Kitt Peak National Observatory. The slightly mottled

surface and tenuous outer edge of the image are thought to be partly genuine features of the star and partly artifacts of the photographic technique.

speck. We take the information from the assemblage of specks in each of many photographed images and feed it into a computer, programming it to analyze and reassemble the information into the unsmeared image of the star. Figure 5.4 shows the reconstructed image of Betelgeuse (Alpha Orionis), demonstrating that speckle photography can resolve minute sources of light, such as the disk of a large star.

Other nuisances hamper our view of the heavens. The night sky's transparency varies as smog, dust, and atmospheric haze cloud it. Also, the upper atmosphere is suffused with a faint light called *airglow*. Molecules of hydrogen, nitrogen, and hydroxyl (OH), and atoms of oxygen, sodium, and calcium are responsible for the airglow. These gases absorb the ultraviolet photons in sunlight and reradiate the energy in a few wavelengths of the green, red, and infrared spectral regions. On long exposures airglow fogs a photograph and reduces the contrast between the faintest images and the sky background.

Another problem is that starlight entering the atmosphere is bent increasingly toward the vertical so that we see a star slightly closer to the zenith (the point directly above the observer) than it really is (see Figure 5.5). This refraction effect is greatest near the horizon (about 1/2°), for there the light's path through the air is the longest. When we observe the rising or setting sun, it is really below our horizon, but refraction raises the sun's image above the horizon by an amount equal to its apparent diameter, which is 1/2°.

Other viewing problems are related to the geographical location of the observatory. An ideal site

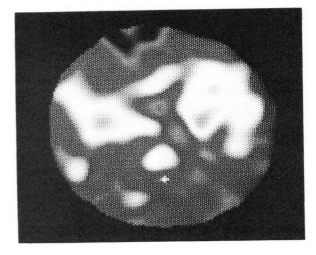

FIGURE 5.5
Bending of starlight or sunlight in earth's atmosphere due to refraction.

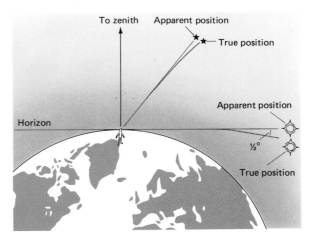

FIGURE 5.6
Panorama of the Kitt Peak National Observatory. The dome for the 4-meter Mayall reflector is in the foreground (right) and the McMath solar telescope building can be seen in the background (left).

for an optical observatory is a mountaintop where the air is dry, transparent, and steady. Mountaintops are also advantageous because the sky is generally darker at high altitudes. However, an observatory also needs a minimum amount of wind and relatively easy access. The southwestern part of the United States satisfies many of these conditions and, in addition, has many clear days and nights. Kitt Peak National Observatory is located there, about 65 miles southwest of Tucson, Arizona. A panoramic view of Kitt Peak is shown in Figure 5.6.

REFRACTING AND REFLECTING TELESCOPES

Telescopes that use lenses are known as *refracting telescopes*. In the seventeenth century, defects in lenses marred the images they presented. The objective of the simple refracting telescope then used could not form a sharp image due to a condition known as *spherical aberration*, and the lens failed to bring all the colors to a common focus, called *chromatic aberration*. A compound lens or two lenses of different types of glass joined together, known as an *achromatic lens*, can bring violet and red light to approximately the same focus and can also minimize spherical aberration.

Spherical aberration also occurs in a *reflecting telescope*, which uses a spherical mirror rather than a lens to produce the image. If the surface of the mirror is parabolic rather than spherical, this aberration is eliminated, although some minor deficiencies remain.

The first reflecting telescope, built in 1668 by Isaac Newton, is shown in Figure 5.7. Why are the big modern telescopes of the reflecting type? Reflecting telescopes have many advantages over refractors. The reflecting telescope is completely free from chromatic aberration, making it ideal for all-purpose photography and spectroscopy. Also, since a lens must be supported by its edges, there is a limit to how large a lens system can be. But a mirror, such as the one shown in Figure 5.7, can be held both at its edges and from the back, allowing a wide range of sizes for mirror systems. The largest refractor has an aperture slightly larger than 1 meter, but the largest reflector is 6 meters in diameter.

There are other advantages to reflectors as well. The glass for the mirror in a reflecting telescope need not be as optically pure and homogeneous as that required for a large lens, because the light reflects off the front surface and does not pass

FIGURE 5.7

Evolution of the telescope. Shown are Newton's reflector and a variety of modern versions.

Top row:
Newton's reflector. The mirror is nearly 5 centimeters in diameter.

Hale reflector (5.1 meters). The horse-shoe collar on the right slowly rotates the great telescope about its polar axis. The observer's cage is the cylindrical centerpiece at the top of the tube.

The mirror for the 5.1-meter Hale telescope as it appears before the surface is coated with a reflecting aluminum. The ribbed structure, which minimizes the weight of the mirror, can be seen through the front surface.

Center row:
Mayall reflector, Kitt Peak National Observatory, Arizona (4 meters).

Lick refractor (0.9 meters). A hydraulic system moves the observatory floor up and down for the observer's convenience.

A new concept in telescope design is the multi-mirror telescope (MMT) on Mount Hopkins in southern Arizona. It uses six small, independent mirrors to form an image rather than one large mirror.

Bottom row:
Fixed dish, Arecibo, Puerto Rico (305 meters). The movable overhead antenna can pick up signals within 20° of the vertical.

The Very Large Array (VLA) radio interferometer near Socorro, New Mexico. The telescope contains 27 dishes in a large Y-shaped configuration.

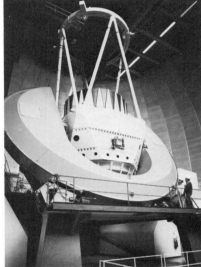

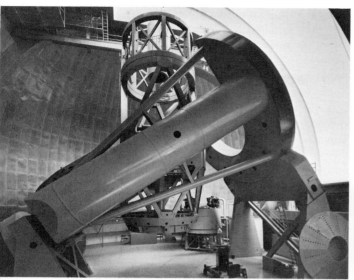

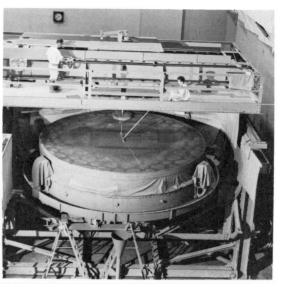

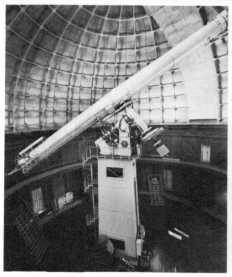

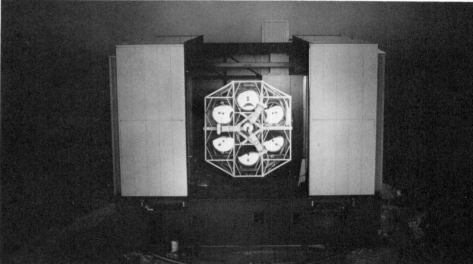

through the mirror, as it does through a lens. And the mirror has only one surface that must be painstakingly ground—the achromatic lens has four. To avoid changes in temperature that would affect the focal length of the reflector, large mirrors are constructed of fused quartz or of a zero-expansion pyroceramic material. The mirror's surface is coated with a thin layer of highly reflecting aluminum.

Reflecting telescopes can be designed for many kinds of astronomical work by choosing the focal arrangement (see Figure 5.8) to suit the type of observation. For photography, photometry, and spectroscopy of faint objects, the *prime focus* is best because its small focal ratio lessens the exposure time required. The *Newtonian focus,* most useful for small telescopes, is now little used by professional astronomers. In both these arrangements the observer works at a considerable distance above the observatory floor.

In the *Cassegrain* version a secondary convex mirror positioned at the top in front of the focus slows the rate at which the light rays converge, effectively increasing the telescope's focal length. It reflects the converging rays to the bottom of the telescope, an arrangement much more convenient for the observer. We might think that putting the secondary mirror and its supports, or the observer's cage, into the path of the light rays would cut down the light reaching the objective, but the loss is small and the quality of the images is not affected. Equipment that is too heavy

FIGURE 5.8
Focal arrangements of a large reflector. The focal ratios are those of the 5.1-meter Hale reflector.

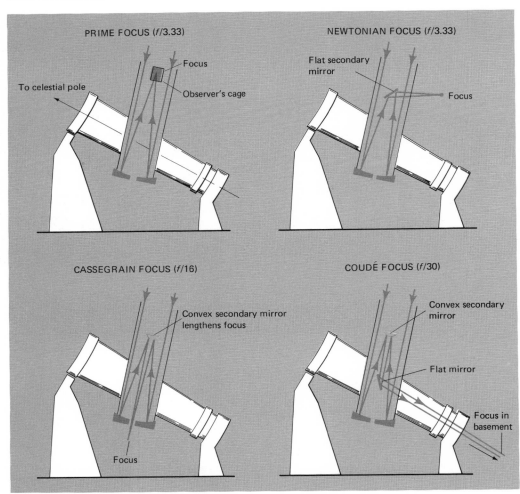

GALILEO GALILEI (1564–1642)

Galileo was born in Pisa in 1564, the year that Shakespeare was born and Michelangelo died. He was one of the last great figures of the Renaissance. He gave up the study of medicine in favor of mathematics. In school he annoyed his professors by refusing to accept without proof the dogmatic statements of past authorities. It was this same questioning in later years that led to his well-known difficulties with the Church. From 1589 to 1610 he held teaching posts as professor of mathematics at the Pisa and the Padua universities. He left in 1610 to become mathematician to the grand duke of Tuscany.

As the father of mechanics, Galileo experimented with the pendulum, with light and mirrors, with balls rolling down inclined planes, and with falling bodies. Out of his researches he formulated in precise mathematical terms the principles of their behavior. It remained for Newton later to incorporate and generalize Galileo's ideas into the well-known laws of motion (see page 35).

In 1609 Galileo was the first to survey the heavens with telescopes of his own construction after having heard about its invention by a Dutch spectacle maker. His observations of the phases of Venus and the motions of Jupiter's four largest satellites, which he discovered, were instrumental in helping to discredit the Ptolemaic theory. One discovery that eluded Galileo was the rings of Saturn. He noticed that the planet had small appendages on either side, which, over a period of time, varied in size and finally disappeared. "Has Saturn devoured his own children?" he queried. He was so disgusted he gave up observing Saturn. Although a confirmed Copernican, Galileo refrained from speaking out in favor of the new system for fear of being ridiculed. Later, his telescopic observations induced him to publicly advocate the heliocentric system.

---

and bulky to be attached to the back side of the primary mirror, or that is sensitive to changing gravitation as the telescope moves, can be placed in a room below the observatory floor. An auxiliary flat mirror diverts the long converging beam down the hollow polar axis around which the telescope rotates, and with this *coudé arrangement* the focus can remain stationary no matter which way the telescope points.

SCHMIDT TELESCOPE

In 1930 a German optician, Bernhard Schmidt (1879–1935), devised a telescopic system using both a mirror and a lens. Aberration introduced by the spherical mirror used by the Schmidt telescope is eliminated by putting a thin corrector lens at the center of the mirror's curvature (see Figure 5.9). The mirror's diameter is appreciably larger than the aperture of the objective correcting lens, so that excellent images can be obtained over a large field of view. Because the light rays come to a focus approximately midway between the mirror and the corrector lens on a spherically shaped surface, the photographic plate must be slightly curved to keep it all in focus. Several variations of the original design are in use.

The Schmidt telescope is the ideal instrument for surveying the sky so that astronomers can identify objects for later study with the big reflectors. It can cover extensive areas, and it has superb light-gathering ability with short exposure times because of its small focal ratio (which also reduces the scale of its photographs).

EQUATORIAL TELESCOPE MOUNTING

An optical telescope, in order to follow an object as the earth's rotation carries it across the sky, must be free to move. To track the stars accurately and to permit the telescope to be conveniently pointed in any direction, the *equatorial mounting system* is used for most telescopes (Figure 5.10). This system has two axes of rotation. The telescope can rotate in an east-west direction, called hour-angle, around its *polar axis*,

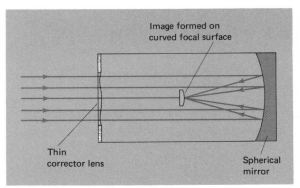

FIGURE 5.9
Schmidt telescope. Light rays entering
the corrector lens from additional directions than that shown in the diagram
fall on different parts of the spherical
mirror so that the entire surface of the
mirror is utilized in producing bright
images over a wide field of view.

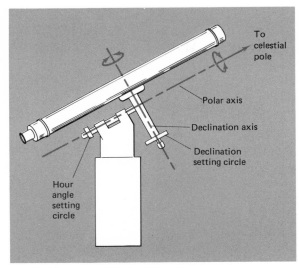

FIGURE 5.10
Schematic of an equatorially mounted
refracting telescope.

which is aligned with the earth's axis of rotation. Another axis allows the telescope to swing in a north-
south direction about the *declination axis,* which is
perpendicular to the polar axis. (For a discussion of astronomical coordinates, see Appendix 3.) Larger telescopes are usually positioned electronically from an
operating console and guided to the exact location
with hand controls. Once the large telescope is
properly set, an electronic clock drive slowly turns it
westward around its polar axis at the same rate as the
earth turns eastward, keeping the stellar images locked
in position in the field of view.

5.2
Accessory Instruments for
Telescopes

RADIATION DETECTORS

Before discussing the accessory instruments used with
optical telescopes, let us consider briefly the most important component of these instruments: the *radiation
detector.* The telescope is capable of collecting light
over a very wide range of wavelengths, but it is the
radiation detector that determines what the telescope
sees. One radiation detector with which we are all
familiar is the human eye. It possesses most of the

properties of radiation detectors in general and is thus
illustrative of the points we wish to make about them.

The properties of interest are the wavelength
region to which the detector is sensitive; the differing
response of the detector over that wavelength region;
and the nature and range of detector response (see
Figure 5.11). Using the human eye, we can briefly illustrate each of these properties.

The eye is sensitive to the narrow wavelength
region between about 3500 angstroms and 7000
angstroms, as shown schematically in Figure 5.11.
However, the eye does not respond equally to all
colors in the visible spectrum. It is most sensitive
to the middle of the wavelength region, the green
wavelengths, and the sensitivity drops to zero toward
either the violet (short wavelengths) or the red
(long wavelengths).

Finally, the nature and the range of detector response are the ways in which the eye responds to one
photon and to a tremendous flood of photons. Common experience tells us that the eye does not respond
the same way for both. There is some threshold
number of photons, depending upon their wavelength, necessary to make the eye respond as shown
in Figure 5.11. In other words, there is a limit to
how faint a light source we can see, and that limit
depends upon whether we are looking at a violet,
green, or red light. Also, all of us have experienced the
loss of response of the eye when trying to look at too

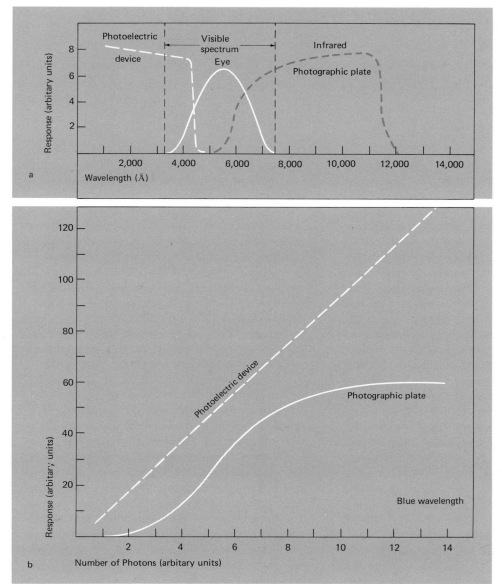

FIGURE 5.11
Illustration of the basic properties of radiation detectors, using a fictitious photographic emulsion and photoelectric device. Part a shows the wavelength region to which the detector is sensitive and the difference in response to different wavelengths in that wavelength region. Part b shows the nature of the response for a photographic emulsion and for a photoelectric device for a blue wavelength to increasing numbers of photons.

bright a light. That is, the eye saturates—it no longer responds to very large numbers of photons. To be a useful radiation detector, the dynamic range between threshold and saturation should be quite large, say a factor of a hundred or a thousand. Now we may ask,

What is the response of the eye to doubling the number of photons in between the lower and upper limits of threshold and saturation? If we double the number of photons, do we observe that the light is twice as bright? The answer in general is no. By and large, over the dynamic range of response of the eye, doubling the stimulus does not double the response; in other words, we say that the response is *nonlinear*. This concept is important, because we would like to know the relative amounts of radiant energy being emitted by different astronomical sources either at one wavelength or over a range of wavelengths.

Let's look now at two other radiation detectors, the photographic emulsion and the photoelectric device.

The *photographic emulsion* records photons by producing a chemical change (a photochemical effect) that will ultimately deposit silver on a glass plate or acetate film. The photographic plate can be made to respond to different wavelength regions within and beyond either end of the visible spectrum, which makes it much more versatile than the eye. Also, its response over a wavelength interval can be made much more uniform than that of the eye. A simulated wavelength response is shown in Figure 5.11 for a photographic emulsion sensitive to infrared photons. However, the photographic plate, like the eye, is nonlinear in its response; it has a rather complicated response depending upon the position in its dynamic range. The response is simulated in Figure 5.11 for a fictitious photographic emulsion.

The photographic plate has a strong advantage over the eye since it will build up a response by storing up the image. Thus time exposures allow the astronomer to collect information on a photographic plate about very faint light sources that can not be detected by the eye through the same telescope. How faint a star can we photograph? The telescope's aperture sets the initial limit. Ultimately, however, the limit is set by the weak illumination in the night sky. This background interference comes from starlight scattered by the earth's atmosphere and from diffuse radiation in the atmosphere (*airglow*). Unfortunately, the photographic plate's photon-capturing efficiency is low. The emulsion can record only 1 or 2 percent of the incident photons (those that activate the light-sensitive coating). Facing this inefficiency, astronomers have found other techniques to improve the telescope's performance.

The *photoelectric device* is an application of the photoelectric effect. The basic principle is to liberate electrons from a metal surface by exposing it to photons and then to measure the number of electrons with electronic circuitry. The photoelectric device, like the photographic emulsion, can be made to respond to different wavelength regions by varying the metals used in making the surface of the device. The biggest advantage of the photoelectric device is that it can be manufactured to have a very large dynamic range of response; in addition, its response is linear to the number of incident photons, as shown in Figure 5.11 for a fictitious device. With modern electronics it is possible to adapt the photoelectric device to count individual photons or to use a mosaic of devices to form a picture much like a photographic plate does.

Electronic *image intensifiers* are our most efficient way of recording electromagnetic radiation. In one such system, photons from the telescope are focused onto a photocathode surface, which ejects electrons. The electrons are increased in number, accelerated, and focused by means of electric and magnetic fields onto a phosphorescent screen, which emits a spark of light for each electron that strikes it. Thus the faint light from the astronomical source is amplified by the device into sufficient light to record the image on a photographic plate. Alternatively, a computer circuit can be used to count the electrons during the exposure. Still other image-intensifying techniques are in use or in developmental stages; these techniques can reduce exposure times by factors of fifty to a hundred over those for photographic systems.

SPECTROGRAPHS

The photographic plate and the photoelectric device enhance our ability to detect light from different astronomical sources. We can equip an accessory instrument with either of these detectors and attach it to the telescope to analyze that light. The two basic accessory instruments are the spectrograph and the photometer.

The *spectrograph* disperses the composite light from the source into its component wavelengths, so we can, for example, determine the elements that compose the light source. Spectroscopy is astronomy's most fundamental interpretive tool.

A *prism* or *grating spectrograph* (see Figure 5.12) receives the concentrated light coming from the telescope's objective on an entrance slit. The light diverging past the slit enters a *collimator*, which is placed so that a beam of parallel rays is delivered to the dispersing device. Then these rays pass through either a prism or a grating, which separates the light into its constituent wavelengths. The dispersed light is focused by a camera system onto a light detector (a photographic plate or a photoelectric device) as individual color images of the entrance slit. Each wavelength forms a distinct image of the slit. The

images of the slit in the different wavelengths are arrayed in an orderly progression of colors from red to violet to create the observed spectrum of the composite light falling on the entrance slit.

PHOTOMETERS

The *photometer* is an accessory device that the astronomer attaches to the telescope at the focal position of the objective to measure the amount of radiation coming from the astronomical object. Where the spectrograph is used to examine the spectral composition of radiation, the photometer measures the amount of radiant energy, either on a relative or an absolute scale, at one wavelength or in a band of wavelengths. The photometer works much like an exposure meter on a camera; basically, incident light is measured by a current. One can use a variety of techniques to define the wavelength region, such as a spectrograph or color filters. And the radiation detector could be either a photographic plate or a photoelectric device. However, almost all modern photometric work is done with a photoelectric photometer.

The photoelectric photometer does have one disadvantage. The brightness of many star images can be

FIGURE 5.12
Schematic drawings of the optical arrangement in a prism spectrograph and a grating spectrograph.

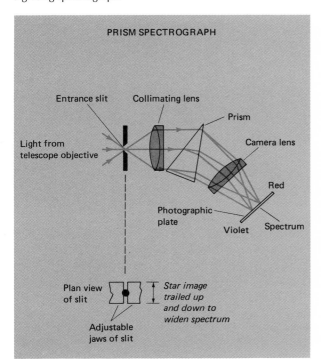

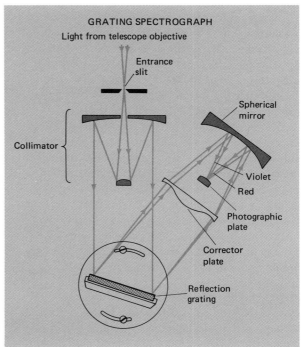

measured on a single photographic plate, but the photoelectric photometer measures only one object at a time. However, the photoelectric photometer has a much greater accuracy than the photographic photometer. Because of its quick response to changes in light, the photoelectric photometer is particularly useful in continually monitoring the change in brightness of an object whose light varies during the observing period.

5.3
Radio Astronomy

DISCOVERY OF CELESTIAL RADIO WAVES
In 1931 a Bell Telephone engineer, Karl Jansky (1905–1950), was trying to find where the interference disrupting the transatlantic radiophone circuits came from. He discovered that some of the radio noise was not from the earth; it was extraterrestrial. The primary source was the center of the Milky Way in the constellation of Sagittarius. In 1936 an Illinois radio engineer, Grote Reber (1911–   ), pursued the phenomenon further. He built the first parabolic radio telescope, 9.5 meters in diameter, and made the first radio map of the sky. The strongest signals he found came from the star clouds in Sagittarius and from several discrete sources toward the center of the Galaxy. Then in 1942 British radar operators and scientists, tracking down suspected radar jamming during World War II, discovered that the interference was radio emission from the sun.

At first astronomers did not grasp just how significant Jansky's work was—they were preoccupied with their observations of the universe through the optical window of the earth's atmosphere. After the war, radio astronomy accelerated as physicists, radio engineers, and astronomers joined to build larger and more efficient radio telescopes with which to study the cosmos. Radio astronomy since then has led to startling discoveries, such as quasars, radio galaxies, pulsars, and interstellar molecules.

RADIO TELESCOPE DESIGN
Because the physical nature of a radio wave is exactly the same as that of a light wave, the problem of designing a radio telescope is similar to that of designing an optical telescope. However, there are some practical differences. Radio waves pass through most materials without any interaction with it. Thus it is not feasible to design a "lens" for radio waves that will focus them in a refracting telescope. But any metal will reflect radio waves, so that a dish-shaped metal mirror will focus radio waves, just as a glass mirror focuses light waves. The reflecting surface of the dish can be an open, fine-wire mesh or a solid metal with a parabolic shape. Radio waves are reflected from the surface and converge toward a focal point, where a small collector aerial absorbs the concentrated energy, as shown in Figure 5.13. From there the signal is carried by an electrical cable to the receiving equipment, which processes the signal just as in your home radio receiver.

FIGURE 5.13
Simplified diagram of a radio telescope system. Radio waves are reflected from the antenna and focused on the collector aerial. There the radio waves are converted to an electric current, which is carried by cables to the receiver. The output of the receiver is displayed on a chart recorder or recorded in digital form on punch cards, paper tape, or magnetic tape.

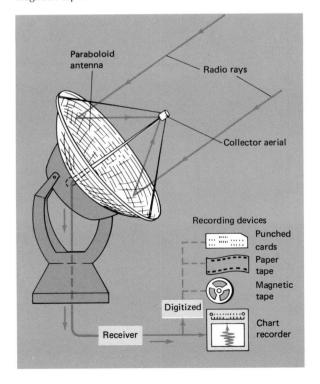

After amplification, the signal variations are recorded in one of several ways (see Figure 5.13). The signal changes can then be fed into a computer for analysis. When the computer has done its job, a formerly invisible view of the universe is revealed in the radio map (see Figure 16.16).

Radio telescopes can be made more sensitive, with better accuracy in pointing and with higher resolution, by expanding the collection area of the dish or by improving the capabilities of the receiver. With the largest radio telescopes we can obtain a resolution of about 1 arc minute, comparable to that of the eye. That is like seeing a penny 65 meters away. Radio energy can be detected by the most powerful radio telescopes from sources whose power is comparable to that of a terrestrial FM broadcast station many light years away.

The radio telescope is remotely controlled by the radio astronomer from an electronic console. Moderate-sized radio dishes, up to about 100 meters or so in diameter, are steerable and have equatorial mountings that follow the rotating sky just as optical telescopes do. Larger and heavier dishes employ an *altazimuth mounting*—that is, one that rotates about a vertical axis and a horizontal axis. This minimizes the distortion in the shape of the dish due to changing the orientation of the dish in the earth's gravitational field while tracking an object. A computer directs the rotation about both axes. Bigger, unwieldy antennas are fixed, pointing upward while the rotating earth sweeps much of the sky by the antenna's field of view. Arecibo, Puerto Rico, has the biggest fixed antenna, a metal dish 305 meters across contoured out of a natural bowl in the ground (see Figure 5.7). It can survey the sky within 20° of the zenith, allowing it to cover about 40 percent of the entire sky.

RADIO INTERFEROMETRY

Astronomers have searched endlessly for better resolving power. It can be achieved by building bigger telescopes or by observing at shorter wavelengths for a chosen aperture or by using the phenomenon of interference (see Figure 5.14). *Interferometry* is a technique involving two or more radio telescopes. Radio radiation from an astronomical source received at the individual telescopes is combined to obtain data that have a spatial resolving power equal to that of a single telescope as large as the distance between the individual receivers. With interferometry an astronomer

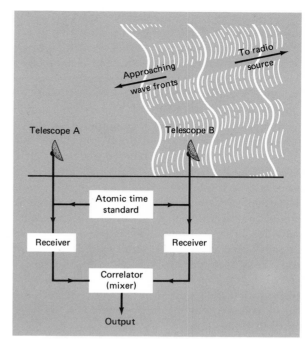

FIGURE 5.14
Simplified schematic of a radio telescope interferometer. Incoming wave crests from the radio source arrive at slightly different times at antennas A and B. They may be in phase, and, if so, the signal is reinforced. Or the crest of a wave may arrive at antenna A while the trough of the same wave or a succeeding wave arrives at B, so that they are out of phase and the signal is canceled. As the sky rotates, the path lengths to each antenna constantly change, resulting in variations in signal strength with time. The signals from each antenna are mixed in an electronic device called a correlator for proper analysis.

can obtain details about the spatial structure of a given celestial object that a single radio telescope could never reveal.

For many years the separation between the individual telescopes of the interferometer was limited by the lengths of the cables connecting the antennas, since the technique depends upon combining, at the same instant, the signals received by the separate telescopes. With the advent of the atomic clock (a clock governed by the vibrations of certain atoms), it became possible to record the signals received by the different telescopes, along with the precise time, and compare them later. This allowed the individual telescopes to be separated—even on opposite sides of the earth. The technique is called *Very Long Baseline Interferometry* (VLBI). It has been used with a geo-synchronous or geostationary satellite as the link in the communications channel between the telescopes.

In the *Very Large Array* (VLA) radio interferometer recently put into operation in New Mexico (see Figure 5.7), signals from each of 27 individual radio telescopes are combined by a computer. Each dish is 25 meters in diameter, and the 27 individual telescopes are moved along railroad tracks arranged in the shape of an enormous 21-kilometer Y. Nine dishes will be located on each branch of the Y and the system can provide a total of 351 interferometer pairs of antennas. The VLA will be able to achieve spatial resolution of about 1 second of arc in ten hours of observing or about that of the 5.1-meter Hale optical telescope. This ability makes it comparable to large optical telescopes and will allow the construction of extremely detailed maps of portions of the sky, that are much better than the one shown in Figure 16.16.

5.4 Infrared Astronomy

INFRARED DETECTORS

In 1800 William Herschel's detection of the infrared component of solar radiation, by positioning thermometers beyond the red end of the sun's visible spectrum, foreshadowed the astronomy of invisible spectral regions. Although astronomers have known for some time of the existence of a few atmospheric

> The giant Hale 200-inch telescope on Mount Palomar could easily record the images of a million galaxies in the bowl of the Big Dipper.
>
> Harlow Shapley

windows in the infrared region of the electromagnetic spectrum, infrared astronomy has become a vital component of astronomy in only the past few years. The limiting problem was the lack of good infrared detectors. In the past, astronomers studied infrared radiation with the thermocouple and *bolometer*. The bolometer originally consisted of a blackened metal foil that absorbed heat radiation. The resulting rise in temperature increased its electrical resistance, which then could be measured in an electrical circuit. If the detector is cooled to the temperature of liquid helium, which is a few degrees above absolute zero, it can be made responsive to even the faintest trace of heat radiation. Modern bolometers, by shielding them against extraneous heat radiation, can measure a change in temperature of 1×10^{-7} degree Celsius. Another type of infrared detector has a photoconductive crystal that releases electrons when it absorbs infrared photons. The flow of the electrons, which is proportional to the intensity of the source emitting the photons, is then measured.

We can subdivide the infrared spectrum into three segments, as shown in Figure 5.15. As you can see in the figure, a large part of the infrared spectrum is not visible at ground level because of absorption by water vapor, carbon dioxide, and molecular oxygen, which lie between the ground and 15 kilometers altitude. Consequently, airplanes, balloons, rockets, and satellites are used extensively to supplement ground-based infrared observations.

FIGURE 5.15
Atmospheric windows in the infrared. The windows appear in the light areas; the opaque portions are in the dark areas.

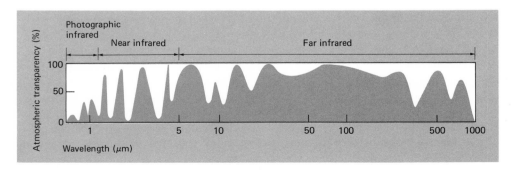

FIGURE 5.16
A panorama of Mauna Kea, the NASA infrared telescope facility in Hawaii.

INFRARED TELESCOPES

The liquid-helium-cooled infrared detector can be used with the appropriate accessory instruments on an ordinary optical telescope to study the cosmos. However, because of the longer wavelength of infrared radiation, the image-producing quality of the telescope objective need not be as fine as it must be for the visible region. Thus a number of new telescopes have been designed and built for infrared astronomy only. A national observatory for infrared astronomy is being built high on the 4200-meter inactive Hawaiian volcano Mauna Kea (see Figure 5.16). A 3.0-meter infrared telescope constructed by NASA and the University of Hawaii is in operation along with a 3.8-meter infrared telescope belonging to the United Kingdom.

A new telescope design called a *Multiple Mirror Telescope* (MMT), which is well suited for infrared observations (see Figure 5.7), has been installed on Mount Hopkins in Arizona. Its design uses a mosaic of independent mirrors of small size to collect and focus light in order to simulate the collecting ability of a large-aperture single mirror. The MMT consists of a circular array of six identical 1.8-meter mirrors on a altazimuth mounting; the array has the light-gathering power equivalent to a 4.5-meter single mirror. The six images may either be superimposed to form a single image or aligned along a spectrographic slit to take full advantage of slit geometry. The pointing directions of the six individual telescopes are locked together by the use of laser beams. If this instrument is successful in demonstrating the practicality of the multiple-

mirror concept, it may be the forerunner of telescopes that are equivalent to a 25-meter telescope.

5.5
Ultraviolet, X-Ray, and Gamma Ray Astronomy

INTRODUCTION

While much useful and important observational work remains to be done from ground-based observatories, an increasing portion of future astronomical research will be carried out from platforms outside the major portion of the veil that is the earth's atmosphere. Up to the middle of this century nearly all our knowledge about the cosmos had come from studying the visible light of astronomical objects through an ever-changing earth atmosphere. The visible and the radio windows still constitute our most heavily used sources of information. However, today's research is centered on the invisible portions of the electromagnetic spectrum, which embraces the infrared, ultraviolet, X-ray, and gamma ray regions. To explore these regions, new techniques and equipment are being developed, which can be flown above the atmosphere in various types of space vehicles.

SPACE VEHICLES

High-altitude aircraft and balloons are the least expensive way of exploring invisible extraterrestrial radiation. Jet aircraft can ascend to about 15 kilometers,

THE WORLD'S LARGEST TELESCOPES

The largest optical telescope in the world is a Russian 6-meter reflector, located at an altitude of 2100 meters in the Caucasus Mountains. Because of its bulk, it has an altazimuth mounting rather than an equatorial mounting. Next in size is the 5.1-meter Hale reflector on Mount Palomar in southern California, at an altitude of 1700 meters. The highest major observatory (4200 meters) is Mauna Kea on the island of Hawaii, housing a 2.2-meter reflector. It will soon be joined by a 3.6-meter French Canadian reflector. Several large reflectors in the range of 3 to 4 meters are located on high mountains in both hemispheres.

The two largest refractors are the 1-meter at the Yerkes Observatory and the 0.9-meter at the Lick Observatory, constructed many decades ago. The largest Schmidt telescopes are the 1.4-meter in East Germany and the 1.2-meter Schmidts in California, Chile, and Australia. Two large infrared telescopes (3.8 meters and 3.0 meters) are located on Mauna Kea in Hawaii.

Among the largest steerable radio telescopes are the 100-meter dish in Bonn, West Germany; the 91-meter (semisteerable) at the National Radio Observatory in Green Bank, West Virginia; the 76-meter at Jodrell Bank in England; and the 64-meter in Australia.

All these utilize an altazimuth mounting. The world's largest dish, 305 meters (not steerable), is in Puerto Rico (see Figure 5.7). Antenna installations with multielement arrays spread over a large acreage or interferometers extending several kilometers long are the biggest of all. Approximately twenty are in the United States, Western Europe, England, Australia, and the Soviet Union. In Socorro, New Mexico, the Very Large Array (VLA) interferometer with 27 individual dishes, each of 25-meter diameter, has been put into operation and will be in full use by 1981.

---

where there is an infrared window relatively free of atmospheric absorption (caused mainly by water vapor at the lower levels). Balloons are useful up to about 30 kilometers, above which only 5 percent of the atmosphere remains.

Rocket flights are easy to maneuver in space and can be launched from many different sites. Though their flights are short compared with balloon flights, lasting for minutes instead of hours, rockets can climb five times higher than balloons. Artificial satellites cost much more than rocket flights, but satellites can continuously monitor events over different regions of extraterrestrial space for long periods, an advantage that outweighs their additional cost. Since 1958 the United States has launched hundreds of instrumented satellites.

An important group of space vehicles, the observatory satellites, has been placed in orbits several hundred kilometers above the earth. These spacecraft carry equipment for experiments in ultraviolet spectroscopy and X-ray and gamma ray measurements. Two of the most sophisticated and costly observatory satellites are the *Orbiting Astronomical Observatories: OAO-2* and *OAO-Copernicus*. (A drawing of the *OAO-Copernicus* is shown in Figure 7.5.)

The *Skylab* program cost $6 billion. Three 3-man crews spent 171 days in *Skylab* between May and November 1973. These astronauts carried out dozens of carefully selected scientific, biomedical, and technological experiments. For example, the Apollo Telescope Mount (ATM), a solar observatory, recorded 180,000 film frames of ultraviolet and X-ray emissions from the solar atmosphere—enough data to keep astronomers busy for years. An added bonus came from the observations of Comet Kohoutek around the time of its perihelion passage in late December 1973. Also on board *Skylab* was an objective-prism camera, which photographed the ultraviolet spectra of several hundred star fields. The abandoned station was to remain in orbit several years; it was hoped that space shuttle astronauts would be able to reuse the station in the future. However, *Skylab* was dragged down by the atmosphere to a fiery demise over the Indian Ocean scattering pieces over western Australia on 11 July 1979, before efforts could be mounted to elevate it into a higher orbit.

In late 1983 NASA expects to place a 2.4-meter unmanned reflecting telescope in orbit at an altitude of 500 kilometers, called the *Space Telescope* (see Figure 7.5). Its optics and instrumentation will be enclosed in a cylindrical tube 13 meters long and 4.3 meters wide equipped with a sunshield and interior baffles to eliminate stray light and to protect it from meteoroids. Auxiliary apparatus include an electronic camera, ultraviolet spectrograph, photometers, polarimeters, infrared detectors, and other specialized devices. Data from the telescope will be radioed in digital (number) form to the Goddard Space Flight Center in Greenbelt, Maryland, for processing. Out in space no atmospheric absorption or turbulence will distort the images produced by the telescope. Thus the telescope should see astronomical sources up to fifty times fainter than those visible from the earth's surface. With proper maintenance from the space shuttle, the telescope could operate for at least a decade. Unlike its ground-based counterparts, the *Space Telescope* will scan the electromagnetic spectrum from the deep ultraviolet to the far infrared.

Other space observatories are the planetary probes (see Figure 7.5), which are the most exotic. Their role is to go to a planet to photograph and to analyze from a close flyby or to orbit the planet—or in some cases to land. As examples, the *Viking 1* and *Viking 2* spacecraft landed on the surface of Mars (we will discuss them later). Other examples are the *Voyager 1* and *Voyager 2* spacecraft launched to encounter Jupiter, Saturn, and perhaps Uranus and Neptune. Much of our understanding of the nature of the universe will change—rapidly and dramatically—because of these space observatories.

ULTRAVIOLET ASTRONOMY

The ultraviolet portion of the electromagnetic spectrum has been divided by astronomers into three segments, more or less dependent on the time in which serious research in them began. First, there is the ground-based ultraviolet, from 4000 angstroms to the atmospheric cutoff at 3000 angstroms; next, the far ultraviolet from 3000 angstroms to 1000 angstroms; and last, the extreme ultraviolet from 1000 angstroms to 100 angstroms. Ultraviolet observations began with the sun after World War II, in October 1946, when a captured German V-2 rocket carried a small ultraviolet grating spectrograph to a height of 100 kilometers. During its ascent a camera made a running record of the never-before-photographed ultraviolet portion of the solar spectrum down to 2200 angstroms. This event was followed in succeeding years by instrumented rockets that recorded the solar spectrum to much shorter wavelengths (see Figure 11.18). Between 1962 and 1975 eight *Orbiting Solar Observatories* (*OSO-1* through *OSO-8*) were launched to carry out experiments in far ultraviolet spectroscopy of the sun, along with some solar X-ray and gamma ray measurements.

In December 1968 the first *Orbiting Astronomical Observatory* (*OAO-2*) began sampling the ultraviolet and far ultraviolet radiation from many different sources. By the time *OAO-2* ended its useful life in February 1973, it had carried out photometry on more than a thousand objects from planets to galaxies and had made spectral scans of hundreds of stars. Its successor, *OAO-Copernicus*, launched in August 1972, carried an 0.8-meter ultraviolet telescope and three small X-ray telescopes. In April 1972 the striking ultraviolet airglow, the *geocorona*, surrounding the earth out to nearly 100,000 kilometers, was photographed from the moon by the Apollo astronauts. An ultraviolet telescope flown during the U.S. Apollo/U.S.S.R. Soyuz mission in July 1975 detected the first extrasolar extreme ultraviolet object—a very blue "white dwarf" star in Coma Berenices.

Until 1975 the extreme ultraviolet spectral region was not seriously considered a likely place to conduct astronomical research. It was originally thought that all stellar ultraviolet light from 912 angstroms, where the Lyman continuum begins,[1] to about 20 angstroms would be strongly absorbed by an abundant number of hydrogen atoms present in interstellar space. The hydrogen atom absorbs all these wavelengths when it is ionized from its lowest energy level. Hence it was supposed that starlight in the extreme ultraviolet region on its way to the earth would not be observed. However, there are gaps between the hydrogen clouds, or regions of very low hydrogen density in some directions, where the extreme ultraviolet photons get through to the earth. In many directions extreme ultraviolet observations are possible out to several hundred light years, encompassing a volume of space containing thousands of candidate stars.

In January 1978 the *International Ultraviolet Explorer,* an orbiting observatory, was launched by NASA. This is a joint undertaking by NASA and several Western European countries involving about 200 astronomers from 17 different nations. Its facilities have been used for studies of planets, stars, galaxies, and the interstellar medium in the wavelength range from 1150 angstroms to 3200 angstroms. Astronomers conduct their experiments from an elaborate console of controls located at the Goddard Space Flight Center (see Figure 1.8).

---

[1]Just as there is a Balmer continuum for the Balmer line series beginning at 3646 angstroms (see page 71), so there is a Lyman continuum for the ultraviolet Lyman series, which begins at 912 angstroms.

X-RAY ASTRONOMY

X-ray astronomers divide their portion of the electromagnetic spectrum into two categories: *soft X rays*, from 100 to 1,000 electron volts (120–12 angstroms), and the more penetrating *hard X rays,* from 1,000 to 100,000 electron volts (12–1.2 angstroms). Both X rays and gamma rays are emitted by regions of space characterized by very high temperatures, low density, and high-speed subatomic particles—that is, wherever there are extreme conditions involving nuclear and atomic reactions. The observed radiation is in part either thermal (blackbody) radiation or nonthermal radiation (see page 181).

Astronomers first began using X-ray detectors in balloons, rockets, and in a few unmanned satellites during the 1960s. By 1967 they had discovered some thirty discrete X-ray sources. Then in December 1970 NASA's *Explorer 42* satellite named *Uhuru*, which means "freedom" in Swahili, was launched off the coast of Kenya, Africa. By the end of its useful life in 1973 it had scanned nearly the whole sky and had located and measured nearly 200 X-ray sources (see Figure 16.18). The newly discovered X-ray objects were named after the constellation in which they appeared, followed by X-1, X-2, and so on ("X" for X ray) in the order of discovery. For example, Taurus X-1, the first X-ray object discovered, is the Crab Nebula (see Section 14.3). Today with a growing number of X-ray satellites in operation by about a half dozen countries, and with a large increase in X-ray discoveries, it is convenient to designate the source by a catalog number.

A second generation of NASA satellites known as the *High Energy Astronomy Observatories*, designated *HEAO-1, HEAO-2,* and *HEAO-3*, was launched in August 1977, November 1978, and September 1979, respectively. The three *HEAO* satellites are designed specifically to study X rays, gamma rays, and subatomic particle radiation (called cosmic rays). The satellites shown in Figure 7.5 are each about 6 meters long and weigh approximately 3000 kilograms. Instruments aboard these satellites are designed to search the sky for discrete and diffuse background sources of X rays and gamma rays; to measure their total energy output and how that varies with wavelength; and to measure the ranges of energy, the composition, and the numbers of cosmic rays. The *HEAO* satellite project opens up new avenues of research not previously accessible with any great detail or sustained viewing.

GAMMA RAY ASTRONOMY

This new field of astronomy has burgeoned in the last decade or so from modest beginnings employing balloons and rockets to satellites (European and American) carrying highly sensitive gamma ray detectors. In 1967 Vela satellites, launched by the United States to monitor violations of the treaty banning atmospheric nuclear tests, first detected gamma ray bursts in outer space. Since then several such events have been recorded each year by different satellites. Many times there is an observed connection between gamma ray and X-ray bursts, and occasional radio outbursts. The present series of three large *High Energy Astronomy Observatories* (*HEAO*) and the forthcoming *Gamma Ray Observatory* to be launched in the early 1980s should significantly expand this new window to the universe.

Gamma ray photons, which carry the highest energy of any photon, range in energy from 10^5 to 10^9 electron volts. In 1978 came the first evidence of several gamma ray spectral lines corresponding to energies ranging from 0.5×10^6 to 6.3×10^6 electron volts. The search for additional discrete lines has been greatly extended with the launch of *HEAO-3*, as it will be later with the launch of the *Gamma Ray Observatory*.

The universe is highly transparent to gamma rays because of their great penetrating power. One would therefore expect them to be capable of carrying the imprint of their origin from far distant places because they pass so easily through interstellar matter. Thus the unique penetration of gamma rays serves as a probe to provide us with new insights into the structure of the cosmos.

SUMMARY

Modern astronomers rarely study the heavens by looking through a telescope. Instead, they use the telescope to photograph celestial bodies with photographic and electronic cameras or to feed celestial radiation into various analyzing instruments such as photometers; spectrographs; infrared, ultraviolet, and

X-ray detectors; and radio receivers. These devices are much more sensitive to light and responsive to a broader range of wavelengths than the human eye. Coupled with automatic recording and processing of observational data with the aid of minicomputers, these accessories are greatly extending the scope of astronomical research.

Optical astronomers working in the visible wavelengths employ refracting telescopes, reflecting telescopes, and Schmidt telescopes. Early refracting telescopes suffered from spherical and chromatic aberrations—handicaps that the achromatic refractor or the parabolic mirror largely overcome. To analyze the light that the telescope collects, it is equipped with a spectrograph or other analyzing instrument.

For investigation of the shorter ultraviolet and longer infrared wavelengths, conventional or specially designed reflecting telescopes are the most efficient collectors of light. The shortest wavelengths are explored with modified laboratory X-ray and gamma ray detectors mounted in balloons, rockets, and space vehicles.

Radio astronomers use radio telescopes of various configurations: single large parabolic-shaped antennas, rows of interconnected smaller antennas, and various other antenna arrays. Increasing the diameters of optical and radio telescopes not only increases the amount of radiant energy collected but also sharpens the resolution of detail.

REVIEW QUESTIONS

1. What is an achromatic lens?

2. What is meant by the "resolving power" of a telescope?

3. The famous star Mira (Omicron Ceti) reaches a minimum brightness about twenty times fainter than naked-eye visibility. Assuming the aperture of the pupil of the eye is 1/2 centimeter, would a 2-centimeter telescope be capable of showing the star at minimum brightness?

4. Which optical arrangement of a modern reflector should be employed to obtain the largest picture of a planet? The faintest image of a star?

5. What kind of telescope would you recommend to photograph a faint comet in as short an exposure as possible?

6. The night sky was clear, but after developing the photographic plate, the astronomer noticed that the star images were larger than usual. What could have happened?

7. Describe a faster way of recording the spectrum of an object than on a photographic plate.

8. Why doesn't the parabolic figure of a radio telescope have to be as accurate as an optical telescope?

9. Why don't radio waves pass through the wire mesh of a radio dish?

10. Why is a very dry atmosphere necessary for an infrared telescope?

11. How does light pollution affect an astronomer's work?

12. Why can't we use an X-ray telescope on a very high mountain site?

13. Why are large modern telescopes of the reflecting type rather than of the refracting type?

14. How can better resolution of celestial sources be achieved with radio telescopes?

15. In what ways are the six 1.8-meter mirrors of the Multiple Mirror Telescope not equivalent to a 6 × 1.8 or 10.8-meter telescope?

16. How are computers used in astronomy?

17. Why do X rays and gamma rays penetrate through space more easily than visible light?

18. Which properties of an image formed by a telescope interest astronomers?

19. Discuss the properties of radiation detectors.

20. If you wanted a radiation detector whose response was linear, which of those discussed would you choose?

6
The Earth-Moon: A Double Planet

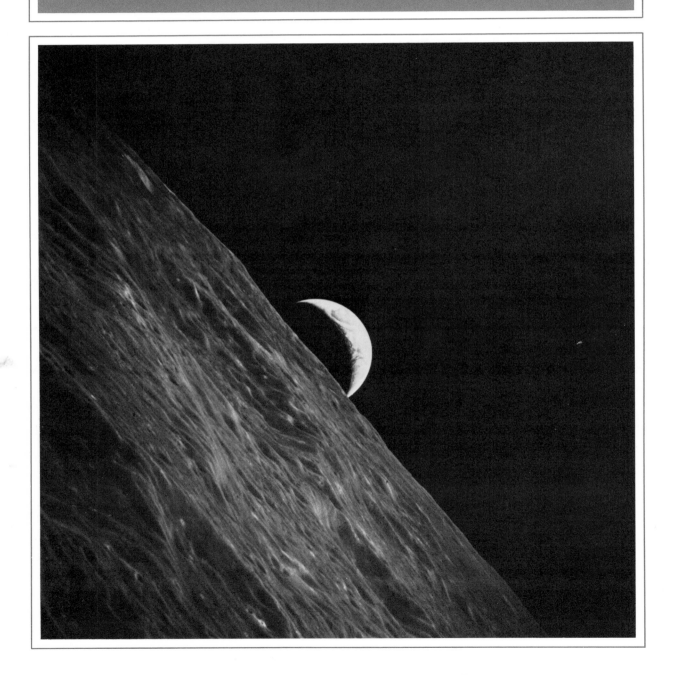

6.1
How Earth Became a Planet

The title of this chapter suggests that the earth-moon system is more than a planet and its satellite. It is probably not unreasonable to classify the system as being more like a double planet for two reasons. First, as we will discuss in more detail later, the moon is not significantly smaller than the planet Mercury. Second, the ratio of the moon's mass to that of the earth's is considerably larger than any other planet-satellite combination in the solar system. Thus the earth-moon system is somewhat unique and its origin and evolution may not be similar to any other planet in the solar system.

One of the most significant events in our view of the world began only in the eighteenth century and in a period of about a hundred years it completely changed our thinking. What was that event? It was the concept of change or evolution. Today our world view is one dominated by a knowledge that the universe, the stars, the earth, and life upon the earth's surface are gradually, sometimes rapidly, evolving in directions shaped by natural processes governed by the laws of physics. Throughout this chapter and those that follow, we will try to sketch the processes of evolution, as well as its results.

EARLY HISTORY

Here we give only a bare outline of our planet's origins; more details are given in Chapter 7. About 4.6 billion years ago—about as long as the sun has been shining—the earth and the other planets were formed. How did it all begin? The prominent theory suggests that there was a contracting cloud of cosmic gas and dust lying somewhere between the stars, called the *solar nebula*. Ultimately shaped by its contraction, its central portion became the sun, with the other material in a surrounding disk of gas and dust eventually forming the planets. This process in all likelihood has and will occur millions of times throughout our Galaxy as stars like the sun are formed. Within a relatively short time the young earth had collected most of the

◀ Earthrise of a crescent-phase earth as seen from the backside of the moon. This photograph was taken by the crew of *Apollo 17* in December 1972.

matter that composes it today. Matter attracted by the growing earth collided with it, giving up its kinetic energy as heat. This energy, along with the energy resulting from the earth's gravitational contraction and emission by radioactive nuclei, heated the earth's interior.

> Here is the world, sound as a nut, not the smallest piece of chaos left, never a stitch nor an end, not a mark of haste, or botching, or second thought; but the theory of the world is a thing of shreds and patches.
>
> Ralph Waldo Emerson

In a few tens of millions of years the earth became molten. Chemical differentiation followed—the heaviest elements, iron in particular, separated from the lighter elements oxygen and silicon (primarily silicates and oxides of iron and magnesium) and sank toward the center. The silicates (molecules containing both silicon and oxygen) and the oxides (containing oxygen) rose to form the mantle surrounding this iron-rich core. The lightest materials rose to the top. During the Archeozoic era, about 4.5 and 4.0 billion years ago, the earth as a whole was cooling even though volcanic activity on the surface was intense. The original atmosphere of hydrogen and helium, if it even existed, was apparently swept away by the young sun's intense outpouring of atomic particles, called the *solar wind*. We believe that this was replaced by a secondary atmosphere whose composition, of which we are not sure, was drawn probably from the gases carbon dioxide, methane, ammonia, nitrogen, hydrogen sulfide, and water vapor. As the earth cooled, gases escaped from the interior and water condensed, forming the oceans. From granitic slag cooling on a bed of molten basalt (a dark lavalike rock containing crystallized minerals), the earliest continents formed.

HOW OLD IS OUR PLANET?

In 1654 Ireland's Archbishop Ussher calculated that the earth was born on 26 October in the year 4004 B.C. He had worked out the date of creation from the

Bible's chronology of the generations of patriarchs. Late in the nineteenth century, however, scientific estimates of the earth's age rose to 50 million years. By the century's end even this figure was far too low to match growing geological evidence. The discovery of natural radioactivity at the start of the twentieth century freed the dating of the earth from theology and from other rational but still inadequate scientific methods (see bottom of page). Estimates of the age of the earth continued to go up from 1900 till now, where it has stabilized at about 4.6 billion years.

6.2
The Geosphere

EARTH'S SIZE AND SHAPE

How do we know the earth's dimensions? One way to find them is to measure the distance along its surface between two parallels of latitude. Measurements in different places reveal that the number of kilometers in 1° of latitude increases slightly from the equator (110.6 kilometers) toward the poles (111.7 kilometers). These measurements indicate that the earth is shaped like an oblate spheroid, with the longer diameter in the equatorial plane and the shorter one in the polar direction (see Table 6.1). The rotation of the earth is primarily responsible for its shape in that the rotation causes material from high latitudes to float toward lower latitudes forming an equatorial bulge.

Because the earth is not a perfect sphere, its gravitational field is not uniform across its surface. These variations affect the motions of artificial satellites in low orbits. From the changes in their orbits, called orbital perturbations, we know that our planet is slightly pear-shaped. The stem portion at the North Pole is about 19 meters farther, and the bottom portion at the South Pole about 26 meters closer, to the center of the ellipsoidal figure.

DATING THE EARTH

The most accurate method we now have to date the earth is based on knowledge of radioactivity. Nuclei of radioactive elements such as uranium and thorium spontaneously break down into lighter nuclei at definite rates, emitting alpha particles (helium nuclei), electrons (beta particles), and gamma radiation. Their decay ends in a stable isotope (see table below) of a lighter element such as lead. The time required for half of the original material to decay into the final products is known as the element's *half-life*. Some examples are given in the accompanying table.

How old is a rock? We start by determining how much of the final stable nucleus has formed, comparing that figure with the remaining parent nucleus from which the daughter nucleus is descended. We combine that figure with the known rate of disintegration and some assumptions about the initial ratios. For example, half a sample of uranium 238 decays in 4.51 billion years

into lead 206. If we found in some rocks a 50-to-50 proportion of uranium 238 to lead 206, the age of the rocks would be about 4.5 billion years (if no lead 206 were present at the beginning). The oldest meteorite specimens show that some primitive lead 206 probably was present when the planetary system was formed. The oldest rock formations we know, in Greenland and Labrador, are 3.7 to 3.8 billion years old. Allowing for some primordial abundance of lead in the earth's rocks, geophysicists obtain 4.6 billion years for the age of the earth.

Scientists use *radioactive dating* of fossil-bearing rock to piece together the story of the earth's earliest inhabitants and their evolution. Another method is *relative dating*, which uses the fact that more recent layers of the earth are on top of older sedimentary rock beds. Also, the history of successive forms of fossils in sedimentary layers of known age can be used to date similar sedimentary layers.

Radioactive Nucleus	Half-life (10^9 years)	Final Stable Nucleus in Decay Sequence
Uranium 238	4.51	Lead 206
Uranium 235	0.71	Lead 207
Thorium 232	13.9	Lead 208
Rubidium 87	49.8	Strontium 87
Potassium 40	1.31	Argon 40

TABLE 6.1
Physical Data for the Earth

Quantity	Value
Equatorial radius	6,378 km
Polar radius	6,357 km
Mean radius	6,371 km
Mean circumference	40,030 km
Volume	1.08×10^{27} cm³
Mass	5.98×10^{27} g
Mean density	5.5 g/cm³
Surface gravity	980 cm/s²
Escape velocity	11.2 km/s
Equatorial rotational velocity	0.46 km/s (1,040 mi/h)
Lengthening of day	1.5×10^{-3} second/century
Surface area	5.10×10^{18} cm²
Land area	1.49×10^{18} cm² (29.2%)
Ocean area	3.61×10^{18} cm² (70.8%)
Geomagnetic poles' locations	
North Pole	76°N 101°W
South Pole	66°S 140°E
Magnetic field	≈0.5 g

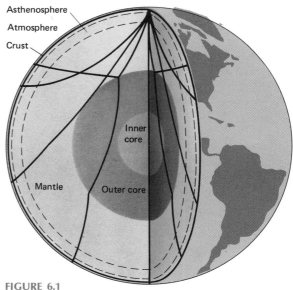

FIGURE 6.1
The earth's internal structure and seismic waves. The principal layers are labeled. Information about these layers is obtained by analyzing data from seismic recordings of earthquake vibrations passing through the earth. The velocities of the waves and their paths change abruptly at the interfaces between adjoining layers, as shown in the diagram.

INSIDE THE EARTH

Geophysicists have a natural tool for probing the planet's internal structure: *seismic waves*, which are generated by earthquakes and spread out in all directions from the site of the quake. From the manner of their propagation, their periods and amplitudes of vibration, and their arrival time at various stations, scientists can deduce much about the earth's structure.

Two kinds of seismic waves, *pressure (P) waves* and *shear (S) waves*, propagate inside the earth. The speed with which these waves travel through the earth (5 kilometers per second to 14 kilometers per second) depends on the material's density, compressibility, and rigidity. The particles of the earth that transmit the *P* waves vibrate back and forth in the direction in which the wave propagates, similar to the way sound waves are propagated through air. The *S* waves, which move at about half the speed of the *P* waves, cause the particles that transport the disturbance to vibrate perpendicular to the direction of the waves' propagation, like waves on a string. Unlike the *P* waves, *S* waves cannot propagate through liquids, which smother their vibrations.

As the *P* and *S* waves move downward through different layers, their speed increases and they are refracted or reflected on reaching a boundary between two distinctly different layers. By tracking the path of these waves, we have a picture of the earth's interior, which is the layered structure left by the melting that separated the constituents according to weight into stratified zones.

A model of the earth's structure is shown in Figure 6.1. The central portion is a hot, highly compressed *inner core*, presumably solid, nearly 2800 kilometers in diameter. It is believed to be mainly iron and nickel. The inner core is surrounded by an *outer core*, a molten shell primarily of liquid iron and nickel, about 2100 kilometers thick, with lighter liquid material on the top. The outer envelope, the *mantle*, is some 2800 kilometers deep. Its upper portion is mostly solid rock in the form of olivine, an iron-magnesium silicate, and its lower portion is chiefly iron and magnesium oxides. A thin coat of granitic composition called the *crust* forms the outermost skin. A summary of the physical data for the earth's layered structure is given in Table 6.2.

TABLE 6.2
Earth's Interior

Layer	Depth Below Surface (km)	Fraction of Mass Interior	Density Range (g/cm³)	Pressure (× 10⁶ atm.)	Approximate Temperature (°K)	Main Composition
Crust	0–30	1.00–0.99	2.6–3.3	0–0.01	290–700	Continents: granite or water
Mantle	30–2900	0.99–0.32	3.4–9.7	0.01–1.3	700–4300	Basaltic rocks: silicates and oxides of iron and magnesium
Outer core	2900–4980	0.32–0.03	9.7–12.5	1.3–3.2	4300–6000	Molten iron and nickel
Inner core	4980–6370	0.03–0.0	12.5–13	3.2–3.7	6000–6400	Solid iron and nickel

OUTER LAYERS

Earth's crust and the uppermost part of the mantle are known as the *lithosphere*. This is a fairly rigid zone about 100 kilometers deep. The crust extends 60 kilometers or so under the continents and only about 10 kilometers below the ocean floor. The continental blocks are primarily a light granitic rock rich in the silicates of aluminum, iron, and magnesium. On top of the igneous strata (molten rock that has hardened) lies a thin veneer of material such as sedimentary rocks (rocks formed by sediment and fragments that water deposited) and soil deposited during past ages in the parts of continents that had no recent volcanic activity or mountain building. In the coastal region on the continental margin, the sedimentary material is mud and clay washed down from the land by rivers and streams.

Sandwiched between the main body of the lithosphere and the rest of the mantle is the partly molten *asthenosphere*, about 150 kilometers thick. It consists primarily of iron and magnesium silicates derived from lava oozing out of the earth's mantle.

As far as the surface geography is concerned, there appear to have been two major terrain-shaping mechanisms at work on the earth (and, for that matter, also presumably on the moon, Mars, Venus, and Mercury, the terrestrial planets). These are cratering by meteoric impact and volcanic activity with its accompanying crustal strains and slippages. Erosion and *tectonic activity* (deformations and motions of the crust), which are the dominate mechanisms now, have all but erased the results of a cratering phase in the earth's history. However, remnants of the last of the cratering phase remain in nearly a hundred ancient impact structures, some of which are as large as the largest visible ones on the moon. We estimate that on the earth tectonic activity with its accompanying volcanic activity dominates better than 90 percent of the present terrain, with not more than 10 percent of the cratered terrain remaining. Present evidence suggests that the pattern of evolution on the other terrestrial planets, as revealed by various space missions, is not the same as that for the earth.

CONTINENTAL DRIFT

The idea that the continents drift relative to each other was one that few hurried to accept when German geologist Alfred Wegener proposed it in 1912. Yet recent research has revealed a variety of evidence showing that the lithosphere is indeed segmented into about a dozen or so major *plates* of different sizes. Floating on the earth's mantle, they move slowly, carrying the continents with them at an average rate of several centimeters each year. This motion is known as *continental drift*.

Explorers of the ocean bottoms have discovered a number of midocean ridges that rise several kilometers above the ocean floor and are thousands of kilometers long. They mark one type of plate boundary. For example, the Mid-Atlantic Ridge separates the North and South American plates from the Eurasian and African plates, while the East Pacific Ridge separates the Pacific plate from the North American, Cocas, and Nazca plates (see Figure 6.2). It appears that molten matter, forced upward from the asthenosphere into a midocean rift, pushes laterally from the ridge to form new plate material, which gradually cools, thickens, and solidifies at the trailing edge of the older plate material (see Figure 6.3). Deep core samples from as far down as 8 kilometers below sea level verify that earth's youngest volcanic rocks are those found near these midocean ridges.

We have further confirmation that the plates do move—the shape, geological structure, and paleon-

FIGURE 6.2

The drifting continents. Two hundred million years ago a single supercontinent—which geologists now call *Pangaea*, "all lands"—was washed by a universal ocean, *Panthalassa*. In the last 65 million years almost half the oceans were created. Greenland split from North America, which completed its cleavage from Eurasia as the Atlantic widened, and northward-drifting India collided with Eurasia, thrusting up the Himalayas. Australia was then joined to Antarctica. Thirty million years ago the eastward movement of Africa, after splitting from Gondwana, stopped. South America was still moving westward, widening the South Atlantic. The Atlantic and Indian oceans will continue to widen and the Mediterranean to shrink. The Pacific Ocean will narrow and California west of the San Andreas Fault will be detached from the mainland and slide toward the Aleutian Trench.

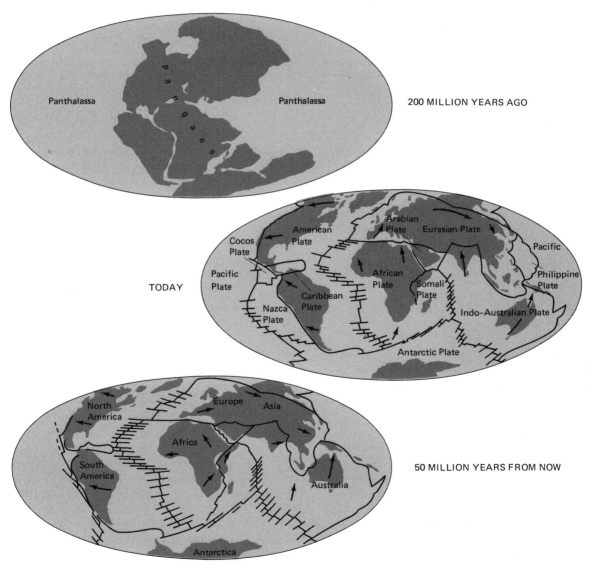

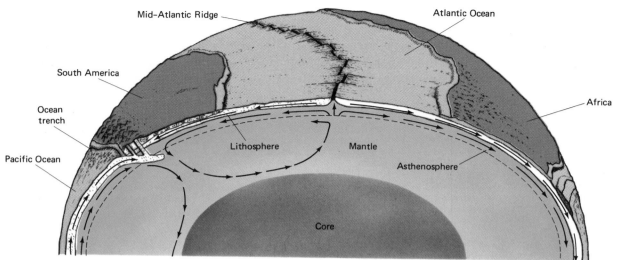

FIGURE 6.3
Schematic view illustrating tectonic activity. The diagram shows one model of convection currents proposed to explain how the plates are moved. Warmer material moves upward from the asthenosphere toward the ridge and spreads out laterally as it cools, forming new crustal plate material. The plates, carrying the continents with them, ride on this convection flow. The cooler material descends at the trench and is reheated within the asthenosphere, completing the cycle of convection.

tology of the continents. Still more evidence comes from rocks. Igneous rocks with similar magnetic fields, which were frozen at the time the rocks solidified, have been found at continental margins that are now widely separated.

Another line of evidence comes from the heat flow out of the earth's interior. Compared to the energy coming into the earth from the sun, the flow of heat from the interior is scarcely a trickle. The heat conducted through an area the size of a football field is roughly equivalent to the energy given off by three 100-watt light bulbs. However, over the 4.6-billion-year history of the earth, this trickle of energy has contributed to the work of making continents drift, opening and closing ocean basins, building mountains, and causing earthquakes. The geographic variation in the heat flow from the interior is not great, but the global variation shows that the major oceanic ridges are high-heat-flow zones while the older continental shields and sedimentary regions are low-heat-flow zones.

How are the plates transported across the mantle? A favored explanation is that they are driven by the horizontal flow of *convection currents* within the mantle, circulating in the upper, softer portion of the mantle, as illustrated in Figure 6.3. The leading edge of one plate encounters a second plate, and one of the plates is pushed downward and forced into the mantle, to create a deep trench. This process can form a coastal mountain belt like the Andes on the overriding plate. As the other plate descends over millions of years, it heats up and becomes part of the general circulation in the asthenosphere. Plates separate along midocean ridges. Most of the great geologic processes—volcanic activity, mountain building, formation of ocean trenches, earthquakes—are concentrated on or near plate boundaries.

About 200 million years ago the last mass movement of the continents began (see Figure 6.2). Earth then had only one consolidated land mass, today called Pangaea. It is believed this supercontinent accumulated from migrations produced by previous drifting. Some 20 million years later seafloor spreading had separated the supercontinent into two segments: Laurasia in the north and Gondwana in the south. About 45 million years later the North Atlantic and Indian oceans had widened, South America had begun to separate from Africa, while India had begun drifting northward. During the next 70 million years the South Atlantic Ocean widened into a major ocean, the Mediterranean Sea began to open up, and North America just began to separate from Eurasia.

A computer-generated projection for the next 50 million years suggests that the Atlantic and Indian oceans will enlarge and the Pacific will contract. Australia will continue to drift northward toward a possible collision with Eurasia. Africa's northward movement will doom the Mediterranean. In 10 million years Los Angeles, which is part of the Pacific plate, will have come abreast of San Francisco, which is sitting on the North American plate, and it will eventually slide into the Aleutian Trench. Recent measurements suggest that the south side of the San Andreas Fault south of the San Gabriel Mountains has moved westward relative to the north side between 12 and 20 centimeters during a three-year period. Average plate motions are on the order of 5 to 6 centimeters per year, so the California change is quite dramatic.

In about 2 billion years the gradual cooling of the earth from heat loss will mean that the asthenosphere will flow less readily and that the plate-motion phase of the earth's evolution will probably come to an end. Thus the earth will enter a new phase in which the plate motions of the earth's lithosphere are not responsible for most of the large-scale terrain features. Larger mountain ranges, like the Himalayas, will no longer be uplifted, and they will erode away over millions of years.

LUNISOLAR TIDES

The moon's gravitational pull on the oceans produces two tidal bulges on opposite sides of the earth in line with the moon. Why two high-water tides? Take an idealized earth entirely surrounded by water, as in Figure 6.4. The moon's gravitation pulls harder on the water closest to it than it does on the water on the opposite side of the earth. This *tidal force* causes a larger acceleration of the oceans at point A than at B compared with the earth's center at P. This causes water to move toward the moon at A and to recede from it at B, compared with the earth as a whole. Consequently, water piles up in the form of an ellipsoid whose long axis is directed toward the moon. Midway between the high tides are the low tides.

Earth's 24-hour rotation underneath the tidal bulges results in alternating high and low tides twice each

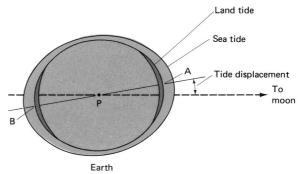

FIGURE 6.4
Polar view of the earth's tidal bulges produced by the moon's gravitational pull. The rotating earth carries the lagging tidal bulge slightly forward of the moon's position in its orbit.

day. Because there is a slight lag before the earth's waters fully adjust to the moon's tidal force, the tidal bulges are dragged by the rotating earth somewhat ahead of the line joining the centers of moon and earth (Figure 6.4).

The sun contributes to the tides, too, but only half as much as the moon does because of its much greater distance, despite its large mass. When the sun and moon are in line, as at new or full moon, their combined gravitational pull is strongest.

If the sun and moon are pulling the earth, why doesn't the land move, too? It does, because the land is not absolutely rigid. However, land has a greater internal strength than water. Therefore, the land tides are very small, but approximately every twelve hours earth's solid surface rises and falls a few centimeters at any given place.

EARTH'S ROTATION

There are several phenomena that show the rotation of the earth. One of the most vivid is the pendulum experiment devised in 1851 by the French physicist Jean Foucault (1819–1868). He hung a 28-kilogram iron

That gentle Moon, the lesser light, the Lover's
    Lamp, the Swain's delight
A ruined world, a globe burnt out, a corpse
    Upon the road of night.

Sir Richard Burton

ball on a 70-meter-long wire from the dome of the Pantheon in Paris. Underneath it was a large circular table with a ridge of sand on its edge. As the pendulum swung, a pin attached to the bottom of the ball would make a mark in the sand. After the pendulum was carefully set into motion, it was apparent from the marks in the sand that the pendulum was deviating slowly in a clockwise direction. In actuality, the spectators and the building were turning underneath the plane of oscillation of the pendulum due to the earth's rotation. Once a plane of oscillation has been established for a pendulum, an external force is required to change the plane's orientation. Figure 6.5 illustrates what would have happened if the experiment had been performed either at the geographic pole or the equator.

The constant friction generated by the lunisolar tides (mainly near the shores and in the shallow seas) has slowed the earth's rotation. As a result the day has lengthened in several billion years from an estimated several hours to our present 24 hours. This conversion of the earth's rotational energy into heat by friction will continue indefinitely. The slowdown in the length of the day is not uniform. Irregularities have been uncovered, particularly once the rotation rate could be measured with atomic clocks.

As the earth turns, it wobbles slightly on its rotational axis over a roughly circular path at the poles about 15 meters in diameter during a 14-month period. Known as the *Chandler wobble*, this movement is believed to be sustained by major earthquakes; otherwise, according to calculations, the motion would die

FIGURE 6.5
Foucault pendulum experiment. A scientist sets pendulums in motion in part a, one at a pole and the other at the equator. The only force acting on the pendulum is gravity, which is directed toward the center of the earth. After six hours (part b), the earth has turned through an angle of 90° relative to you as an observer outside the earth. For the scientist the plane of oscillation for the pendulum at the pole is now perpendicular (90°) to the observer's meridian of longitude. The rate at which the two planes change orientation relative to each other is 15° per hour clockwise in the Northern Hemisphere and counterclockwise in the Southern Hemisphere. However, for the pendulum at the equator the plane of oscillation rotates with the meridian plane—that is, the rate of change between the two planes is 0° per hour. For latitudes between 0° and 90° the rate of rotation is between 0° per hour and 15° per hour. For example, the rate at San Francisco is 9.2° per hour and requires 39 hours for a full rotation.

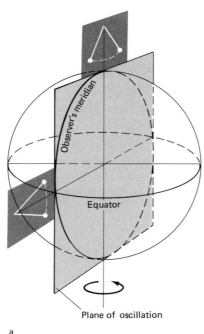

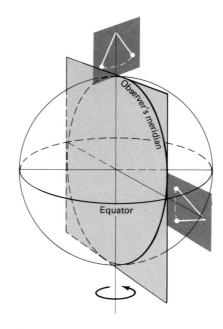

down in the short time of 20 years. There appears to be some correlation between long-term changes in the length of the day, variations in the Chandler wobble, and global seismic activity.

MAGNETOSPHERE

Another puzzle: Where does the earth's magnetic field come from? We now believe it is caused by circulation of liquid metal in the outer core. If friction can ionize the metal atoms, then the flow of ionized material is an electrical current. The current produces the magnetic field. Such a mechanism is a *dynamo,* a device that converts mechanical energy of motion into electrical energy. In appearance the magnetic field of the earth is similar to that of a bar magnet inclined slightly to the earth's axis of rotation. The magnetic lines of force run between the north and south polar regions of the earth, much like the pattern formed by iron filings sprinkled around a bar magnet (see Figure 6.6). The magnetic field loses strength away from the earth's surface, but it can still be measured many tens of thousands of kilometers out in space.

However it began, the earth's magnetic field has changed polarity (the north magnetic pole becomes the south magnetic pole and vice versa) many times over geologic ages. Scientists trace the history of changes by studying the magnetism frozen into rocks

FIGURE 6.6
Cross section of the magnetosphere in the plane of the earth-sun line and earth's magnetic axis. Earth's magnetic field presents an obstacle to the plasma flow of the solar wind and is shaped by the encounter with the solar wind, which gives rise to the frontal bow shock wave. The magnetosheath is the region of compressed subsonic plasma flow behind the shock. The magnetopause marks the thin boundary that separates the magnetosheath from the magnetosphere. Trapped charged particles are concentrated in the inner and outer Van Allen radiation belts (dark orange regions). Within the lighter shaded areas the particles are less stable and tend to be precipitated into the upper atmosphere, where they produce the auroras.

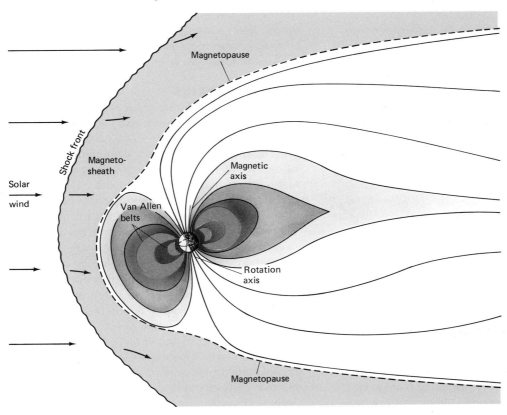

of different ages. Iron particles in molten lava beds align themselves along the lines of the existing magnetic force. After the rocks solidify, they retain the orientation of the magnetic field indefinitely. Such rocks show that magnetic reversals have come at intervals as short as 35,000 years. Why the reversals? We do not know. One suggestion is that they may be related in some way to changes in the earth's rotation or in the fluid state of its outer core.

The *magnetosphere* is that part of the magnetic field surrounding the earth in which it exerts a force strong enough to control the motions of charged subatomic particles entering the field. Exerting a sufficiently strong force even as far as 50,000 kilometers away from the earth's surface, this magnetic field protects us from bombardment by many of the charged subatomic particles traveling at speeds through space that are a significant fraction of the speed of light.

From satellites monitoring the magnetic field, we have learned much about the magnetosphere's strength, direction, and composition. A cutaway section of its structure is shown in Figure 6.6. It has several concentric zones; the principal zones are the *Van Allen radiation belts* (named after American physicist James Van Allen, who discovered their existence in 1958 from *Explorer* satellite data). The Van Allen belts encircle the planet in two doughnut-shaped regions lying about 3,000 kilometers and 17,000 kilometers from the earth's surface, respectively.

Charged particles, mainly protons and electrons, populate the magnetosphere's zones. Most of these subatomic particles are ejected from the sun as the *solar wind*. When they encounter the earth, they are either diverted away from it or trapped by its magnetic field. The collision of solar wind particles with the earth's magnetosphere creates a shock wave that distorts and compresses the magnetic field on the sunlit side and stretches it into a long tail on the nighttime side. A *shock wave* is a large-amplitude compression, such as the sonic boom made by a jet plane.

6.3
The Atmosphere

If the earth had no atmosphere, life as we know it would not exist here. The insulating blanket of air surrounding us maintains a temperature range favorable for life. This happens as follows: Sunlight passing through the air is absorbed by the ground. The sun's radiant energy, which is primarily in the visible wavelengths, warms the ground, which in turn reradiates energy in the infrared region of the electromagnetic spectrum. Its passage outward into space, however, is restricted by carbon dioxide and water vapor molecules in the atmosphere. They absorb the energy and reradiate much of it back to the earth. This is called the *greenhouse effect*, after the similarity of the action to that of the glass in a greenhouse, which prevents heat radiation produced inside the greenhouse from escaping. Without its atmosphere the earth's average temperature would be about 20 to 30 degrees lower than its present value of 15°C. Since water would be frozen at that temperature, it would not be as effective in the development and maintenance of life as it has been in a liquid form.

The upper atmosphere is also important for our survival. It protects us from harmful ultraviolet and X-ray radiation from the sun, vaporizes meteoroids entering the atmosphere, and absorbs most of the incoming highly energetic subatomic particles (mostly protons) that we call *cosmic rays*. Finally, the atmosphere creates the soft blue appearance of the sky: atmospheric gases scatter the photons of sunlight in the blue region much more efficiently than they do photons of longer wavelengths.

PHYSICAL PROPERTIES

The total mass of the atmosphere is about one-millionth of the total mass of the earth. It has several distinct layers (shown in Figure 6.7), with unique thermal, physical, chemical, and electrical properties. Approximately half the atmosphere is contained in the first 5.6 kilometers, and 99 percent is below 35 kilometers.

Our weather takes place in the bottom layer called the *troposphere*. Eleven kilometers up, the temperature drops to −55°C. Above this region lies a 40-kilometer-thick layer, the *stratosphere*, where the temperature slowly rises, reaching a maximum of about 0°C at 50 kilometers. Somewhat below this altitude an absorbing layer of ozone screens out most of the incoming ultraviolet radiation. Within the next layer, the *mesosphere*, the temperature rapidly drops to a minimum of −85°C at its upper limit, 90 kilometers.

Above the mesosphere is the *thermosphere*. Here the still more dangerous X rays and gamma rays are effectively filtered out by molecular oxygen and nitrogen and by their dissociated atoms at even higher altitudes. The temperature climbs steadily throughout the temperature into the *exosphere*—the atmospheric fringe several hundred kilometers above sea level.

Up to about 90 kilometers there is no diffusive separation of the atmospheric gases due to the mixing of the atomic and molecular constituents by air currents and random thermal motion. The chemical composition of the atmosphere therefore remains nearly uniform, with 78 percent nitrogen, 21 percent oxygen, nearly 1 percent argon, 0.03 percent carbon dioxide, and water vapor (which varies up to several percent in the troposphere). The atmosphere has minute traces of other gases, including sodium, methane, ammonia, nitrous oxide, neon, and krypton. Above about 90 kilometers the constituents are not well mixed and the heavier molecules and atoms settle to the bottom with the lighter ones diffusing to the top. At extreme heights a rarefied layer of helium extends from about 600 to 1000 kilometers, topped by a very tenuous hydrogen layer merging into interplanetary space.

Charged particles constantly spill out of the outer Van Allen radiation belt and fall into the auroral latitudes of the earth's atmosphere. There they collide with atoms of oxygen and nitrogen, stimulating these gases to radiate pale greens and occasional bright reds in patches or across the whole sky. These are the *auroras*, called the northern lights in our hemisphere (see Figure 6.8). They are most often seen in zones between 65° and 70° north and south magnetic latitudes. Because the subatomic particles enter the atmosphere more easily when the solar wind is most intense, more auroras color our night skies during the height of the eleven-year sunspot cycle (see Section 11.3), when the sun is emitting more subatomic particles.

IONOSPHERE

Within the earth's atmosphere are layers in which the concentration of free electrons is above the average atmospheric value. These layers constitute the *ionosphere*, identified as the D, E, F_1, and F_2 ionospheric levels shown in Figure 6.9. The electrons come from solar ultraviolet and X-ray photons, which ionize atmospheric atoms and molecules. Some radio waves transmitted by ground stations are reflected between the ionosphere and the earth's surface, making possible long-distance communication between stations

FIGURE 6.7
Structure of the earth's atmosphere. The various regions are differentiated by their distinct thermal, physical, chemical, and electrical properties.

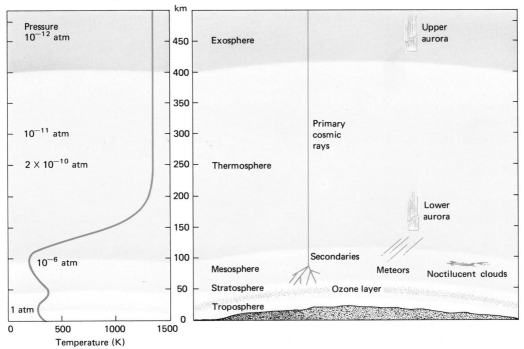

FIGURE 6.8
Northern lights over Canada. Charged particles from the sun crash through the upper atmosphere near the magnetic poles and cause it to glow like a neon sign.

that are not along a direct line of sight because of the earth's curvature. These waves enter the ionosphere, causing the free electrons to vibrate and to reradiate energy at the same frequency as the incident waves. As the waves penetrate deeper into denser portions of the ionosphere, the oscillating electrons propagate more energy in a backward direction toward the earth's surface than in the direction of the incident wave, reflecting the waves toward the earth.

Eventually the higher ionospheric density of free electrons reduces the wave energy so that it no longer propagates forward. Radio waves of frequencies less than about 30 megahertz (millions of hertz) are turned back by the ionospheric layers. Frequencies higher than about 30 megahertz pass through the ionosphere into space with little or no bending.

MOON'S DISTANCE
A lively, variegated earth and a lifeless, barren moon—these two dissimilar worlds exist in close proximity. The mean distance between the moon's center and the earth's is about four hundred thousand kilometers. Recently that distance has been precisely measured within several centimeters by timing the round-trip interval of a laser beam reflected from an array of reflectors that the Apollo astronauts placed on the moon.

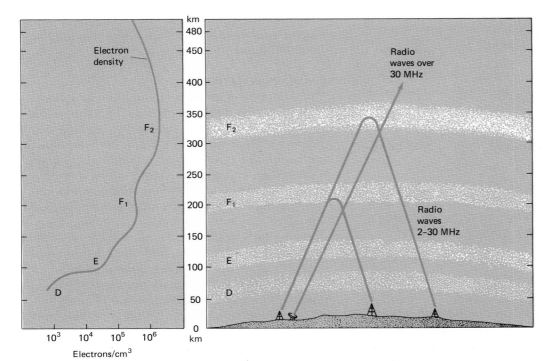

FIGURE 6.9
Earth's ionosphere. This is the region in the atmosphere where radio waves longer than about 10 meters are bent back to the earth. The numbers of electrons vary with the height, as the electron density curve shows. They are concentrated in the D, E, F_1, and F_2 layers. The paths of the radio waves are influenced by the ionosphere and vary with the intensity of ionization—that is, the electron density—and the wavelength.

LUNAR ORBIT

Our view of the moon is determined by its orbit and rotation and by the earth's movements. The moon travels around the earth in an ellipse of small eccentricity, with the earth at one focus, as shown in Figure 6.10. Some of the physical parameters of the lunar orbit are given in Table 6.3. It moves 13° east in twenty-four hours, so to bring the moon back to our eastern horizon, the earth must rotate an additional 13°, which takes fifty minutes. Thus the moon rises an average of fifty minutes later every day, or about six hours later from week to week.[1]

The point that orbits the sun annually according to Kepler's laws is not the geographic center of the

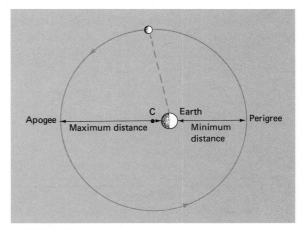

FIGURE 6.10
Moon's orbit relative to the earth. C denotes the center of the orbit; the earth is off center by 21,000 kilometers.

---

[1] Around full moon in September, the moon's path makes so small an angle with the eastern horizon that the harvest moon rises with very little delay on successive nights.

TABLE 6.3
Statistics for the Lunar Orbit

Orbital Parameter	Value
Maximum distance from earth (apogee)	406,700 km
Minimum distance from earth (perigee)	356,400 km
Mean distance from earth	384,401 km
Sidereal period (sidereal month)	27.32 d
Synodic period (synodic month)	29.53 d
Average orbital velocity	1 km/s
Inclination of orbit to ecliptic	5°8′
Eccentricity	0.055

line with the earth and the sun and to appear again as the new moon. The period of lunar phases, the *synodic month,* is 29 1/2 days.

MOON'S ROTATION

Why does the moon always show us the same face? It turns once on its axis in the same time that it completes one orbit around the earth, so the same hemisphere is always toward us. This is relatively easy to demonstrate to yourself. Walk around a stool, continually facing it. Next walk around the stool keeping your head and body pointed toward the same direction. In the first instance you rotated once while you revolved once, just as the moon does. In the second you did not rotate about your axis. If the moon did not rotate, we could see all its sides during the month.

We can, in fact, see about 59 percent of our satellite's surface, although we never see more than 50 percent at any one time. We get the 9 percent bonus because the moon's axis is tilted 6.5° from the perpendicular to its orbital plane; this lets us look over the poles as they are alternately tipped toward us every two weeks. Also, the moon's orbital speed varies while its rotation is uniform, allowing us to peek around the east and west edges alternately during the month.

The fact that the moon's rotation coincides with the period of its orbital revolution (27.32 days) is not accidental. Tidal forces between the earth and the moon (see Section 6.2) over the eons have set up a spin-orbit synchronization that causes the moon always to present the same hemisphere to the earth. Originally both bodies were much closer, perhaps only 5 to 10 percent of their present distance, and were rotating more rapidly. The earth's day was then a few hours long and its month, or the moon's or-

earth. It is a point on the line joining the earth and the moon and is known as the *center of mass* or *barycenter.* The center of mass for the earth-moon system lies inside the earth and corresponds to a center of balance of an imaginary rod supporting the earth at one end and the moon at the other, as shown in Figure 6.11.

If we track the moon's movement against the stars, we find that it takes the moon 27 1/3 days to complete its orbit; this is the *sidereal month* (see Figure 6.12). But because we are also moving around the sun, the time between each cycle of lunar phases is longer than the sidereal month. Although the moon has completed its revolution around the earth at the end of 27 1/3 days, it takes about 2 more days of travel to bring the moon in

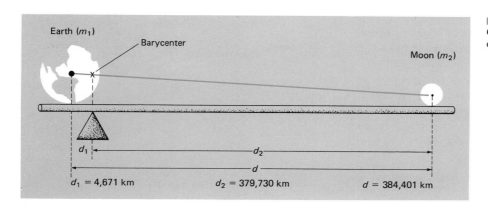

Earth (m_1)

Barycenter

Moon (m_2)

d_1

d_2

d

$d_1 = 4{,}671$ km $d_2 = 379{,}730$ km $d = 384{,}401$ km

FIGURE 6.11
Center of mass, or barycenter, of the earth-moon system.

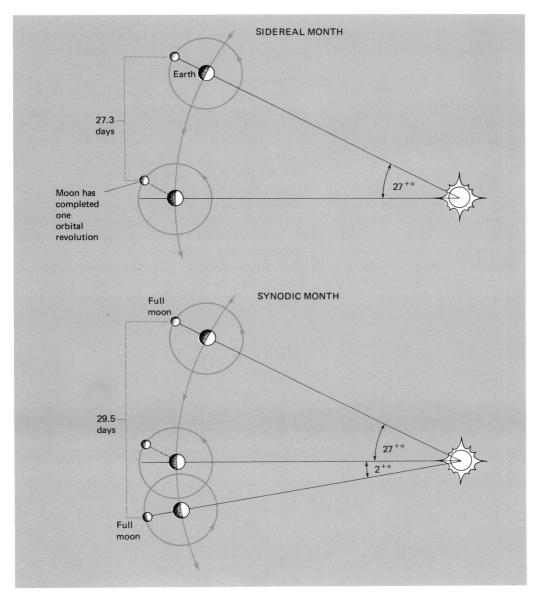

FIGURE 6.12
Sidereal and synodic months compared.

bital period, much shorter than now. Because of the earth's greater tidal force, the moon's rotation has slowed more rapidly than the earth's. Some of the earth's rotational energy is gradually transferred by the lunar tides to the orbiting moon, so that the moon recedes from the earth several centimeters every year. Why? The moon's tidal force has a braking effect on the earth, which decreases its angular momentum, or quantity of rotational motion. To conserve the total angular momentum of the earth-moon system, the angular momentum in the moon's orbital motion must be increased. Hence it is accelerated ever so slightly in its orbit, spiraling outward from the earth. As the moon recedes, the month must lengthen, according to Kepler's third law. Eventually the earth and the moon will face each other with equal periods of rotation and revolution (about 47 days) at a distance of about 560,000 kilometers. But the calculated time for this event to happen, several tens of billions of years, far exceeds our estimates of the earth-moon system's probable life span.

6.5
Nature of the Moon

As stated at the beginning of this chapter, we could think of the earth-moon system as a double planet. Compared with other planets and their satellites, the moon is more nearly comparable in size to the earth than the other satellites are to their planets. Yet the moon does not quite span the width of the United States, and its mass is roughly 1 percent of that of the earth. By comparison, the next largest satellite

mass relative to the planet is about a tenth of a percent. The planets Mercury and Pluto are about 6 percent and 1 percent of the mass of the earth, respectively. Table 6.4 summarizes the moon's physical properties.

The lunar exploration program that began in 1964 with unmanned craft and culminated in six manned Apollo landings between 1969 and 1972 has provided us with a priceless legacy of lunar materials and data. Lunar rocks (see Figure 6.13) have been collected from nine different locations, six by the United States and three by the Soviet Union, amounting to more than 2,000 individual samples weighing about 382 kilograms (843 pounds). Five instrument packages were left on the lunar surface, which operated until October 1977. The seismometers in these packages detected about 1,800 meteoric impacts, 36 shallow quakes, and 10,000 deep ones during their operating life span of about eight years. The Apollo program also carried out an ex-

TABLE 6.4
Physical Data for the Moon

Physical Datum	Value or Description
Mean radius[a]	1738 km (27% of earth's equatorial)
Rotation period	27.32 d
Inclination of axis	6.5°
Mass	7.35×10^{25} g (1/81.3 that of earth)
Average density	3.3 g/cm³
Surface gravity[b]	162 cm/s² (1/6 that of earth)
Escape velocity[c]	2.4 km/s
Atmosphere	Nearly absent; traces of helium, argon, possibly neon
Water	None present in lunar samples; interior may hold some
Average surface reflectivity[d]	7%; lowest in maria (approximately 5%); highest in cratered regions (approximately 10% to 15%)
Maximum day temperature[e]	130°C
Minimum night temperature[e]	−170°C
Magnetic field[f]	No general field; weak local fields over the lunar surface but less than 1% of earth's general field

[a]Found by trigonometry from the moon's known mean distance d and the average angle it subtends in the sky.

[b]The gravitational attraction of a body of mass M and radius R on an object resting on its surface is equal to GM/R^2, where G is the constant of gravitation. This result is derived from Newton's law of gravitation.

[c]This is the minimum velocity an object needs to become free of the gravitational attraction of a body of mass M and radius R. It is derived from orbital gravitational theory and is equal to $\sqrt{2GM/R}$.

[d]Derived from photographic and photometric color measurements of the sunlight reflected from the moon's surface.

[e]Found from terrestrial observations and from temperature gauges placed on the moon's surface.

[f]Obtained from magnetometers placed on the lunar surface, from portable magnetometers carried by Apollo astronauts from place to place, and from magnetometers installed on spacecraft orbiting the moon.

tensive effort to photograph and analyze the lunar surface. The result is better maps of some parts of the moon than some areas on earth. Also, X-ray and radioactivity studies from orbit have yielded estimates of the chemical composition of about one-quarter of the lunar surface, an area about the size of the United States and Mexico together. Thus in less than a decade the moon has changed from being a strange and distant world to one about which we are quite knowledgeable.

Because of its small mass, the moon's history has been vastly different from the earth's. With a small mass comes a weak gravitational force; as a result the moon retains almost no atmosphere. It has no surface water, either free or chemically combined in the rocks (as in earth rocks), although some water may be trapped under its surface. Also, it has no general magnetic field, but its rocks suggest that a strong one existed in the very distant past. However, the moon is far from a simple, featureless satellite.

FIGURE 6.13
Astronaut David Scott on the moon, preparing to photograph a lunar rock sample.

SURFACE OF THE MOON

Galileo's subdivision of the lunar surface into *maria*, the low-lying, almost circular dark regions in Figure 6.14, and *terrae*, the rough cratered highlands, is still significant in terms of lunar history and processes. The kinds of surface features on the moon are outlined in Table 6.5.

A tremendous number of craters pit the moon, evidence of cataclysms that altered the crust during its past. More than 30,000 are visible by telescope. The total, down to bushel-basket size, may well exceed a million. The great walled plains or super craters with low profiles like Clavius or Grimaldi (see Figure 6.15) have a structure similar to the maria, but on a smaller scale. Their diameters are about 240 kilometers.

Next in size on the moon's front side are some three dozen craters from 80 kilometers to 190 kilometers in diameter. A third of them have conspicuous light-colored streaks called *rays* radiating outward in all directions up to several hundred kilometers long, such as the well-known ray craters—Tycho, Copernicus, Kepler, and Aristarchus. Many of the small secondary craters, as well as the ray systems, apparently were formed by a rain of debris ejected from the primary crater after a large body struck the surface. The impact-produced craters are reasonably circular, with the interior rim steeper than the outer rim. The larger craters have terraces on their inner walls and frequently have a fairly smooth floor from which a few low peaks rise (Figure 6.15). Beyond the craters the terrain is hummocky and overlain with the ejected material from the cratering activity. Even moderate-sized craters, like the larger ones, have high walls. Some craters have a central peak believed to have been created by the elastic rebound of rock from below the surface after the initial impact. Others have bare floors, presumably because they were flooded with lava; the crater Plato is a good example. Volcano-produced craters, formed mostly during the moon's early history, are also present in smaller numbers. They are not always circular. Their rims slope at about the same angle both inside and outside, and their rims and floors are shallower than those of impact craters.

Although fractures are observed in the lunar crust, there is no evidence of folded mountain belts as on the earth or other indications of tectonic activity. Lunar mountain ranges apparently were produced in conjunction with the formation of impact basins. Most mountain chains are on or near the periphery of the circular maria. The mountains bordering the maria rise more steeply on the side facing the seas than on the other side. Many have lofty peaks, occasionally rising to 7500 meters above the surrounding plains. We have precise elevations and

TABLE 6.5
The Moon's Surface

Surface Feature	Description	Surface Feature	Description
Maria (seas)	Dark-colored, relatively smooth basins, roughly circular, diameter about 300 km to 1000 km; wrinkled by serpentine ridges; indented with a few shallow craters; sixteen on side facing earth, including one ocean; four on back side of moon; cover about 17% of surface	Rilles (crevices)	More than 1000 on front and back, both straight and winding
		Ridges	Narrow, wrinkled projections rising to hundreds of meters and many kilometers long, extending across the basins
Craters	Millions, from microscopic size up to about 250 km in diameter; more on the back side of moon	Scarps	Occasional abrupt discontinuities in the form of steep slopes
Ray systems	Several radiating from prominent craters, mostly on front side of moon	Domes	Volcanic blisters up to about 15 km in diameter dotting the maria here and there
Mountains	Several dozen on side facing earth, including mountain ranges as well as isolated peaks		

FIGURE 6.14
Full moon. Note that the moon as it is photographed by a telescope is inverted.

Mare Crisium—Sea of Crises
Mare Foecunditatus—Sea of Fertility

Mare Frigoris—Sea of Cold
Mare Humorum—Sea of Moisture
Mare Imbrium—Sea of Showers
Mare Nectaris—Sea of Nectar
Mare Nubium—Sea of Clouds

Mare Serentatis—Sea of Serenity
Mare Tranquillitatus—Sea of Tranquillity
Mare Vaporum—Sea of Vapors
Oceanus Procellarum—Ocean of Storms

FIGURE 6.15
Some famous lunar craters. The many craters that pockmark the moon's surface attest to the violence of the moon's past, when it underwent a series of asteroidal bombardments that petered out about four billion years ago. Plato is the large smooth-bottomed crater in the top left photograph. It is immediately below the center of the photograph here, which shows the eastern section of Mare Imbrium. Close to the terminator, near the left edge center, is the Alpine Valley, appearing as a long dark streak across the Alps Mountains. Clavius (top right) is the largest crater on the near side of the moon, about 240 kilometers in diameter. The crater Copernicus (bottom left) is about 80 kilometers across and clearly shows the diverging ray system associated with it. The close-up view (bottom center) of its floor covers about 27 kilometers. From its floor mountains rise 300 meters with slopes up to 30°; cliffs on the crater's rims are 300 meters high. The somewhat smaller crater Tycho (bottom right) has a long, striking ray system that is easily seen with a small telescope at the time of full moon. Notice the terraced walls on the left side of the photograph.

Top row: Crater Plato; crater Clavius

Bottom row: Crater Copernicus; floor of Copernicus; crater Tycho.

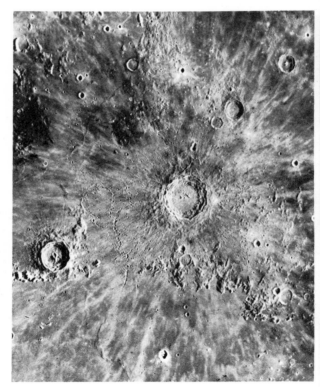

depths for many topographic features from stereoscopic photographs taken by lunar orbiters and from radar measurements.

Beyond the eastern edge of Mare Imbrium a narrow valley cuts across the lunar Alps Mountains. This feature has long been known from photographs taken from the earth. Now, from photographs taken by a lunar orbiter spacecraft, we know that the Alpine Valley is a deep trough 3 kilometers to 10 kilometers wide and about 130 kilometers long (see Figure 6.16). Narrow channels called *rilles*, which resemble chasms or gorges, cut many kilometers across the lunar terrain, frequently without interruption. Running lengthwise down the middle of the Alpine Valley is a very conspicuous rille (see Figure 6.16). Many shorter rilles take tortuous paths. Some may be lava channels or collapsed tubes partly filled with rubble from lava flows. The walls of Hadley Rille, a 365-meter-deep canyon near the base of the Apennine Mountains, and their lower slopes are stratified into parallel tilted layers. Perhaps the large object that created Mare Imbrium heaved up blocks of earlier lava-flow substrata that had impregnated the lunar crust, creating the layering.

Structures that look like wrinkles on the mare surface turn out to be sinuous ridges. They may have been pushed up by compressional waves spreading

from the giant blows that created the maria. Evidence of past volcanic activity is clear in the small domes and cinder cones and their associated lava flows in the mare flatlands (see Figure 6.17). Some scarps also may have been made by fractures opened by faults.

Topography on the far side of the moon is strikingly different from that on the near side (see Figure 6.18). Craters are everywhere but few have steep slopes. This face has no extensive mountain ranges and no large lava-flooded basins comparable to Mare Imbrium on the near side, though some small basins are there. The far side also lacks the near side's extensive lava flooding. The moon's center of mass is displaced from its geometric center about 2 kilometers earthward. One consequence is that the lunar crust facing the earth is about half as thick as that of the far side. Perhaps this variation explains why the near side has more volcanic activity, accounting for its large deposits of dark mantling material. It may also help explain why the basins on the far side are only partially filled.

CHEMICAL COMPOSITION OF THE MOON

The moon appears to have accreted from the same chemical elements as those that formed the earth, although in somewhat different proportions. It has less iron, more of the substances that are hard to melt (refractory materials)—such as calcium, aluminum, and

FIGURE 6.16
The Alpine Valley. Note the rille running centrally along its floor. The valley is about 130 kilometers long and up to 10 kilometers wide. Compare Figure 6.15.

FIGURE 6.17

Domes in Oceanus Procellarum. The domes are best seen on the left of the picture. Varying from 3 to 16 kilometers in diameter and from 300 to 460 meters in height, they resemble the volcanic domes of northern California and Oregon. You can see several ridge systems in the photograph. The large crater in the upper right is Marius.

titanium—and less of the easily vaporized atoms and molecules (volatile compounds) than the earth does, such as sodium and potassium. The most common surface rocks are anorthosites (silicates mainly of aluminum and calcium), iron-rich basalts from the maria, thorium-rich and uranium-rich rocks containing potassium (K), the rare earth elements (REE), and phosphorus (P), together labeled KREEP. No traces of water and no organic compounds, the indicators of living processes, were discovered in any of the lunar samples. In fact, the Apollo lunar rocks contain only tiny amounts of carbon and the carbon-based compounds from which life originates.

The lunar highlands, which cover about four-fifths of the lunar surface, are the oldest preserved terrain. They consist mostly of material rich in aluminum and silicon and lighter than the mare basalts, having

been formed during the first half billion years of the moon's existence during violent igneous activity and bombardment by bodies like asteroids. Meteorites constantly hitting the moon for over 3 billion years made most of the large and small craters. The craters Copernicus and Tycho are relatively recent, about 600 and 200 million years old, respectively. Meanwhile, small meteoroids pulverized the lunar surface, leaving a dusty layer some 1 to 20 meters deep, the *regolith,* on which the astronauts left their footprints. Since this lunar soil contains no water or organic matter, it is totally different from soils formed on the earth by water, wind, and life. It could only have been formed on the surface of an airless body.

The lunar basins are overlain by dark volcanic basalts enriched with iron, magnesium, and titanium

. . . the moon is nothing
But a circumambulating aphrodisiac
Divinely subsidized to provoke the world
Into a rising birth rate . . .

Christopher Fry

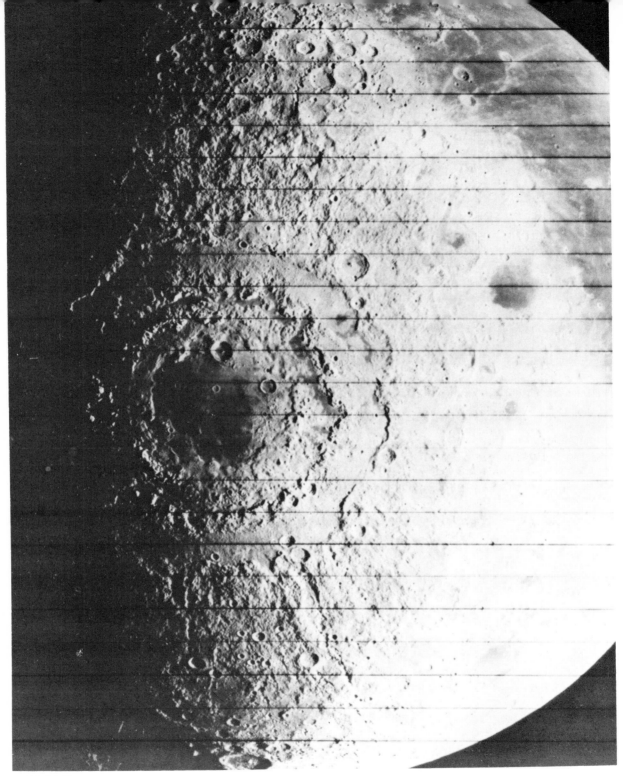

FIGURE 6.18
Mare Orientale basin photographed on the back side of the moon. The basin's outer scarp is about 930 kilometers in diameter. The bull's-eye basin is in an excellent state of preservation with very little encrustation, erosion, or disfigurement. The large dark portion in the upper right is Oceanus Procellarum, which is visible on the moon's near side (see Figure 6.17).

formed by extensive flooding of the impact basin with lava. The lava is apparently due to radioactive heating and partial melting in the interior following the moon's heaviest bombardment. For example, the great Mare Imbrium basin was probably formed about 3.9 billion years ago by an object that may have been as big as Rhode Island. The upheaval splashed out a thick layer of rocks and rubble, and molten basalt welled from below to flood the ravaged area. The Apennine Mountains around the basin may have been created by material pushed upward by the tremendous internal pressure caused by the impact. During the moon's violent cratering phase, catastrophes of different intensities modified its surface, often recurring in the same regions. Evidence for this buildup in stages is the successive lava flows inundating the mare basins at different times, set off by intervals of quiescence with subsequent cooling and solidification.

THE MOON'S INTERNAL STRUCTURE

Radioactive dating of the lunar rocks points to an origin much like that of the earth, about 4.6 billion years ago. Most of the lunar material probably came from accreting planetesimals (small solid bodies) orbiting within the contracting solar nebula. Nearly all the original crust was lost when the moon underwent a global melting followed by chemical differentiation or separation shortly after the moon's formation. A few of the rock samples returned to the earth are about 4.6 billion years old, the same age as the earth. The highland areas are apparently about 4.0 to 4.3 billion years old. The global crust of the moon was continually modified after its formation by the impact of material from elsewhere in the solar system. The cratering record preserved in the early crustal units represents a distinct phase of intense cratering that began to decline rapidly about 3.8 billion years ago (see Figure 6.19). Although volcanic processes may have operated during this early period, the surface history of the moon is primarily that due to cratering. As mentioned earlier, this phase in the earth's history has been almost completely erased.

The next stage in lunar history was dominated by the formation of dark mare plains that cover about 17 percent of the lunar surface, favoring the earthward side. These structures are relatively thin ponds of basaltic lava totaling less than 1 percent of the volume of the crust. Apollo mare rocks suggest that the major outpouring of lava occurred between 3.9 and 3.2 billion years ago. Although some mare deposits may be

as young as 2 billion years, there appears to have been no extensive igneous activity on the lunar surface for the last 3 billion years. Thus the shaping of the present lunar terrain is almost the opposite of that of the earth's—the moon being dominated by cratering and the earth by volcanic and tectonic activity.

There is no general lunar magnetic field as large as approximately 1/10,000 that of the earth, which seems to indicate that the moon does not now have the molten iron-nickel core believed necessary to produce a magnetic field comparable to that of the earth. However, the evidence suggests that the moon may have had a stronger magnetic field early in its history. Random magnetic fields up to about 0.6 percent of the earth's field intensity were detected at different sites by the Apollo astronauts, but we do not know the reason for such magnetic anomalies.

> Man makes a great fuss
> About this planet which is only a ball-bearing
> In the hub of the universe.
>
> Christopher Morley

From seismographs left on its surface we know that seismic events on the moon follow patterns different from those here on the earth. Moonquakes, rare meteorite impacts, and artificially produced vibrations (grenade explosions and crash landings of discarded spacecraft) are transmitted very slowly through the lunar material. They build gradually, then take up to an hour to subside. Some seismic disturbances have been traced to geologic movements in the rilles; others, to occasional impacts of meteoroid swarms. Moonquakes frequently coincide with tidal stresses triggered by the varying distance between moon and earth. They occur at depths of 600 to 900 kilometers, much deeper than earthquakes. About eighty or so sources for these deep moonquakes have been discovered so far. But compared with the earth's seismic activity, the moon's is fairly subdued—the whole moon releases less than one ten-billionth of the earth's earthquake energy. About thirty-five or so shallow quakes, presumably tectonic events, have been detected. Thus if the moon is expanding or contract-

FIGURE 6.19
Evolution of the moon's surface. The moon 3.9 billion years ago (left), when it was undergoing collisions with fragments of material left over after the formation of the solar system. The largest of these impact features were like Imbrium Basin in the upper left. Most of the surface was saturated with craters.

The moon 3.1 billion years ago (above) after the cratering barrage ended. Shattering of the crust allowed magma from within the moon to reach the surface filling large basin floors. This continued over the next half a billion years or so forming the maria.

The moon today (left) with newer craters such as Copernicus and Tycho whose bright rays overlay the ancient surface. Material ejected from recent cratering events thrown out into the maria tend to lighten their contrast with the highlands.

ing, it is doing it extremely slowly. In the last 3 billion years thermal and geological activity has been relatively tranquil. Most volcanic activity appears to have ceased about 3 billion years ago, but some minor activity may still be going on.

Seismic data tell us that the crust is about 65 kilometers thick, twice as great as the earth's crust. The basaltic material of the maria was formed by the impact of large bodies, which provided the energy to

melt partially the iron-rich regions below the crustal layer. The moon's mantle, nearly 1000 kilometers deep under the lunar crust, is uniformly structured. Most of it may be pyroxene, a mineral rich in the silicates of calcium, magnesium, and iron. The seismic data reinforce the view that the moon's core is unlike earth's metallic core. Probably the lunar core consists of partly molten silicates, with a small metallic center, as shown in Figure 6.20. Lunar orbiters passing over the major maria brought a surprise: their speed increased unexpectedly. A strong gravitational irregularity seemed to be the cause and apparently comes from a concentration of mass or *mascon* beneath the surface. The mascons are believed to have formed from magma (molten rock) welling up beneath the original mare surface and spreading widely. They have a higher proportion of iron, magnesium, and titanium than the surrounding highlands, which have larger concentrations of the lighter elements such as calcium and aluminum. This difference in composition and density probably accounts for the gravitational anomaly.

6.6
Origin of the
Earth-Moon System

We have all this new information, yet the moon's origin, like that of the solar system itself, is shrouded in mystery. No proposal either of a catastrophic or

FIGURE 6.20
Schematic diagram of one proposed model of the moon's interior. The moon is not uniform, but is apparently divided into layers as is the earth. There is probably a crust of about 65 kilometers and a mantle of some 800 to 1000 kilometers. Although still unknown, there may be a small iron core surrounded by a partly molten zone.

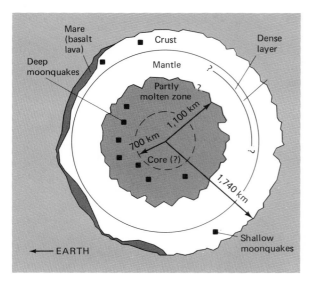

noncatastrophic beginning for the earth-moon system is universally accepted.

The earliest concept, the *fission theory* (see Figure 6.21), was that eons ago a rapidly spinning earth was flattened to dumbbell shape by rotational instability, perhaps because movement in the earth's molten core was uneven. The smaller end of the dumbbell broke away to become the moon, separating ever more from

FIGURE 6.21
Various theories for the origin of the earth-moon system. The fission theory presumes that because of rapid rotation the earth broke apart forming the moon that spiraled outward. In the binary theory the earth and the moon formed in orbit about each other. In the capture theory the moon formed in another part of the solar system and was later captured by the earth.

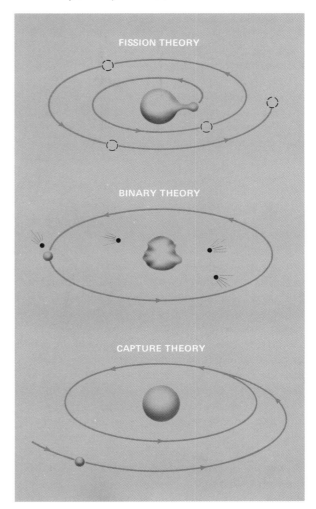

Art thou pale for weariness
Of climbing heaven, and gazing on the earth,
Wandering companionless
Among the stars that have a different birth?

Percy Bysshe Shelley (referring to the moon)

the earth because of tidal forces. The major objection to this theory is that the primitive earth could not have spun rapidly enough to promote fission through rotational instability.

Earth and moon may have formed by accretion from chemically related primordial planetesimals condensing out of the gas and dust of the solar nebula; this is the *binary theory* (see Figure 6.21). Colliding debris trapped in orbit around the growing earth accreted to form the moon. The two bodies are of comparable ages (4.6 billion years), a fact strongly backing this theory; the similar isotopic composition of oxygen on the moon and the earth is further evidence supporting this theory. The most recent chemical analysis of the lunar rock samples indicates that the moon and the earth did indeed evolve from the same parent material.

Another suggestion is that a collision between the already differentiated earth and a large planetesimal ejected part of the earth's outer layers into an orbit around the earth, where they condensed and eventually coalesced into the moon (see Figure 6.21). And finally, another possibility is that a protobody originally was moving in a highly eccentric orbit around the sun. As it approached the earth almost on a collision course, it was disrupted by strong tidal forces and part of the fragmented body became the earth's satellite—the capture theory.

It seems to us appropriate to close this section with a quotation from Mark Twain's *Life on the Mississippi*: "There is something fascinating about science. One gets such wholesale returns of conjecture out of such a trifling investment of fact."

SUMMARY

About 4.6 billion years ago, it is believed, in a contracting cloud of gas and dust that formed the solar nebula, this planet of ours was begun. Analysis of radioactivity in terrestrial rocks and meteorites leads to this figure. Earthquake vibrations tell us about the earth's composition: a core mainly of iron and nickel, solid at the center and liquid farther out; a rocky mantle topped by a thin granite crust. From exploration of the oceans' bottoms we find that earth once had just one huge land mass, which broke into separate plates 200 million years ago as the seafloors spread. The plates, on which the continents ride, are still moving.

The rotation of the earth has given the earth its ellipsoidal figure. Apparently the earth's rotation is gradually slowing down. Earth's magnetic field depends upon its rotation and the presence of a liquid outer core. Geological evidence points to a number of reversals of the magnetic field over the earth's history.

The outer parts of the magnetic field, or magnetosphere, act as a barrier to incoming subatomic particles. The atmosphere also acts as a barrier both for incoming particles and electromagnetic radiation. Our atmosphere behaves like a greenhouse to trap heat and maintain a reasonably stable environment for life on the earth's surface.

We can think of the earth-moon system as a double planet. These are close but dissimilar worlds: a lively, variegated earth and a lifeless, barren moon, each of which evolved in different ways. The moon and earth were formed at about the same time, 4.6 billion years ago. Because it does not have enough mass to retain an atmosphere, the moon's surface has changed little during most of its history. In contrast, earth's topographical features have changed constantly.

Astronomers know a great deal about the moon's general behavior, its complex orbital motions, its rotation, and the mutual gravitational interactions between it and the earth. The moon had been thoroughly observed with earthbound telescopes—revealing its maria, craters, mountains, and valleys—but surprises came when extraterrestrial exploration of the moon was possible. Rock samples brought back by Apollo astronauts reveal certain chemical, physical, and mineralogical differences from terrestrial rocks. The youngest rocks returned are slightly more than 3 billion years old. From our extraterrestrial evidence we think the moon passed through great geological and thermal activity during its first 1.5 billion years. Since then it has been more or less physically inactive.

In the last few years we have learned more about the moon's history and how its surface was shaped than we learned in all preceding centuries. One major mystery that still remains is the cause of the different appearance of the terrain on the moon's far side. In addition, we know a great deal about the moon's geology and chemistry, but we have yet to obtain a firm understanding of how the earth-moon system formed.

REVIEW QUESTIONS

1. Explain how the age of the earth is derived from its oldest radioactive rocks.

2. How have geophysicists been able to arrive at a comprehensive picture of the earth's interior?

3. Describe the phenomenon of seafloor spreading. How is continental drift accounted for?

4. How are tides produced by the gravitational attraction of the sun and moon on the earth? How do tides affect earth's rotation?

5. What is the lithosphere? The magnetosphere? The ionosphere?

6. Describe the magnetosphere and its interaction with the solar wind.

7. Name the principal divisions of the atmosphere and discuss their physical and chemical properties.

8. What role does the ionosphere play in the propagation of radio waves between transmitted and received signals from ground stations?

9. Name and briefly describe the various kinds of lunar features.

10. How does the far side of the moon differ from the near side?

11. How does the chemical composition of lunar surface material differ from that of earth?

12. How is it that some lunar rock samples are older than any terrestrial rocks if the earth and the moon were formed at the same time?

13. What are the ray systems observed around some craters?

14. What kind of activity is still present on the moon?

15. How were the lunar craters formed?

16. Would a human body buried on the moon decompose? Explain your answer.

17. Discuss the three theories covered in the text for the origin of the earth-moon system.

18. Why do we find short-lived radioactive elements still existing on earth when their half-lives are a minute fraction of the earth's age?

19. Why can't the radio frequencies of the AM broadcast stations (550–1600 on a radio dial) be used in communication satellites?

20. Why does the clear sky sometimes have a whitish appearance instead of its usual blue color?

7

The Solar System

7.1 Surveying the Solar System

THE SUN

What is it about the sun that forces planets, asteroids, comets, and other bodies of the solar system to orbit around it? It is its immense gravitational reach, since the sun contains 99.86 percent of the solar system's mass. The natural satellites of the planets, on the other hand, are gravitationally bound to the parent planet because of their closeness to it.

Like other stars, the sun is a gas from center to surface and generates its radiant energy within its hot core. This giant gaseous sphere has a diameter 109 times greater and a mass 333,000 times greater than that of the earth. The sun's family of planets intercepts only a minute amount of the radiation that streams from the sun, flooding the solar system.

◀ Artist's illustration of the flight of the *Voyager* spacecraft to Jupiter and Saturn. The 1800 pound spacecraft after leaving Saturn will journey on to Uranus and possibly Neptune for a complete survey of all of the Jovian planets.

THE PLANETS

Outside the sun, the nine planets contain the next major share of the mass of the solar system (see Figure 7.1). Although the planets had a common origin, they have significant chemical, physical, and geological differences. This diversity stems mostly from their different masses and distances from the sun. These factors determined their ability to retain matter and its chemical composition at the time of their formation. We know of enough chemical and physical similarities among the planets, however, to divide them into two well-defined categories, summarized in Table 7.1. One category is an inner group composed mostly of rocky material: the *terrestrial planets,* Mercury, Venus, Earth, and Mars. Though like the moon in size, Pluto probably has an icy composition and thus clearly does not belong to either group. The second group, the *Jovian planets*—Jupiter, Saturn, Uranus, and Neptune—are farther from the sun, are larger, and consist of the lighter elements, primarily hydrogen and helium. Jupiter and Saturn apparently have the same chemical composition as the sun. Uranus and Neptune seem to have less hydrogen and helium and presumably more of the icy materials (frozen gases such as H_2O, NH_3, etc.). (Table 7.4 contains a summary of the specific physical properties of the planets.)

Astronomers have employed several independent techniques in deriving the scale of the solar system.

TABLE 7.1
Comparing Terrestrial and Jovian Planets

Average Characteristic	Terrestrial	Jovian
Distance from sun (AU)	0.9	16
Sidereal period	Hundreds of days	Tens of years
Spacing between members (AU)	0.4	8
Mass (earth = 1)	0.5	110
Diameter (earth = 1)	0.7	7
Mean density (g/cm³)	5.0	1.2
Rotation period	Slow, in days	Rapid, in hours
Shape	Spherical	Oblate
Mean temperature	450 K	85 K
Atmosphere	Thin: CO_2, N_2, O_2	Thick: H, He, CH_4, NH_3
Surface	Cratered,[a] volcanic	None
Interior composition	Iron, silicates, oxides	Hydrogen, helium
Satellites	0.5	8

[a]Except earth, whose extensive tectonic activity has obliterated its early cratered history.

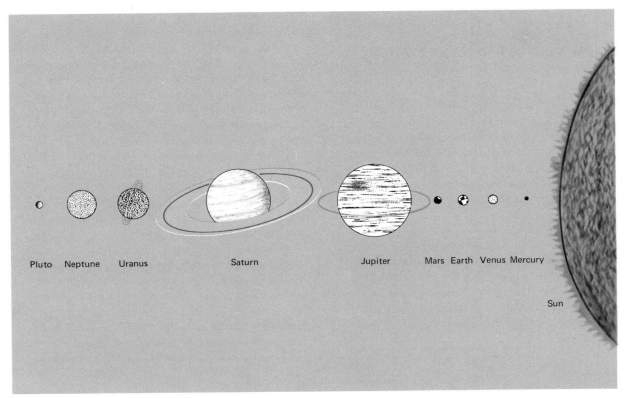

FIGURE 7.1
Relative sizes of the planets and the sun.

The scale is based on the earth's average distance from the sun, which is known as the *astronomical unit* (1 astronomical unit = 149,597,871 kilometers).[1] It is an important measurement, for it is the yardstick by which we measure distances within the solar system. The most accurate method for determining the scale is by timing the round trip of a pulsed radio signal reflected from a planet. Combining this information with the planets' distances, using Kepler's third law (in astronomical units), leads to the absolute size of the solar system (in kilometers).

German astronomer Johann Bode (1747–1826) called attention in 1772 to a rule originally discovered by Johann Titius in 1766 that seemed to predict the mean distances of the then-known planets (see Table 7.2). It is known as *Bode's law*, and both Uranus, discovered in 1781, and the first asteroid, Ceres, found in 1801, adhered fairly well to the rule. But the rule broke down later when Neptune and Pluto were discovered; the law may have no unique physical basis.

[1]Earth's distance from the sun ranges between 147,097,000 kilometers and 152,086,000 kilometers.

ORBITS OF THE PLANETS

All the planets have some similarities in orbital characteristics. They revolve around the sun in the same direction in roughly circular orbits that lie nearly in the same plane (see Figure 7.2). Mercury and Pluto, the innermost and outermost planets, depart most from this regularity. Between the terrestrial planets the average spacing is much smaller than that separating the Jovian planets. All the planets orbit the sun at mean distances ranging from 2/5 of earth's distance from the sun to 40 times earth's distance, with orbital periods between 1/4 year and 248 years. Orbital and physical data for the planets are in Tables 7.3 and 7.4.

THE SATELLITES

Of the 40 or so satellites, all but 4 of them belong to the Jovian planets. It seems likely that more will be discovered in the future, since there is an unconfirmed discovery for Jupiter. Very small and faint satellites of the Jovian planets could escape detection given our present technology. Table 7.5 lists information about the ones we know. Three of them—

TABLE 7.2
Bode's Law of Planetary Distances

Planet	Titius-Bode Rule	Actual Distance from Sun (AU)	Spacing Ratio
Mercury	$0 \times 0.3 + 0.4 = 0.4$	0.39	
Venus	$1 \times 0.3 + 0.4 = 0.7$	0.72	1.3
Earth	$2 \times 0.3 + 0.4 = 1.0$	1.00	1.4
Mars	$4 \times 0.3 + 0.4 = 1.6$	1.52	1.5
Minor planets[a]	$8 \times 0.3 + 0.4 = 2.8$	2.81 (average)	1.8
Jupiter	$16 \times 0.3 + 0.4 = 5.2$	5.20	1.9
Saturn	$32 \times 0.3 + 0.4 = 10.0$	9.54	1.8
Uranus[a]	$64 \times 0.3 + 0.4 = 19.6$	19.18	2.0
Neptune[a]	$128 \times 0.3 + 0.4 = 38.8$	30.07	1.6
Pluto[a]	$256 \times 0.3 + 0.4 = 77.2$	39.46	1.3

[a]These bodies had not been discovered when the rule was formulated.

FIGURE 7.2
Orbits of the planets. The scale of the lower diagram is 1/20 that of the upper diagram. The numbers give the planets' distances from the sun in light minutes, the distance light travels in a minute of time or about 1.799×10^7 kilometers.

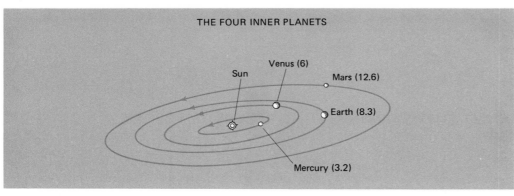

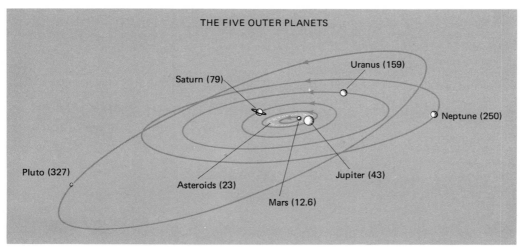

TABLE 7.3
Planetary Orbital Data

Data	Mercury	Venus	Earth	Mars	Jupiter	Saturn	Uranus	Neptune	Pluto
Mean distance from sun (AU)	0.39	0.72	1.00	1.52	5.20	9.54	19.18	30.06	39.44
($\times 10^6$ km)	57.9	108.2	149.6	227.9	778.3	1427.0	2869.6	4496.6	5900
Minimum distance from sun ($\times 10^6$ km)	45.9	107.4	147.1	206.7	740.9	1347	2735	4456	4425
Maximum distance from sun ($\times 10^6$ km)	69.7	109.0	152.1	249.1	815.7	1507	3004	4537	7375
Sidereal period (yr)	0.241	0.615	1.00	1.88	11.86	29.46	84.01	164.79	247.69
Synodic period (d)	115.9	583.9	—	779.9	398.9	378.1	369.7	367.5	366.7
Inclination of orbit to ecliptic plane[a]	7.0°	3.4°	0.0°	1.85°	1.3°	2.5°	0.77°	1.8°	17.2°
Orbital eccentricity[a]	0.206	0.007	0.017	0.093	0.048	0.056	0.047	0.009	0.250
Average orbital velocity (km/s)	48	35	30	24	13	9.6	6.8	5.4	4.7

[a]These are two of the six parameters that uniquely define the planet's elliptic orbit. A third gives the angle measured eastward along the ecliptic plane from the vernal equinox to the ascending node of the planet's orbit; a fourth, the angle between the ascending node and the perihelion point measured in the direction of the planet's motion; a fifth, the date of closest approach to the sun (time of perihelion passage); and the sixth, the semimajor axis (mean distance from the sun), or the sidereal period, which is related to the mean distance by Kepler's third law.

PRECESSION OF MERCURY'S ORBIT: TESTING RELATIVITY THEORY

Einstein's relativity theory cleared up a long-standing problem that had plagued astronomers for decades—the advance of the perihelion of Mercury's orbit in the direction of the planet's revolution around the sun, that is, the precession of Mercury's orbit. Perturbations by the other planets were the cause (see Figure 7.3), but when astronomers had applied Newtonian theory, observations still left them with 43 arc seconds per century more than the calculated value of the precession. They faced the unwanted choice of increasing the mass of Venus by an inadmissible one-seventh or postulating the existence of a never-discovered planet called Vulcan within Mercury's orbit. Einstein removed the difficulty in 1915.

The general theory of relativity says that the orbit of a planet rotates in its own plane, and the classical theory of Newton says the same. In either theory the change in perihelion is most pronounced for Mercury's orbit. But Einstein's equations for the elliptic motion of a planet orbiting the sun include a term not present in the Newtonian equations. Its contribution is a tiny fraction of the total amount of the orbital precession. It adds up to one extra revolution in 3 million years. Looked at relativistically, the planet's eccentric orbital motion periodically moves it into stronger and weaker gravitational fields, where it encounters varying properties of the space-time structure surrounding the sun.

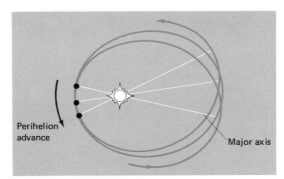

FIGURE 7.3
Precession of the major axis of Mercury's orbit. The observed advance in the perihelion of Mercury's orbit is 573 arc seconds per century.

Perihelion advance

Major axis

TABLE 7.4
Planetary Physical Data

	Mercury	Venus	Earth	Mars	Jupiter	Saturn	Uranus	Neptune	Pluto
Equatorial diameter (earth = 1)	0.38	0.95	1.00	0.53	11.19	9.47	4.15	3.88	0.20
(km)	4,878	12,100	12,756	6,795	142,800	120,800	52,900	49,500	2,800
Mass (earth = 1)[a]	0.055	0.815	1.00	0.11	318.0	95.2	14.6	17.2	0.01?
Mean density (g/cm³)	5.4	5.2	5.5	3.9	1.3	0.7	1.2	1.6	$\approx$1.0?
Rotation period[b]	58.6 d	−243 d	23.94 h	24.62 h	9.84 h	10.23 h	−12.3 h	16 h	6.4 d
Surface gravity (earth = 1)	0.37	0.88	1.00	0.38	2.64	1.15	1.06	1.43	0.2?
Inclination of axis to orbital plane	7°	6°	23.5°	24°	3.1°	26.7°	97.9°	28.8°	?
Oblateness	0	0	0.003	0.005	0.067	0.105	0.007	0.025	?
Velocity of escape (km/s)	4.26	10.30	11.18	5.02	59.6	35.5	22.1	24.8	2.5?
Albedo[c]	0.06	0.77	0.39	0.16	0.50	0.50	0.66	0.62	Low
Solar energy received (earth = 1)	6.7	1.91	1.00	0.043	0.037	0.011	0.003	0.001	0.0006
Magnetic field strength	Weak	Not detectable	Moderate	Weak	Very strong	Moder-ate	?	?	?
Approximate mean temperature (°C)	350 (day) −170 (night)	480 (surface) −23 (clouds)	15 (average)	−23 (average)	−150	−180	−210	−220	−230
Atmospheric gases observed (principal constituents)	H, He Trace	CO_2	N_2 O_2	CO_2 N_2, Ar	H_2 He	H_2 He	H_2, He	H_2, He	?
Number of known satellites	0	0	1	2	15	13	5	2	1

[a]If the planet has no satellite, the mass is found by mathematical analysis of the gravitational disturbances. These disturbances would be the forces that the planet exerts on comets, asteroids, and space vehicles during near approaches. If the planet has satellites, the planetary mass may be found by applying Newton's modification of Kepler's third law (page 37) or from a spacecraft flyby.

[b]Minus sign indicates reverse rotation.

[c]Measured percentage of reflectivity of sunlight from the planet's surface or atmosphere.

Ganymede, Callisto, and Titan—are as large or larger than Mercury. The nearer satellites move in circular orbits in the plane of their primaries' equators (a *primary* here refers to the planet about which the satellite orbits) and in the same direction as their primaries rotate. The outer satellites have more eccentric orbital motions, which are more highly inclined to the equatorial planes of their primaries. The four outer satellites of Jupiter, the most distant satellite of Saturn, and the inner satellite of Neptune have orbits that are reversed from the direction of their primaries' rotation. The outer satellites may have been captured by the primaries at the time that the planets and their inner satellite systems were formed.

One of the most exciting developments in planetary research in recent years is the discovery of ring systems for Uranus and Jupiter. (Saturn's rings were discovered with the introduction of the telescope in astronomy.) The rings are actually individual particles in orbit about the planet and are thus very small satellites of the planet. Uranus's and Jupiter's rings do not contain as many particles as does Saturn's, which makes them much fainter than the ring system about Saturn. Thus they managed to escape detection until now. Although no ring system has been found for Neptune, it is probable that it also has a faint set of rings similar to those around the other three Jovian planets.

TABLE 7.5
Natural Planetary Satellites

Planet and Satellite	Average Distance from Planet's Center (km)	Period of Revolution (d)	Orbital Eccentricity	Orbital Inclination	Diameter (km)	Mass (relative to moon)	Discoverer	Year
Earth								
Moon	384,410	27.32	0.055	5.1°	3,475	1	—	—
Mars								
Phobos	9,350	0.32	0.021	1°	25	—	A. Hall	1877
Deimos	23,500	1.26	0.003	2°	13	—	A. Hall	1877
Jupiter								
V Amalthea	181,000	0.42	0.003	0.5°	150	—	E. Barnard	1892
I Io	422,000	1.77	0.000	0°	3,680	1.21	Galileo	1610
II Europa	671,000	3.55	0.000	0.5°	3,100	0.65	Galileo	1610
III Ganymede	1,071,000	7.16	0.002	0°	5,300	2.04	Galileo	1610
IV Callisto	1,884,000	16.69	0.008	0°	4,840	1.46	Galileo	1610
XIII Leda	11,090,000	238.7	0.148	27°	8?	—	C. Kowal	1974
VI Himalia	11,500,000	250.6	0.158	28°	≈100	—	C. Perrine	1904
VII Elara	11,750,000	259.6	0.207	26°	≈25	—	C. Perrine	1905
X Lysithea	11,800,000	263.5	0.130	29°	≈20	—	S. Nicholson	1938
XII Ananke[a]	21,000,000	625	0.169	33°	≈15	—	S. Nicholson	1951
XI Carme[a]	23,000,000	714	0.207	17°	≈20	—	S. Nicholson	1938
VIII Pasiphae[a]	23,500,000	735	0.378	34°	≈25	—	P. Melotte	1908
IX Sinope[a]	23,700,000	758	0.275	29°	≈15	—	S. Nicholson	1914
XIV 1979J1	129,000	0.12	—	—	≈35	—	*Voyager 2*	1979
Saturn								
X Janus	168,700	0.75	0.000	0°	200	—	A. Dollfus	1966
I Mimas	185,400	0.94	0.020	2°	400	0.0005	W. Herschel	1789
II Enceladus	238,200	1.37	0.004	0°	560	0.001	W. Herschel	1789
III Tethys	294,800	1.89	0.000	1°	1,000	0.009	G. Cassini	1684
IV Dione	377,700	2.74	0.002	0°	800	0.014	G. Cassini	1684
V Rhea	527,500	4.52	0.001	0°	1,500	0.031	G. Cassini	1672
VI Titan	1,223,000	15.94	0.029	0°	5,500	1.80	C. Huygens	1655
VII Hyperion	1,484,000	21.26	0.104	1°	≈400	0.001	W. Bond	1848
VIII Iapetus	3,563,000	79.32	0.028	15°	≈1,300	0.015	G. Cassini	1671
IX Phoebe[a]	12,950,000	550.37	0.163	30°	≈250	—	W. Pickering	1898
XI—	150,000	0.68	—	—	≈170	—	*Pioneer 11*	1979
Uranus								
V Miranda	130,000	1.41	0.000	0°	550	0.001	G. Kuiper	1948
I Ariel	191,800	2.52	0.003	0°	1,500	0.018	W. Lassell	1851
II Umbriel	267,300	4.14	0.004	0°	1,000	0.007	W. Lassell	1851
III Titania	438,700	8.71	0.002	0°	1,800	0.058	W. Herschel	1787
IV Oberon	586,600	13.46	0.001	0°	1,600	0.035	W. Herschel	1787
Neptune								
Triton[a]	353,600	5.88	0.000	20°	4,500	1.90	W. Lassell	1846
Nereid	5,570,000	359.42	0.750	28°	500	—	G. Kuiper	1949
Pluto								
Charon	≈20,000	6.4?	≈0.000	75°	850?	?	J. Christy	1978

[a]Retrograde motion.

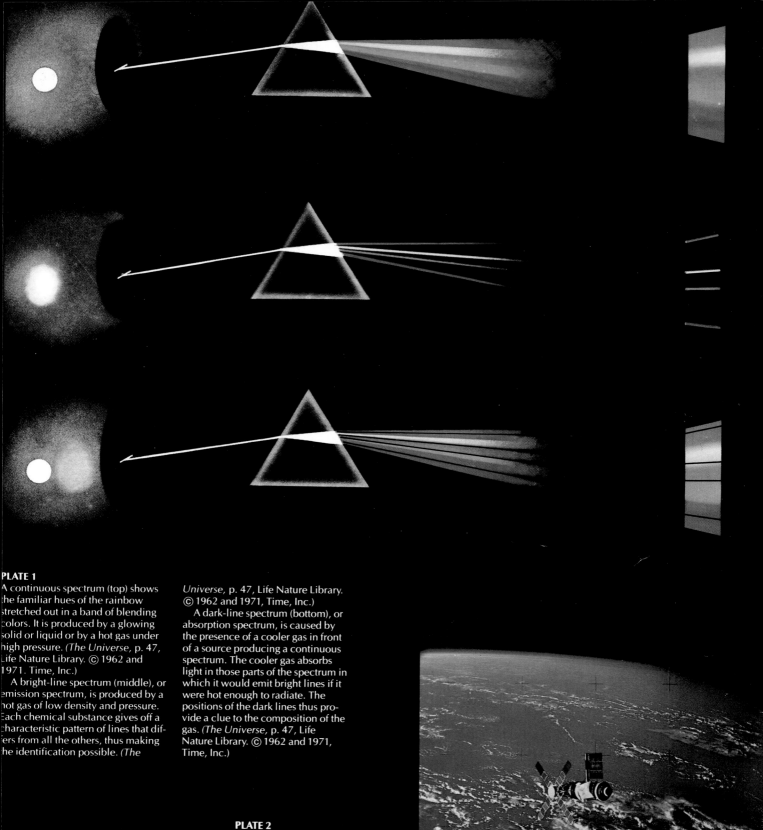

PLATE 1

A continuous spectrum (top) shows the familiar hues of the rainbow stretched out in a band of blending colors. It is produced by a glowing solid or liquid or by a hot gas under high pressure. *(The Universe,* p. 47, Life Nature Library. © 1962 and 1971. Time, Inc.)

A bright-line spectrum (middle), or emission spectrum, is produced by a hot gas of low density and pressure. Each chemical substance gives off a characteristic pattern of lines that differs from all the others, thus making the identification possible. *(The Universe,* p. 47, Life Nature Library. © 1962 and 1971, Time, Inc.)*

A dark-line spectrum (bottom), or absorption spectrum, is caused by the presence of a cooler gas in front of a source producing a continuous spectrum. The cooler gas absorbs light in those parts of the spectrum in which it would emit bright lines if it were hot enough to radiate. The positions of the dark lines thus provide a clue to the composition of the gas. *(The Universe,* p. 47, Life Nature Library. © 1962 and 1971, Time, Inc.)

PLATE 2

Skylab space observatory as seen from the command module during the second mission in June 1973. Both clouds and stars can be seen in the distance. (NASA)

PLATE 3
Earth, photographed from *Apollo 16* (April 1972), showing extensive cloud cover. Much of Mexico and the southern half of the United States, however, are clearly visible. (NASA)

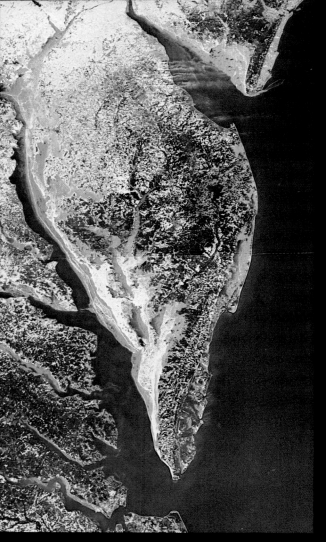

PLATE 4
Landsat photograph of the frozen Chesapeake Bay. Two images recorded in February 1977 were used to produce this picture ice patterns along the western shore of the Delaware-Marylan Virginia Peninsula. (NASA)

PLATE 5
Surface of the moon. Scientist-Astronaut Harrison H. Schmitt shown examining a huge, split boulder at the Taurus-Littrow landing site of *Apollo 17*. (NASA)

PLATE 9
Jupiter as seen by *Voyager 2* in June 1979. This picture, taken from about 24 million kilometers, shows the Galilean satellite Io to the right; the shadow on the left of Jupiter is Ganymede, the largest of the satellites. Clearly visible are the intricate and colorful cloud patterns in the atmosphere of Jupiter. (NASA)

PLATE 10

Volcanic eruption on the Galilean satellite Io. This view, taken by *Voyager 1* on 4 March 1979 from about 490,000 kilometers, shows an enormous volcanic eruption silhouetted against the blackness of interstellar space. Although the intensity of the volcanic plume has been exaggerated by a computer, the color and size have been preserved. This erupting volcano is one of several observed by *Voyagers 1* and *2*. (NASA)

PLATE 11

Saturn, photographed on February 19, 1973, with the Catalina Observatory telescope. (NASA)

PLATE 12
Total eclipse of the sun, photographed on
July 10, 1972, by David Baysinger at Tuk-
toyaktuk, Northwest Territories, Canada.
(David Baysinger)

MINOR MEMBERS

The *asteroids*, or minor planets, are rocky bodies whose diameters vary from a few between 150 kilometers and 1000 kilometers down to thousands less than a kilometer across. Most asteroids are found between Mars and Jupiter, traveling around the sun in the same direction as the planets. However, many of them orbit the sun in the vicinity of the earth's orbit, with some in fairly elliptical orbits.

The *meteoroids* range in size from irregular solid bodies, called *meteorites* when they strike the ground, to tiny particles, called *meteors* if they merely flash through the atmosphere. As we go down the scale in size, the number of meteoroids increases very rapidly. All the meteoroids are satellites of the sun and are apparently moving in a wide variety of orbits, as best we can determine.

Unlike the planetary bodies, most *comets* move around the sun in highly eccentric orbits with very long periods of revolution and at all angles of inclination. There are, however, some comets with short periods that are regular visitors to the vicinity of the earth. Their small masses mean that it is possible for the larger planets, Jupiter in particular, to alter the orbits of comets. We believe a comet is an icy conglomerate of light molecular compounds mixed with meteoritic matter.

The *interplanetary medium* is primarily gas particles—mostly protons and electrons—that are ejected from the sun's atmosphere at several hundred kilometers per second. These subatomic particles form the *solar wind*. Some dust is there too, most of it being cometary debris. Despite huge numbers of gas and dust particles, interplanetary space has fewer bits of matter and is a better vacuum than can be made in a terrestrial laboratory.

All around me:
  planet, moon, sun, riverbed, marsh:
    grew out of cataclysms galore;
  nothing ever sprang whole, stays put.
  I feel the earth beneath my feet
    suddenly shale away....

Diane Ackerman

OBSERVING THE PLANETS

Very little naked-eye study of the planets with a telescope is done by astronomers these days. The primary use of a telescope is for either direct photography or photoelectric work, because these techniques leave a record that we can study as needed. But sometimes these instruments do not record minute surface details sufficiently well, because of blurring from the unsteadiness of our own atmosphere. In some cases, and when observing conditions are excellent, eye observation can be more effective for determining specific surface details.

Photography through color filters, which restricts the light to a narrow spectral region, and conventional spectroscopy give us clues about the chemical composition of a planet's surface, clouds, and atmosphere. With Polaroid filters and other devices for measuring the polarization of light, astronomers can analyze a planet's surface and atmosphere by the manner in which reflected sunlight is polarized. To understand the significance of polarization of light in this context, you should remember what happens when sunlight reflected from, say, a car windshield is observed through Polaroid sunglasses. The Polaroid lens does not transmit all the reflected light that is polarized and thus the glare is less.

Analyzing a planet's light spectroscopically provides us with information about its atmosphere. Atmospheric constituents may be revealed by absorption lines or bands that are superimposed on the spectrum of the sunlight reflected from within the atmosphere. These planetary absorption features are sometimes difficult to separate from very similar absorption lines originating in the sun's atmosphere. *Thermal radiation* from the planets, which is due to the fact that the planets are hot, can be studied far into the infrared with today's heat-sensing instruments. This radiation is *blackbody radiation*, caused by the random thermal motion of the particles that compose the outer parts of the planet. The data obtained from infrared radiation provide important additional information on the surface and atmospheric temperatures and, indirectly, its chemical composition.

The planets' normal thermal radiation and any nonthermal radiation present are also observable in the millimeter, centimeter, and meter wavelength regions with radio telescopes. *Nonthermal radiation* is radiation due to physical processes other than that involved in producing blackbody radiation (see page 181). For example, the light produced in lightning is nonthermal radiation. Pulsed signals sent and received by radar have given us relief maps of Mercury, Venus, and Mars. This is all-important and somewhat remarkable for Venus since a thick cloudy atmosphere prevents visual observations of the surface. The signals are reflected from the planet's surface and return to the earth as an echo modified by the surface terrain (see Figure 7.4). The returning wave train from different portions of the surface is analyzed by a computer and then can be used to construct a topographic map of the planet's surface.

Analyzing the Doppler shift (Section 4.1) in the reflected radar signal caused by the rotation of the planet gave us the period of rotation for Mercury and Venus, which cannot be found accurately for Venus by optical methods because of its cloud cover. The radio wave reflected from the part of the surface that is approaching the earth has its frequency slightly increased because of the Doppler effect. The echo from the side of the planet that is receding from us is slightly decreased in frequency.

SPACE EXPLORATION

It is difficult in a few short paragraphs to do justice to the drama, excitement, and intellectual achievements of the space program (see Figure 7.5). We have all become somewhat blasé about its accomplishments. But we should remember that only a little over thirty-nine years have elapsed between the discovery of Pluto and *Voyager 1* and *2*'s close-up views of Jupiter and its major satellites. It has been a little less than twenty years since the first Venus probes by Russia in 1961 and the United States in 1962 propelled astronomy into the realm of physical exploration of the planets. In that interim we have landed spacecraft on Mars and Venus and carried out experiments on their surfaces. We have brought back samples from the moon and could do so from Mars,

FIGURE 7.4

Mapping by radar. At left the signals are shown bouncing off a smooth surface. Only the center of the curved surface reflects a strong echo (colored arrow) back to earth. Where the surface curves away, the rays bounce off at an angle, so that the planet's edges return only weak signals to the earth. If, however, the radar rays bounce off irregularities (right diagram) such as mountains, the patterns are changed. In what is otherwise an area of weak echoes, strong echoes bounce back from a surface projecting outward, as shown on the left of the diagram. Radar rays striking a surface projection at an oblique angle, as shown to the right, bounce their strongest signals off to one side, and only weak scattered signals return to the earth.

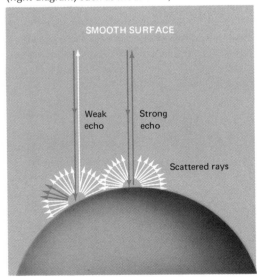
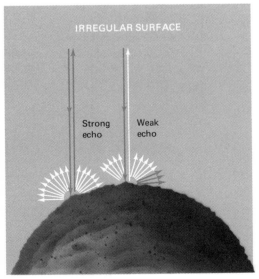

asteroids, and comets. We have photographed and mapped the surface of Mars and are in the process of mapping the surface of Venus with an orbiting radar system. This is in spite of our inability to actually see the surface of Venus because of the cloud cover. (A more extensive discussion of these space accomplishments is given in the next two chapters.)

Spacecraft are the most promising instruments of the future for planetary research. These include instrumented flybys, entry probes, orbiters, and landers. For example, they can carry infrared and ultraviolet spectrometers to send back clues on chemical composition, cloud structure, atmospheric and surface densities, pressures, and temperatures. Television cameras have been developed for spacecraft that can produce color images of the surface at close range in different wavelengths. Also, magnetometers can be included in order to record the magnetic field strength of the interplanetary medium as well as that of the planet. Spacecraft also carry particle detectors for measuring the energies and numbers of subatomic particles in the vicinity of the planet. From the influence of the planet on the spacecraft's orbit, we can derive information on the mass and internal structure of the planet. Future space probes should yield new insights into the physical, chemical, and possibly biological properties of the planets and help us better understand how our planetary system formed and evolved.

Today and for some time into the future the most practical method we have for launching a spacecraft toward a planet is by chemical propulsion. Just to escape from the earth, the rocket used to launch the spacecraft will consume almost all its propellant. Then the spacecraft becomes a satellite of the sun as it is directed into a long, unpowered, coasting orbit—a *transfer orbit*—toward an encounter with the planet. The *minimum-energy orbit* is the one that uses the least amount of rocket propellant, a precious commodity, to inject the spacecraft into its transfer orbit. Deviations from the minimum-energy orbit may be desired in order to reach the planet sooner, depending on the supply of propellant and the payload weight. A small amount of fuel is kept in reserve to provide midcourse corrections for a more precise trajectory while the vehicle is en route to the planet or for an orbit around the planet or for a soft landing.

Aiming a spacecraft, such as *Voyager 1* and *2*, toward a planetary target on a long, curved, coasting path (the transfer orbit) is a delicate and difficult task. Both the earth and the planet are moving around the sun at different speeds and in different directions. The problem is similar to that of a hunter on a rotating merry-go-round shooting at a fast-flying duck. The planet and the earth must be in the correct relative positions when the rocket firing takes place, which must be in the proper direction and at the correct velocity. *Voyager*'s trip to Jupiter was approximately twenty months, yet the spacecraft arrived within seconds of its projected schedule.

For the flight to Venus the departure must be opposite to the earth's orbital motion to curve the path inward toward Venus. This permits the vehicle's orbital velocity, which is acquired from the earth's orbital velocity at escape time, to be reduced by the velocity required to escape from the earth. The resultant velocity of about 19 kilometers per second relative to the sun causes the craft to fall toward the sun to encounter Venus about four months after launch.

For the flight to Mars the departure must be in the same direction as earth's orbital motion. The resultant velocity relative to the sun after escape from the earth is approximately 40 kilometers per second. This forces the vehicle to swing outward from the sun along its transfer orbit, to encounter Mars about seven to eight months after launch. The spacecraft may then be placed in an orbit around Mars (or any other planet) by firing retro-rockets to slow the spacecraft, allowing it to be captured by the planet as a satellite.

PLANETARY SURFACES AND INTERIORS

We can examine directly only the surfaces of the moon, Mercury, Mars, and several of the larger satellites because of their transparent atmosphere or lack of one. These studies show that intense bombardment by meteorites and planetesimals that apparently pounded the terrestrial planets some 4 billion years ago left them with a heavily cratered terrain. In addition, surface features on Mercury, Mars, and the moon are not symmetrical. For example, Mercury and the moon have more maria on one side than the other, which is a great puzzle. The northern hemisphere of Mars has few craters and is somewhat more depressed than the southern hemisphere, which is elevated and heavily cratered—another mystery. Cratered terrain is generally believed to be the oldest type of surface on a terrestrial planet. Its preservation implies a relatively stable crustal history. Volcanic material can only reach a planet's surface to develop plains and other formations if the crust can be broken, allowing the lava to

FIGURE 7.5

Space observatories for the exploration of the cosmos.

Top row:
OAO-Copernicus satellite was placed into earth orbit in August 1972. The large flat surfaces on either side of the vehicle are solar panels that convert incident sunlight into electrical power. The satellite was used for exploration in the ultraviolet wavelengths.

Voyager spacecraft experiments are shown in this drawing. Each *Voyager* spacecraft will use eleven different instruments as shown for studies of the Jovian planets, including extensive photographic surveys, the satellites of Jupiter, the rings of Saturn, the magnetosphere surrounding Jupiter, and interplanetary space.

Center row:
Viking I's lander touched down on the surface of Mars on 20 July 1976. Its array of scientific instruments varied from meteorological equipment to life-detection equipment. The work of the landers from *Viking I* and its sister ship *Viking II* have immensely extended our knowledge of the surface and atmosphere of Mars.

The *High Energy Astronomy Observatory* (HEAO) project of NASA includes three earth-orbiting satellites. The satellites launched between 1977 and 1979 were designed to study some of the most perplexing astronomical mysteries —pulsars, black holes, supernovae, and quasars. These three space observatories are designed to collect data in the gamma ray, X-ray, and ultraviolet regions of the electromagnetic spectrum, as well as on cosmic ray particles.

Project Galileo spacecraft is shown in this artist's conception. The mission is proposed for 1982, and would include both an orbiter vehicle to be put into orbit of the giant planet and a probe to be launched to enter Jupiter's sunlit hemisphere.

Pioneer Venus mission as conceived by an artist. The orbiter and the multiprobe were launched in the summer of 1978, and arrived at Venus in late December of that year. The mission has been very successful in enhancing our knowledge of the cloud-shrouded Venus.

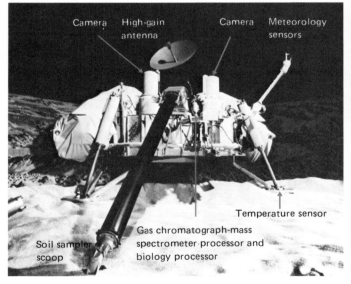

Bottom row:

Space Telescope as visualized by an artist being placed into earth orbit by the *Space Shuttle Orbiter.* The *Space Telescope* is an orbiting astronomical observatory with a 2.9-meter aperture telescope as the principal research instrument. It is scheduled to be used as a national observatory with astronomers able to come to a ground station to operate the telescope just as in an earth-based observatory.

Infrared Astronomical Satellite (IRAS) is an earth-orbiting observatory for infrared studies of the cosmos. It is scheduled to be launched in mid-1981, into a 563-mile high polar orbit. IRAS is an international program between the United Kingdom, the Netherlands, and the United States.

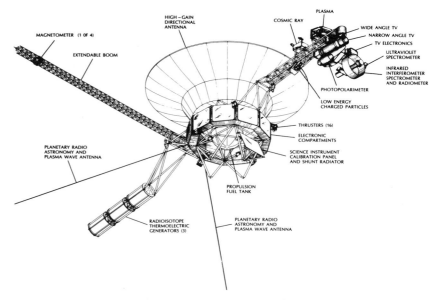

MAGNETOMETER (1 OF 4)

EXTENDABLE BOOM

HIGH – GAIN DIRECTIONAL ANTENNA

COSMIC RAY

PLASMA

WIDE ANGLE TV

NARROW ANGLE TV

TV ELECTRONICS

ULTRAVIOLET SPECTROMETER

INFRARED INTERFEROMETER SPECTROMETER AND RADIOMETER

PHOTOPOLARIMETER

LOW ENERGY CHARGED PARTICLES

THRUSTERS (16)

ELECTRONIC COMPARTMENTS

SCIENCE INSTRUMENT CALIBRATION PANEL AND SHUNT RADIATOR

PLANETARY RADIO ASTRONOMY AND PLASMA WAVE ANTENNA

PROPULSION FUEL TANK

RADIOISOTOPE THERMOELECTRIC GENERATORS (3)

PLANETARY RADIO ASTRONOMY AND PLASMA WAVE ANTENNA

seep onto the surface. Of the terrestrial planets the earth shows far and away the greatest volcanic activity, with Mars next, followed by Mercury and the moon. Somewhat surprisingly, Jupiter's satellite Io also shows current volcanic activity. The limited radar data for Venus point to a primarily cratered terrain more Mars-like than earthlike. Thus the surface of the earth has evolved more than the surfaces of other terrestrial planets, but they do show evidence of some evolution.

We construct models of planets' interiors from their observed physical properties and from theory. The physical facts are their known masses, densities, shapes, rotation rates, gravitational and magnetic field strengths, surface temperatures, and chemistry. Our models are naturally more precise when we have seismic data and rock samples, as we have for the earth and the moon.

The mean density of a planet, combined with other physical data, supplies clues to its internal chemistry and distribution of mass. From these we can construct a theoretical model of a planet's interior, starting with these assumptions: (1) the planet has a stable configuration—it is neither contracting nor expanding; (2) the downward weight of material caused by gravity is balanced by upward pressure from the compressed material beneath. When we have constructed the model, we can use it to help us predict how the temperature, pressure, and density vary from the planet's center to the surface. Examples of some interior models for the terrestrial and Jovian planets are shown in Figure 7.6.

PLANETARY ATMOSPHERES

We are all aware of the importance of the earth's atmosphere in maintaining life. Mercury and Pluto, in contrast, have almost no atmosphere. How did the planets' atmospheres get their diverse physical characterisics? These characteristics result from several factors:

1. The planet's distance from the sun along with its size and mass, which influence its ability to retain an atmosphere.

2. Its chemical composition, which determines what processes go on in the atmosphere.

3. Geological and chemical evolution of the planet's surface layers.

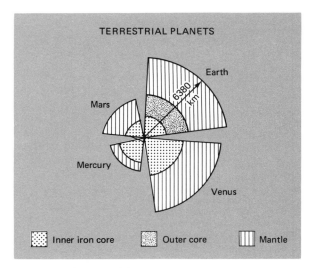

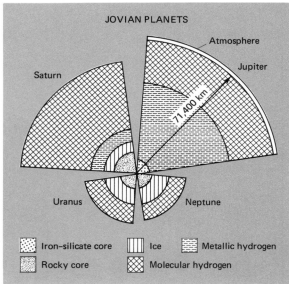

FIGURE 7.6
Models of planets' internal structure. The cross-sectional views of the terrestrial and Jovian planets are drawn to different scales.

4. The atmosphere's interaction with biological life, if living organisms are, in fact, present.

The first factor can be explained as follows. The higher the temperature or the smaller the mass of a molecule (or both), the greater the average velocity of gas particles in the atmosphere will be. A particle moving outward might gain enough velocity after colliding with another particle to escape into space.

When can a gas escape the planet's gravitational field? We make calculations by using our knowledge of the planetary temperatures, masses, and radii, and the masses and thermal velocities of their atmosphere's constituents, and then we come up with some conclusions. If a molecule's velocity is near a third of the velocity needed for escape, about half of that chemical species will escape from the atmosphere within weeks. For a planet to preserve its atmosphere indefinitely, the mean velocity of the gases must be less than a tenth of the velocity of escape.

The massive Jovian planets with their large escape velocities (20 to 60 kilometers per second) have held their primeval atmospheres of hydrogen and helium, while the less massive terrestrial planets with smaller escape velocities (2 to 12 kilometers per second) have lost these light gases. Venus, Earth, and Mars have managed to retain atmospheric water molecules, whose molecular weight is 18, as well as some heavier gases. Mercury, the moon, and Pluto (apparently) lack appreciable atmospheres. The surprising scarcity of the noble gases such as neon, argon, krypton, and xenon, in spite of their abundance in the sun, suggests that the terrestrial planets did not retain their original atmospheres. A secondary atmosphere for Venus, Earth, and Mars was apparently formed out of gases escaping from their interiors through volcanic action.

The second factor affecting a planet's atmosphere is the atmosphere's chemistry, which is very different for the terrestrial from that of the Jovian planets. If the planets were formed by planetesimals accreting in a contracting solar nebula (most believe they were), we would expect those originating closer to the sun to lose more of the highly volatile materials—such as hydrogen, helium, methane, ammonia, and water ice—than those formed farther out where temperatures were lower. Thus the terrestrial planets would have formed with larger percentages of the refractory materials—such as silicates and metal oxides—than the Jovian planets. The accreted planetesimals in the outer solar system that formed the Jovian planets would have retained more of the various volatiles in addition to large amounts of hydrogen and helium. For example, if the ratio of the abundance of hydrogen to that of silicon were the same in the earth as in the sun, then the mass of the earth would have ended up being about the same as Saturn's. Planets like Jupiter and Saturn are more nearly like the sun in chemical composition than like the terrestrial planets.

The third factor, the ways geological and chemical evolution affect a planet's atmosphere, is important in the case of the Earth, Venus, and Mars. Outgassing from these planets' interiors in their early history consisted mainly of water vapor, carbon dioxide, and nitrogen, in approximately the same proportions as we observe in volcanic gases today. On the earth the water vapor condensed to form the oceans but nitrogen remained in the gaseous state. Most of the carbon dioxide combined with the silicate rocks of the earth's crust to form the carbonate rocks such as limestone or calcium carbonate, a reaction that occurs most efficiently in the presence of liquid water. If it could be released from earth's rocks along with the small amount in the oceans, carbon dioxide in the earth's atmosphere would equal the amount in the dense atmosphere of Venus. Also, if water vapor at some earlier time did condense on Venus, the high surface temperature due to the greenhouse effect prevented water from remaining in a liquid form, keeping the carbon dioxide in its atmosphere. Most of the water vapor apparently dissociated into hydrogen and oxygen by absorbing ultraviolet sunlight, while a small amount may have combined with volcanic sulfur products to produce the sulfuric acid droplets found in Venus's atmospheric clouds. Hydrogen escaped, and the heavier oxygen may have combined with crustal material to form oxides. On Mars the outgassing of water vapor, carbon dioxide, and nitrogen was less complete than on the earth. Carbon dioxide forms the largest part of the atmosphere of Mars; argon is also present. Mars appears at present to be in a cold phase, and a large amount of water apparently is stored beneath the frozen carbon dioxide polar caps and elsewhere as permafrost.

Finally, we need to consider a planet's biological life, if any. Living organisms on a planet are bound to affect its atmosphere if the interaction between its biosphere and atmosphere is anything like that on earth. Large expanses of liquid water will moderate a planet's climate and can provide an environment conducive to the development of life if there is adequate protection from ultraviolet solar radiation. Most of earth's free oxygen, so necessary to animal life, comes from photosynthesis (see Section 10.1). It is constantly replenished by green plants, plankton, and some bacteria. When living organisms became able to extract carbon dioxide from the atmosphere, they helped save the earth from the heat death that Venus has apparently experienced.

The biological picture on Mars has been drawn from the biochemical experiments on the *Viking* landers, as we will discuss later. Since Jupiter's atmosphere contains large amounts of H_2, CH_4 and NH_3, the lower, warmer atmospheric levels may be suitable for synthesizing the complex carbon molecules found in living organisms. Thus within another decade entry probes may detect simple organisms in the warmer regions of Jupiter's atmosphere.

Even at this early stage, exploration of the solar system has already paid dividends in the solution of terrestrial problems. For example, we now have a better understanding of the effect of variations in the solar wind on the earth's magnetosphere, short wavelength electromagnetic and particle radiation on the level of ionization of atmospheric constituents, and the solar constant on the earth's total environment. These factors will undoubtedly become an essential ingredient in forecasting terrestrial weather. The full impact of our increased knowledge and understanding of the solar-terrestrial connection, the interplanetary environment, and the comparative study of the planets on earth's problems is still to be realized. It is not an accident that we are here on the earth, rather than on Mars or Venus, and we would do well to recognize this fact when we tamper with our environment.

7.3
Origin of the Solar System

ARCHITECTURE OF THE SYSTEM

One of the most challenging questions the astronomer can ask is: "How did the solar system begin?" Two kinds of theories have been proposed in the past: an accidental catastrophic event such as a near collision between a star and the sun, or a natural noncatastrophic event such as nebular condensation. A

historical look at the ways of explaining planetary genesis is a good beginning, but first let us re-examine the arrangement of planets in the solar system.

A progressive sequence of natural forces evidently created and shaped the system's destiny somewhat along the lines revealed by the following facts. These clues suggest that the solar system's design most likely did not materialize through a sequence of unrelated, random events.

1. The planets are well-isolated from each other without bunching, and they are placed at orderly intervals as Table 7.2 shows.

2. The planets' orbits are nearly circular except for those of Mercury and Pluto.

3. Their orbits are nearly in the same plane (Mercury and Pluto again excepted.)

4. All the planets and asteroids revolve around the sun in the same direction as the sun rotates (from west to east).

5. Except for Venus and Uranus, the planets rotate around their axes in the same direction as the sun turns.

6. The terrestrial planets have high densities, a relatively thin or no atmosphere, slow rotation, and few or no satellites, points that are undoubtedly related to the fact that they occupy the inner regions of the solar system.

7. The giant planets have low densities, relatively thick atmospheres, many satellites, and rapid rotation, which derive from the fact that they are in the outer regions of the solar system.

8. With some exceptions, the satellites revolve around their primary in the same direction as the primary rotates.

9. Studies of chemical composition suggest that the temperature at which the material formed producing the planets decreased with distance from the sun.

NEBULAR HYPOTHESIS

German philosopher Immanuel Kant speculated in 1755 that the solar system was formed out of a huge, rotating gaseous nebula slowly contracting and condensing. Pierre Laplace (1749–1827) celebrated French mathematical astronomer, expanded the idea in 1796, and it became the *nebular hypothesis*. Laplace theorized that as the large, slowly rotating solar nebula contracted, it rotated faster and faster, which meant it flattened into an equatorial ring. Involved here is the principle of *conservation of angular momentum,* which requires a spinning body to rotate faster as it shrinks. The angular momentum of a rotating body, a measure of its quantity of rotation, remains constant unless energy is taken out of rotation and put into some other form. If the radius decreases, the rotational velocity must increase to compensate for the reduced radius in order to conserve angular momentum. This is what we observe when a spinning ice skater rotates faster as outstretched arms are brought close to the body.

Laplace supposed that when the centrifugal force acting on the outer rotating edge of the solar nebula exceeded the inward gravitational force of the nebular mass, a ring of gaseous matter was split off, eventually coalescing into a planet. The splitting process repeated itself, making concentric rings that formed into planets, while the central portion condensed to become the sun. The satellite systems might have materialized within the ring structure before the planets were fully developed, and the asteroids could be the shattered remains of a disrupted ring.

The theory has two major defects. First, 98 percent of the angular momentum of the solar system resides in the planets' orbital motions. The central mass could not have transferred this much momentum to the planets. Second, a hot gaseous ring of this type would disperse into space, not pull itself together gravitationally to form a planet.

ENCOUNTER THEORIES

At the beginning of this century, attempts to reconcile the nebular hypothesis with physical principles were temporarily abandoned. A different approach, the so-called *encounter theory,* was first speculated on in 1745 by the French naturalist Georges Buffon (1707–1788), when he proposed that material that was ripped off from the sun by a collision with a comet condensed into the planets. The theory was revived in a different form early in this century by the American geologist Thomas Chamberlin (1843–1928) and the American astronomer Forest Ray Moulton (1872–1952) who together developed this theory. They suggested that giant eruptions were pulled off the sun by the gravitational attraction of a passing star. The stellar visitor imparted a curved lateral motion to the strung-out hot material that later fragmented and solidified into small bodies (planetesimals), which coalesced into planets by collisional accretion. Somewhat later another geologist-astronomer pair, Harold Jeffreys (1891–   ) and James Jeans (1877–1946) in England concluded that such an ejection could not have taken place. They theorized that a cigar-shaped gaseous filament was pulled out of the sun by the sideswiping action of a passing star (see Figure 7.7).

FIGURE 7.7
One version of the encounter theory.

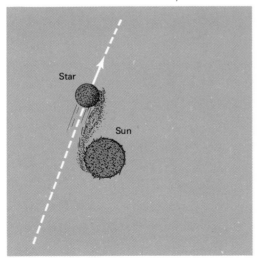

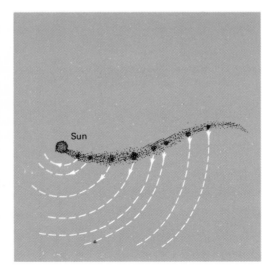

The middle section condensed into the Jovian planets and the ends into the smaller planets.

Both theories account for the common direction of the planets' orbital motion and the sun's rotation, as well as for the planets' nearly circular and coplanar orbits. The encounter theory in either version, however, has serious failings. The ejected planetary material could not have acquired sufficient angular momentum nor would the hot gas have condensed into planets. Besides, the probability of a near encounter in our region of the Galaxy is vanishingly small, less than one in many millions.

PROTOPLANET THEORY

By midcentury astronomers once more turned their attention to possible improvements in the nebular hypothesis. A new factor was introduced in the form of the existence of a small amount of dust in the cool gaseous nebula, providing nuclei for the condensation of gas particles into larger aggregates that could accrete and solidify into the embryo planets. (The existence of dust particles in the interstellar gas clouds out of which stars are formed was accepted in the 1930s). This modern version of the nebular hypothesis is called the *protoplanet hypothesis*. It was first formulated independently by Carl von Weizsacker (1912–   ) and Gerard Kuiper (1905–1973) in 1945, then extended and modified over the years by others.

The hypothesis begins with a fragment separating from an interstellar cloud composed mainly of hydrogen and helium, with trace amounts of the other elements. With other fragments of the interstellar cloud presumably following a similar evolution, its central region, being somewhat more dense, collapsed more rapidly than its outlying parts. This formed the central portion of the *solar nebula*, whose outer portion contained a thin disk of solids within a thicker disk of gases. The solar nebula grew by accretion as material continued to fall inward from its surroundings (see Figure 7.8). Large-scale turbulence from gravitational instabilities ruptured the thin disk into eddies, each containing many small particles (see Figure 7.9). These particles gradually built up into larger bodies by accretion. Repeated encounters between them resulted in accretion of still larger aggregates called planetesimals, orbiting the center of the solar nebula. Mutual gravitational disturbances led to further encounters and gradual coalescence into the planets. As the central portion of the solar nebula contracted, the temperature rose to around 2000 K,

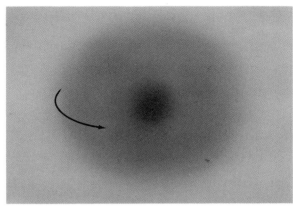

The central region of an interstellar cloud fragment collapses faster than the rest.

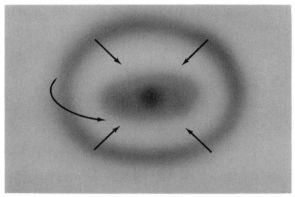

This fragment forms a small solar nebula: a central mass that is not yet the sun, surrounded by a disk of gas and dust grains, with more gas and dust concentrated around the periphery.

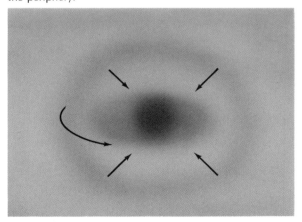

The small nebula then grows by accretion over a long time.

FIGURE 7.8
Protoplanet theory: accretion model of the solar nebula. Formation of the planets is shown in Figure 7.9.

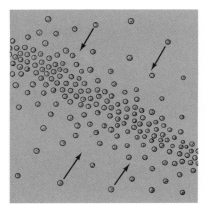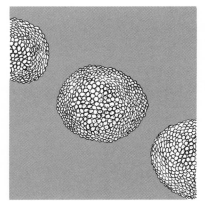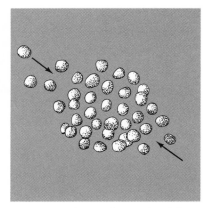

Planets begin to form when interstellar dust grains collide and stick to one another, forming ever larger clumps that fall toward the midplane of the nebula (left) and form a diffuse disk. Gravitational instabilities collect this material into millions of bodies of asteroidal size that collect into gravitational clusters (center). These clusters collide and intermingle (right).

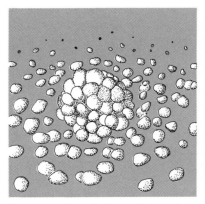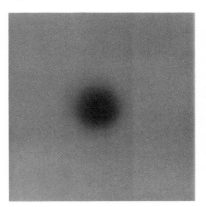

The clusters' gravitational fields relax, and they coagulate into solid cores (left); some bodies perhaps orbit the cores. Continued accretion and consolidation may create a planet-size body (center). If the core gets larger, it may concentrate gas from the nebula gravitationally (right). A large enough core may make the gas collapse into a dense shell that constitutes most of the planet's mass.

FIGURE 7.9
Protoplanet theory: the formation of planets.

hot enough to vaporize any compounds in the dust grains within the inner portion of the disk, while the outer disk remained relatively cool. Planets that formed closer to the center, such as the terrestrial planets, would be expected to lose more of the volatile compounds and thus be left with relatively more of the refractory compounds (the silicates and metal oxides) than those formed farther out, such as the Jovian planets, where the temperature was lower. The terrestrial planets and the cores of the giant planets accreted from condensed matter in the solar nebula.

During and following the formation of the terrestrial planets, there was a catastrophic bombardment of asteroidal-meteoric fragments that cratered

these planets. The impacting material, coupled with intense radioactivity and subsequent gravitational concentration, produced sufficient heat to melt and chemically differentiate the planets into their presently layered structure (core, mantle, and crust). The atmospheres of the terrestrial planets were formed afterward by outgassing from their hot interiors.

Within the outer, cooler regions of the solar nebula, the volatile substances such as ammonia and methane, largely in combination with water ice, formed thick condensates of complex mixtures, including organic compounds, around the small rocky cores of the Jovian planets. These bodies grew massive as they drew in more hydrogen and helium from the surrounding interplanetary gas. Because of their great mass, they have kept very nearly the same relative proportion of hydrogen and helium to the heavier elements as the sun and the interstellar medium have. The comets are probably a fossil relic of the primordial ices that existed within the outermost regions of the solar nebula.

When the protoplanets were all formed, the nebula's central bulge rapidly collapsed into the protosun. Continued contraction raised its internal temperature from a few tens of thousands of degrees to several million degrees when the first stages of nuclear burning were initiated. (The nuclear burning processes will be discussed in Chapter 13.) In the last stages of formation the sun may have had a much more intense solar wind, which presumably blew away much of the primordial gas and dust left over from the original interstellar cloud. However, this point is still pretty much of a mystery.

A weakness in the protoplanet hypothesis is that it does not provide a completely satisfactory explanation for the observed distribution of angular momentum in the solar system. If the angular momentum of the planets could somehow be returned to the sun, its present slow rotation (like other stars similar to it) of 2 kilometers per second would be increased to about 100 kilometers per second. The primitive sun apparently transferred most of its rotational energy to the planets as they were forming. To explain this transfer of energy, astronomers have proposed a braking action caused by magnetohydrodynamic forces on the sun as its magnetic field interacted with the ionized nebular gas in the disk. The magnetic lines of force spiraling outward from the rotating sun into the surrounding nebula would act like a magnetic drag on the spinning sun and serve as conduits, transferring angular momentum to the planetary disk. Another possibility is that much of the sun's original angular momentum was carried off by the matter leaving the sun as part of a vigorous solar wind and was transferred to the protoplanets.

We have moved a small part of the way toward understanding how the solar system began. Further progress will come as geologists, mineralogists, and physicists study meteorites, lunar rock samples, and other planetary surfaces, and as astronomers develop improved theories on the formation of stars. For example, there is a recent discovery of a relatively high abundance of some rare—by earth standards—isotopes in primitive meteorites. It has been proposed that the isotope anomalies are due to the injection into the solar system of matter from a supernova explosion a few million years before the meteorites solidified. (A *supernova* is the explosion of a star in the last stages of its life.) Possibly the concussion from the explosion triggered the collapse forming the solar nebula. Regardless of the means of starting the formation process, planetary systems are believed to grow naturally from physical events that develop after an interstellar cloud has begun to contract. Finally, in recent years astronomers have discovered large, cool dust envelopes around infrared stars in the interstellar clouds of the Milky Way. Some of these objects may be in the early stages of nebular condensation visualized in the protoplanet theory.

> Comets are the nearest thing to nothing that anything can be and still be called something.
>
> *National Geographic,* 31 March 1955

SUMMARY

We study the planets to learn more about the origin and evolution of the solar system of which the earth is a part. How did two distinct families of planets, the terrestrial and the Jovian, evolve from a common origin, yet with each planet stamped with its own individuality? The dominant influence of the sun, over regions near and far, when the solar system was formed must have largely determined the fate of other members of the system. Diversity among the planets resulted mainly from their distances from the sun and from their differences in mass.

The terrestrial planets are Mercury, Venus, Earth, and Mars. They are the inner planets. Smaller and denser than the outer planets, they are rocky and metallic and rotate more slowly. Jupiter, Saturn, Uranus, and Neptune are the outer, Jovian planets; we know too little about far-off Pluto to fit it definitely in either category. The Jovian planets are large, low-density planets, made mostly of hydrogen and helium; they rotate rapidly.

Astronomers now generally believe that the solar system was born from a gravitationally contracting solar nebula whose central portion became the sun and whose surrounding disk of planetesimal fragments accreted into planets. The inner planets, because of their smaller mass and proximity to the sun, lost most of the volatile matter present when the solar system was formed. Their atmospheres apparently formed by outgassing from their interiors; because of its small mass, Mercury's atmosphere must have escaped. The Jovian planets, because of their larger mass and cooler environment, retained the light, abundant hydrogen and helium and the ices of water, ammonia, and methane. The icy comets are believed to have formed in the cool, outer fringes of the solar nebula; the asteroids from the fragmentation of planetesimals orbiting between Mars and Jupiter. The details of how a planet forms are still obscure, but we believe that planets have formed often in our Galaxy.

REVIEW QUESTIONS

1. Describe briefly the techniques employed in studying the planets.

2. How is Bode's law of planetary distances arrived at?

3. Why is the numerical value of the astronomical unit so important? How is it derived?

4. What factors are responsible for the retention of a planet's atmosphere?

5. How could the presence of life alter the composition of a planet's atmosphere?

6. Discuss the differences between the terrestrial and Jovian planets in size, mass, rotation, and surface features.

7. How do the terrestrial and Jovian planets differ in composition and structure? How do we account for their differences?

8. Describe the satellite systems of the different planets.

9. List and briefly describe the minor members of the solar system.

10. What is there about the planetary system that informs us that it is not a chance aggregation of bodies or a "fortuitous concourse of atoms," as one astronomer put it?

11. Why are the atmospheres of the terrestrial and Jovian planets so different?

12. How did photosynthesis save the earth from the heat death of Venus?

13. What percentage of mass in the solar system does the sun have? What percentage of angular momentum? How do we account for these?

14. Describe in general terms the catastrophic and noncatastrophic theories of the origin of the solar system.

15. How does the protoplanet theory account for the two kinds of planets: terrestrial and Jovian?

16. Describe how a spacecraft is launched so that it can rendezvous with Mercury or Venus; with Mars or Jupiter.

17. Explain how a spacecraft on a flyby mission to Jupiter can be given a gravitational assist that will direct it to Saturn.

18. How do the orbits of the minor members compare with the orbits of the planets?

19. Why is Pluto considered an unusual planet?

20. Describe the gravitational contraction of an interstellar cloud to form a planetary system.

8

Inner Solar System:
The Terrestrial Planets

More has been learned about the planets and some of their satellites since the space age began in the late 1950s than in all of preceding history. It seems likely that we will have explored, as part of the space program, much of our solar system by the end of the century. Even though curiosity has been one of the prime motives that fueled exploration of the earth, it has been the abundant economic rewards that have maintained exploration. Similarly, our justification for the grand, but costly, enterprise of space exploration is not just curiosity but also usable knowledge.

By studying different planets, each with its own characteristics, we hope to learn how they have evolved to their present state through the same set of dynamic processes. The planets are the laboratories needed for observing processes beyond the range, in both time and extent, of our terrestrial environment. To better understand our own planet, we need a perspective that can only be acquired by comparative study of the other terrestrial planets.

8.1
Mercury

GENERAL PROPERTIES

Although it is one of the brighter objects in the heavens, Mercury is difficult to study from the earth because it is so close to the sun. Its maximum angular separation is only 28° on either side of the sun. Swift orbital motion keeps the planet visible low above the horizon for only a few days each year, immediately after sundown or before sunup. It is best seen when it is an evening star during March and April or a morning star during September and October for the same reasons as were given for the harvest moon (footnote 1 in Section 6.6).

Of the terrestrial planets, Mercury is the least massive and smallest in size. Mercury's rotation period is two-thirds of its orbital period; thus the planet completes three rotations during two orbital revolutions. This synchronization of its rotation and revolution, like that between the earth and the moon, is not ac-

◀ Sunrise on the planet Venus from the *Pioneer Venus* spacecraft in orbit about the planet.

cidental. It was apparently set up by the strong tidal pull exerted by the sun, which slowed the planet's spin, trapping it in a spin-orbit lock in the ratio of 3:2. As a result the sun takes 88 days after rising on the eastern horizon to cross Mercury's sky and set on the western horizon; meanwhile the planet completes one orbit of the sun.

SURFACE STRUCTURE

Observations from the earth had hinted that Mercury might look like our moon. The *Mariner 10* flyby on 29 March 1974 (and the two subsequent flybys on 21 September 1974 and 16 March 1975) showed that the planet does have some rugged terrain and is heavily cratered (see Figures 8.1 and 8.2). Over twenty-seven hundred useful pictures of Mercury were taken in the *Mariner 10* mission, covering about 50 percent of its surface, at resolutions varying from 100 meters to 5 kilometers. Although the surface of Mercury is remarkably similar to the surface of the moon, there are significant differences—differences that suggest a somewhat different surface evolution from that of the other terrestrial planets.

In general, the surface of Mercury is pockmarked with craters ranging from at least 100 meters up to about 1000 kilometers. Some of the bright craters have extensive ray systems like those on the moon. Twenty or so of its maria or basins are more than 200 kilometers across. Caloris, the largest basin (Figure 8.1), is 1300 kilometers wide, and its interior surface resembles the famous Orientale Basin on the moon (see Figure 6.18). The basins were probably created during bombardment early in the history of the solar system, followed by a period of vulcanism that filled them with lava. Mercury is also like the moon in a way not yet fully explained for either body: The craters cluster in one region and the maria in another.

But there are some conspicuous differences between Mercury and the moon. Craters on the lunar highlands are densely packed, with rims of young craters overlying old craters, and the mare regions are sharply bounded. On Mercury, by contrast, craters are often interspersed with relatively smooth plains, giving the terrain a speckled appearance. It has few craters as large as 20 kilometers to 50 kilometers across. Scarps or cliffs at least 3 kilometers high and often 500 kilometers long are distributed widely over the old and heavily cratered regions. They cut across plains and craters alike. Unlike the scarps on the

FIGURE 8.1
Caloris. The semicircle of cratered mountains in the left half forms the boundary of Caloris, which was the largest basin on Mercury photographed by *Mariner 10,* the first spacecraft to fly by Mercury in late March 1974. The ring of mountains is 1400 kilometers in diameter and up to 2 kilometers high. The basin floor consists of highly fractured and ridged plains.

FIGURE 8.2
Views of Mercury taken by *Mariner 10* from about 80,000 kilometers. A shadowed, eastward-facing escarpment at the top extends southward several hundred kilometers from the planet's northern limb. The frame shows a 600-kilometer wide portion of Mercury's surface.

moon, they may have been formed when the crust of Mercury cooled and shrank, like wrinkles on the skin of an old apple.

The atmosphere on Mercury is very tenuous and is constantly replenished by the solar wind. In fact, it is probably much like the outermost portion of the earth's atmosphere called the exosphere right down to Mercury's surface. This means there are probably no collisions to speak of between the atmospheric molecules. Helium and atomic hydrogen have been identified as its principal constituents. Other atomic or molecular species, if present, are insignificant in abundance. No evidence has been found for atmospheric modification of any landform.

INTERNAL STRUCTURE AND EVOLUTION

The values of the mass and radius imply that Mercury must contain a large fraction of iron, the only heavy element sufficiently abundant to account for the high mean density (see Figure 7.6). By analogy with terrestrial, lunar, and meteoritic chemical abundances, we presume that silicates of iron and magnesium are also prevalent.

One discovery by the *Mariner 10* mission was unexpected: Mercury has a detached bow wave close to the planet, caused by the onrushing solar wind colliding with the planet's magnetic field. From this it is apparent that Mercury has a magnetic field, which is about 1 percent as strong as that of the earth. The magnetic axis almost coincides with Mercury's axis of rotation.

The basic facts that must be accounted for in any evolutionary history for the planet Mercury can be briefly summarized in the following way. First, the surface, on the basis of *Mariner 10* data, appears to be the product of melting of at least the outer layers, if not the entire planet, accompanied by chemical differentiation. Second, the magnetic field is intrinsic to the planet and is most likely the result of an internal mechanism that continuously generates the field in much the same way as is the earth's (see page 107).

This is additional evidence for an iron core. Third, differentiation appears to have occurred very early in the planet's history, probably the first half billion years. Finally, the surface has been largely undisturbed by thermal and tectonic processes.

8.2
Venus

GENERAL PROPERTIES

Venus, the second closest planet to the sun, is a yellow color and it is the third after the sun and moon in brightness in our night sky. Like Mercury, Venus goes through all the lunar phases, as Galileo first observed in 1609. Since it has a larger orbit than Mercury, Venus swings outward more slowly from the sun as viewed from the earth, about 47°, or twice as far as Mercury. Venus remains visible as an evening star in the western sky or as a morning star in the eastern sky for weeks at a time. Although Venus comes slightly closer to the earth than Mars does, we cannot see surface or atmospheric features then because its dark hemisphere is turned toward us.

Venus's diameter, mass, and density are slightly less than those of the earth. Like Mercury and the earth, it probably has an iron-rich core because of chemical differentiation (see Figure 7.6). The very high surface temperature of 460°C is due to Venus having retained a large amount of carbon dioxide in its atmosphere. This gas is fairly transparent to incoming solar radiation, but it is opaque to the reradiated infrared energy from the planet's surface, which is the greenhouse effect. Thus thermal radiation is trapped, similar to the way it is on the earth, but on a much larger scale. The night surface temperature differs little from the daytime temperature, apparently because of a strong atmospheric circulation that transports heat efficiently to all parts of the planet.

Venus's rotation was a mystery that eluded solution by optical or spectroscopic observations because of its slow rate. But Doppler shifts noted in radar observations solved it: The planet rotates in a retrograde direction, with its axis of rotation inclined only 6° from the perpendicular to its orbital plane. (*Retrograde* here means a direction of rotation reversed from that of the other planets, excepting Uranus.) The period of rotation as determined from radar measurements is 243 days, 18 days longer than its orbital period. The Venusian day is 117 days long, with 58.5 days of sunlight and 58.5 days of darkness. Thus the sun rises on the western horizon and sets approximately twice during the Venusian orbit with respect to the earth. The planet is locked in a 4:1 synchronous rotation period relative to the earth, which means that with this spin-orbit relationship the same hemisphere of Venus faces the earth at every inferior conjunction. Venus may originally have had direct rotation with a period comparable to that of the earth, but the prolonged tidal influence of the earth and sun slowed the rotation past the point when its rotation period was equal to its orbital period. Then the planet began to rotate in the reverse direction until it became locked in resonance with the earth.

SPACE MISSIONS TO VENUS

The Soviet Union has launched twelve *Venera* spacecraft toward Venus since 1967, with ten of them descending through the atmosphere to the surface. *Venera 7* in 1970 and *Venera 8* in 1972 made the first successful, although short-lived, landings. *Venera 9* and *10* landed on the daylight side and transmitted data for about one hour before succumbing to the hostile environment. These two crafts were designed to study the atmosphere below about 65 kilometers and down to the surface. *Venera 11* and *12* landed in late December 1978 about 800 kilometers apart. Apparently no pictures were transmitted. However, both did transmit data on surface conditions for about 100 minutes before going silent.

The heavens themselves, the planets, and this center
Observe degree, priority, and place,
Insisture, course, proportion, season, form,
Office, and custom, in all line of order

Shakespeare

The United States has launched three Venus flybys: *Mariner 2* in 1964, *Mariner 5* in 1967, and *Mariner 10* in 1973. However, our most ambitious effort was the two *Pioneer Venus* craft launched on 20 May 1978 and 8 August 1978. The first member of the *Pioneer Venus* fleet was an orbiter, shown in Figure 7.4, which went into orbit around the planet on 4 December, with an orbit varying from as close as about 150 kilometers above the surface to as far away as 66,000 kilometers and a period of about 24 hours. The rest of the armada, which consisted of one large and three small probes and the mother ship that carried the probes, arrived on 9 December. They went immediately to the surface, as shown in Figure 8.3, to widely separated landing points in both the northern and southern hemispheres and on both the day and night sides.

The array of scientific experiments aboard the *Pioneer Venus* mission was extensive. In summary, the orbiter carried instruments to measure neutral particle and ion densities in the atmosphere, the planet's magnetic field, the atmospheric temperature, and the movement of clouds as seen in ultraviolet light, and radar to map the surface terrain.

FIGURE 8.3

Artist's conception of day probe having landed on the Venusian surface. This probe transmitted data from the surface for about 67 minutes before failing. At lower right is the approximate position at which the probes landed in December 1978.

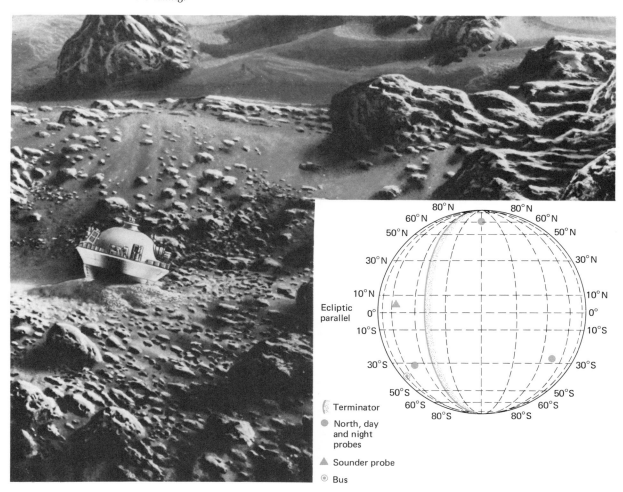

The probes carried instruments to measure the atmospheric structure—temperature, pressure, and density—as well as its chemical composition, cloud composition, heat balance, and wind velocities. Although none of the probes were designed to survive on the surface after impact, one small probe on the daylight side did survive one hour and eight minutes.

VENUS'S HIDDEN SURFACE

Venus's surface is hidden by a total cloud cover, whose visual markings observed from the earth are indistinct and transient. Thus in visible light the planet appears bland and featureless, with a light yellow color. However, in the ultraviolet part of the electromagnetic spectrum, the reflected solar radiation reveal Y-shaped cloud formations, as shown in Figure 8.4.

Earth-based radar observations of the hidden surface disclose some almost circular structures between 30 and 1000 kilometers in diameter, which may be impact craters and basins. Other large features include a 1400-kilometer-long trough, 280 kilometers wide and 4.6 kilometers deep, which is similar in scale to the Martian Valles Marineris (partly shown in Figure 8.13). Also detected is a large, low circular dome about 700 kilometers across, with a central depression 90 kilometers in diameter, similar in many respects to a volcanic peak. If truly a volcano, then it is about 25 percent larger than Olympus Mons on Mars (see Figure 8.14).

At the time of this writing, only the most preliminary data from the *Pioneer Venus* mission had been received and analyzed. However, fascinating new findings are already evident. Radar results from the orbiter suggest a topography on Venus similar to the earth's, with high mountainlike features and extensive flatlands. One northern hemisphere area appears to be

FIGURE 8.4
Venus photographed by *Pioneer Venus*. The top four photographs show the evolution of the dark, horizontal Y-shaped feature over two periods, 10 and 11 February 1979, left two images, and 19 and 20 February 1979, right. The Y feature is not a single feature, but a combination of recurring cloud formations. The clouds move rapidly, right to left or east to west, around the planet at speeds of about 100 meters per second. Convective motions produce the mottled appearance near the center of the images. The bottom four photographs were taken from 2 February to 2 May 1979.

a huge uplift, 3 to 5 kilometers high and three times the area of the Colorado plateau.

The two Soviet landers, *Venera 9* and *10,* sent back the first photographs of the surface on 22 and 25 October 1975. Sunlight filtering through the cloudy atmosphere supplies enough light to make the surface look like a dark overcast day on the earth. The *Venera 9* view in Figure 8.5 shows a rock-strewn region. Many of the stones are rather slablike, with conspicuously sharp edges. They are about 50 to 70 centimeters across and 15 to 20 centimeters high. This could be debris resulting from meteoric cratering or local crustal movements. The visible horizon is estimated at several dozen meters with possibly a cliff in the distance. This view roughly resembles Martian terrain. *Venera 10* landed about 2000 kilometers from *Venera 9.* The view is of a rather smooth surface with fewer rocks visible, whose edges seem to be more worn. The elevations that do show appear to be covered with a relatively darker, fine-grained sand and there are some suggestions of exposed bedrock. Apparently the surface possesses a variety of terrain and is still geologically active.

ATMOSPHERE AND CLOUDS

Measured atmospheric pressure on the surface of Venus is about ninety times greater than that of the air in your room. Analysis of the lower atmosphere suggests that it is about 96.4 percent carbon dioxide,

FIGURE 8.5
Venera 9 photograph of the first view of Venus's surface, showing rocky terrain (October 1975). Perhaps vulcanism formed these rocks; this would be in accord with radar views that show evidence of tectonic activity.

with 3.4 percent nitrogen, 0.14 percent water vapor, and traces of molecular oxygen, argon, neon, and sulfur dioxide. Because of the high surface temperature (about 730 K), the carbon dioxide was apparently not reduced, as it was on the earth, by reacting with the primitive rocks to form carbonates and limestones and by absorption by surface water. Above 150 kilometers, atomic oxygen is the most abundant species; other gases present are carbon dioxide, carbon monoxide, and molecular nitrogen. And finally, far above the atmosphere a huge cloud of hydrogen surrounds the planet.

More than almost any other aspect of Venus, the mysterious clouds that perpetually obscure the surface have been the subject of extensive speculation. In 1973 it was suggested that the clouds are composed of sulfuric acid droplets. In fact, it appears that more than five-sixths of the clouds' composition by weight is sulfuric acid. The clouds begin around 46 kilometers above the planet's surface and seem to be confined to a fairly distinct layer rising up to about 55 kilometers. Thin haze regions lie above and below the cloud layer. The lower one has a surprisingly abrupt cutoff some 32 kilometers above the surface.

From the *Pioneer Venus* results we think that the clouds are indeed composed of sulfuric acid droplets and other particles, possibly free sulfur, so thick that during the probes' descent they appeared to be passing through a blizzard. Early analysis of the data also suggests that a wide range of possible sulfur compounds such as sulfur monoxide, sulfur dioxide, carbonyl sulfide, and hydrogen sulfide also exist in the atmosphere. The falling probes found the atmosphere's particles and droplets to be grouped into several discrete size ranges and to abruptly disappear at about

46 kilometers above the surface. From the bottom of the haze at 32 kilometers, down to the surface, the atmosphere appears to be surprisingly clear.

The *Mariner 10* flyby, first to photograph Venus from space, passed within 5800 kilometers of its surface on 5 February 1974 before its encounter with Mercury. The atmospheric circulation that you see in Figure 8.4 is the same in both hemispheres. A vigorous equatorial east-west jet stream is quite evident in the upper atmosphere, moving around the planet in only four days, opposite to the direction of the planet's slow spin. The wind velocity decreases at lower altitudes until at the surface it slows to a gentle breeze. The lower atmosphere apparently circulates because of differences in solar heating between the equatorial and polar regions. Clouds rise near the equator, spiral toward the poles, and descend in what appears to be an almost continuous flow. But why such high winds reverse in the upper atmosphere we do not know. The planet does have an extremely weak magnetic field, possibly due to its slow rotation, which allows solar wind particles to penetrate well down into the atmosphere.

Finally, two of the *Pioneer Venus* probes parachuting into the dark side appear to have picked up a glow on the surface or in the low atmosphere. In what may be related to the American findings, *Venera 11* and *12* detected what Soviet astronomers believe are intense atmospheric lightning flashes. *Venera 11* data detected as many as 25 lightning flashes per second. The *Pioneer Venus* orbiter detected radio signals on the night side, which may be triggered by lightning flashes.

In spite of this wealth of new information, there is still far more we do not know than we do know about Venus.

8.3
Mars

GENERAL PROPERTIES

Mars is a little more than half the earth's size, has about 1/9 of the earth's mass, and therefore has a mean density about 3/4 that of the earth. The Martian day lasts 24 hours and 37.4 minutes. Its axis of rotation tilts from the perpendicular to its orbital plane by 24°, giving the red planet seasons like those of

the earth, but they last twice as long because the Martian orbital period is nearly 2 years.

When Mars is closest to the sun, the south pole is inclined toward it. As a result the large southern polar cap recedes during the Martian summer in the southern hemisphere leaving behind a residual cap some 300 kilometers in diameter. On the other side of the orbit, the north pole is inclined toward the sun when the planet is farthest from the sun. A somewhat smaller and warmer residual northern polar cap remains during the Martian summer in the northern hemisphere. No one yet knows why the southern residual polar cap is larger and colder than the northern one.

Mars has such an eccentric orbit that the closest approach at opposition between earth and Mars comes every fifteen or seventeen years when Mars is near perihelion in its orbit. At the last favorable opposition in August 1971, Mars came within 56 million kilometers of the earth. At a time like that, even the most casual observers of the heavens are struck by the planet's brilliant ruddy color, far outshining the brightest stars.

GROUND-BASED AND SPACE EXPLORATION

Visual observers have made countless detailed maps of the numerous and varied surface features of Mars. Most of the maps are marked with Greek and Latin names. Two observers are especially notable: the Italian astronomer Giovanni Schiaparelli (1835–1910) and the American astronomer Percival Lowell (1855–1916), who named many of the Martian features (see Figure 8.6).

Through the telescope from the earth the red planet appears to have earthlike characteristics, such as white polar caps and large dark areas, which vary with the Martian seasons.

Large grayish equatorial shadings streaked with darker mottlings also go through seasonal changes in color and intensity (see Figure 8.7). Once many astronomers thought these seasonal changes were simply growing and declining vegetation. However, the photographs transmitted to the earth by the *Mariner* and *Viking* flybys, orbiters, and landers between 1965 and 1976 revealed a waterless, cratered planet with some large extinct volcanos. They tell of a Mars that in many ways is similar but in many other ways is different from both the moon and the earth.

In the dry Martian environment the prevailing sea-

PERCIVAL LOWELL (1855–1916)

Percival Lowell was descended from a distinguished New England family. His younger brother, Abbot, became president of Harvard University and his sister, Amy, was a well-known poet and critic. After his graduation from Harvard in 1876, with distinction in mathematics, he traveled for a number of years throughout the Far East before settling down to a career in astronomy. He was particularly interested in Mars and its "canals," whose drawings by the Italian astronomer Schiaparelli had received wide public acclaim.

In 1894 he founded the Lowell Observatory at Flagstaff, Arizona. Its altitude of some 7000 feet and its dry desert air made it an excellent observing site for the study of Mars, which was then close to the earth. During fifteen years of intensive studies of Mars, whose surface markings he drew in intricate detail, there was one distinguishing feature on his maps: a network of several hundred fine straight lines he called "canals," crisscrossing in a number of oases. Lowell concluded that the bright Martian areas were deserts and the dark ones, patches of vegetation; and that water released from the melting polar cap would flow down the canals, supposedly constructed by intelligent Martians who once flourished on the planet, toward the equatorial region to revive the vegetation. Lowell published his views in two books: *Mars and Its Canals* (1906) and *Mars As the Abode of Life* (1908). The canals, as it turns out, are mostly chance alignments of dark patches that the eye, at the limit of resolution, tends to form together into lines.

Lowell's greatest contribution to planetary studies came during his last eight years, which he devoted to the search for a planet beyond Neptune. He first analyzed the discrepancies between the observed and the calculated positions of Uranus after making allowance for the perturbations of Neptune. But on examining photographs of the region of the sky where he predicted the missing planet might be, he found no such object. The search continued for a number of years after his death at Flagstaff in 1916; the new planet, named Pluto, was discovered by Clyde Tombaugh in 1930 with the newly acquired 13-inch astrograph (see page 191).

sonal winds periodically shift the dusty surface material to form the recurring patterns of light and dark markings. There is no evidence in the space mission photographs of a seasonal wave of soil darkening that would account for the telescopic observations. The fine delicate streaks called *canals*, sketched by observers on early Martian maps, are illusory. The *Mariner* and *Viking* pictures reveal these canals to be nothing more than dark-floored craters or irregular dark patches aligned by chance and linked unconsciously by the observer into a line that looked like a canal.

Viking 1 and *2* were two of the most sophisticated pieces of technical hardware in the space effort. Both arrived in the vicinity of Mars in the summer of 1976 after a ten-month journey. Each detached a lander from an orbiter, which set down on the surface of Mars and began to collect information. In late July 1978, the *Viking 2* orbiter was turned off because it could no longer be oriented in space. The remaining orbiter and the two landers continue to function at the time of this writing. However, it is expected that they will probably be turned off in late 1979 or early 1980. All outlived their expected useful life and provided us with a wealth of information about the Martian atmosphere, surface terrain, biological activity, and satellites. This information will continue to be analyzed for many years to come.

MARTIAN ATMOSPHERE

Carbon dioxide is the most abundant constituent of the thin Martian atmosphere, amounting to about 95 percent. We know the atmosphere also contains about 3 percent nitrogen, about 2 percent argon, lesser amounts of atomic and molecular oxygen, and traces of ozone, carbon monoxide, nitrogen oxide, neon, krypton, and xenon. Table 8.1 compares the abundance of the principal atmospheric constituents for Venus, Earth, and Mars. There are similarities between

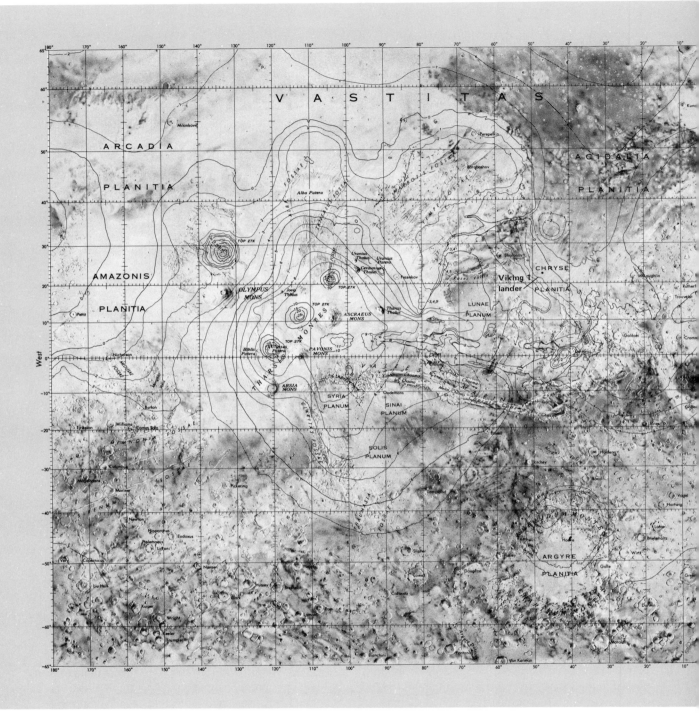

FIGURE 8.6
Topographical map of Mars. The photograph, taken by *Mariner 9,* was enhanced by computer.

Viking 1's landing site was 22.3°N
48.0°W; *Viking 2*'s was 48°N 226°W.

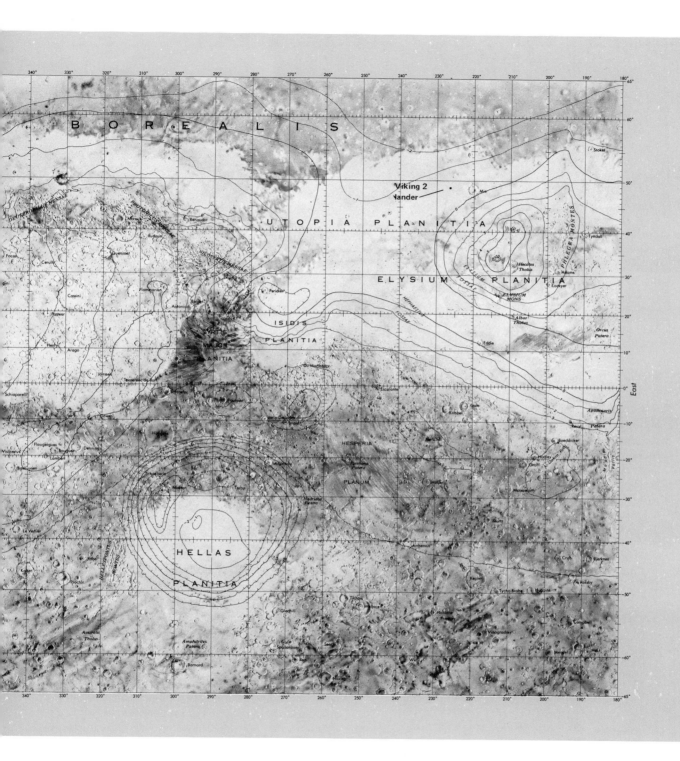

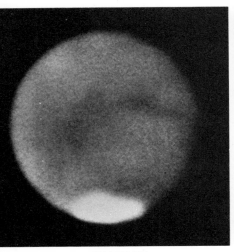 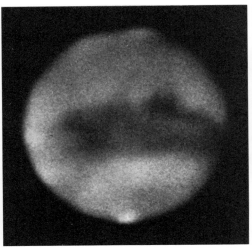 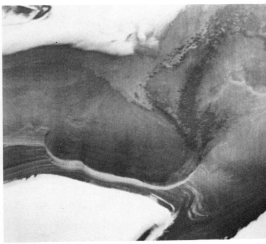

FIGURE 8.7

Martian south polar cap. The two earth-based photographs at left show the seasonal changes in the polar cap and darkening of the equatorial area. At the right is a close-up view of part of the residual north polar cap of Mars. The bright areas are apparently water ice, while the terraces are some kind of geological formation. Photograph is by *Viking* orbiter.

the three planets, but there are also distinct differences in the composition of their atmospheres.

A small, daily, and seasonally variable amount of water vapor also has been detected on Mars. The abundance of water in the atmosphere, however, is far too low for rain. Because the atmospheric pressure on Mars's surface is low (0.7 percent of the earth's sea-level atmospheric pressure), water vapor cannot exist as a liquid on the open, flat ground. The appearance of early morning fog lying in craters and other low places is probably evidence of an exchange of water vapor between subsurface or surface ice and the atmosphere. The Martian atmosphere also possesses clouds, which are most probably water ice and carbon dioxide ice condensations (see Figure 8.8). In addition, the sky is always hazy up to about 50 kilometer altitudes, as shown in Figure 8.9. The warmest daytime temperature is around 30°C at the Martian equator, while the nighttime temperature drops to −130°C. Over the polar regions it is even colder. During summer the north polar ice cap gets up to only about −70°C. Though still very cold, it is not cold enough for the residual north polar cap to be made of carbon dioxide ice. Thus it appears that it is water ice, which is consistent with the finding of more water vapor in the atmosphere at high latitudes near the poles.

There is speculation that water ice may be a remnant of a denser atmosphere that Mars may have had in the first billion years or so of its existence. If that early atmosphere was a denser carbon dioxide and water vapor one, or had a somewhat different composition than at present, then it could have acted to trap infrared radiation in the greenhouse effect, and it would have been warm enough to contain substantial amounts of water vapor. For some reason, possibly having to do with the removal of CO_2 by water activity, the atmosphere became more transparent to infrared radiation. Consequently, it cooled, and the water vapor condensed and precipitated onto the surface. The runoff of rain from this ancient dense

TABLE 8.1
Comparison of Principal Atmospheric Constituents

Component	Molecular Weight	Percentage by Volume		
		Venus	Earth	Mars
CO_2	44.0	96.4	0.03	95
N_2	28.0	3.41	78.1	2.6
O_2	32.0	$< 10^{-2}$	20.9	0.15
H_2O	18.0	0.14	≈ 1, Varies	≈ 0.03, Varies
Ar	39.9	$< 10^{-3}$	0.9	1.6
CO	28.0	0.002	?	0.6
Kr	83.8	?	10^{-4}	Trace
Xe	131.3	?	$< 10^{-5}$	Trace

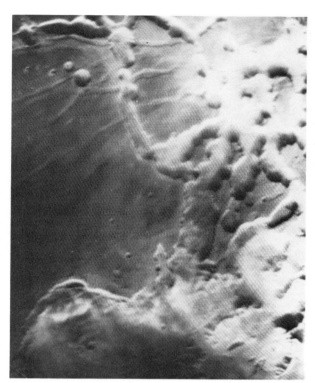

FIGURE 8.8
Late afternoon sunlight illuminates clouds in the Martian atmosphere. Clouds have been observed in a number of locations in the photographs taken by *Viking* orbiters.

FIGURE 8.9
Atmospheric phenomena on Mars. *Viking 1* took this picture in August 1978, which is mid-winter in the southern hemisphere. On the horizon cloud streamers and a faint haze are visible.

atmosphere may have carved some of the treelike network of channels seen on the surface. But where is the water now if it is not in the atmosphere or standing on the surface? There are indirect arguments for the presence of water ice in the crust and regolith of Mars down to depths of several meters. Thus water is suspected as being an important agent of change in Mars's past.

The skies at the locations of *Viking 1* and *2* are yellowish brown in color and seem to remain that way over the course of the Martian year. This color is probably due to dust particles suspended up to altitudes of 40 kilometers or so in the atmosphere. Surface winds can stir the atmosphere sufficiently to hold dust particles. From the *Viking* data we know that the prevailing winds are westerly as on earth, with velocities of 36 to 70 kilometers per hour at the surface and over 360 kilometers per hour at altitudes above 10 kilometers. Winds of 400 kilometers per hour on Mars are comparable to winds of 40 kilometers per hour on

the earth, where the air density is a hundred times greater (because the force of surface winds scales as the air density times the square of the velocity). Notice the effect of wind on the sand inside a crater in Figure 8.10. Undoubtedly the winds come from unequal solar heating between different portions of the Martian surface driving air from high-pressure areas to low-pressure areas as on the earth.

The winds are strong enough to create major dust storms. This is particularly true when Mars is near perihelion and is receiving the maximum amount of solar radiation. Major dust storms appear to be rare when the planet is near aphelion. The *Viking* landers have obtained data during at least two major dust storms in February and May of 1977 when Mars was near perihelion. Both storms began in the southern hemisphere and were fairly quickly distributed around the planet by high-altitude winds. Each dust storm took more than two months to completely subside.

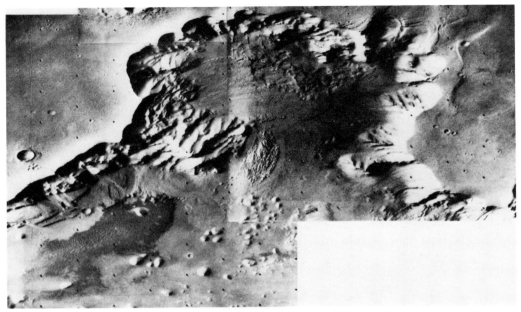

FIGURE 8.10

A field of sand dunes (lower left) some 50 kilometers long. The general view, taken by *Viking 1,* is of the north wall of Gangis Chasma, one of the branch canyons along the equator. The walls of the canyon are fluted by wind erosion.

MARTIAN TERRAIN

Like the moon and Mercury, Mars has a different topographic pattern in each hemisphere. The northern hemisphere is generally depressed with few craters. Its crust has been altered by intense volcanic activity and subsequent lava flooding. The southern hemisphere has a densely cratered and elevated crust that has not changed appreciably throughout its history. The division between the two types of terrain is roughly along a great circle inclined to the equator by about 30°.

There are only a few smooth circular basins on Mars. One that has long been observed from the earth is Hellas (see Figure 8.6), an almost craterless basin about 1600 kilometers wide or about one and a half times the size of the largest lunar sea, Mare Imbrium. Variations in the brightness of Hellas probably are caused by shifting windblown dust or sand, which either covers or erodes the surface.

The abundance of craters in some of the more heavily cratered regions of the southern hemisphere is comparable to that in the bright highlands of the moon. The similarities between the cratered Martian southern hemisphere and lunar highlands has prompted the speculation that the two are about the same age. Thus almost half the surface of Mars is ancient terrain, with many of its landforms having

remained essentially unchanged over the last 4 billion years. On the other hand, the northern hemisphere is composed of plains of lava that extensively flooded the surface at various times after the cessation of the heavy bombardment of the planet during the first half billion years of the solar system. Although it is difficult to estimate absolute ages for the plains, the variation in the abundance of craters they possess suggests they vary from a few hundred millions to several billions of years. However, since lava flows can be seen in *Viking* orbiter photographs in even the most primitive cratered terrain, it may well be that early in Mars's history even much of the ancient cratered terrain in the southern hemisphere was flooded with lava. Such thermal activity would explain why the highly cratered terrain on Mars is relatively smooth compared with the mountainous highlands of the moon. However, thermal and tectonic activity on Mars is still much weaker than that on the earth.

The area photographed by the *Viking 1* lander in the Chryse region (22.5°N latitude, 47.8° longitude) is a gently rolling landscape, yellowish brown in color, strewn with rocks and dotted with drifts of fine-grained material. Within about 30 meters of the lander, several outcrops of bedrock can be seen, which in many ways is similar to the semidesert regions of the American Southwest, but without vegetation (see Figure 8.11). Inorganic analysis by the lander found iron, calcium, silicon, titanium, sodium, and aluminum in the soil—the common elements on the earth. Organic analysis failed to detect any organic compounds (those containing carbon, one of the essential ingredients for life). From orbit the dominant features of the region are craters. From the ground there are only a few obvious craters to be seen in the immediate vicinity of the lander. If Mars were like the moon, then there should be several small craters, tens of meters in diameter, visible. Their absence indicates that the Martian atmosphere is dense enough to burn up small meteoroids before they reach the surface. Thus there is not the profusion of small craters as seen on the moon.

The soil at the *Viking 1* and *2* landing sites has a low cohesion and a consistency about the same as talcum powder. Its density is about 1.2 to 1.7 grams per cubic centimeter and is about 80 percent iron-rich clay minerals, 10 percent magnesium sulfate, 5 percent carbonate minerals, and 5 percent iron oxides. The soil contains about 1 percent water by weight, some of which is probably in hydrated minerals.

South of Chryse in Figure 8.6 is a plateau with small channels, sinuous gorges, and even larger canyons many kilometers wide (see Figure 8.12). Since both residual polar caps are presumably composed of water ice, hypotheses involving water as an active agent are currently accepted. The orbiter photographs certainly show that the small Martian channels form an extensive drainage system. Thus it seems reasonable to suppose that these features may have been carved out by great torrents of water.

In the equatorial region lies the spectacular canyon Valles Marineris, cutting across the middle of the plateau. It is nearly 4000 kilometers long, up to 250 kilometers wide in some places, and at least 6 kilometers deep. A small portion appears in Figure 8.13. At its western end lies a complex pattern of intersecting fault valleys.

Another terrain feature known as Amazonis (see Figure 8.6) is lightly cratered and resembles the moon's Oceanus Procellarum basin. This area is different because of three large volcanos, running diagonally along the crest of a ridge called Tharsis, and the spectacular isolated volcanic structure, Olympus Mons (see Figure 8.14). It is similar to, but much larger than, Mauna Loa and Mauna Kea in Hawaii as seen from the bottom of the Pacific Ocean. Close-up, the sides of Olympus Mons are striated from past lava flows on its slopes. In addition to these large-shield volcanos rising some 25 kilometers or so above the surrounding plains, there are flattish saucer-shaped volcanos. One of them is 1500 kilometers across but less than 3 kilometers high. The lack of impact craters on their slopes and surrounding area suggests that they are relatively young. South of Alba in Figure 8.6 lie the remains of what is possibly an eroded volcanic area.

Six weeks after *Viking 1* landed, the *Viking 2* lander touched down on the rock-littered plains of Utopia (48°N latitude, 225.6° longitude). There the atmosphere has ten times more water vapor than Chryse. The general appearance of most of the rocks at both

FIGURE 8.11

View from the *Viking 1* lander, taken 3 August 1976. This remarkable picture shows a dune field with features similar to many seen on the earth's deserts. Cutting through the picture's center is the meteorology boom, which supports *Viking's* miniature weather station. The photograph covers 100° in azimuth. Just beyond the far right edge of the picture are several areas of exposed bedrock. The left side of the picture shows a large field of drifted material, probably deposited when the wind was blowing from north to south. The picture is looking toward the east.

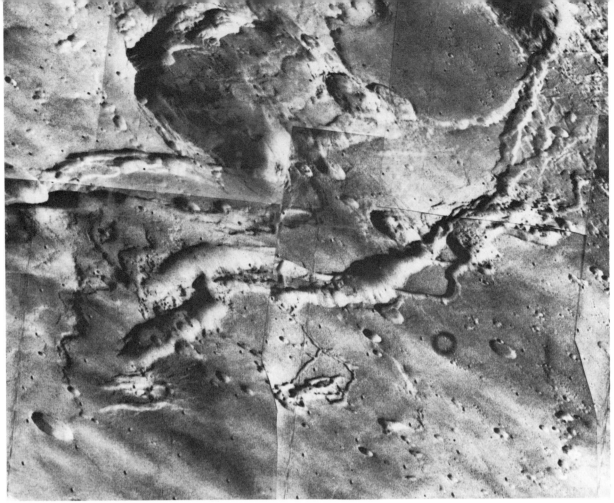

FIGURE 8.12
Fault zones (above) breaking the Martian crust. The view, taken by *Viking 1,* shows an area 2° south of the equator.

FIGURE 8.13
This super chasm, known as Tithonius Chasma, is part of the system of canyons called Valles Marineris. The volcanic activity in Tharsis split the planet's surface in all directions. Along one of the radial faults a mighty chasm developed, the large part of which is shown here.

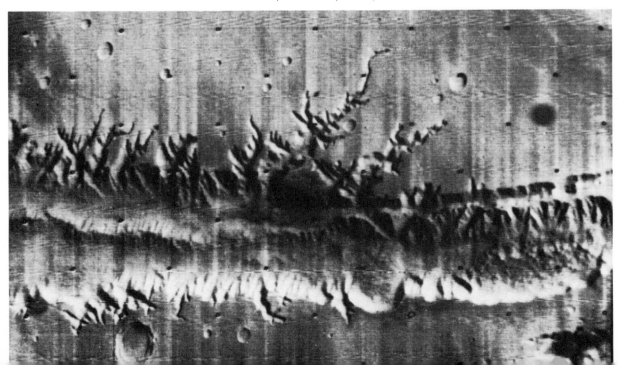

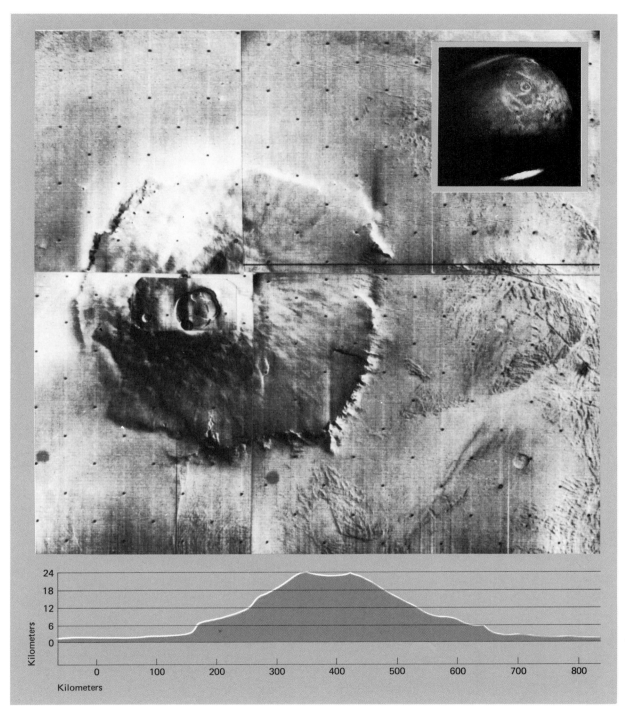

FIGURE 8.14
Mosaic of Olympus Mons on Mars. The super volcano is nearly 600 kilometers wide at the base and 25 kilometers high and is capped by a crater 65 kilometers in diameter. It is about five times larger than the most massive volcanic cone on earth, Mauna Loa in Hawaii. At a great distance it appears as the small bright ring in the upper right of the inset photograph, which was taken by *Mariner 7*. The south polar cap is very conspicuous at the bottom of this picture. (The elevation diagram and mosaic of Olympus Mons are shown at different scales.)

landing sites suggests that they are volcanic in origin. However, some show significant wind erosion (see Figure 8.11). Many of the rocks photographed show pit marks, which are due to gas pockets formed in gas-rich lavas. The reddish ground has scrambled patches suggestive of alternating freezing and thawing of ice. A meandering, troughlike feature is similar to earth features which are caused by running water. Evidence that plenty of water was present in the distant past is the hard rocky surface, resembling calcium carbonate deposits, formed by the evaporation of water after leaching salts from the ground.

One of the most exciting things to happen since the landing of the *Viking*s was the photographing of frost on the surface at the *Viking 2* site (see Figure 8.15). The frost occurred during the northern winter between May and November 1977. The composition of the frost is not known; the air temperature was too warm for it to have been carbon dioxide ice and the air was too dry for it to have been pure water ice. The best speculation is that it is some kind of mixture of CO_2 and H_2O.

The multilayered polar regions are still another type of Martian topography. A region near the south pole appears in Figure 8.7. The layered deposits probably hold appreciable quantities of frozen water mixed with dust beneath a carbon dioxide coating. Periodic changes in Martian climate may be responsible for the deposition of the successive layers of material. It appears that the bulk of the polar deposits were formed early in Martian history and have since undergone wind erosion.

Viking's life-detection experiments are described in Chapter 10. They revealed that certain minerals in the soil show a surprising amount of exotic chemical activity. There was no firm evidence for biological activity of the kind familiar to us on the earth.

Finally, both *Viking* landers carried instruments to record quakes on Mars. Unfortunately, only the one on *Viking 2* worked, and on 24 November 1976 it appears to have detected the first quake. If real, it suggests that the Martian crust has an average thickness of about 40 kilometers with a maximum thickness of about 75 kilometers under the Tharsis ridge and a minimum thickness in the Hellas basin of about 10 kilometers. By comparison, the earth's average crust thickness is about 33 kilometers and it covers a planet with nearly twice the radius of Mars. Hence the crust of the earth is about half of a percent of the radius, while that of Mars is about 1.2 percent of the radius.

FIGURE 8.15
Frost at the *Viking 2* landing site. This photograph taken 13 September 1977 when the temperature was about 174K shows patches of frost on the ground. It is thought that the frost is probably a mixture of water and carbon dioxide.

The moon's crust is about 4 percent of its radius. Like the moon, Mars appears to have a mascon, or mass concentration, in its crust at the site of a well-worn northern hemisphere basin named Isidis Planitia (12°N latitude, 271° longitude). Such features are very difficult to identify on earth and it is not obvious they have even been found.

MARTIAN SATELLITES

The two little satellites of Mars—Phobos, the inner, and Deimos, the outer—are potato-shaped bodies with cratered surfaces. Phobos orbits eastward, just as our moon does, and in the same direction that Mars rotates, in a period of 7 1/2 hours at a distance of 5950 kilometers from the surface of Mars. This gives it an angular size, as seen from the surface of Mars, of about half that of our moon. Since it revolves about Mars much faster than the planet rotates, it rises on the western horizon and sets on the eastern horizon 5 1/2 hours later. This is counter to any other natural satellite in the solar system as observed from its primary. Phobos (see Figure 8.16) is about 27 kilo-

FIGURE 8.16

Phobos as photographed by *Viking 1*. The surface of Phobos is covered with many impact craters, the largest being Stickney (not in this picture). The most striking feature is the numerous long, parallel grooves that are 100 to 200 meters wide and 20 or so meters deep. The picture is about 12 kilometers across from top to bottom.

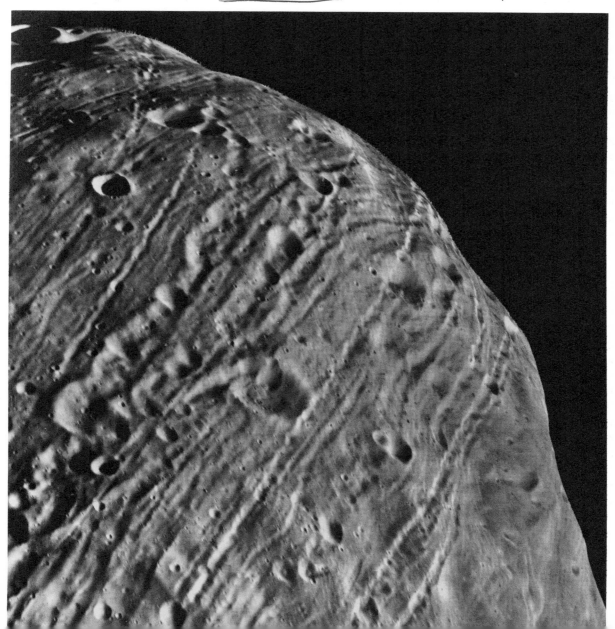

meters long, 21 kilometers high, and 19 kilometers wide. Both Phobos and Deimos have been shaped by high-velocity impacts, which appear to have sheared off large sections of each satellite. In addition, both have many craters but no ejecta, or craters with central peaks. This feature is reasonable since their gravitational attraction is very small due to their small masses. Both satellites are darkish and possess a regolith. However, there are some distinctive differences. Phobos seems more heavily cratered than Deimos, with the largest crater, Stickney, being about 10 kilometers across. It also has mysterious long parallel grooves across a large part of its surface, as shown in Figure 8.16. They are from 100 to 200 meters wide and 20 to 90 meters deep and may have been formed in the same process as the crater Stickney.

Deimos is about half the size of Phobos, being about 15 kilometers by 12 kilometers. It orbits Mars 20,000 kilometers from the planet's surface in a period of 30.3 hours. Its angular size, as seen from Mars, is equivalent to a quarter viewed at a distance of 37 meters. Its orbital period is somewhat longer than the rotational period of Mars, so it rises on the eastern horizon and sets on the western horizon nearly three days later, while going through its phases twice. Deimos appears to have a much smoother surface than Phobos; many of its craters have apparently been filled in with a light-colored material. This material may be debris from other cratering events. Evidence to date suggests that Deimos probably has the same composition and physical structure as Phobos.

8.4
Asteroids

DISCOVERY

On 1 January 1801 Sicilian astronomer Giuseppe Piazzi (1746–1826) accidentally discovered a faint object whose orbital motion was that of a body 2.8 astronomical units from the sun and about where a major planet would be expected according to Bode's law (see Section 7.1). The object was named Ceres, after the Roman goddess of agriculture. Three more objects were discovered with orbits near 2.8 astronomical units: Pallas in 1802, Juno in 1804, and Vesta in 1807. Since photographic techniques were introduced in astronomical research in the 1890s, many more have been discovered.

Instead of one planet in the slot at 2.8 astronomical units, more than two thousand small bodies have been discovered orbiting in the region between Mars and Jupiter. William Herschel called these objects *asteroids* because in a telescope they looked like stars. Almost 95 percent of these bodies have orbits between 1.6 and 3.3 astronomical units, with periods from two to six years. Their orbits are more elliptical than those of the planets and more inclined to the ecliptic. They also move in the same direction as the planets around the sun.[1] For some time the asteroid Hidalgo was thought to have the largest orbit. Its orbital period is fourteen years, with an aphelion just outside the orbit of Saturn. However, in October 1977, a new asteroid, named Chiron, was discovered, which travels in a highly eccentric orbit (eccentricity = 0.38) at an angle of 6.9° to the plane of the ecliptic. It ranges between 8.5 and 18.9 astronomical units, or roughly between the orbits of Saturn and Uranus, with a period of 50.7 years. Because of its great distance from the asteroid belt, there is a question whether it might be the first discovery in an outer zone of asteroids. Or possibly it is not even an asteroid but is something related to a comet or an escaped satellite of Saturn.

Asteroids vary in size from Ceres (1025 kilometers), down to some one hundred thousand of them that are a kilometer in diameter, and countless numbers of even smaller ones. All the asteroids together may add up to no more than a few ten-thousandths of the mass of the earth. The six largest asteroids are listed in Table 8.2.

CHARACTERISTICS

Photometric studies of the asteroids show how they differ in size, shape, and rotation. From the variation in their brightness, we gather that most have somewhat irregular shapes. Their periods of rotation are measured in hours. All but the largest of the asteroids are too small to show a measurable disk. Their colors put nearly all asteroids in two categories. Some are bright reddish, a sign of the silicates and metals, and they populate mostly the inner part of the asteroid belt. However, most asteroids have the darker neutral color of material containing various carbon compounds (carbonaceous) and occupy the outer part of the belt.

[1] An asteroid discovered in 1977 was found to be orbiting in the opposite direction.

TABLE 8.2
The Largest Asteroids

Asteroid	Diameter (km)	Mean Distance from Sun (AU)	Rotation Period (h)	Year Discovered	Type
Ceres	1025	2.77	9.1	1801	Carbonaceous
Pallas	583	2.77	7.9	1802	Peculiar carbon
Vesta	555	2.36	5.3	1807	Basaltic
Hygiea	443	3.15	18	1849	Carbonaceous
Interamnia	338	3.06	8.7	1910	Unknown
Davida	335	3.19	5.2	1903	Carbonaceous

Collisions between two asteroids may produce effects ranging from craters—if a small one collides with a large one—to fragmentation of the two asteroids—if they are of comparable size. For example, if the body producing crater Stickney on Mars's satellite Phobos (see Figure 8.16) had been a little larger, Phobos might have been broken into many small pieces. As it is, the grooves on Phobos may be large cracks produced by the impact.

Minor planets that come close to the earth are valuable sources of information on the characteristics of asteroids. The first one discovered was Apollo in 1932 and they are consequently called the Apollo asteroids. They cross the earth's orbit and have perihelia inside the earth's orbit, with periods of around a year. So far more than two dozen have been discovered. However, it is estimated that as many as a thousand larger than one kilometer in diameter pass close to the earth at some time or other.

In June 1968 Icarus passed within 6.4 million kilometers of the earth. Radar signals bounced off its surface showed that Icarus has an uneven pitted surface a kilometer or two in diameter. Hermes, less than 1.5 kilometers in diameter, passed within twice the moon's distance from the earth (about 800 thousand kilometers) in 1937 (see the orbits of several in Figure 8.17). The well-known asteroid Eros last approached the earth in January 1975 (at 22.5 million kilometers). It has a shape similar to a football, is about 36 kilometers by 14 kilometers, and spins around its short axis in 5.3 hours (as judged from its changing brightness).

ORIGIN OF ASTEROIDS

Where are the asteroids from? Some think that they might be debris left over from an exploding planetary body, but this conjecture is difficult to prove. No

known internal source of energy is great enough to rupture a planet and scatter its fragments against its own gravity. The evidence suggests instead that while planetesimals were accreting in the solar nebula, possibly several dozen in the size range from 100 to 1000 kilometers orbiting the protosun near Jupiter were unable to coalesce into anything larger perhaps than Ceres. Too close to Jupiter and its perturbing gravitational field, they could form no appreciable planetary mass. Jovian gravitation probably would have accelerated the asteroids, increasing the chances of colli-

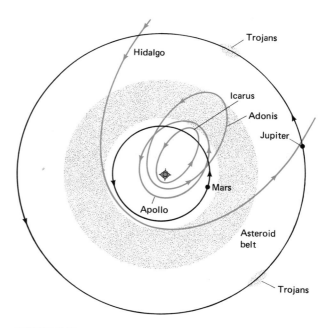

FIGURE 8.17
Orbits of unusual minor planets. There is always a remote possibility that an asteroid whose orbit intersects the orbit of the earth may collide with it.

sion. Only a few differentiated asteroids, such as Vesta, have survived intact.

The asteroids are one key to the solar system's past. Their number, size, density, composition, and distribution are important links in this chain of understanding. They may be the best specimens of solidified primordial matter still moving about the inner solar system.

8.5 Meteoroids, Meteors, and Meteorites

As much as 1000 tons of cosmic debris—billions of microscopic particles—pepper the earth daily. We are aware only of those weighing a significant fraction of a gram, which produce the shooting stars that flash across the sky. All but a few are too small to leave luminous trails. These solid particles are called *meteoroids* before they encounter the earth. Those large enough to survive flight through our atmosphere and land are called *meteorites*. And the luminous trails of the smaller particles that are completely vaporized in the atmosphere are called *meteors*. In order of increasing size and brightness, meteors are classified as (1) telescopic and radio meteors; (2) visual meteors; and (3) fireballs or bolides.

> I would rather be a superb meteor, every atom of me in magnificent glow, than a sleepy and permanent planet. The proper function of man is to live, not to exist.
>
> Jack London

A meteoroid passing through the atmosphere leaves a dense column of electrons stripped from the atoms and molecules in its path. As the ionized atoms regain their electrons, they de-excite, emitting photons, which makes the momentary luminous trail we see from the ground as a meteor or shooting star. Timing radar signals reflected from the trail of ionized particles yields our best estimates of meteor heights and velocities.

When meteoritic particles encounter the earth, they are moving from about 12 to 72 kilometers per second, depending on their direction and the angle at which they strike the earth. The velocities convince us that meteoroids belong to the solar system, moving in independent orbits around the sun.

The normal observed rate for meteors is about ten per hour over the entire sky. However, we see fewer meteors before midnight than after midnight. Why? During the evening hours we are on the back side of the earth facing the direction opposite to earth's orbital motion, and we see only those swift meteoroids overtaking us from the rear. During the morning hours earth's rotation has turned us so that we are facing in the same direction as its orbital motion. Hence we see those meteoroids that we overtake and those that meet us head-on. The effect is analogous to raindrops hitting the windows of a moving car: more raindrops hit the front windshield than hit the rear window.

Our atmosphere slows the particles and transforms their kinetic energy into radiant and thermal energy. Anything that remains slowly filters down through the air as dust and solidified droplets of melted meteoroid. Spectroscopically analyzed, meteor trails show emission lines of iron, sodium, magnesium, calcium, silicon, and several less abundant metals. Many of the physical samples have a porous structure with densities from about a half to several grams per cubic centimeter, suggesting a stony composition of metallic silicates. The average particle weighs a fraction of a gram and is microscopic in size.

METEOR SHOWERS

Several times a year we can see *meteor showers*, the swarms of shooting stars that dart from a small area in the sky and persist for hours or days. On such occasions the earth is passing through a large group of particles moving in ribbonlike fashion along an orbit around the sun. Perspective makes their tracks seem to diverge from a small spot in the sky called the *radiant*. The shower is named after the constellation in which the radiant appears. Some of the better-known showers are listed in Table 8.3.

Long ago astronomers found that some meteoroids travel in orbits much like those of some comets. They

TABLE 8.3
Meteor Showers and Their Associated Comets

Name of Shower	Approximate Date of Maximum Display	Approximate Maximum Visual Hourly Count	Associated Comet	Period of Comet (yrs)
Lyrids	21 April	12	1861 I (Thatcher)	415
Eta Aquarids	4 May	20	Halley	76
Perseids	11 August	70	1862 III	105
Draconids	10 October	Up to 500	Giacobini-Zinner	6.6
Orionids	20 October	30	Halley	76
Taurids	1 November	12	Encke	3.3
Andromedids	14 November	Low	Biela	6.5
Leonids	16 November	10 to 140,000	1866 I (Tempel)	33
Ursids	22 December	15	Tuttle	13.6

had spotted a link between meteor showers and the short-period comets (to be discussed in Chapter 9). The particle swarms may be debris left by evaporation and tidal disruption of comets. For example, on the night of 13 November 1833 watchers in the southern part of the Atlantic seaboard were awestruck as over a hundred thousand shooting stars per hour plummeted from the constellation Leo for three hours. The great display was produced when the earth encountered a swarm of meteors orbiting the sun in a period of 33 years and associated with Comet Tempel (1866 I), which has long since vanished. The meteoric displays of 1866, 1899, and 1932 were progressively weaker; then on 17 November 1966 a fairly spectacular meteor shower was observed in the southwestern part of the United States. With the passage of time, the meteor stream—which is made up of conglomerates of fine dust, ices, and ice-covered particles—is strung out along the comet's orbit. This ribbon of particles typically averages about 50,000 kilometers in cross section. Thus the earth must come fairly close to the meteor stream in order for us to see a meteor shower.

METEORITES

Most meteorites are discovered accidentally years after they fall. Of some three dozen meteorite falls weighing more than a ton, only a few were seen descending. Not many of the falls are ever recovered. Most meteorites land in the oceans or in unoccupied places where their fall is not likely to be observed. No known record tells of a community destroyed or an individual killed by a meteorite, in spite of some close

calls. Approximately two thousand meteorite specimens have been recovered.

Meteorites striking the earth probably have formed thousands of craters, but only two hundred or so have been identified at this time. One great collision in 10,000 years is a conservative estimate, and at that rate at least 50,000 giant meteorites must have fallen on the earth in the past 500 million years. But the fossil craters left by many of these may lie buried and unnoticed in the earth's crust. Probably most of them have been obliterated by weathering, erosion, and geological processes. One that we know about, near Winslow, Arizona, is the Barringer meteorite crater (see Figure 8.18), created by a meteorite weighing at least 30,000 tons. It hit the earth about 24,000 years ago and must have devastated all plant and animal life within a large area. The crater is 155 meters deep and is crowned with a raised rim 1300 meters wide. Thirty tons of shattered iron fragments have been picked up within 6.5 kilometers of the crater.

At 7 A.M. on 30 June 1908 a tremendous fireball flashed across the sky in Siberia. A great ball of flame brighter than the sun was seen leaping from a forested region near the Tunguska River. The sight of the fire was followed by the sound of an explosion powerful enough to level trees within 50 kilometers. Earth tremors were recorded on seismographs throughout Europe. The most plausible explanation for the event is that a small comet (possibly part of Comet Encke) or a large, fragile, stony meteorite struck the earth, dissipated its kinetic energy on the forest and the ground, and completely vaporized.

Three classes of meteorites have been established by their chemical and metallurgical properties.

1. *Stones* are composed primarily of silicates of iron, magnesium, aluminum, and other metals. These generally have a relatively smooth, brown or grayish fused crust indented with pits and cavities. Buried inside all but a small fraction of them are small pieces of glassy minerals called *chondrules* that apparently formed from molten droplets, presumably during the early formation of the solar system. One subgroup of the stones are the *carbonaceous chondrites*, which contain large amounts of carbon, water, and other volatiles that would have been driven off with the slightest heating above about 500 K. Therefore, these are the most elemental samples of matter that we have, dating back to the primeval solar nebula. They are doubly interesting because they contain organic compounds, such as hydrocarbons, amino acids, and lipids. These biologically important compounds evidently formed in the primordial solar nebula without the assistance of living organisms.

2. *Stony irons* are a matrix of stone and iron. Their brownish crust is sometimes broken by yellow olivine cavities. Inside the iron may have a veinlike or globular structure.

3. *Irons* are almost exclusively composed of iron, with some nickel. They are easily identified by their characteristic pitted brownish exterior and high density. Cut, etched, and polished, they usually have a peculiar crystalline pattern unlike any in terrestrial iron. They show evidence of melting and signs of other igneous processes.

Stones are the most brittle kind of meteorite and they are more fragile than the irons. Even though most falls are stones, more of the recovered meteorites are irons because they are relatively easy to identify and they resist weathering. All the meteorites carry signs of atmospheric ablation (melting of the surface layers of the meteorite) from their descent through the air, like a space capsule that has reentered the atmosphere.

Many meteorites that have been radioactively dated average tens of millions of years for the stones and 600 million years for the irons. These are their ages only since the larger mass broke up. The most ancient specimens are about 4.6 billion years old, the same age as the earth. The chemical and mineralogical sequences in the different classes of meteorites strongly suggest that they share the same heritage as that of the rest of the solar system.

We are still not sure of the origin of meteorites. Are

FIGURE 8.18
Aerial view of Barringer meteorite crater.

they the remains of comets? Perhaps, but the supporting evidence for this idea is not strong. Another line of speculation is that most meteorites may be descended from a few chemically differentiated asteroids, whittled down by repeated collisions early in the planetary system's history. In such a case, stony meteorites come from the original crusts, the stony irons from the intermediate parts, and the irons from the core. Regardless of our ability to understand their origins, it is evident that asteroids and meteorites are the discarded building material from which the inner solar system was fabricated.

8.6
Interplanetary Medium

INTERPLANETARY DUST

The space between the planets is a vacuum by terrestrial standards, but it is not devoid of gas and small solid particles. The particulate matter consists of particles blown out from the sun's atmosphere by the solar wind, micrometeoric debris scattered by comets, and, perhaps less plentiful, granular powder strewn about by asteroid and meteoroid collisions.

We have learned about interplanetary dust from several sources. One is the *zodiacal light*, which is most easily observed in our Northern Hemisphere in spring after sundown in the west and in fall before dawn in the east. It appears as a faint pyramidal band of light tapering upward from the horizon along the line of the ecliptic. The spectrum of zodiacal light is a faint replica of the solar spectrum, produced by small particles lying in the plane of the planets' orbits, which scatter the solar photons in our direction.

More direct evidence of interplanetary dust comes to us from spacecraft experiments. Sensors on the skin of the spacecraft arranged to electronically count micrometeorites striking the surface. From the numbers of impacts registered, it is estimated that the average spacing between interplanetary particles is in the range of many meters. The total mass of dust particles is estimated to be about 10^{20} grams, or about a hundred-millionth of the mass of the earth.

INTERPLANETARY GAS

Most interplanetary matter is in the form of the gas comprising the *solar wind*. It consists of an almost continuous stream of particles, mostly protons and electrons, flowing out from the sun's corona. As the solar wind moves forward, it forms an expanding spiral pattern due to the sun's rotation, and its velocity increases until it equals the speed of sound in the plasma, several solar radii from the sun. Its velocity continues to increase as it flows outward, much as rocket gases are accelerated to supersonic velocities in a rocket nozzle. Near the earth the solar wind reaches a velocity of about 450 kilometers per second. Beyond the earth its speed remains very nearly constant. At earth's distance, the wind's density is down to about five protons and five electrons per cubic centimeter on the average, but it can rise on occasion to a hundred particles per cubic centimeter. Compare that with the number of molecules in your room—about 3×10^{19} per cubic centimeter. The temperature of the wind particles is about 200,000 K in the vicinity of the earth. This is less than their 1,000,000 K temperature when they were in the outer parts of the corona. The density is so very low, however, that the wind transfers no appreciable quantity of heat to the earth (see page 68).

SUMMARY

Although telescopic observations of the inner planets from ground-based observations have been most fruitful, explorations of these planets by unmanned instrumented space vehicles have greatly enlarged our knowledge. Flybys have scanned Mercury, Venus, and Mars; orbiters have surveyed and mapped Mars; and landers have studied the surface of Venus and Mars and have sought to find life on Mars.

Among the four terrestrial planets and our moon, there are both similarities and differences. All, except the earth, show evidence for an extremely intense cratering phase early in their evolution. The moon and Mercury still possess most of the effects of that cratering period. Cratering is one major terrain-shaping mechanism, the other is thermal and tectonic activity. Earth, and to some extent Mars and Venus, have had their surfaces continuously modified by such activity over the past two or three billion years. The atmospheres of Venus and Mars are somewhat similar to each other in that their chief constituent is carbon dioxide; their atmospheres may well resemble an earlier phase in the earth's atmosphere. Mercury and the moon are too small to hold any appreciable atmosphere. Venus is unique in having a cloud cover over the entire planet, while the earth is unique in having a low, surface atmosphere in which water is a significant agent of change. There is reason to believe that the chemical compositions of these five bodies (considering the entire planet) are similar and they also have somewhat similar internal structures.

The asteroids—small, irregularly shaped solid bodies, most under a kilometer in diameter—number in the thousands. Matter orbiting the sun in the form of meteoroids, dust, and gas constitutes a highly rarefied interplanetary medium. A grain-sized or smaller particle in this medium, when captured by the earth, produces a meteor trail as it passes through the earth's upper atmosphere. Larger bodies that survive the atmospheric flight and land are called meteorites.

How were the planets and their satellites formed? How did the planets each evolve so differently from the same physical processes under somewhat different initial conditions? The difference in the way planets have evolved allows us to see more clearly the nature and origin of our past and where we are going. Global phenomena on earth may best be understood by comparing them with those on other planets.

REVIEW QUESTIONS

1. In what ways are the moon and Mercury alike or different?

2. What are the surface and atmospheric conditions on Venus?

3. Discuss the differences in the atmospheres of Venus, Mars, and earth.

4. How is the Martian terrain similar to and different from that on the earth?

5. How are asteroids discovered? How many have we found so far? How big are they? What limits the discovery of others?

6. How may the asteroids have originated? The meteorites? Is there a connection between the two?

7. Suppose you were with a friend one night and both of you saw a shooting star. How would you explain what actually happened to your friend, who has never taken a course in astronomy?

8. Describe each of these: (a) a meteoroid; (b) a meteor; (c) a bolide; and (d) a meteorite.

9. How did we discover that there was a physical connection between certain comets and meteor swarms?

10. Why are the carbonaceous chondrites believed to be the most primitive material in the solar system? What else makes them interesting?

11. What evidence do we have that there have been actual collisions between the earth and large meteorites?

12. Why can't Mercury or Venus be seen at midnight from the earth?

13. Describe the orbital motions of the Martian moons, Phobos and Deimos. How big are they and what is their physical appearance?

14. Why is the surface temperature of Venus so hot (460°C)? Why does it not vary much over the entire surface?

15. How do we know that Mars has no canals of the kind depicted by Lowell?

16. What causes the seasonal changes of Mars, once thought to be due to the growth and decline of vegetation?

17. Why does the earth possess large expanses of water while neither Mars nor Venus does?

18. Describe the solar wind. What is its origin? How does the earth's atmosphere protect us from the onslaught of the solar wind?

19. What are the physical processes that have shaped the surfaces of the terrestrial planets and the moon? Have these bodies followed identically the same path in the evolution of their surfaces? What are the differences, if any?

20. List as many features in common between the terrestrial planets as you can think of. List the differences between them.

9
Outer Solar System: The Jovian Planets

Our knowledge of the inner part of the solar system (Chapter 8) is considerably greater than that of the outer region: there are little or no data on many aspects of the outer parts of the system. And while the boundaries of the inner solar system are reasonably well defined (even if not well studied), the outermost limits of the solar system are poorly defined. For example, is it possible that there are small, faint, distant planets beyond Pluto, awaiting discovery? What is the icy realm of the comets like? As we will indicate, major space efforts are in progress or proposed that will greatly improve our knowledge of the Jovian planets and their environments. However, there is still a great deal to be learned beyond what even these voyages of exploration are expected to reveal.

The general information on the nine planets presented in Chapter 7 shows that there are distinctive differences between the terrestrial and the Jovian planets. For example, the four giant planets—Jupiter, Saturn, Uranus, and Neptune—contain 99.6 percent of the total mass of the sun's planets. Also, Jupiter and Saturn, with their large complements of satellites, are like miniature solar systems. And certainly the differences in the compositions of these bodies compared to the terrestrial planets indicate differences in the details of their formation. We will discuss these differences and other details of the outer solar system in this chapter.

9.1 Jupiter

GENERAL PROPERTIES
Fifth planet from the sun, Jupiter is the largest and most massive of the planets in the solar system. In our nighttime sky it glows with a bright, steady yellow light, outshining the stars. The mean diameter of Jupiter is about eleven times greater than the earth's, and Jupiter is more than a thousand times larger in volume than the earth. However, Jupiter's mass is barely more than three hundred times that of the earth, even though it exceeds the combined masses of all the other bodies orbiting the sun. Thus its

◄ Composite NASA photograph of Jupiter and satellites.

mean density is about one-fourth that of the earth. Because its axis is tilted only 3 degrees from the perpendicular to its orbital plane, the planet has little seasonal change.

All portions of the layers of the planet visible to us do not rotate in unison. The equatorial region completes its rotation several minutes sooner than adjacent higher latitudes. Slight speedups and slowdowns in some areas have also been observed. Jupiter's rapid ten-hour rotation and low density have combined to flatten the planet about 6 percent in its polar diameter. Both sunlit and dark sides have the same temperature. During the 1960s one of the most important discoveries in planetary infrared astronomy was that Jupiter radiates twice as much heat as it receives from the sun. The excess apparently comes from internal heat that was generated by gravitational contraction at the time of Jupiter's formation and that may be continuing today at a very slow pace. (As a material body such as a forming planet contracts, it converts gravitational potential energy into thermal energy.)

Saturn also appears to emit about twice as much energy as it receives from the sun. Somewhat surprisingly, Neptune also appears to have an internal heat source, while Uranus alone among the giant Jovian planets appears to have no major heat source. It was expected that Uranus and Neptune might differ from Jupiter and Saturn, but not from each other.

SPACE MISSIONS TO THE OUTER SOLAR SYSTEM
The highly successful *Pioneer 10* and *11* spacecraft flew a first reconnaissance mission by Jupiter in December of 1973 and 1974. With gravitational assistance from Jupiter, *Pioneer 11* went on its way to a rendezvous with Saturn in September 1979. *Pioneer 10* and *11* will cross the orbit of Pluto sometime between 1987 and 1990 on their way out of the solar system. Both spacecraft carry a variety of scientific equipment plus a message from the earth engraved on a plaque mounted on the vehicle. Although the likelihood of either spacecraft being discovered by an intelligent civilization in the neighborhood of the sun is remote, they are our first attempts to send written messages to other occupants of our Galaxy. If these spacecraft were headed for the nearest star, Alpha Centauri—which they are not—they would take a little more than a hundred thousand years to reach its vicinity.

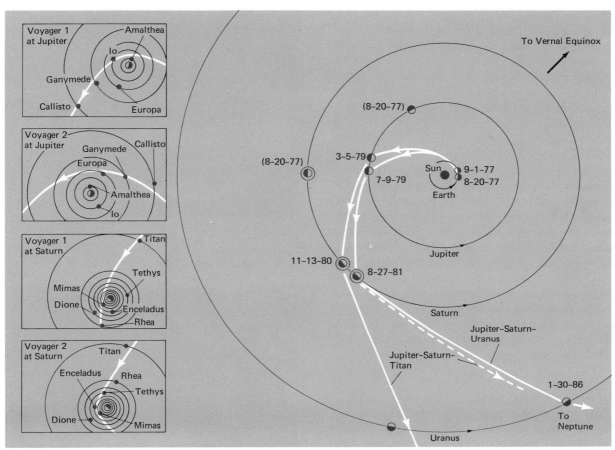

FIGURE 9.1
Space mission of *Voyager 1* and *2* to the Jovian planets. Relative distances are approximately to scale, although the sizes of the planets and the sun are not to scale. The four insets show the details of the encounter with Jupiter and Saturn by *Voyager 1* and *2*.

In the late summer of 1977, *Voyager 1* and *2* (see Figure 9.1) were launched toward Jupiter, Saturn, and their satellites. These spacecraft carried eleven different scientific instruments, which included two television cameras, spectrographic equipment for the ultraviolet and the infrared spectral regions, devices for measuring magnetic fields, instruments for studying the interplanetary plasma and other charged subatomic particles, and antennas for studying Jupiter's radio emission.

Because of Jupiter's intense radiation field, which might damage measuring instruments, *Voyager 1* was targeted no closer than about five Jovian radii in its March 1979 encounter and *Voyager 2* no closer than about ten Jovian radii in July 1979. Besides studying

Jupiter, both spacecraft looked at the innermost satellite Amalthea and the Galilean satellites (Io, Europa, Ganymede, and Callisto). For two and a half days *Voyager 1* was within 2 million kilometers of Jupiter, traveling at speeds between 15 and 35 kilometers per second. *Voyager 1* was able to pay special attention to Io, innermost of the Galilean satellites, as the spacecraft passed below Io's south pole within 22,000 kilometers. The Galilean satellites (shown in the photograph at the beginning of this chapter) were photographed with a resolution as good as that in *Mariner 9*'s photographs of Mars.

After departing Jupiter, *Voyager 1* and *2* headed for an encounter with Saturn. The *Voyager* spacecraft used the *gravitational assist* of Jupiter for their acceleration toward Saturn. *Voyager 2* will attempt to obtain a gravitational assist from Saturn for a rendezvous with Uranus in January 1986 and possibly Neptune in 1989 or 1990. By then *Voyager 2* will have traveled a total distance of 30 astronomical units, or about 4.5 billion kilometers, in twelve years before the spacecraft eventually makes its exit from the solar system.

A future NASA space mission, Project Galileo, is currently scheduled for launch in 1984. This mission will send a 2-ton orbiter with a detachable probe to Jupiter. About a hundred days before reaching Jupiter, the orbiter is to release the probe in such a way that it will enter the planet's atmosphere on the sunlit side, taking measurements through the atmosphere during its thirty-minute-long descent. The orbiter will then go into orbit about the planet for at least twenty months, studying the planet, its largest satellites, and the interplanetary space near Jupiter. It is proposed that the spacecraft be carried into earth orbit by the *Space Shuttle* and then launched from there on a roughly thousand-day journey to Jupiter, arriving in early 1986.

JUPITER'S ATMOSPHERE

Observing Jupiter through a telescope, we see a yellowish, somewhat flattened disk crossed by alternating light and dark atmospheric cloud bands parallel to the equator. The predominant colors of the dark *belts* are gray or brown, occasionally interspersed with blue, green, and red blotches; the lighter *zones* are yellowish. The entire banded structure is constantly undergoing changes in color and intensity, probably because of the formation or dissolution of clouds of differing chemical composition at different altitudes and latitudes.

The general structure of the bands (see Figure 9.2) appears to be due to underlying currents flowing alternately east and west in both hemispheres. The observed belts and zones are the outward manifestations of these currents. The gas convected upward near the slower polar regions moves around Jupiter in the direction of its rotation. In general, the dark belts are at the lower altitude and are possibly warmer descending regions; the bright zones are at a higher altitude and are slightly cooler ascending regions of cloud material. However, the cloud motions on a small scale are by no means orderly, as is evident in Figure 9.3. *Voyager* scientists were un-

FIGURE 9.2
North polar region of Jupiter. This view, as seen by *Pioneer 11,* is not visible from the earth.

FIGURE 9.3
Close-up of Jupiter's clouds. This photograph was taken by *Voyager 1* on 1 March 1979 from a distance of about 5 million kilometers. The smallest features resolvable in the picture are about 95 kilometers across. The Great Red Spot is in the upper right corner along with a white oval. Note what appears to be a very turbulent region immediately to the west (left) of the Great Red Spot. There is also an intricate and involved structure that can be seen within the spot itself.

prepared for the diversity and sometimes high state of turbulence in the cloud motions as photographed by the spacecraft. Surprisingly, photographs failed to reveal cloud features smaller than about 100 kilometers across. Narrow bands appear to coalesce and widen, while wide bands break apart. Material seems to be transferring between bands.

Conspicuous in the southern hemisphere is the oval Great Red Spot, which has varied both in size and intensity since its telescopic discovery three centuries ago. It measures about 14,000 kilometers by 40,000 kilometers. A close-up view of the Red Spot is shown in Figure 9.4. It may well be a cyclonic region of swirling gas, a mammoth storm. White plumes of clouds race by the Spot in a rather chaotic fashion; the interior is relatively calm by comparison. There are, in addition to the Great Red Spot, smaller light- and dark-colored spots, which are also apparently circulation cells.

The modified spectrum of the sunlight reflected from the Jovian atmosphere shows ammonia, methane, helium, and atomic and molecular hydrogen.

However, the dominant constituents are hydrogen and helium. Small amounts of water vapor, carbon monoxide, acetylene (C_2H_2), ethane (C_2H_6), and phosphine (PH_3) have been discovered in the infrared spectrum. Traces of two uncommon isotopes of hydrogen and carbon, $^2H_1$ and $^{13}C_6$, are also present. Other compounds, such as hydrogen sulfide (H_2S), are suspected but have not yet been spectroscopically identified.

One of the more startling results of the *Voyager* mission was the observation of a 30,000 kilometer stretch of what is thought to be auroral activity (see Figure 9.5). Photographs of the dark side of the planet, taken while the spacecraft headed away from Jupiter, reveal the auroral display in the atmosphere along the limb. Also observed were bright flashes on the dark side, which were coming apparently from above the clouds and are thought to be lightning.

JUPITER'S STRUCTURE

One, but certainly not the only, proposed model of Jupiter's structure, shown in Figure 7.6 begins with a

FIGURE 9.4
Great Red Spot of Jupiter. This *Voyager 2* photograph was taken on 3 July 1979 from a distance of about 6 million kilometers. The large white oval next to the red spot is different from one visible in *Voyager 1* photographs of the same region.

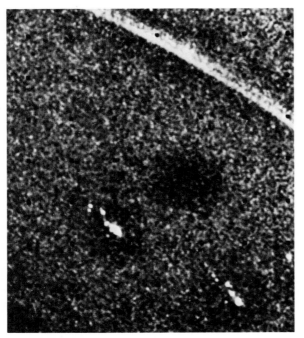

FIGURE 9.5
Auroral glow on Jupiter's dark limb. This photograph was taken by *Voyager 1* of Jupiter's dark hemisphere. What appears to be an auroral glow traces out the dark limb of the planet. The bright spots toward the center of the planet's darkened disk are thought to be lightning flashes.

top layer of clouds 240 kilometers deep. The atmosphere is 84 percent hydrogen, 15 percent helium, and 1 percent methane, ammonia, and other molecules. Below the top layer of clouds, in successive strata, are clouds of ammonia crystals suspended in the atmosphere, ammonium hydrosulfide droplets, water ice, and liquid water droplets down to a depth of a thousand kilometers. Farther down is an exten-

sive shell of liquid molecular hydrogen at a pressure of 3 million earth atmospheres to a depth of 25,000 kilometers (about one-third of the radius). Below this level the liquid hydrogen is compressed into metallic hydrogen nearly all the way to the center, where the pressure is on the order of 100 million earth atmospheres. At the center is a small iron silicate core of ten to twenty earth masses. The temperature of the core is estimated to be greater than 30,000 K. Thus in this speculative model Jupiter is more liquid than gaseous or solid. It is because of the tentative nature of this model that more space missions, such as Project Galileo, are needed.

Since the calculated abundance ratios of hydrogen, helium, carbon, and nitrogen are nearly the same as for the sun, it appears that Jupiter has retained the primordial chemical composition that was in the solar nebula. In its composition Jupiter resembles the sun more than it does the earth. Possibly, had it been several dozen times more massive, it would have become a star.

JUPITER'S MAGNETOSPHERE

Jupiter is the strongest radio emitter in the solar system after the sun. It emits both thermal or blackbody radiation and nonthermal radiation, which is synchrotron radiation (see page 181). At times its radio emission exceeds even the sun's in intensity. Jupiter emits a short-wavelength radio radiation often called *Jupiter's decimeter radiation,* since its wavelength is in the decimeter (0.1 meter) range. This radiation is synchrotron radiation, showing that Jupiter has a magnetic field and energetic, free electrons in radiation belts analogous to the earth's Van Allen radiation belts (see Section 6.3).

There is also a long-wavelength *decameter* (10 meters) *radiation,* with occasional bursts having energies up to 10 million kilowatts. The bursts are strongest

Lo' from the dread immensity of space,
Returning, with accelerated course,
The rushing Comet to the Sun descends;
And, as he shrinks below the shading earth
With awful train projected o'er the heavens,
The guilty nations tremble.

James Thompson

NONTHERMAL ELECTROMAGNETIC RADIATION AND POLARIZATION

There are several ways in which a gas may produce electromagnetic radiation. The most important of these is simply because the gas is hot, and therefore the gas is known as a *thermal source* of radiation. As we discussed in Chapter 4, the atoms, ions, and free electrons in a hot gas continually collide with each other, billions of times a second, and in so doing, kinetic energy of motion is converted into energy that excites the bound electrons of the atoms or ions. The spontaneous de-excitation of the atom or ion produces the photons that are observed coming from the gas. The type of spectrum in this case is continuous if the gas is reasonably dense and is an emission line if it has a low density.

The free electrons of a hot gas, moving about at random, are another thermal source. In a hot gas the electrons can emit energy in the form of photons when their paths are altered by protons (hydrogen nuclei) or other types of nuclei. Thus they also are emitting part of their kinetic energy of motion in the form of electromagnetic energy. This chance encounter is called a *free-free transition* because the electron is free before and after the encounter. The spectrum is continuous and is related to the temperature of the gas since the electrons have a range of kinetic energies, with the average kinetic energy determining the temperature.

However, the emission of radiation does not always signify that the gas is hot. For example, we are all familiar with the drastic difference in temperature between the ordinary incandescent light bulb, which is a thermal source of radiation and is hot, and the fluorescent bulb, which is cool by comparison. Sources of radiation for which the nature of the radiation does not directly indicate the temperature of the gas are called *nonthermal sources*. Examples of nonthermal sources are those instances in cosmic space, of gases in which free electrons and possibly nuclei are moving at speeds close to the speed of light. These electrons, called *relativistic electrons*, can have their paths altered—for

example, by the presence of magnetic lines of force embedded in the gas. The electrons will be forced to spiral about the magnetic field lines, if present, as shown in Figure 9.6. Since this type of radiation was first observed in the laboratory coming from relativistic electrons being accelerated to very high energies in a device called a synchrotron, it is called *synchrotron radiation*. For typical values of the magnetic field and electron energy, synchrotron radiation will be primarily in the radio portion of the electromagnetic spectrum. However, for strong fields and very energetic electrons, the radiation can be primarily in the visible portion of the spectrum and is thus a measure of these quantities rather than of the kinetic temperatures of the gas particles. Each type of nonthermal radiation has a quite different spectrum, which allows the astronomer to distinguish one mechanism from another.

Another property of electromagnetic radiation, which conveys significant information about the source of radiation and the environment through which it passes, is its state of *polarization*. The electric field in electromagnetic radiation can be thought of as oscillating in one particular direction (see Figure 4.1),

which is at right angles to the direction of propagation. Radiation from most natural sources contains a mixture of waves for which the oscillations occur equally in all possible directions about the direction of propagation. In this case the radiation is said to be *unpolarized*. However, if there is a preferred direction, then the radiation is said to be *polarized* in that direction. Thus in Figure 4.1 the radiation is polarized in the up-and-down direction. Light waves can also be polarized by various processes after the radiation is emitted, such as selective absorption by certain crystals, reflection from nonmetallic surfaces, and scattering by small particles.

The important point for us to note here is that synchrotron radiation is polarized, while, in general, thermal sources emit unpolarized radiation. Thus studies of the polarization of electromagnetic radiation can provide the astronomer with clues about the nature of the source of radiation—whether or not a magnetic field is present, for example. They also provide information about the environment through which the radiation has passed, such as reflection by planetary surfaces or scattering by particles in a planetary atmosphere and the interplanetary medium.

FIGURE 9.6
Synchrotron radiation, depicted by the wiggly lines, emitted continuously along the spiral path of the electron. (It is shown here only for one loop of the electron's path.) The double arrow indicates the vibrating plane of polarized light, which is perpendicular to the magnetic line of force and to the direction of the radiation.

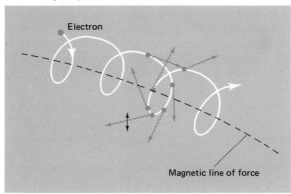

Electron

Magnetic line of force

around 30 meters wavelength and are more intense when the nearest Galilean satellite Io appears on one side of Jupiter as viewed from the earth. Why should the position of Io make a difference? We suspect that it is due to the motion of Io through Jupiter's magnetic field, disturbing the field and the electrons trapped in it, which normally produce the decimeter radiation. Some of these electrons are dumped out of the radiation belts and into the atmosphere, producing the decameter radiation in much the same way that aurora are produced in the earth's atmosphere. Jupiter's magnetic field may also be focusing some of the less energetic cosmic rays, which the earth intercepts.

Pioneer space probes ran into the bow shock wave formed by the solar wind interacting with the Jovian magnetic field as far out as 108 Jovian radii. The planet's inner radiation region has high-energy protons, electrons, and a thermal plasma. They are like earth's Van Allen belts but from five to ten thousand times more intense. Farther out, the magnetic field flattens into a disk extending several million kilometers from the planet. Its shape is influenced by the large centrifugal force due to the planet's rapid rotation. The outer part of the field constantly undergoes fluctuations and distortions, apparently produced by variations in the solar wind as particles move in and out of this region. Its long tail, flowing out opposite to the direction of the sun, extends an unknown distance beyond the orbit of Saturn. Within the rounded inner portion of the field, the particles are trapped more or less permanently. Jupiter's magnetic axis is inclined to its spin axis by about 11°, making the magnetosphere wobble up and down about 22° during the planet's ten-hour rotation period.

SATELLITES OF JUPITER

At least 14 known satellites orbit Jupiter. The thirteenth was discovered in 1974; a fourteenth was found by *Voyager 2* and a fifteenth has been suggested. Among the four largest, discovered by Galileo in 1610, Io and Europa are about the size of our moon, while Ganymede and Callisto are larger than Mercury. These four Galilean satellites and the small innermost satellite, Amalthea, orbit within Jupiter's magnetosphere.

In recent years astronomers have given the Galilean satellites (shown in Figure 9.7) a great deal of attention. Their mean densities in grams per cubic centimeter are as follows, in order of distance from Jupiter: Io, 3.41; Europa, 3.06; Ganymede, 1.90; Callisto,

1.81. Hence Io and Europa, with size, density, and mass comparable to the moon, probably have a composition and structure similar to the moon, while Ganymede and Callisto are lighter and made of icy materials. From both spectroscopic and radar reflection studies, we conclude that Europa and Ganymede are covered with ice, or a mixture of ice and rocks, many meters thick.

The satellite Io shown in Figure 9.7 surprised *Voyager* scientists by its surface appearance, which is a collage of mottled yellows, reds, and blackish browns. Passing very close to Io, *Voyager 1* was able to photograph the satellite's surface, resolving features as small as a few kilometers. Evidence suggests that the satellite does have a thin atmosphere, with even a suggestion of lightning occurring somewhere in the atmosphere or near the vicinity of the satellite. However, the greatest excitement about Io is the positive identification of active volcanos on its surface. Figure 9.7 shows an eruption occurring on the limb, with material being thrown up to altitudes of about 150 kilometers at velocities of about 2000 kilometers per second. At least eight active volcanos have tentatively been identified in the *Voyager 1* photographs. In this case the orange color of the surface could well be due to sulfur compounds from volcanic activity. Related to the discovery of volcanic activity is the fact that there are virtually no impact craters on the surface. Thus the surface is active enough so that craters are either eroded away or filled in by volcanic debris in time periods as short as 10 million years. Io must possess the youngest surface of any solar system body we have examined, and it is the only body besides the earth to show significant volcanic activity, which is actually greater than that of the earth.

There is also an extended banana-shaped cloud of gaseous sulfur partly surrounding Io. It possibly comes from sulfur atoms driven off the surface by a rain of high-energy particles from Jupiter's magnetosphere or is thrown off by volcanic activity. Potassium and sodium have also been detected spectroscopically in the vicinity of Io. Not yet satisfactorily explained is the suspected brightening of Io as it leaves Jupiter's shadow after an eclipse by the planet. Finally, atomic hydrogen fills all the orbit in the shape of a torus (doughnut). Interaction between this hydrogen and the plasma in Jupiter's magnetosphere may produce the observed meter-wavelength radio bursts.

In stark contrast to Io is the next Galilean satellite out from Jupiter, Europa, shown in the *Voyager 1* pho-

Hast thou ne'er seen the Comet's flaming flight?
Th' illustrious stranger passing, terror sheds
On gazing Nations from his fiery train
Of length enormous, takes his ample round
Thro' depths of ether; coasts unnumber'd worlds,
Of more than solar glory; doubles wide
Heav'n's mighty cape; and then revisits earth,
From the long travel of a thousand years.

Edward Young

tograph in Figure 9.7. The satellite is moonlike in size, density, and mass but has a much higher reflectivity than the moon, which suggests an ice-rich surface. The surface color is a lightly orange-hued off-white and has the least contrast of the four Galilean satellites. There appear to be few if any craters or bright spots, but there are vast numbers of crisscrossing light and dark markings. Some of these features are hundreds of kilometers wide and thousands of kilometers long. In fact, some appear to extend halfway around the satellite and may be up to 3500 kilometers in length, if they are truly continuous.

A *Voyager 1* photograph of Ganymede (Figure 9.7) shows a round disk with light and dark mottlings. There are some craters and several lunarlike maria. In comparison to the lunar maria, there are more craters, but, nevertheless, fewer than in the lunar highlands. Ganymede is a low-density body with presumably a high composition of the icy materials. However, signs of surface or subsurface ice have not been found which is puzzling. Ganymede's surface, like Europa's, is crisscrossed with linear features, as shown in Figure 9.7. However, unlike Europa's features, those on Ganymede are definitely a system of grooves and ridges superimposed on each other. They may be the result of transverse faulting similar to the earth's plate tectonics. The satellite's period of rotation, 7.16 days, is the same as its orbital period, so it keeps approximately the same face to Jupiter. The other Galilean satellites and little Amalthea also have a spin-orbit lock.

The outermost of the Galilean satellites is Callisto, which is shown in a *Voyager 1* photograph in Figure 9.7. It is a bit smaller and a little less dense than Ganymede, and it too is probably an ice-rock composition.

Callisto has about ten times as many craters as Ganymede. However, it appears to lack fractures in its crust, which Ganymede has. It does have two large basinlike features surrounded by an almost concentric sequence of rings. The rings are raised features, over 1500 kilometers in diameter for one basin and 500 kilometers for the other. There does not appear to be a significant difference in elevation for any features on Callisto, which is somewhat puzzling. This characteristic may be an indication of a relatively weak surface material, which, either because of composition or because of structure, is unable to support much vertical relief.

The idea that Jupiter possessed a ring like that of Saturn was proposed some twenty years ago. *Pioneer 11* data were interpreted as consistent with the existence of a system of tiny satellites forming a ring about Jupiter. However, this was at best speculation, and it was *Voyager 1*'s photograph of the Beehive star cluster that actually revealed the ring, which is shown in Figure 9.8. The figure shows the ring extending to the right of the planet (not shown), some 57,000 kilometers above the cloud tops of Jupiter. At most, the ring is about 30 kilometers thick and 7,000 kilometers wide. We are still not sure about the size of the particles composing the ring—whether they have typical diameters of tens, hundreds, or thousands of meters.

As far as Amalthea and the other eight satellites are concerned, not a great deal is known about their physical structure and surface appearance. As shown in Table 7.5, all were discovered after 1892. With the exception of Amalthea, these non-Galilean satellites have eccentric orbits and on the average have orbits inclined to Jupiter's orbit by about 28°.

FIGURE 9.7

The Galilean satellites of Jupiter as photographed by *Voyager 1* and *Voyager 2*.

Top row (page 185):
Io is the innermost of the four Galilean satellites. This photograph was taken on the morning of 5 March 1979, at a distance of 377,000 kilometers by *Voyager 1*. Smallest visible features are about 10 kilometers across. The light-colored areas are thought to be surface deposits consisting of mixtures of various salts, sulfur, and material from volcanic activity. The black areas are apparently the volcanic cones and their associated lava flows. In the middle of the picture is a heart-shaped basin or plateau. The lack of a significant number of impact craters on the surface is a good indication of the youth of Io's surface as compared to the other Galilean satellites.

Voyager 2 took this picture of Io on the evening of 9 July 1979, from a distance of 1.2 million kilometers. On the limb of the satellite are two erupting volcanos. The plumes extend about 100 to 200 kilometers above the surface. *Voyager 2* found six of the eight volcanos discovered by *Voyager 1* still erupting four months later. The largest plume viewed by *Voyager 1* appears no longer to be erupting in *Voyager 2* photographs. And the eighth plume was not covered by the *Voyager 2* survey. Material sprays out of the volcanos in dome-like formation due to the lack of any significant atmosphere on the satellite. Io certainly seems to be the most active object in the solar system as far as volcanic activity is concerned.

Center row (pages 184, 185):
A comparison of the sizes of the Galilean satellites with the planet Mercury and the Moon. Also shown are their relative orbital spacings along with the innermost satellite Amalthea. *Voyager 1* and *2* missions proved to be a great revelation to astronomers about the diversity of the Galilean satellites.

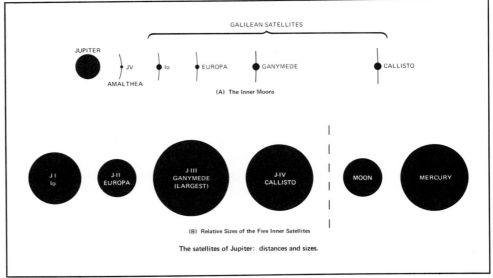

GALILEAN SATELLITES

JUPITER · JV · Io · EUROPA · GANYMEDE · CALLISTO

AMALTHEA

(A) The Inner Moons

J-I Io · J-II EUROPA · J-III GANYMEDE (LARGEST) · J-IV CALLISTO · MOON · MERCURY

(B) Relative Sizes of the Five Inner Satellites

The satellites of Jupiter: distances and sizes.

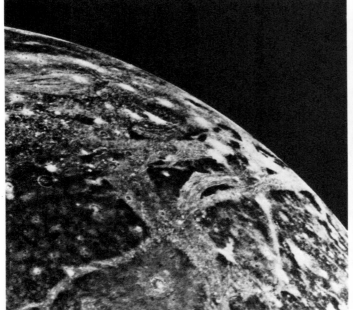

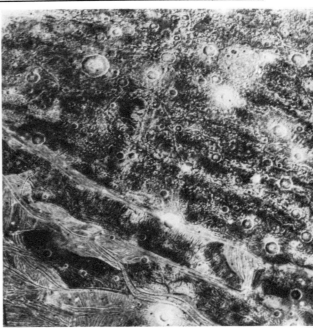

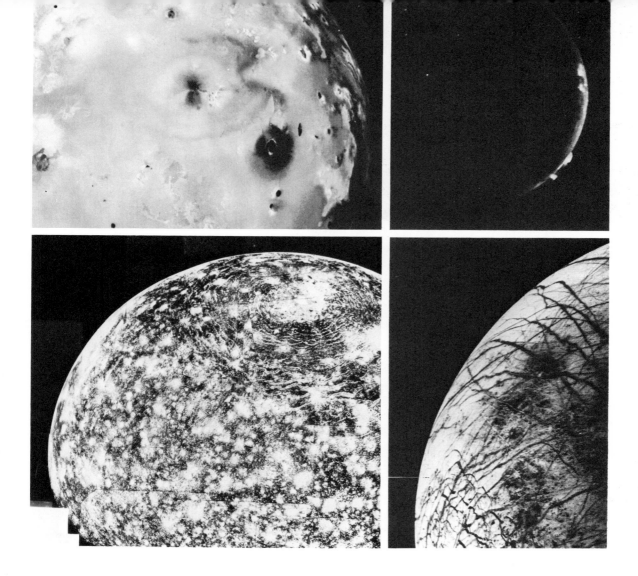

On the morning of 6 March 1979, *Voyager 1* obtained this photograph of Callisto, the fourth and most distant of the Galilean satellites, when 202,000 kilometers from the satellite. The dark surface has been very heavily cratered and thus is probably the oldest of those of the Galilean satellites. The bright spot near the limb is an impact basin with concentric rings around it. This probably shows the response of the icy surface to shock waves moving through the crust by the impact.

This picture of the second Galilean satellite Europa was taken by *Voyager 2* during a close encounter on the morning of 9 July 1979. The surface of Europa appears to be a crust of ice perhaps 100 kilometers thick. The complex array of streaks appear to be cracks in the crust which have been filled by material pushed up from below. Note that there is no indication of hills and valleys along the limb indicating a relatively smooth surface. There is also a lack of impact craters in the crust. Thus Europa's surface is also fairly young, but probably for a different reason than that of Io. No volcanic activity was seen on Europa by either *Voyager 1* or *2*.

Bottom row (page 184):
Ganymede, Jupiter's largest satellite, was photographed in this picture by *Voyager 1* on 5 March 1979, from about 253,000 kilometers from the surface. The smallest features visible are about 2.5 kilometers across. Numerous impact craters can be seen on the surface, many with bright ray systems. The striking features are the light-colored bands traversing the surface that contain alternate light and dark streaks.

Voyager 2, when about 150,000 kilometers from Ganymede, took this picture on 9 July 1979. In the foreground is typical grooved terrain as seen by *Voyager 1*. Note that it consists of intersecting bands of closely-spaced, parallel ridges and grooves. This grooved terrain is somewhat less cratered than the dark terrain. Therefore it is probably younger. Many of the craters have light-colored halos and rays, while some craters have dark halos. No good explanation exists for Ganymede's surface features.

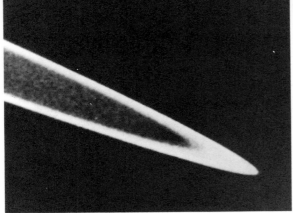

FIGURE 9.8
Jupiter's ring system. This photograph was taken by *Voyager 2* on 10 July 1979 when the spacecraft was 1.5 million kilometers from the planet. The brightness of the ring suggests that the particles composing it are small. Seen within the inner edge of the bright ring is a fainter ring that may extend all the way down to Jupiter's cloud tops.

9.2
Saturn

GENERAL PROPERTIES AND TELESCOPIC APPEARANCE

Its rings make Saturn, sixth planet from the sun, one of the most remarkable objects in the heavens. Brighter than all the stars but Sirius and Canopus, it shines with a steady ashen color. Saturn is second among the planets in mass and size.

It is twice as far from us as Jupiter, but the markings that we can see on the noticeably flattened disk of Saturn faintly resemble the banded cloud structure of Jupiter's atmosphere (see Figure 9.9). However, the coloration is more restrained and the details are less distinct. On rare occasions a bright spot may appear.

The mean diameter of Saturn, minus its ring system, is 9.5 times that of the earth and its mass is 95 times greater. Its density is the lowest of any planet, 0.7 times that of water. Rapid rotation and an unusually low density give it more polar flattening than any other planet, about 11 percent. We can detect weak radio emission in low-frequency bursts synchronized with the planet's 10.2-hour rotation period. Like the planet Jupiter, Saturn radiates about twice as much heat as it receives from the sun.

Saturn's spectrum, unlike Jupiter's, has stronger absorption lines for gaseous methane than for ammonia. Ammonia's visibility is reduced because of the lower atmospheric temperature on Saturn. At the measured cloud temperature of −180°C, the ammonia has probably formed small solid crystals, leaving exposed the molecular hydrogen and the gaseous methane that are still easily observable spectroscopically. Minute amounts of combinations of hydrogen and carbon like those on Jupiter have been detected in the infrared spectrum. Saturn's atmospheric chemistry and internal structure are believed to be similar to those of Jupiter.

RINGS OF SATURN

The circular rings lie in a plane coinciding with Saturn's equator. During the 29.5-year period of the planet's revolution around the sun, the rings are observed obliquely at different angles from the earth, as shown in Figure 9.9. At intervals of a little over 7 years, the rings change orientation relative to the earth from about 45° to parallel to the line of sight, at which point they practically disappear. They may be only a few kilometers thick.

Three concentric rings have been known for some time and are labeled A, B, and C in order of decreasing distance from Saturn. The outside diameter of the ring system is 274,000 kilometers, and the inside diameter is 140,000 kilometers. The bright ring B is separated from ring A by a space of 4000 kilometers, called *Cassini's division*. Next is the semitransparent ring C, the so-called crepe ring. An exceptionally faint D ring has been reported inside of the C ring based on ground-base observations. Outside the A ring, two other faint rings, E and F, have been found by the *Pioneer 11* spacecraft.

The rings vary in width and are believed to be composed of particles consisting of or at least covered by water ice, which vary in size from a few centimeters up to several meters. Each particle pursues its independent orbit around Saturn in accordance with Kepler's third law. This is shown by a Doppler shift in the spectral lines of reflected sunlight. The farther out from the planet, the lower are the particles' speeds. (A solid ring structure, on the other hand, would rotate fastest at its outer rim.) The entire ring system lies within the critical distance called the *Roche limit*, equal to about 2.4 Saturnian radii. This limit is named after the nineteenth-century French mathematician Edouard Roche, who found that inside this limit the gravitational attraction exerted by a planet on two ad-

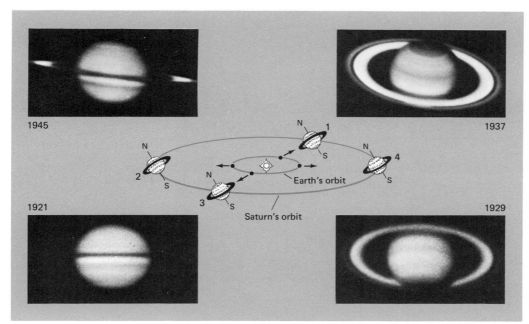

FIGURE 9.9
Changing views of Saturn's rings. The rings are inclined 27° to the plane of Saturn's orbit.

jacent orbiting particles is larger than the attraction of the two particles for each other. Whether the rings were formed inside the *Roche limit* by a satellite breaking up or whether Saturn's gravitational force prevented primordial particles from coalescing to form a satellite is unknown.

Cassini's division, in which there is a relative absence of particles corresponds to a place in the ring structure where a hypothetical particle would orbit in a period that is a simple fraction of the orbital periods of the inner satellites. A particle moving in this region is subjected by any one of these satellites to repeated gravitational disturbances that may pull the particle out of the region.

SATELLITES OF SATURN

All eleven satellites lie outside the Roche limit. Two of the satellites are particularly interesting. Titan is 1.6 times larger than our moon and in fact is larger than all the satellites in the solar system; it is even larger than the planet Mercury. It has a rotation period of 16 days. It is unusual because of its reddish brown atmosphere, which contains primarily methane. Its atmospheric pressure possibly exceeds that of the earth, while its surface temperature is probably less than 100 K. From a distance of 370,000 kilometers, the *Pioneer 11* photograph shows Titan as a smooth and hazy body. Another satellite, Iapetus, may be partially snow-covered. One side is very bright and the other is rough and dark. The remaining satellites appear bright, perhaps because of their light, icy composition.

EXPLORATION OF SATURN

After *Pioneer 11* sped by Jupiter, the immense gravitational attraction of the planet accelerated the spacecraft into a new trajectory. Traveling at more than 170,000 kilometers per hour, it headed back across the solar system for a rendezvous with Saturn in September 1979. In June 1978, a few months short of four years since leaving Jupiter, it again reached Jupiter's orbit on its way to Saturn, having been carried about 16° above the plane of the ecliptic. It became the first spacecraft ever to look down upon the solar system. On 1 September 1979 *Pioneer 11* passed within 3500 kilometers of the outer edge of the visible rings and to within 21,400 kilometers of the planet itself (see Figure 9.10). Saturn's magnetic field is surprisingly weak with its magnetic axis aligned with its rotation axis.

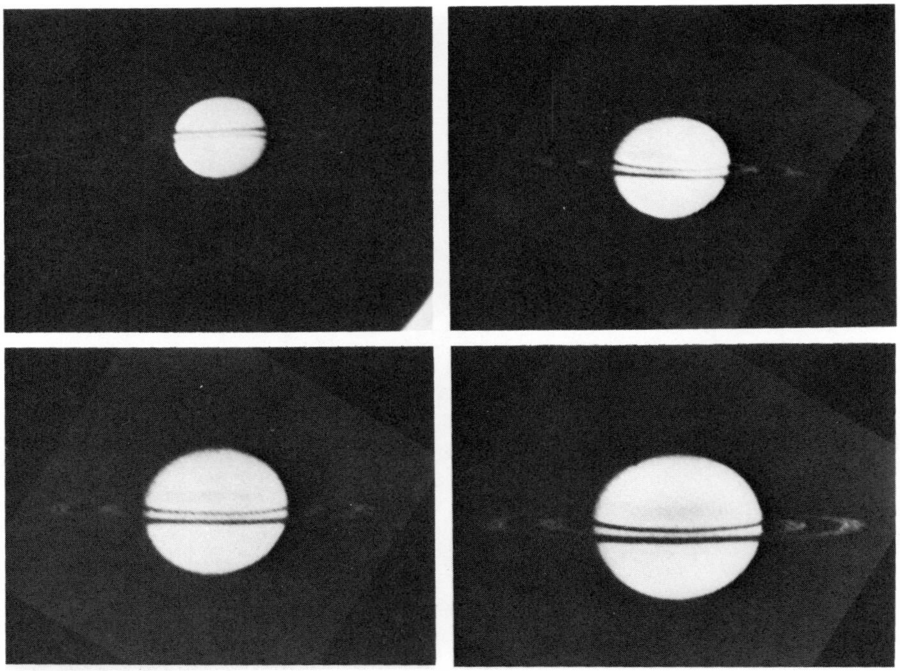

FIGURE 9.10
Saturn and its rings. Four photographs taken by NASA's *Pioneer 11* spacecraft as it approached the planet. The four pictures were taken from distances of 8.4 million kilometers (top left), 5.6 million kilometers (top right), 5.5 million kilometers (bottom left), and 4.3 million kilometers (bottom right). The band structures of the clouds are as distinctive at these distances as are those of Jupiter.

The trajectories of both *Voyager 1* and *2* are such that they will be given a gravitational assist by Jupiter, launching them toward Saturn and its satellite system. *Voyager 1* makes a close approach to the satellite Titan, after which it will come within about 3.3 Saturnian radii of Saturn in early November 1980. It will have a face-on view of the planet's south polar region and the ring system. The present plans call for the *Voyagers* to pass outside the ring system, for reasons of safety to the craft, rather than between it and the planet. The remaining satellites of Saturn will be surveyed before *Voyager 1* leaves the vicinity of Saturn about six weeks later, eventually exiting the solar system and spending the rest of time wandering through interstellar space. Since the orbital planes of Saturn's satellite Titan and the planet Uranus are very different, it is not possible for *Voyager 1* to make a close approach to Titan and then proceed to Uranus.

The flight of *Voyager 2* is scheduled to pass within about 2.7 Saturnian radii of Saturn in August 1981. Its closest approach to Titan will be about 353,000 kilometers. Plans call for *Voyager 2* to repeat the observations of *Voyager 1* but at different distances and angles, giving a somewhat different perspective. After leaving Saturn, it will go on to a rendezvous with Uranus in January 1986.

9.3
Uranus

DISCOVERING URANUS

"In examining the small stars in the neighborhood of H Geminorum I perceived one that appeared larger than the rest; being struck with its uncommon appearance . . . I suspected it to be a comet." So wrote William Herschel on the night of 13 March 1781 in his observing journal. Astronomers vainly tried to derive a cometary orbit for the object; then they realized that this was a new planet moving in a nearly circular orbit approximately twice as far away as Saturn. A check of older records revealed that the planet had been charted as a star many times in the preceding century. Just barely perceptible to the naked eye, it had escaped detection because of its very slow motion relative to a background of stars.

GENERAL PROPERTIES

Uranus has a diameter four times larger than that of the earth and a mass almost fifteen times greater. Its average density is slightly higher than that of water. In a large telescope the slightly flattened disk is seen as apple green. Nearly 3 billion kilometers from the earth, it presents an almost featureless appearance. What few atmospheric features we can see appear to be indistinct bands with occasional vague markings. As with Jupiter and Saturn, we are probably seeing clouds. Both molecular hydrogen and methane have been identified spectroscopically in the atmosphere. Ammonia, unquestionably present, has been frozen out of the atmosphere, accounting for its absence in the spectrum of reflected sunlight. Uranus appears to emit nonthermal radio bursts, which apparently originate deep within its atmosphere. As mentioned earlier, Uranus alone among the Jovian planets appears not to have an internal source of heat. A model of the planet's structure (see Figure 7.6) suggests that the interior is divided into a rocky core and an intermediate shell of ice surrounded by a fairly thick layer of molecular hydrogen.

Uranus's rotation has a peculiarity. Its axis is tilted 98° to the perpendicular of its orbital plane—that is, it lies on its side. We see its rotation in the reverse direction from that of any planet but Venus. For Uranus, the retrograde rotation is due to the peculiar inclination of the axis, while for Venus it is a true reverse rotation. When its axis is in our line of sight every 42 years (half the sidereal period), we observe either its sunlit northern or southern hemisphere, while the opposite hemisphere is dark. One-quarter or three-quarters of its period later (21 years or 63 years), its axis is at right angles to our line of sight, and we observe both the northern and southern hemispheres (see Figure 9.11). Recent evidence suggests that the rotation period may be more nearly 23 hours rather than the 12.3 hours listed in Table 7.4.

URANUS'S SATELLITES AND RINGS

Uranus has five satellites that revolve in the same sense as Uranus, and their orbital planes coincide approximately with Uranus's equatorial plane (see Figure 9.12). Thus the origin of the satellite system is closely related to that of the parent planet. During the 84-year cycle of Uranus's revolution around the sun, we see the orbital motion of the satellites at all possible angles to the line of sight, between edgewise and face-on.

Occasionally a planet will pass between the earth and a star. Such an event is called an *occultation* (from the Latin word meaning "hiding"). In recent years as-

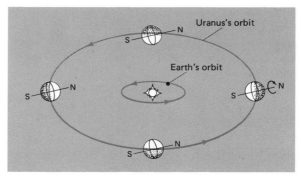

FIGURE 9.11
Hemisphere views of Uranus.

FIGURE 9.12
Uranus and three of its five satellites.

tronomers have carefully monitored these occultations, since the time and place on the earth at which the occultation will be visible can be calculated. It requires a precise knowledge of the planet's orbit to make such a calculation. The preciseness with which the prediction is confirmed by the observation in turn tells us how well we really know the orbit. Photometric observations made at the time of the occultation can also provide the astronomer with information about the planet's atmosphere, or lack of it, as well as the planet's diameter. As the planet begins to occult the star, its atmosphere, which is partially transparent, covers the star first, so that there is a gradual dimming of the star. If there were no atmosphere, the star's brightness would remain constant until the opaque body of the planet cut off all light. The change is sudden, not gradual.

In this manner astronomers in South Africa, Australia, India, China, and particularly aboard the Kuiper Airborne Observatory (see Figure 7.5) flying high over the Indian Ocean, discovered a ring system around Uranus on 10 March 1977. About a half hour before the occultation was to take place, the star's light dimmed unexpectedly for a few seconds, followed by four other dips in brightness minutes later. The sequence was repeated as the star passed beyond the disk of Uranus on the other side. Since the discovery of the original five rings, four less prominent rings have been identified, making a total of nine rings.

The rings appear to be very narrow, not more than 50 to 100 kilometers in width, and they lie close to the planet's equatorial plane. The origin of the rings is closely related to that of Uranus. The outer ring appears to be slightly elliptical, with the radii for all the rings ranging from 42,000 to 48,400 kilometers. All the rings are dark and have sharp edges. They may consist of carbonaceous (carbon-bearing) material similar to that found in certain meteorites and asteroids. Some of the rings have now been detected with the 5.1-meter Hale reflector in the infrared region of the spectrum at 2.2 microns. Less sunlight is reflected by Uranus at this wavelength than at others since there is strong absorption by methane in the atmosphere. Thus the brightness of the planet does not overwhelm that of the very faint rings.

VOYAGER'S ENCOUNTER WITH URANUS

Voyager 2 will approach Uranus in January 1986, at right angles to the satellite system, in a bull's-eye fashion. Such an encounter should offer a dramatic view of the newly discovered dark rings that lie in or close to the planet's equatorial plane. The exact details of the approach will be decided during the long voyage from Saturn to Uranus. Only one of the satellites can be examined at close range if the spacecraft is to make a close approach to the planet.

9.4
Neptune

MATHEMATICAL DISCOVERY

For many years after Uranus was accidentally discovered in 1781, astronomers were perplexed that,

even after allowing for the perturbations of Jupiter and Saturn, Uranus's orbital behavior was less predictable than that of the other planets. The discrepancy was finally resolved in 1845 and 1846 by two astronomers, John Adams in England and Urbain Leverrier in France. By a brilliant application of the law of gravitation, they arrived independently at the same conclusion: a disturbing body beyond the orbit of Uranus was the culprit. Using only pencil and paper, both men succeeded in almost pinpointing this unknown object, Adams a little earlier than Leverrier. Through misunderstanding and lack of proper sky charts, the search conducted in England at Adams's request was prolonged by delays. Leverrier's results were communicated to Johann Galle, the Berlin Observatory astronomer, who received the information on 23 September 1846. Within half an hour, using the observatory's 9-inch refractor, Galle located the new planet among a group of eight stars whose positions had been charted on a recently prepared map. It was found within 1 degree of the predicted position and was about twice as bright as estimated and 3 astronomical units farther out. Though the honor of discovery was first accorded to Leverrier, history now grants recognition to both men.

CHARACTERISTICS AND SATELLITES

Looking at Neptune through a telescope, we see a slightly flattened, bluish green, almost featureless disk. Observers at times have reported irregular, indistinct markings and a bright equatorial zone vaguely similar to that of Uranus. In the infrared part of the electromagnetic spectrum, between 5 and 20 microns, Neptune exhibits emission features that are indicative of an upper atmospheric temperature in excess of 100 K. Also, it radiates about twice as much heat as it receives from the sun, indicating that it must have an internal heat source. Neptune's diameter is about 3.5 times that of the earth. Its mass is 17 times greater and its mean density is one-third that of the earth. Both hydrogen and methane have been spectroscopically detected in its atmosphere. Helium and ammonia undoubtedly are there but are not spectroscopically observable. All the ammonia and part of the methane have crystallized into solid matter at the atmospheric temperature of about −220°C. Internally, Neptune resembles Uranus (see Figure 7.7).

The larger of Neptune's two satellites, Triton (see Figure 9.13), orbits the planet in 5.9 days in a direction

FIGURE 9.13
Neptune and its nearer satellite, Triton.

opposite to the planet's eastward rotation. Triton is 1.3 times larger than the moon, with a mass nearly double the moon's and with a slightly higher density. Analysis of Triton's infrared spectrum indicates the possibility of some frozen or gaseous methane on or near its surface. The smaller satellite, Nereid, takes nearly a year to swing around Neptune in a highly elongated ellipse, from 1.45 million kilometers to 9.6 million kilometers from the planet. The satellite may have been thrust into its eccentric orbit from a more regular orbit during a close encounter with Triton.

If all goes well, *Voyager 2* will be launched toward Neptune following its encounter with Uranus. It should reach Neptune sometime in 1990. Because of the planned trajectory, it will not be possible to have *Voyager 2* rendezvous with Pluto before it leaves the solar system.

9.5
Pluto

DISCOVERING PLUTO

Percival Lowell, founder of the observatory bearing his name in Flagstaff, Arizona, was convinced by his calculations begun in 1905 that minute discrepancies still complicated the orbit of Uranus. (Neptune had not been observed long enough to provide useful data.) He concluded that the irregularities might be caused by a planet beyond Neptune. Several years of intermittent and unproductive search passed, and then the hunt was resumed in January 1929. The Lowell Observatory had acquired a 13-inch photographic refractor and put a young assistant, Clyde Tombaugh, to work on the new search. After a year of photographing star fields along the ecliptic and later all over the sky, Tombaugh made the historic find in January 1930.

Lowell astronomers carefully followed Pluto's slow movement among the stars, holding up the announcement of its discovery for nearly two months. Positive that they had found a new planet beyond the orbit of Neptune, they announced it on 12 March 1930. However, the mass of Pluto is so small that it could not have produced the perturbations of Uranus's orbit used in Lowell's calculations—the discovery was accidental. Nevertheless, Lowell's important contribution was recognized and he was given credit posthumously for being instrumental in discovering the planet.

Pluto moves very slowly. It was first thought that Pluto's relatively high angled eccentric orbit and slow rotation were due to its being a former satellite of Neptune captured by the sun. However, the recent discovery of a satellite to Pluto makes such a possibility less likely. For a small part of its orbit, it is closer to the sun than Neptune. The two planets are not in danger of colliding, though, because their orbits do not now intersect.

CHARACTERISTICS

Pluto's brightness varies slightly, presumably because sunlight reflects unevenly from its rough surface. Photoelectric observations of these variations reveal that the period of Pluto's rotation is 6.4 days. The planet's small disk, measured with difficulty even in a large telescope, is less than one-fourth the earth's diameter. Its mass is at most a few percent that of the earth's, which leads to a best guess for the mean density of about 1 gram per cubic centimeter. Infrared observations suggest a surface composition dominated by ices, which is consistent with the low mean density. Pluto's low surface gravity and extremely low temperature mean that its atmosphere is a tenuous one of possibly methane or neon. The infrared observations are consistent with the prediction that the surface is covered with methane ice or frost. However, its observed reflectivity is not consistent with a prediction of methane ice or frost acting alone.

Pluto appears to have a satellite, as shown in Figure 9.14. It was discovered in June 1978 by James W. Christy at the U.S. Naval Observatory station in Flagstaff, Arizona. The discovery was made during the examination of photographs taken in April and May 1978 as part of the routine task of refining data on the planet's orbital motion. As you can see in Figure 9.14, the image of Pluto is elongated, with a bulge that is hardly a mountain on the surface. Although the photographs do not clearly resolve the satellite, its estimated diameter is around 850 kilometers. If Pluto and its satellite have equal mean densities, then the satellite is about 5 to 10 percent of the mass of Pluto, making it by far the largest satellite in the solar system in comparison to its parent planet (the moon is about 1.2 percent of the earth's mass). The satellite orbits Pluto in an approximately circular orbit at an estimated distance of 20,000 kilometers from the center of the planet. The orbital period is 6.4 days.

9.6
Comets

DISCOVERIES AND APPEARANCE

Comets are some of the most spectacular bodies of the solar system and appear unexpectedly in all parts of the sky. They are often discovered accidentally in photographs taken by professional astronomers for other purposes or by amateur astronomers methodically searching for them with field glasses. Comets are named after their discoverers. A record number appeared in 1977, five new comets and fifteen old ones. Only once every other year or so does a comet become bright enough to be seen with the naked eye. A spectacularly bright comet appears about once or twice each decade.

The usual telescopic comet appears as a small hazy object with a roundish nebulosity called a *coma* (head) and occasionally a short tail. A brighter comet has more interesting features: an enlarged coma surrounding a small bright *nucleus* with possibly a sunward spike and a well-formed *tail* that points away from the sun, as shown in Figure 9.15. The size of the

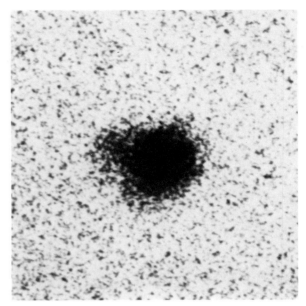

FIGURE 9.14
Presence of satellite orbiting close to Pluto. The bulge at the top of the image of Pluto is probably a satellite. Speckle appearance of photograph is the graininess of the photographic emulsion.

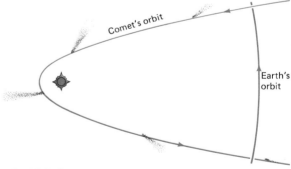

FIGURE 9.15
Changing appearance of a comet's tail. The tail points away from the sun because the comet's material is driven radially outward by the solar wind and the pressure exerted by sunlight. Normally the tail of a bright comet begins to become prominent at about the earth's distance from the sun. It subsequently brightens and lengthens as it nears the sun and its most spectacular near perihelion passage. As it recedes from the sun, its appearance may be different from that during its approach.

coma may vary from about 20,000 kilometers to well over a million kilometers. The comet's mass is concentrated in the nucleus, which is the comet proper and which may be a fraction of a kilometer to several kilometers in diameter. Well-developed tails usually form when the comet is within the earth's orbit and can be millions of kilometers long. Very rarely will the tail be longer than the earth's distance from the sun.

We know that comets are flimsy structures of low density from the following evidence: (1) they cannot be observed upon the solar disk when they pass in front of the sun; (2) we can clearly see stars through the tail and the outer portions of the head; (3) from night to night changes in brightness and size are observed in the head and are even more pronounced in the tail; (4) they are easily perturbed by gusts of solar wind and by tidal, gravitational, or other disruptive forces.

ORBITS OF COMETS

The first positive evidence that comets are extraterrestrial objects was found by sixteenth-century astronomer Tycho Brahe. He tried to find the parallactic displacement of the comet of 1577 among the stars, com-

TYCHO BRAHE (1545–1601)

Tycho Brahe was descended from an old Danish line of noblemen. In 1559 he enrolled at the University of Copenhagen, where he studied rhetoric and philosophy in order to prepare for a career as a statesman. His interest in astronomy was first aroused by a partial eclipse of the sun that occurred on 21 August 1560. So impressed was he with the accuracy of its prediction that he bought a copy of Ptolemy's works in Latin to instruct himself in mathematics and astronomy. To divert him from his preoccupation with astronomy, which his family deemed unworthy of a young nobleman, he was sent to the University of Leipzig, ostensibly to study law.

However, nothing could dissuade Brahe from his pursuit of astronomy. Despite some tightening of the purse strings, he managed to procure a small celestial globe to familiarize himself with the constellations and a number of astronomical tables. On comparing the computed positions of the planets with the actual positions, he discovered, when he was only sixteen years old, that the planetary tables were seriously in error. Thereupon, he resolved to examine more carefully the positions of the planets on the sky by constructing several observing instruments with which he could measure the positions of the planetary bodies more accurately than was ever done before. His instruments employed open sights like those used on a rifle, since the telescope had not yet been invented.

In 1570 he returned to Denmark. On 11 November 1572 his interest in astronomy was dramatically heightened by the sudden appearance of a supernova in the constellation of Cassiopeia, a remnant of which is still visible. The object was nearly as bright as Venus and could be seen in broad daylight. During the year that it remained visible to the eye, Tycho made systematic observations of its position, color, and brightness. His friends persuaded him to publish his results, which bore the title *De Nova Stella*.

In 1576 Tycho received a grant from King Frederick II of Denmark to establish an astronomical observatory on the small island of Hveen about 20 miles northeast of Copenhagen in the Baltic Sea. Two observatories were constructed, and they were equipped with numerous measuring instruments of the finest workmanship. A crew of research assistants carried out the necessary measurements, accumulating a wealth of observational data relating to the sun, moon, planets, fixed stars, and comets. Position accuracy was between one and two minutes of arc, the best that had ever been attained from naked-eye observations.

After twenty years of observing at Hveen, Tycho had a falling out with King Ferdinand's successor, King Christian IV. He left in a huff in 1597 to accept a post as Imperial Astronomer at Prague, taking his records with him. He hired as his assistant the young German mathematician, Johannes Kepler, who served under him for only a year until Brahe's death.

The fame of Tycho rests upon his ability as a skillful practical astronomer whose systematic observations were the most accurate before the invention of the telescope, and which led to Kepler's discoveries of the three laws of planetary motion (see pp. 29–30) and a reaffirmation of the Copernican system.

paring measurements from his observatory and other European centers, and decided the object was more distant than the moon. Isaac Newton demonstrated a century later that comets are members of the solar system and move in elliptical orbits according to his law of gravitation.

Comets fall into two groups, long-period and short-period, depending on the period of orbital revolution around the sun. Those in the first group, which astronomers believe are the majority of all comets, travel in highly elongated ellipses inclined at all angles to the ecliptic plane. Their periods range from hundreds to millions of years. The brighter members are usually among the most magnificent comets, with conspicuously long tails and well-defined nuclei. Some of these comets are approaching the sun perhaps for the first time. Several comets in the long-period category have grazed the outer parts of the sun. Frequently, solar tidal or nontidal forces break them apart during this close encounter, and the fragments travel on as independent comets along nearly identical orbits to the parent's orbit.

Dutch astronomer Jan Oort, studying how cometary orbits are distributed, suggested in 1950 that a cloud of comets of not more than a few earth masses surrounds the sun at an average distance of 50,000 astronomical units. Detached from this great reservoir by perturbations from nearby stars, a few begin to orbit the sun as long-period comets. Around a hundred thousand comets might have come close enough to the sun to be observable. The number of bodies in this cometary cloud may be in the billions.

The second group of comets, now numbering over a hundred, orbit the sun at small or moderate angles of inclination to the ecliptic plane, nearly all in the same direction as the planets. Roughly, every third or fourth new comet discovered has a short period, ranging from 3.3 years to 200 years. Approximately half have their aphelion (the point in the orbit most distant from the sun) somewhere near Jupiter. When their orbital history was traced back mathematically, it was found that these objects were initially moving in long-period eccentric orbits, bringing them on one critical occasion into a chance encounter with Jupiter. The great planet's attraction so modified their orbits that they now are part of Jupiter's short-period family of comets.

The nuclei of short-period comets have lifetimes of a few millennia. They may eventually evaporate down to their basic constituents of gas and dust or they may end up as asteroids (see page 169).

> Then I felt like some watcher of the skies when a new planet swims into his ken.
>
> John Keats

PHYSICAL AND CHEMICAL PROPERTIES

Today most astronomers agree that a comet is an icy core composed mostly of frozen water, methane, ammonia, and carbon dioxide, plus a little particulate matter and dust. This conglomerate forms the nucleus, which is surrounded by material vaporized by solar radiation to form the coma when the comet is close to the sun. As the comet approaches the sun, solar ultraviolet radiation breaks complex molecules down into simpler molecules of hydrogen, carbon, oxygen, and nitrogen, mostly CH, NH, NH_2, CN, OH, and C_2. These molecules are identified by the bright bands they produce in the spectrum of the comet. Emission lines of vaporized sodium are also present in the spectrum, along with some lines of iron, magnesium, and silicon when the comet comes very near the sun. The emission lines and bands are superimposed on the weak background spectrum of sunlight reflected from the cometary material.

A typical bright comet's structure and appearance are illustrated in Figure 9.16. The comet's head plowing through the onrushing solar wind creates a bow shock wave. High-energy electrons in the solar wind ionize the molecular gases in the coma. Chaotic magnetic fields in the solar wind sweep the charged molecules away from the coma at high speeds, forming the narrow, bluish ion tail (item 4 in Figure 9.16). What causes the wide, yellowish curved tail (item 3 in Figure 9.16)? The sun's radiation (not the same as the solar wind) pushes dusty material, which is flowing from the coma at different velocities, away from it at comparatively low speeds.

From their apparent structure and composition, it seems probable that the comets are a link or bridge between the solar system and the interstellar medium. We will defer a complete discussion of the interstellar medium to Chapter 14. However, we note here that many complex molecules are being discovered in dark clouds in the interstellar medium that when frozen could form icy structures, which are or may be similar to comets.

FIGURE 9.16
Schematic drawing of a comet superimposed on an actual photograph. This very bright comet was discovered by the Danish astronomer Richard West in November 1975 at the European Southern Observatory at La Silla, Chile. It was too far south to be observed in the northern latitudes until late February 1976, about the time it was rounding perihelion. Then it was bright enough to be seen with the naked eye near sunset—only the fourth comet to have been so visible this century. By March 1976 the nucleus had split into four different components, which separated from each other.

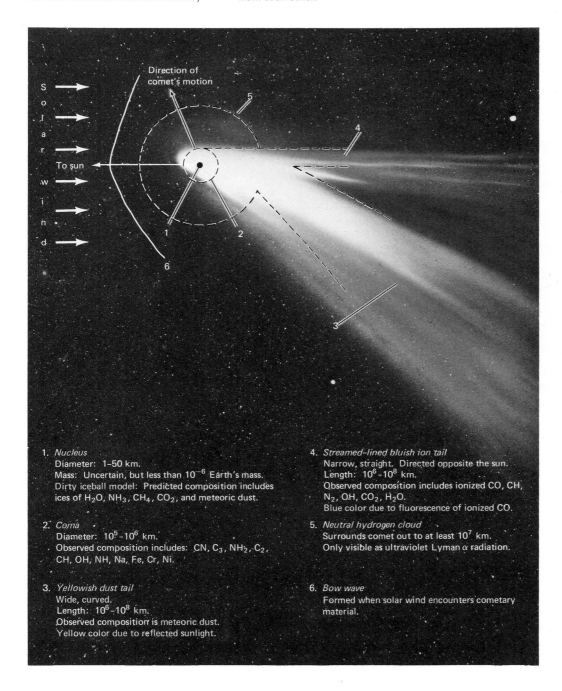

1. *Nucleus*
   Diameter: 1–50 km.
   Mass: Uncertain, but less than 10^{-6} Earth's mass.
   Dirty iceball model: Predicted composition includes ices of H_2O, NH_3, CH_4, CO_2, and meteoric dust.

2. *Coma*
   Diameter: 10^5–10^6 km.
   Observed composition includes: CN, C_3, NH_2, C_2, CH, OH, NH, Na, Fe, Cr, Ni.

3. *Yellowish dust tail*
   Wide, curved.
   Length: 10^6–10^8 km.
   Observed composition is meteoric dust.
   Yellow color due to reflected sunlight.

4. *Streamed-lined bluish ion tail*
   Narrow, straight. Directed opposite the sun.
   Length: 10^6–10^8 km.
   Observed composition includes ionized CO, CH, N_2, OH, CO_2, H_2O.
   Blue color due to fluorescence of ionized CO.

5. *Neutral hydrogen cloud*
   Surrounds comet out to at least 10^7 km.
   Only visible as ultraviolet Lyman α radiation.

6. *Bow wave*
   Formed when solar wind encounters cometary material.

When a bright comet appeared in the heavens late in the summer of 1682, Edmund Halley (1656–1742) calculated its orbit by using the method developed by his good friend Newton. He had done the same for the bright comets of 1531 and 1607 and was struck by the similarity in the three orbits. He concluded that the same comet must have made three revolutions in an elliptical orbit in a period of 75.5 years. He predicted it would return in 1758, but he died 17 years before the comet returned as he had forecast. In March 1759 it passed perihelion, delayed on its approach to the sun by perturbations from Jupiter and Saturn. Because Halley recognized that this comet makes periodic returns, it was named posthumously in his honor. Old records reveal that Halley's comet has been observed at every return since 239 B.C. Its appearance in A.D. 1066 is recorded in the historic Bayeux tapestry. Its orbit is shown in Figure 9.17. Halley's comet will next come near the earth in 1985 and 1986.

A much-publicized member of the long-period group is Comet Kohoutek, discovered in March 1973 when it was 4.7 astronomical units from the sun, ten months before passing perihelion. Early discovery gave astronomers ample time to prepare, so that a rich harvest of observations, extending throughout the electromagnetic spectrum, were made. In December 1973 it passed within 21 ×

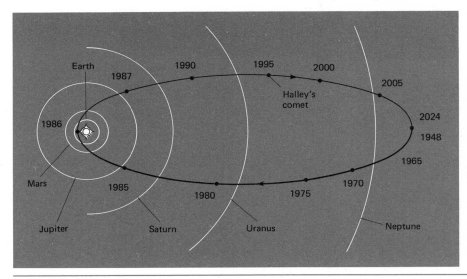

FIGURE 9.17
Orbit of Halley's comet.

SUMMARY

The Jovian planets constitute a family whose physical properties are very different from those of the terrestrial planets. They are much larger, more massive, and are composed chiefly of ten parts hydrogen to one part helium, which accounts for their low density. Furthermore, they rotate more rapidly and possess 32 of the 36 known satellites. Their distances from the sun range from 5 to 40 astronomical units, if we include outermost Pluto, a nonmember.

Jupiter, with its 14 satellites, is the largest and most massive planet, and in some ways it resembles the sun more than a typical planet. Its rapid rotation coupled with vigorous atmospheric convection gives rise to the light and dark parallel bands of various colors encircling the planet. The famous Red Spot, visible for more than three centuries, is a deep-seated, hurricanelike disturbance, four times the width of earth. The planet radiates about twice as much heat as it receives from the sun, implying that it has an internal source of heat. Surrounding Jupiter is an intense magnetic field, which extends out to enormous distances. The interaction of this powerful magnetic field with the charged particles gives rise to strong nonthermal radio emission.

The last of the known planets in ancient times, Saturn—the ringed planet—with its eleven satellites

FIGURE 9.18
Comet Kohoutek and its spectrum in the orange red region, revealing the presence of water vapor in the ionized condition (H_2O^+) (9 January 1974).

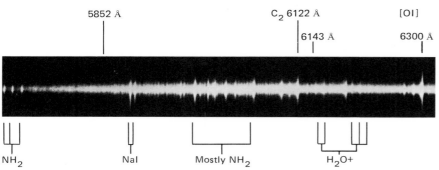

5852 Å C_2 6122 Å   [OI]

6143 Å   6300 Å

NH₂   NaI   Mostly NH₂ H_2O^+

10^6 kilometers of the sun. Astronomers accumulated more data (including *Skylab* observations) on this comet than on any other in history. Besides the expected presence of the common metals, silicates, and fractionated compounds of hydrogen, carbon, nitrogen, and oxygen, ionized water (H_2O^+; see Figure 9.18), hydrogen cyanide (HCN), and methyl cyanide (CH_3CN) were also detected. The expected hydrogen cloud surrounding the coma out to 10 million kilometers was observed in the ultraviolet in early January 1974 from an *Aerobee* sounding rocket.

ranks next in size to Jupiter. It possesses many of Jupiter's physical characteristics. Saturn, too, emits radio waves and radiates more heat than it receives from the sun. The rings appear to consist of small assorted particles like water ice. Saturn's large satellite, Titan, is bigger than Mercury.

The next planet, Uranus, is unique in three respects. Unlike the other Jovian planets, it apparently has no internal heat source; it orbits around on its side; and it is circled by at least nine very thin, extremely dark rings. Its atmosphere is quite clear; it consists mostly of hydrogen, some methane, and probably ammonia. Uranus has five satellites, which orbit in the plane of the planet's equator.

Distant Neptune has physical characteristics somewhat similar to those of Uranus. One of its two satellites, Triton, is larger than our moon, and it orbits Neptune in reverse fashion. Pluto, the planet that conforms to neither the terrestrial nor Jovian characteristics, has a surface consisting partly of frozen methane and has no observable atmosphere. Its diameter is comparable to that of our moon. A satellite has been discovered circling it in the same period as Pluto's rotation.

Our knowledge of the outermost regions of the solar system is very scanty. In this region are the comets, small icy conglomerates of particles, dust, and frozen gases orbiting the sun in highly eccentric orbits of very long periods. As comets near the sun, they evaporate from the heat, forming a halo or coma and long tails of gas or dust particles. The most famous comet, Halley's comet, travels round the sun in a period of from 75 to 76 years, somewhat beyond the orbit of Neptune. It next makes its closest approach to the earth in 1985 or 1986.

REVIEW QUESTIONS

1. What causes the banded appearance of Jupiter and Saturn?

2. What is one interpretation of the Great Red Spot on Jupiter?

3. How do we know that the rings of Saturn are composed of icy particles and are not solid?

4. What makes Titan such an interesting satellite?

5. How were Uranus, Neptune, and Pluto discovered?

6. If there is anything unusual about Venus, Uranus, or Pluto, what might it be?

7. Discuss the structure and radiation of the magnetosphere of Jupiter.

8. What is the presently accepted version of the physical structure and chemical composition of a bright comet's three main components: nucleus, coma, and tail?

9. How were the rings of Uranus discovered? Why are they extremely difficult to photograph?

10. Describe what happens to Io as it orbits around Jupiter.

11. Which planets did *Voyager 1* and *2* visit and when?

12. Describe the observed markings on the Galilean satellites and their interpretation.

13. What could be the source of the internal heat radiated by Jupiter or Saturn?

14. Why is the composition of the Jovian planets, particularly Jupiter and Saturn, similar to that of the sun?

15. Describe the rings of Uranus. Why are they so dark in color?

16. How was the satellite of Pluto discovered?

17. Why is Pluto considered to be a maverick planet?

18. Draw a sketch depicting a model of the interior of Jupiter.

19. Which satellites in the solar system move in retrograde orbits? What is a possible explanation for this anomaly?

20. Relate the history of Halley's comet. Why was this comet named after Halley, since he did not discover it?

10
Life in the Solar System

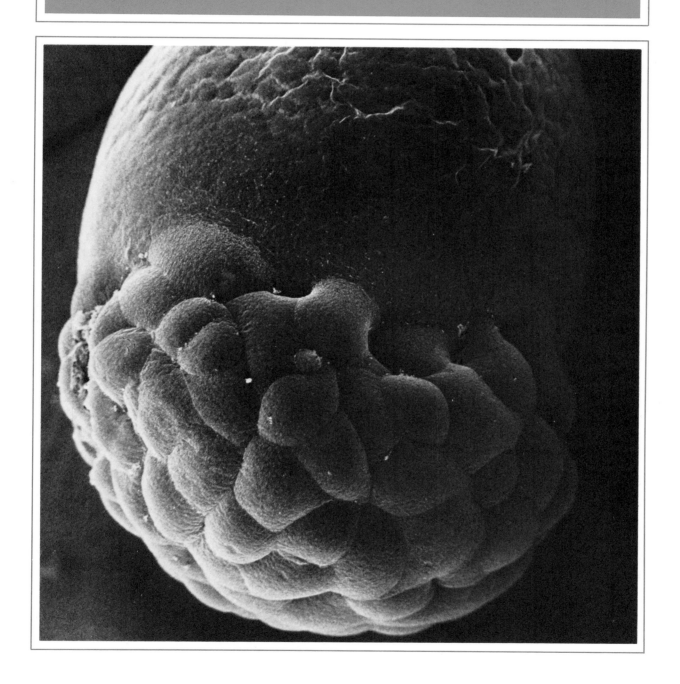

How did life begin on the earth? Are there other habitable worlds in the solar system? Can we find clues to life elsewhere in the space beyond our solar system? Have we any physical evidence on how planetary systems form? Are they common in the universe? These are perplexing questions that past generations have asked themselves. Today, we are somewhat closer to their answers. In fact, a new science—exobiology—is devoted exclusively to the study of extraterrestrial life.

Our exploration of life must begin with the earth and ourselves, since it is the only system of life for which we have any real information. From the earth, our study will move to the entire solar system, and in a later chapter we will come back to the question of life beyond the solar system.

Recall from Time's abysmal chasm
That piece of primal protoplasm
The First Amoeba, strangely splendid,
From whom we're all of us descended.

Arthur Guiterman

10.1
Life on the Earth

LIFE'S ESSENTIALS

A thin zone with the land, water, and air in which the earth's life flourishes—that is the *biosphere*. Had the earth formed somewhat closer to the sun than it did, the greenhouse effect would have dominated, resulting in a hot, sterile surface region similar to that on Venus. On the other hand, had the earth formed farther away from the sun than it did, water would have remained frozen as subsurface ice or polar caps, and the earth would resemble somewhat the cold Martian landscape. Our earth, therefore, is at the right distance from our star, the sun, for development of an active biosphere. The chemistry of the only kind of life we know is built on the element carbon and the solvent water, supplemented by the biologically important atoms nitrogen, oxygen, phosphorus, and sulfur.

The carbon atoms bond easily with other carbon atoms, producing long chains of carbon atoms to which other biologically significant atoms can bond—hydrogen, oxygen, and nitrogen. These molecules, whether or not they are a part or product of living matter, are called *organic molecules*. Water, be-

cause it can flow readily and remain in liquid form through a range of conditions, is the ideal solvent for organic compounds. From water come the hydrogen bonds, which give structural stability to long strings of proteins, nucleic acids, and other long-chain carbon compounds. A liquid water environment and moderate temperature make it possible for such long-chain carbon molecules to form, and they are the basis of life as we know it.

Even with this chemical basis, life would not have been able to develop and sustain itself without proper temperature, a supply of nutrients, self-regulating mechanisms, and the sun's energy. The sun is the prime source of energy in the life cycle.

HOW DID LIFE BEGIN?

But how did life come to be in this biosphere? How did it grow from nonlife? To discuss life elsewhere, we cannot start unaware of the chemical and biological evolution that apparently led to life on our planet. Biologists are not unanimous on all the factors in the definition of life, but most agree that life is different from nonlife: it evolves or changes (by chance mutations or otherwise) as time goes on while interacting with its environment in a unique way. For terrestrial life, evolution has led to intelligence, the aspect of extraterrestrial life in which we are most interested.

Until the mid nineteenth century, most people believed in spontaneous generation: that life rose from nonliving matter—worms from mud, mice from refuse, maggots from decaying meat, and so forth. By careful experimentation, Pasteur proved otherwise. Then Darwin set forth his theory of biological evolution. Darwin held that the more complex forms of life evolved from simple organisms over very long stretches of time, thus suggesting a unity for all of earth's life. Considering the problem of how life came from nonlife, he speculated that in some "warm little pond," such as the tidal ponds in Figure 10.1, proteins and more complex organic molecules could have been

◄ The nucleus of a cell—the minimum organized unit of matter that displays life.

FIGURE 10.1
Tidal ponds in which complex organic molecules may have synthesized from more elementary molecules such as ammonia, methane, water, and so forth.

synthesized from ammonia and phosphoric salts energized by sunlight, heat, or electricity.

Our contemporary ideas on *chemical evolution* actually began in 1924 when the Russian biochemist Alexander Oparin reintroduced chemical evolution as a necessary forerunner to biological evolution. In 1928 the English biologist John Haldane independently suggested an outline for chemical evolution, differing somewhat in details from that of Oparin. His is still the basis of our understanding of chemical evolution, which we will briefly outline in the remaining paragraphs of this section.

Earth's cooling and solidifying crust was wracked by volcanic activity that presumably vented water vapor, carbon dioxide, nitrogen, and smaller amounts of other molecules that are easily vaporized. These probably formed earth's early atmosphere. Today active volcanos discharge large quantities of water vapor, carbon dioxide, nitrogen, some sulfur, and traces of other gases (see Figure 10.2).

An alternative possibility is that hydrogen and methane were produced when carbon dioxide and water were reduced by the free iron present in earth's early crust. Some ammonia was formed by the combination of hydrogen and nitrogen present in earth's primitive atmosphere. However, opinion seems to be shifting away from a methane-ammonia atmosphere toward one that retained CO_2, N_2, and a small amount of H_2. Earth's magnetic field, which probably developed after the interior differentiated into its present layered structure, helped to keep this primitive atmosphere from being swept away by the solar wind. Out of the inorganic compounds containing chiefly hydrogen, carbon, nitrogen, and oxygen were synthesized the long-chain carbon molecules, such as the amino acids, and the simpler organic compounds; these were formed by energy from solar ultraviolet and visible light, electric discharges, and heat from radioactivity, volcanos, and meteoric impacts.

Subsequent cooling of the earth condensed water vapor, forming the warm seas and the shallow lagoons and pools that were destined to provide a haven for the development of organic compounds. From this primordial soup, over a long time, the more complex organic molecules or biological macromolecules, such as proteins and nucleic acids—the basic ingredients of life—were fashioned. The probable course that biochemical evolution followed from the primordial raw materials to the first living organisms is outlined in Figure 10.3. The cell is the basic unit of life from which complex organisms such as humans are built. We will investigate the cell and its components next.

FIGURE 10.2
Outgassing can be seen at current volcanic sites.

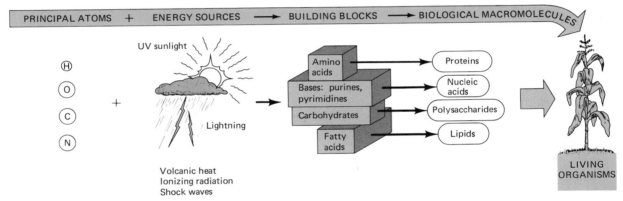

FIGURE 10.3
Chemical evolution leading to life. The chemical content of living matter is approximately 78 percent water, 15 percent protein, 5 percent fat, and 2 percent carbohydrate.

THE CHEMISTRY OF LIFE

An individual *cell* is sketched in Figure 10.4. Modern biology has shown that the cell is in general the minimum organized unit of matter that displays those properties we collectively refer to as *life*. Collections of cells go to make up the organism. Cells vary in size from bird eggs and nerve cells several feet long to bacterial cells one-tenth of a micron in diameter. The typical human cell is about 10 microns across.

The cell has a well-ordered, water-based substance, *cytoplasm*, surrounding the *nucleus*, which holds coiled, threadlike strands, the *chromosomes*. The chromosomes, which transfer hereditary characteristics to each generation of new cells, occur in pairs, with a fixed number in every cell of every species. (Human cells have 23 pairs.) The nucleus controls the cell's activities, structure, maintenance, and repair. It contains instructions for manufacturing the specialized cells (muscle, bone, liver, and so forth) and maintaining their functions to keep the organism alive. The cytoplasm outside the nucleus contains the complex organic molecules used in synthesizing amino acids and proteins, as directed by the genetic code in the DNA (*deoxyribonucleic acid*) molecule in the nucleus. The DNA molecule specifically determines the type of organism (human or elephant, for example).

Amino acids are small, chemically reactive organic molecules composed mostly of hydrogen, carbon, nitrogen, and oxygen, of which 20 are found in living organisms. They are arranged like beads and strung into long molecular sequences called *proteins*. A protein molecule may contain a hundred to a thousand amino acids. If each amino acid corresponds to a letter of the alphabet, then each protein is a sentence in a book of instructions. Because the 20 letters (amino acids) can be arranged in almost limitless ways to form a row (the protein), the possible number of proteins is enormous. Of these, only a small percentage appear in living matter. The sequence of amino acids in the protein molecule determines its specific function. Proteins serve both as structural material and as *enzymes*, or catalysts, that govern chemical reactions in the cell. (A catalyst is a chemical that speeds up the reaction without being consumed by it.) These chemical reactions give the cell properties that collectively can be called life. The proteins give any organism its distinguishing characteristics; synthesis of protein molecules is controlled by the genetic code in the DNA molecule.

The nucleic acid molecule, DNA, is very large and complex and is in the chromosomes of every living cell. One human cell has about eight hundred thousand DNA molecules. Since there are many varieties of living organisms with many chromosomes, there are many different forms of DNA. The DNA has two functions: it carries the heredity instructions for manufacturing proteins in the cell and it passes genetic information on to daughter cells during cell division by making copies of itself (replication). Its structure was first worked out jointly by the American biologist James Watson and the English chemist Francis Crick in 1953. They found that the molecule takes the form of a double helix. A short untwisted segment of the DNA molecule is drawn in Figure 10.5.

Essentially, DNA is a string of four kinds of subunits, called *nucleotides*, linked like a twisted ladder or a double helix. The nucleotides in turn are made up of bases (carbon-nitrogen compounds), sugars (carbon-hydrogen compounds), and phosphates

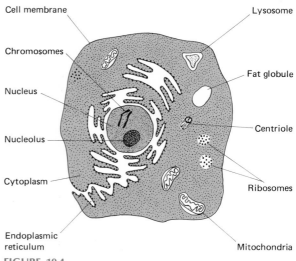

FIGURE 10.4
Structure of a cell. A million human cells could be placed on the head of a pin. See the nucleus on page 199.

(phosphorus-oxygen compounds). The sugars and the phosphates form the two handrails of the spiral ladder, with the bases, of which there are four, appearing together in definite pairs, forming the tread. The spiral is a right-handed spiral in which each tread is of the same size, at the same distance from the next, and turns at a rate of about 30° between successive treads. The genetic code is carried on the treads of the ladder, and the sequence of the four different nucleotides specifies the hereditary message. Thus the language of heredity is written in an alphabet of only four letters. For each protein potentially capable of being formed, a specific segment of the DNA molecule carries the information by which the 20 kinds of amino acid subunits in that protein are properly ordered during its synthesis. The number, type, and arrangement of the nucleotides in DNA determine the kind of organism that is created. A human cell has about 5 billion nucleotide pairs in the DNA of the 46 chromosomes. Different organisms have different sequences of nucleotides as well as different amino acid sequences in the proteins.

During cell division, replication of the DNA molecule begins with separation of the two spirals (see Figure 10.6) by breaking the hydrogen bonds between the bases, as shown in Figure 10.5. Each strand of the double helix directs the formation of a complementary strand to pair with it. This replication

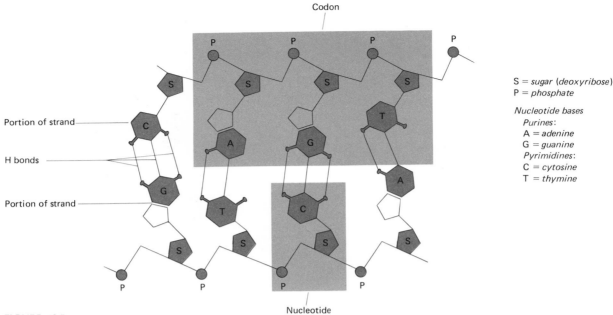

Codon

S = sugar (deoxyribose)
P = phosphate

Nucleotide bases
  Purines:
    A = adenine
    G = guanine
  Pyrimidines:
    C = cytosine
    T = thymine

Portion of strand

H bonds

Portion of strand

Nucleotide

FIGURE 10.5
Portion of a DNA molecule. Nucleotide bases form the treads of the ladder held together by the hydrogen bonds. The rails of the ladder contain alternating sugar and phosphate stiffeners.

continues until the entire double-twisted configuration is complete. Each of the two newly formed helices contains one old strand and one new one. They migrate to opposite ends of the cell before cell division. In the daughter cells are two daughter DNA molecules, with nucleotide sequences identical to those of the parent DNA molecule in the parent cell.

This, then, is the biological process that defines terrestrial life. The genetic instructions for all organisms are written in the same chemical language. Indeed, such a shared hereditary language is one reason to believe that all organisms on earth are descended from a single ancestor, one single instant some 4 billion years ago when all of earth's present life began.

BIOCHEMICAL EXPERIMENTS

Laboratory experiments copying the presumed composition of the earth's primitive atmosphere have chemically synthesized, with fair success, the building blocks of life—the organic molecules. In his pioneering investigation of 1953, Stanley Miller subjected a gaseous mixture of methane, ammonia, water, and hydrogen to an electrical discharge for about a week. He found the gas had turned into several amino acids, urea, and some other organic molecules. Varying the original experiment, other researchers have produced in the laboratory many of the life-associated molecules that are essential components of the nucleic acids.

More recent experimenters have had some success in the steps that link amino acids to form proteins, imitating the step from building blocks to biological macromolecules (see Figure 10.3). The most difficult laboratory experiment, synthesizing the nucleic acids under conditions that are supposed to have predated life, has not yet been achieved.

FREE OXYGEN, PHOTOSYNTHESIS, AND THE CARBON CYCLE

If free atmospheric oxygen was not here in the beginning, how did it become the second most abundant constituent? Some oxygen may have been released from the dissociation of the water vapor molecule into oxygen and hydrogen by the sun's ultraviolet radiation. The lighter, more rapidly moving hydrogen atom eventually escaped into space, but earth's gravitational field retained the heavier, slower-moving oxygen. Ox-

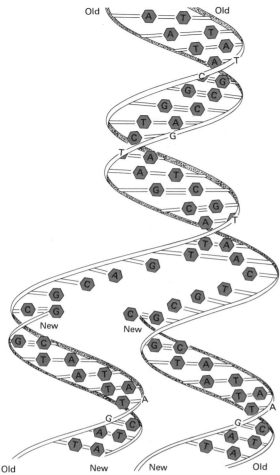

FIGURE 10.6
Unzipping of the DNA molecule to form a new DNA molecule (replication).

(O_3) formed from the combination of molecular oxygen (O_2) and atomic oxygen (O). This layer, 12 kilometers to 50 kilometers up, absorbed most of the solar ultraviolet light, keeping the radiation from being absorbed in the surface layers of the primordial seas, where it had been the primary source of energy for synthesizing organic compounds.

As the supply of organic compounds was depleted, the earliest organisms, which depended on them for food, went into a wholesale decline. Presumably the organisms that survived could utilize the visible wavelength radiation that penetrated the ozone shield. Green plants and plankton in the oceans, relying on photosynthesis, were favored in this new environment and became dominant in later development. The first organisms using photosynthesis appeared about 3.5 billion years ago. They have had the upper hand ever since.

Plants produce and store energy from sunlight, which animals then consume. By photosynthesis, plants and plankton absorb water, carbon dioxide, and solar energy to form the organic compounds they need for growth, and they release oxygen, their waste product. Respiration by animals and transpiration by decaying organic matter and groundwater in turn consumes the oxygen and releases carbon dioxide as a waste product.

10.2
Biological History of the Earth

ygen combines so readily with many atoms that some of the early oxygen atoms must have been rapidly depleted in oxidizing the crustal material of the young earth. Apparently most of the oxygen now in our atmosphere was produced by *photosynthesis*. This is the name for the process in which plants generate carbohydrates, using carbon dioxide and water as raw materials and sunlight as an energy source, with the release of free oxygen as a by-product. We have evidence of predecessors of photosynthesizing organisms in the microscopic one-cell algaelike fossils deposited in sedimentary rocks about 3.5 billion years ago.

Eventually, when enough free oxygen had accumulated, a protective atmospheric layer of ozone

Evolution in plants and animals is controlled by two forces: limits that the environment sets for the organisms and changes in their hereditary material. Organisms are constantly modified by chance mutations, sexual selection, and natural selection. Together these produce members of a species that can survive in a changing environment. Mutations occur at random. Unfavorable ones are eliminated because they limit the organism's ability to cope with its environment, to reproduce, or to compete with rival species. Relentless experimenting with cumulative mutation and natural selection have produced the great diversity in life on our planet. Alfred Wallace, an English contemporary of Charles Darwin, hit upon the theory of natural selection at about the same time as Darwin did, but Darwin's *Origin of Species* appeared in 1859 before Wallace had decided to publish his book. Darwinian evolution was the initial step toward our ideas on evolution.

Do we find enough evidence from paleontology and geology to follow biological evolution from the first living organisms to humans? A summary of the progress of evolution is shown in Figure 10.7. There

FIGURE 10.7
Evolution of living systems on the earth.

TIME (millions of years)	GEOLOGIC ERA	LIVING SYSTEMS EVOLVE	EARTH CHANGES
Present		Human beings (consciousness)	Ice Age (most recent)
70	Cenozoic		
	Mesozoic	First mammals	
220		Creatures with spines	
	Paleozoic		
600		Invertebrates	
1000		First multicelled creatures	Close to present atmosphere
2000			Oxygen increases
	Precambrian		
3000			Atmosphere enriched with oxygen by algae
		First microfossils (algae, bacteria)	Atmosphere water, methane, ammonia, and possibly hydrogen
4000		Evolution of carbon compounds (the building blocks of living systems)	A crust and water oceans
	Earth formation	Too hot and chaotic for living systems	An earth sphere forms
			Particles Gases A formless void
5000			

is reason to believe, on the basis of physical evidence, that the sun's radiation has not changed appreciably during earth's biological age. If the sun's radiation had drastically changed, the gradual evolutionary sequence would not have been preserved more or less unbroken in the biological record. However, contemporary theoretical ideas in stellar evolution predict that the sun's radiation was 25 percent less 4.6 billion years ago than it is today. This means that the early atmosphere of the earth must have been able to provide a more efficient greenhouse effect than it does at present. This in turn requires a different density, or composition, or both. Either a methane-ammonia or a CO_2-rich atmosphere would do it and thus perserve the biological record. This divergency in scientific opinion is typical of many questions in science in which experimental evidence and our conceptual thinking differ. In this case, as in many other examples, the majority favor theoretical understanding rather than experimentation.

Earth's geologic history divides into four eras of time: Precambrian, Paleozoic, Mesozoic, and Cenozoic. Each has a dominant biological species, as shown in Figure 10.7. After the primitive oceans formed, the first single-cell organisms of the Precambrian era evolved in the warm seas. The oldest known microscopic fossil remnants are those of a primitive bacteriumlike organism, which dates back 3.5 billion years and was found in the Precambrian sedimentary rocks of Swaziland, South Africa. Shown in Figure 10.8 are 3.5-billion-year-old microfossils of algae, some in the process of cell division.

In the Cambrian period of 600 million years ago, beginning with the Paleozoic era, the invertebrates (lacking a backbone) appeared; the first vertebrates, primitive fishes, appeared some 500 million years ago; and toward the close of the Paleozoic era 220 million years ago, the first reptiles appeared. It was only about 70 million years ago, a small fraction of the earth's age,

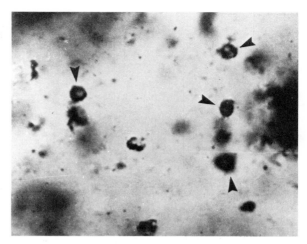

FIGURE 10.8
Preserved microfossils of 3.5-billion-year-old algae embedded in Figtree chert.

that small mammals such as the ancestors of modern horses, rodents, cats, and dogs, and toothless birds with feathers appeared (see Figure 10.9). They were followed by other species of mammals that eventually lead to human beings.

The primitive tree-dwelling mammals, such as the lemur, that developed stereoscopic vision and high mobility, and could oppose the thumb to the hand, were well adapted for survival. From this group, which roamed over eastern and southern Africa, we believe that a precursor of man known as *Australopithecus* evolved (see Figure 10.10). Bones of this creature more than 4 million years old have been found in Tanzania and Ethiopia. It stood about 3 to 5 feet in height and weighed from about 50 to 120 pounds. Life on the ancient grasslands led to further anatomical changes and to adaptation of jaws and teeth to masticate tough animal and plant foods. This creature's ability to hunt

Then wilt thou not be loth
To leave this Paradise, but shalt possess
A Paradise within thee, happier far . . .
They hand in hand with wandering steps and slow
Through Eden took their solitary way.

John Milton

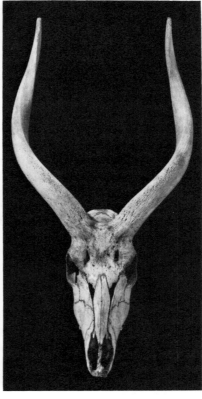

FIGURE 10.9
Fossil and modern nyala horns. Note that the modern horns (right) have changed very little from those of their prehistoric ancestor (left).

and forage for food was later enhanced by speech, which allowed actions to be coordinated among members of the group. As far back as 2 million years ago its descendants, *Homo habilis,* could fashion and use tools. As brain capacity increased, the hominid could profit from experience and choose between alternatives, improving the chances of survival. Modern human beings (*Homo sapiens*) are believed to be descendants of the successor to *Homo habilis,* who is known as *Homo erectus,* one of the two forms that branched out from the ancestral hominids. Two million years ago we were not yet "man," but with *Homo erectus* one million years ago we were.

10.3
Ecology of the Earth

For most of humanity's 2 million years people have struggled just to survive. In the ten thousand years since people learned to domesticate animals, grow crops, and live in settlements, our numbers have bloated. The world's population rose from about 10 million in ancient times to about 500 million by the eighteenth century, and since then it has grown to about 4 billion. By the year 2030, at today's rate of growth, the world's population will be 12 billion. Pressures for food, water, and breathing space will be enormous. Grim prospects of overpopulation and overwhelming demands on the world's resources were described as early as 1798 by Thomas Malthus in *An Essay on Population.* It is interesting to note that the germ of the concept of natural selection came to Darwin and Wallace independently, with each having read Malthus's treatise early in their work.

From the biosphere humans took what they needed for existence and yet tried to preserve a stable physical and chemical environment in which they could function. But trying to manage the biosphere through agriculture and other technological processes is not without an element of risk. By altering the environment in an irreversible fashion, we can reduce the

FIGURE 10.10
Artist's conception of an early human, *Australopithecus.*

chances for our survival. Our species is not guaranteed perpetual existence on this planet—or even continuation of our present level of intelligence. We can last only by carefully considering what human and natural changes are doing to the earth as a whole.

For a hundred years changes in the environment—wrought by humans burning fossil fuels, clearing large forests, and cultivating and excavating the land—have accelerated tremendously. This increase in human activity may cut the biosphere's productivity by loading the atmosphere with dust and carbon dioxide and thereby diminishing photosynthesis. If photosynthesis were to totally cease, animal life in the biosphere would be in jeopardy within about two thousand years, the recycling period of atmospheric oxygen. For carbon dioxide the recycling time is a mere three hundred years. We know that our earth carries only a finite life-support system; we must be particularly careful, therefore, and learn to use its resources wisely.

10.4 Life Beyond the Earth

ORGANIC MOLECULES IN METEORITES

Among meteorites that have been found is a small subgroup of stony meteorites, called carbonaceous chondrites, that contain up to about 5 percent organic molecules. That chondritic meteorites carry organic compounds has been known for more than a century, but only when techniques had been developed for studying the lunar rocks could the organic compounds be definitely ascribed to an extraterrestrial origin. Identification is complicated because contamination by terrestrial organic molecules is possible, though the chance is lessened when the specimen is found soon after its fall.

Interest in these meteorites was renewed when workers at the NASA Ames Research Center in California discovered amino acids in a fresh chondritic specimen that fell near Murchison, Australia, in September 1969 (see Figure 10.11). Extraterrestrial amino acids have since been found in other meteorites, but only a few of these acids are in living cells of earth organisms. Amino acids from meteorites are almost an equal mixture of right- and left-handed molecules. (*Handedness* is the direction in which the plane of polarization rotates when a beam of polarized light passes through the material; see page 181.) Amino acids of earth origin usually are left-handed, supporting the idea that the meteoritic amino acids are of extraterrestrial origin. Biologically produced amino acids would probably be almost exclusively either left- or right-handed; so the mixture of both types in the meteorite specimens suggests a chemical, not a biological, origin.

Along with these acids, several other organic molecules have been found. We cannot at once conclude, however, that these organic precursors to life prove that life exists elsewhere in space. Perhaps the one critical step from chemical to biological evolution that took place on earth 3.5 billion years ago has been duplicated elsewhere, and perhaps it has not; that remains to be determined. However, what we can conclude is that nature can easily take those first steps toward the formation of living entities. Of this point we can be relatively certain. If nature has taken second or third steps, the planet Mars is perhaps the most promising place to look for signs of that fact.

FIGURE 10.11
A piece of the Murchison meteorite that fell in Australia, in September 1969. Chemical analysis revealed the presence of amino acids.

LIFE ON THE MOON?

The moon was the first body outside the earth on which searches were made for the presence or remains of living organisms. As mentioned in Chapter 6, the *Apollo* manned landings returned over 800 pounds of lunar surface material. The procedures and the facilities developed here on the earth for their study were as elaborate as any in the history of chemical and organic analyses. Organic chemists had improved old techniques and developed new ones in anticipation of new findings on chemical evolution and the origin of life that might come from the lunar samples.

But, alas, nowhere in any of the samples was a significant amount of carbon found. Most of the carbon was in the form of traces of carbide, methane, or carbon monoxide. No amino acids, proteins, or nucleic acids were found. Also no water molecules, either free or chemically bound, were found in these surface materials.

These results were not particularly surprising even if somewhat disappointing, since the surface of the moon is completely exposed to bombardment by solar ultraviolet photons, solar wind, subatomic particles, and meteoritic matter. Under such an energetic onslaught, the likelihood of building long-chain carbon molecules is very small. Thus the chance of finding organic matter on the surface of the moon was extremely small to begin with.

MARTIAN LIFE?

The thermally habitable zone in our solar system, the *ecoshell*, where life might flourish on a terrestrial planet, lies between Venus and Mars. The inner limit is roughly the point where water would boil and the outer limit the point where water would freeze. Since Venus does not seem to be a suitable abode because of its high surface temperature, this leaves Mars as the only realistic place for biological exploration. A theoretical chemical analysis tells us that if the earth had formed only 10 million kilometers closer to the sun, it could have been a hot and sterile planet. Conditions would not be so hostile if the earth had formed farther out from the sun by the same amount.

The excellent pictures returned from Mars by the *Mariner 9* spacecraft neither proved nor disproved that Mars has biological activity. However, they are good enough to make very remote the possibilities of intelligent life on Mars that could modify the landscape to the extent we have on the earth.

PROJECT *VIKING*

A billion-dollar undertaking, Project *Viking* was designed in part to search for life on Mars. This dramatic aspect of its mission was part of the other scientific studies described in Chapter 8. Two *Viking* spacecraft were launched in late August and early September 1975 for an eleven-month journey to Mars. On arrival in June and August 1976, respectively, *Viking 1* and *2* were injected into orbit. After looking at the Martian surface for over a month, the capsule portion, with its biochemical laboratory, detached and made a soft landing. The lander and its instruments were sterilized prior to launch to avoid contaminating the planet's surface with terrestrial organisms. Two prime landing sites were originally chosen: one that might have had good drainage or deposit sites for water and other minerals in times past; the other on low (and safer) ground. As it turned out, both were passed up because of the rough surface. *Viking 1* landed on the plain of Chryse; *Viking 2* in the frozen northern region called Utopia (see the topographical map, Figure 8.7).

The biological hunt on Mars searched mainly for possible microorganisms in the Martian soil. A motor activated a 10-foot arm that reached out and scooped up a sample of soil. The arm retracted and deposited the soil sample into a hopper on the lander for au-tomated analysis. Two cameras periodically scanned the immediate surroundings to look for any large-scale biological life. Failure to detect even the simplest organism, as described in the next section, does not mean that such efforts are hopeless ventures. The Martian chemistry of life may be so different that our means for uncovering it are not adequate, or life may be below the surface where the environment is more protective, or we may not have chosen the correct sites.

LIFE-DETECTION EXPERIMENTS

In the *Viking gas exchange* experiment (see Figure 10.12) a sample of Martian soil was inoculated with a rich aqueous nutrient in an atmosphere of helium, krypton, and carbon dioxide. After a number of days of incubation, samples of the atmosphere were analyzed to check for traces of hydrogen (H_2), oxygen (O_2), methane (CH_4), or carbon dioxide (CO_2). These are the most likely by-products of metabolism, a microorganism's way of assimilating food to maintain and reproduce itself. In the experiment both oxygen and carbon dioxide were released, rising rapidly, but then leveling off. If living organisms had been present, a steady rise should have taken place because of metabolic processes.

FIGURE 10.12

Viking's life detection experiments. The pyrolytic release experiment (at left) was designed to detect the carbon cycle. Soil samples were exposed to a simulated Martian atmosphere and then heated to over 600° C. In the labeled release experiment (center) a soil sample was moistened with a nutrient and incubated for about eleven days. Finally, in the gas ex-change experiment (at right) a soil sample was inoculated with an aqueous nutrient in an atmosphere of helium, krypton, and carbon dioxide and was allowed to incubate for several days.

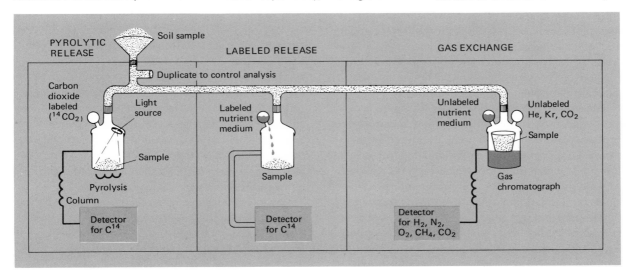

The *pyrolytic release* experiment (see Figure 10.12) was designed to detect the carbon cycle. A soil sample is exposed to a simulated Martian atmosphere of carbon dioxide and a trace of water vapor. Part of the carbon dioxide atmosphere is then replaced by one of carbon monoxide (CO) and carbon dioxide containing radioactive carbon 14 ($^{14}C_6$). A living organism would assimilate some of the radioactive carbon atoms and incorporate them into organic compounds. Artificial sunlight from a xenon lamp illuminated the sample to simulate the conditions necessary for photosynthesis.

> Life, far from being an aberration on the part of Nature, becomes within the field of our experience, nothing less than the most advanced form of one of the most fundamental currents of the Universe, in process taking shape around us.
>
> Pierre Teilhard de Chardin

After several days of incubation the sample was heated to 635°C to drive off organic vapors. Any radioactive carbon 14 in the organic molecules in the exhaust gases would indicate the presence of an organism with a metabolism of the terrestrial type. The result of the experiment was the detection of radioactive carbon, which was apparently fixed by something in the soil. Another soil sample sterilized at modest temperatures still allowed the conversion to occur, suggesting it was an inorganic rather than a biological reaction. The conversion present could have been caused by a reducing agent such as iron in the soil.

The *labeled release* experiment (see Figure 10.12) moistened a soil sample with a nutrient that contained radioactive carbon 14 and incubated it for up to eleven days, enough time for assimilating the nutrient. The rate at which the radioactively tagged gases are released as waste products tells us the reproduction rate and the physiological state of the microorganisms. What was found was a rapid release of gas containing radioactive carbon that resembled biological activity. But the reaction soon slowed down. Had living organisms been present, the biological reactions should have increased as the organisms thrived and multiplied.

All the biological experiments show an unusually active Martian surface. Water and nutrients added to the soil in the experiments appear to cause it to imitate biological activity. The answer may lie in an exotic chemistry of the Martian soil. Of great importance in this interpretation of the biology experiments are the results of the gas chromatograph–mass spectrometer experiments. This instrument searched for organic compounds in the soil and failed to find any at either landing site—even under a rock. This is despite the ease, as we argued earlier, with which nature can make these molecules, even if life is not present. The sensitivity of this instrument was such that many common organic molecules could have been found if they had been present at levels of the order of one part per billion. Possibly living organisms have developed and evolved in an environment so different from that of the earth that we did not formulate the experiments correctly, or we are still not interpreting properly the baffling results. However, at the present juncture, it is unlikely that Mars does have or has had living organisms on its surface.

For the future, a robot lander could collect samples on the Martian surface and return them to the earth for analysis in the sophisticated facilities developed for studying the lunar samples. Such a project is under study at the present. Of course, safeguards would have to be applied to prevent the possibility of a Martian microbe escaping into the earth's environment. But it is by no means obvious that if Martian microbes did exist, they could interact with life on the earth.

PROSPECTS FOR LIFE ON MERCURY AND VENUS

Neither Mercury nor Venus are very promising sites for the development of life. With no protective atmosphere to shield the surface from high-energy photons and subatomic particles, it is very hard to believe that any extensive chemical evolution could have begun on the surface of Mercury. Also, the surface temperature varies from daytime values of 350°C to nighttime temperatures of −170°C, as given in Table 7.4.

On the other hand, Venus has a very dense atmosphere which traps solar energy, so that surface temperatures run about 450°C or so. In spite of the presence of carbon dioxide and traces of oxygen and water vapor, the high temperature suggests a low

> Life can only be understood back-
> wards; but it must be lived forwards.
>
> Sören Kierkegaard

probability for finding any type of living organism on Venus.

THE JOVIAN PLANETS AND THEIR SATELLITES

The Jovian planets lack the clear demarcation between a planet surface and an overlying atmosphere that the terrestrial planets possess. However, Jupiter and Saturn, with large quantities of hydrogen and helium in their atmospheres, probably do have zones in reasonably dense portions of their atmospheres where the thermal conditions are as moderate as on the earth. These zones are below the clouds, and it is possible that conditions are such that chemical processes that occur on the earth may well be happening in Jupiter's and Saturn's atmospheres. Thus chemical evolution may occur, although this does not mean that life is developing. As mentioned in Chapter 9, plans are under way in Project Galileo to enter Jupiter's atmosphere with an instrumented probe that could search for prebiotic compounds. Finally, there is little to suggest that Uranus and Neptune are likely abodes for life.

One of the most likely places in the outer solar system for the evolution of organic molecules is Titan, the largest satellite of Saturn and, at two-thirds the diameter of Mars, it is the largest satellite in the solar system. Recent studies show that its atmosphere contains methane and possibly molecular hydrogen and traces of acetylene and ethane. The reddish coloration of Titan, like Jupiter and Saturn, although it is unlikely, may be indicative of the presence of organic molecules. Missions to Titan are in the planning stage and at some point in the future we may have definitive evidence one way or the other.

SUMMARY

Life came to our planet, we believe, when the primitive atmosphere of carbon dioxide, nitrogen, water vapor, and other gases, vented from the earth's volcanos, was energized by the sun's ultraviolet rays and other sources of energy. Simple organic molecules were created, followed by more complex organic molecular compounds from which the first primitive living organisms were made. The warm watery broth of the ancient seas was a haven in which these organisms could evolve. Over 3 billion years ago primitive organisms, by endless genetic mutations, achieved photosynthesis, and from then on they could sustain themselves in the closed carbon cycle. Highly productive, evolution took life through ever more complex biological orders on toward the human race. Darwin taught us that evolution of species goes on by natural selection and cumulative mutation. The fossil record gives a reasonably unbroken history of living forms from the most primitive one-cell organism to *Homo sapiens*.

We are beginning to render our habitat more unlivable through misuse of earth's natural resources, pollution of our environment, and overpopulation. The consequences of this ecological damage, if not checked, may destroy the quality of our life and possibly life itself.

Considering the prospects for life elsewhere in the solar system, we have now looked for it on both the moon and Mars. In neither place did we find definite evidence of either existing life or the remnant of life. The moon was rated an unlike place for life because of its continual bombardment by high-energy photons and particles—both capable of breaking apart long-chain organic molecules. On the other hand, Mars held some prospects, but proved to be somewhat disappointing after sophisticated remote analysis of soil samples. We appear to have found a rather exotic soil chemistry with only minute possibilities that this chemistry is a sign of life processes.

Elsewhere in the solar system, the prospects for life, certainly above the multicellular category, are remote. However, we will continue to search and constantly wonder if we are asking the right questions in our exploration.

REVIEW QUESTIONS

1. How and when did the earth acquire its atmospheric oxygen?

2. Describe the roles of carbon dioxide, oxygen, and solar energy in sustaining most forms of life by means of the carbon cycle.

3. Describe the early evolutionary steps that are believed to have led to the synthesis of the first living organisms on earth.

4. What is the Darwinian concept of evolution?

5. What caused the demise of the dinosaurs? (The answer to this question will have to be sought outside the text.)

6. What elements and environment would constitute the basic ingredients of life on another biologically suitable planet?

7. Discuss the significance of the presence of the amino acids in meteorites.

8. Describe the experiments undertaken to search for life on Mars in the *Viking* project.

9. Suppose Mars were inadvertently seeded with earth organisms that began to thrive and multiply in the Martian environment. Would that be good or bad? Explain your answer.

10. How are proteins assembled from the amino acids in the cell?

11. Discuss the role of the DNA molecule in cellular activity.

12. Present a possible scenario for the chemical evolution of life on the earth, beginning with the raw, primitive materials.

13. Why would a land creature be expected to evolve into a higher state of technology than a marine or air creature on an alien planet?

14. What did Stanley Miller's experiment of 1953 prove?

15. If life is found on Mars, how would you rate the importance of this discovery in the annals of science?

16. Where did the oxygen not originally present in the earth's primitive atmosphere come from?

17. It has been suggested that the DNA genetic code structure was implanted on the earth when a spaceship landed here for the purpose of seeding the earth. Is it possible for a different kind of genetic code structure to exist elsewhere in space?

11
The Sun: Our Bridge to the Stars

11.1
The Sun as a Star:
An Overview

The sun is a star—a gaseous, self-luminous body—equal in volume to 1.3 million earths. It is the only star whose surface we can study closely. It is, in effect, the source of our inspiration when it comes to understanding a scale of radiation, temperature, pressure, and magnetic fields unattainable on the earth but apparently common to the stars. From this we have learned how nuclear fusion—the process of energy production in the stars—can provide for our earth an almost inexhaustible supply of energy.

THE SUN AND ITS ATMOSPHERE

About half a billionth of the sun's radiation is intercepted by the earth, but that is enough to warm our planet and to sustain life. All the power we now generate on the earth in a year is as much as the earth receives from the sun in about twenty minutes.

Using devices called *pyrheliometers*, we can measure the solar radiation falling on a unit area of the earth within a certain time. When we correct this measurement to allow for absorption by the earth's atmosphere and some other factors, we obtain what is referred to as the *solar constant* (see Table 11.1). How constant it actually is over periods of hundreds of millions of years has not been determined. But for periods of several years, it is constant to within about 0.2 percent.

If we multiply the solar constant by the surface area of a spherical surface whose radius is the earth's mean distance from the sun, we find the rate at which radiation is emitted by the sun in all directions at the earth's distance. This figure is therefore equivalent to the total rate at which energy is radiated from the sun's surface. Thus we find that the sun's total rate of emission—its *luminosity*—is about 500 trillion billion horsepower (the value in metric units is given in Table 11.1).

This flood of radiant energy comes from what appears to be the surface of the sun. However, it is not in reality a distinct surface but is a layer of gas, about 500 kilometers in thickness, called the *photosphere*. It is dotted here and there with sunspots, as shown in Figure 11.1. The photosphere is the bottom level of the sun's atmosphere. Lying above it is a transparent, tenuous layer, the *chromosphere*, which is several thousand kilometers thick. This is topped by an even more rarefied layer, the *corona*, which extends millions of kilometers out from the sun in all directions (see Figure 11.1). These three regions are distinguished by their physical characteristics, but their boundaries are not sharply defined. One region gradually merges into the other.

Why do we see only the photosphere and not the outermost layers? Light from the chromosphere and the corona is usually too weak to be seen off the sun's limb (its edges) against the glare of the photosphere. But they are visible during a total eclipse of the sun, when the moon covers the bright solar disk. The chromosphere and corona are also observable in the short-wavelength regions of the electromagnetic spectrum (the ultraviolet and X-ray regions), or in the longer-wavelength radio region of the spectrum. These observations are possible because the photosphere is not very bright in these wavelength regions in comparison to the chromosphere and the corona. The gases of the corona and chromosphere are transparent in the visible part of the spectrum, and our line of sight normally passes through them and ends somewhere in the photosphere. Below this level the gases that compose the sun become opaque, and radiation coming out of the deep interior of the sun is absorbed and reemitted many times before it reaches the photosphere.

INTERIOR OF THE SUN

The sun is about 150 million kilometers away, yet it is 270,000 times closer to us than any other star. Precise knowledge of the sun's observable properties (see Table 11.1) enables us to calculate, from physical laws, a mathematical model of the sun's internal structure and to see if that model is consistent with the sun's observable properties. The reasonable success astronomers have had with such modeling for the sun encourages us to believe that theoretical models for other stars can be constructed.

The radiant energy emitted by the sun's photosphere is the result of *hydrogen fusion*—the conversion of four hydrogen nuclei into one helium nucleus

◀ Solar eclipse of 7 March 1970, photographed in Mexico.

TABLE 11.1
Properties of the Sun as a Star

Property	Value	How Determined
Mean distance	1.496×10^8 km	Radar reflection from Venus and Kepler's third law
Angular diameter	0.533°	
Radius	6.96×10^5 km	Angular diameter and mean distance
Surface area	6.09×10^{22} cm²	$S_\odot = 4\pi R_\odot^2$
Volume	1.41×10^{33} cm³	$V_\odot = (4/3)\pi R_\odot^3$
Mass	1.99×10^{33} g	Newton's modification of Kepler's third law
Mean density	1.41 g/cm³	$\rho_\odot = $ mass/volume
Solar constant	1.36×10^6 erg/cm²·s	Ground-base, high-altitude rocket and aircraft measurements
Luminosity	3.83×10^{33} erg/s	Solar constant and mean distance
Mean luminosity per unit mass	1.92 erg/g·s	luminosity/mass
Effective temperature	5770 K	Stefan-Boltzmann law, luminosity and surface area ($L_\odot = 4\pi R_\odot^2 \, \sigma T_{eff}^4$)
Spectral type	G2 V	Classifying absorption spectrum (see Section 12.4)
Apparent visual magnitude	−26.74	(see Section 12.3)
Absolute visual magnitude	+4.83	(see Section 12.3)
Color; (B-V)	yellow; +0.65	(see Section 12.3)
Atmospheric layers	Photosphere, chromosphere, corona, in order out from the sun	Direct observation
Rotation period	25 days at equator, 36 days at poles	Motion of sunspots; Doppler shift in photospheric spectrum
Angle between equator and ecliptic	7.2°	Rotation of sun
Magnetic field	$\approx$ 1–2 G averaged over photosphere; hundreds of gauss in active regions; thousands of gauss in sunspots	Zeeman effect
Chemical composition	91% hydrogen, 9% helium, 0.1% heavier elements by numbers of particles	Spectrum analysis of the photospheric absorption spectrum and mathematical models of the interior

in the deep interior of the sun. Every second in the sun's core, at a temperature of about 15 million degrees Kelvin, 600 million metric tons of hydrogen is transformed into 596 million metric tons of helium and 4 million metric tons of matter is converted to energy. This energy is the source of our sunlight, and when first released it is chiefly in the form of gamma ray and X-ray photons and the kinetic energies of the subatomic particles produced in the fusion process. As the radiation works its way through the maze of solar material toward the sun's surface, atoms and ions absorb and reemit the energy. After hundreds of thousands of years of absorption and reemission, the radiant energy reaches the layers about 100,000 kilometers below the sun's surface. There the energy has been so degraded toward longer wavelengths that the photons are readily absorbed by any un-ionized hydrogen atoms. As a result, *convection* (a flow of hot and cold material) takes over as the most efficient and dominant mode of energy transport in this region, known as the *hydrogen convection zone*. This process is like that in a heated room. Cold heavier air descends to be reheated near the floor and then rises, circulating heat in the room. In the sun, hot gases bring energy from the interior to just below the photosphere, and cool gases return to the interior to start the cycle again. While the convection zone occupies the outer 37 percent or so of the

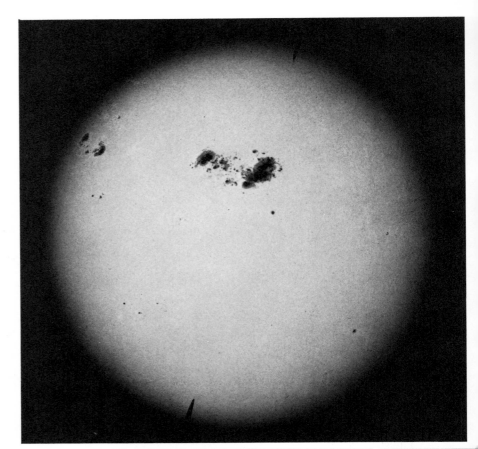

FIGURE 11.1

The sun, its atmosphere, and prominent features.

The sun is the only star that we can study at close range. It generates energy through nuclear fusion, by which hydrogen is consumed to produce helium plus radiation that flows outward and eventually emerges as sunlight.

The photosphere is the visible surface of the lowest level of the sun's atmosphere; its granulation is shown in the high-altitude ballon photograph at the top right. The photosphere is also the region where sunspots occur. A close-up of a large sunspot group that occurred on 17 May 1951 is shown in the middle left photograph. The dark centers are the umbras and the surrounding striated areas the penumbras; photospheric granules are also visible outside the sunspot group. Above the photosphere lies the chromosphere, a thin layer of transparent gases several thousand kilometers thick. The outermost layer of the sun's atmosphere is the corona. During a total solar eclipse, we see the corona's pearly white halo extend outward well over a million kilometers. Extending into the corona and above the chromosphere are prominences. The great eruptive prominence of 4 June 1946 (bottom left) grew to a size almost as large as the sun within an hour; several hours later it disappeared completely. Flares—sudden and short-lived outbursts of highly energetic radiation—emanate from a disturbed region of the sun's surface.

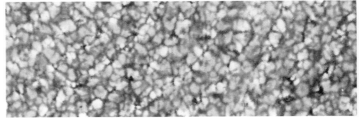

Above: Top, direct view of the sun, showing the photosphere; center, granulation of the photosphere

Facing page, top left: Main features of the sun

Middle left: Large sunspot group

Middle right: Chromosphere and prominences

Bottom left: Prominence of 4 June 1946

Bottom right, upper: Solar flare of 16 July 1959

Bottom right, lower: Eclipsed view of the sun, showing the corona

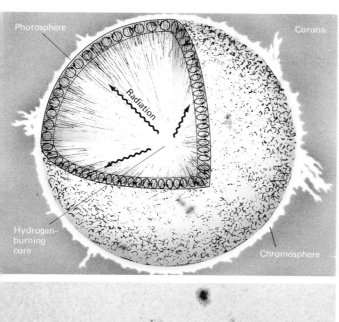

Photosphere

Corona

Radiation

Hydrogen-
burning
core

Chromosphere

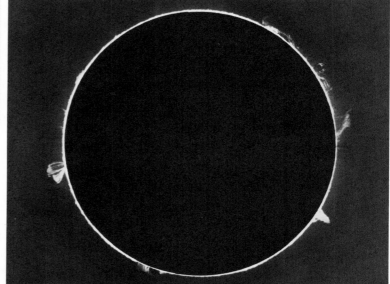

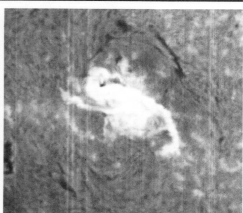

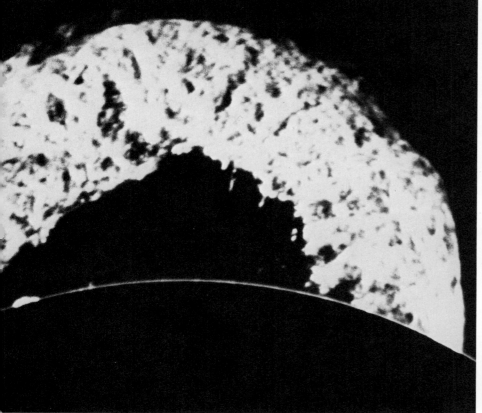

sun's volume, it contains only about 1 to 2 percent of the sun's mass because of the very low density near the surface. At the surface most of the energy that left the deep interior is now in the visible spectral region—the sunlight that we observe about eight minutes after it leaves the sun.

The sun radiates energy away from its surface at the same rate that energy is being created in its interior. The sun is, in effect, in a state of equilibrium, with the temperature declining from the 15 million degrees Kelvin at the center to about 6000 degrees Kelvin in the photosphere. The density also declines from a central value of about 160 grams per cubic centimeter to a mere 10^{-7} gram per cubic centimeter in the photosphere. The drop-off in density is so rapid that most of the mass of the sun is in the deep interior close to the center.

Sparked by speculations on the correct formulation of relativity theory, observations were made to see if the sun were slightly oblate; no oblateness could be found. What was found, however, was a periodic pulsation of the sun, with a fundamental period of about fifty-two minutes. In addition to this fundamental, several higher-frequency harmonics of it have also been detected. Thus the sun vibrates much as a bowl of Jell-O does. Just as the free oscillation of the earth after earthquakes can be used to probe the earth's interior, the pulsation of the sun should help us to verify the theoretical interior structure we have just described.

Glorious the sun in mid-career;
Glorious th' assembled fires appear.

Christopher Smart

In Chapter 13 we will examine in detail how the sun and other stars are structured and how they change in structure as they live out their lives. Briefly, according to present models of the sun, the sun is about midway through its life in its present mode of producing energy. In another 5 billion years or so the hydrogen fuel in the deep interior will be used up and the matter in the core will be primarily helium. As a result the sun will restructure itself by gravitational contraction of its burned-out core. Contraction will raise the internal temperature sufficiently to fuse unprocessed hydrogen outside the core into helium. And over a billion years or so, the sun will expand from a yellow dwarf star to a giant red star with a vastly increased luminosity. The enlarged sun will expand to about the orbit of Mercury. When the contracting core reaches a central temperature of around 100 million degrees Kelvin, helium will be synthesized into carbon and heavier nuclei. The sun will go through further changes in size and luminosity in more advanced stages of evolution, but the time will come when all the nuclear fuel will be exhausted and thermonuclear reactions will cease, about 7 billion years from now. At that point the sun will have become a small, dense dying star, called a white dwarf.

11.2
The Photosphere

PHOTOSPHERIC TEMPERATURE

Large solar telescopes, such as the one in Figure 11.2, are used by astronomers to study the photosphere of the sun. The temperature of the photosphere is an important property of any star. It is a measure of the rate at which radiation is emitted by the star. To find the sun's photospheric temperature, we can use one of three methods: the Stefan-Boltzmann law, Wien's displacement law, and Planck's radiation law. As discussed in Chapter 4, the laws describe the radiation emitted by blackbodies (see Figure 11.3). Because the sun is not a perfect blackbody, the temperatures derived by these three methods differ slightly; the methods yield an approximate value of 6000 K. One means of visualizing how this value is derived from Planck's law is the degree to which the amount of energy measured at various wavelengths in the solar spectrum approximates a blackbody energy curve. This comparison is shown in Figure 11.3, where we can see that the 6000 K blackbody energy curve is a reasonable approximation to the distribution in the sun's continuous spectrum.

The sun's limb looks darker than the center of the disk. Why? At the sun's edge we are viewing radiation obliquely, seeing light that comes from a higher level of the photosphere. Since the higher layers emit less radiation, they must be cooler, as is evident from the blackbody energy curves in Figure 11.3. Radiation visi-

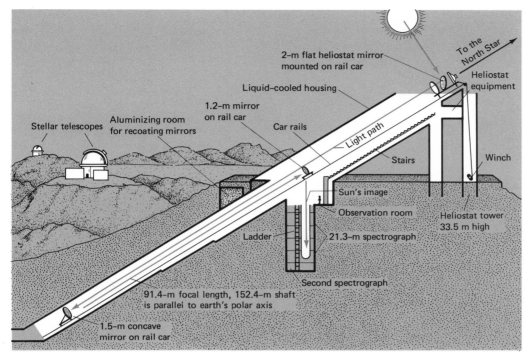

FIGURE 11.2
McMath Solar Telescope, Kitt Peak National Observatory. The telescope is housed in a 152.4-meter concrete tunnel, partly underground, which is inclined 32° to the horizontal and parallels the earth's axis of rotation. A rotating 2-meter flat mirror called a heliostat, mounted on top of the building, tracks the sun. It reflects sunlight down the tunnel onto a 1.5-meter parabolic mirror that focuses light back up the tunnel and down into an underground observing room 91.4-meters away. The image formed by the telescope is nearly a meter in diameter. This is the largest solar telescope now in use.

Labels in figure:
Stellar telescopes
Aluminizing room for recoating mirrors
1.2-m mirror on rail car
2-m flat heliostat mirror mounted on rail car
To the North Star
Heliostat equipment
Liquid-cooled housing
Car rails
Light path
Stairs
Winch
Sun's image
Observation room
Heliostat tower 33.5 m high
Ladder
21.3-m spectrograph
Second spectrograph
91.4-m focal length, 152.4-m shaft is parallel to earth's polar axis
1.5-m concave mirror on rail car

ble to us from the sun's center, however, comes from deeper, hotter layers and is more intense (see Figure 11.4). Thus the temperature declines outward through the photosphere. From this fact the astronomer can deduce the decrease of temperature, density, and pressure to use in determining the abundances of the chemical elements. The sun's limb looks sharp-edged because the layers responsible for most of the emitted white light are too thin to be resolved. They are up to 500 kilometers thick; the typical resolution size for 1 second of arc seeing corresponds to about 750 kilometers.

SPECTRUM OF THE PHOTOSPHERE

The spectrum of the visible solar disk displays a continuous band of colors from red to violet crossed by many absorption lines, which can be easily seen in Figure 11.5. German physicist Joseph von Fraunhofer (1787–1826) first mapped nearly six hundred of the most prominent lines in 1814. He designated by letters the strongest absorption lines, from A in the red to K in the violet. We can interpret these lines according to Kirchhoff's third law of spectral analysis (see Section 4.2). The radiation coming up from the interior of the sun has a continuous spectrum. As it passes through the photosphere, certain wavelengths of this radiation are absorbed by atoms and ions of different chemical species in the photosphere's cooler layers and in the adjoining low chromosphere, causing the observed dark lines. The uninterrupted wavelength regions, between the absorption lines, are regions of continuous radiation that pass into space without being absorbed.

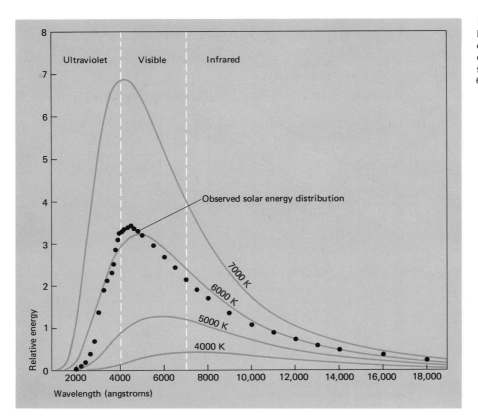

FIGURE 11.3
Blackbody energy curves. The observed distribution of energy in the sun's continuous spectrum (dots) matches the 6000 K theoretical curve best.

CHEMICAL COMPOSITION

By precisely measuring wavelengths of absorption lines in the solar spectrum, astronomers have identified in the sun nearly seventy of the ninety-two naturally occurring elements and about twenty molecules. The identifications are made by comparing the wavelengths of lines in the solar spectrum with those obtained from laboratory analyses of the spectra of the elements. The elements missing from the spectra, mostly the heavier ones, are probably present in the sun's atmosphere, but either they are not abundant enough to be detected spectroscopically or their spectral lines are not in the visible region, the only region thoroughly explored. We can estimate the abundances of the photosphere's elements (see Table 11.2) by combining our theoretical knowledge of the probability that an atom will absorb radiation at the wavelength in question and at a specified temperature with measurements of the line's blackness and width. In Section 12.4 we will discuss these measurements further.

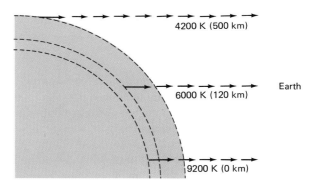

FIGURE 11.4
Darkening of the sun's limb. Temperatures at different depths of the photosphere are indicated; the numbers in parentheses are the heights above the lowest level from which solar radiation reaches the earth. Radiation visible to us from the center of the solar disk comes from deeper in the photosphere, where the temperature is higher, than does the radiation from the sun's limb. Drawing is not to scale.

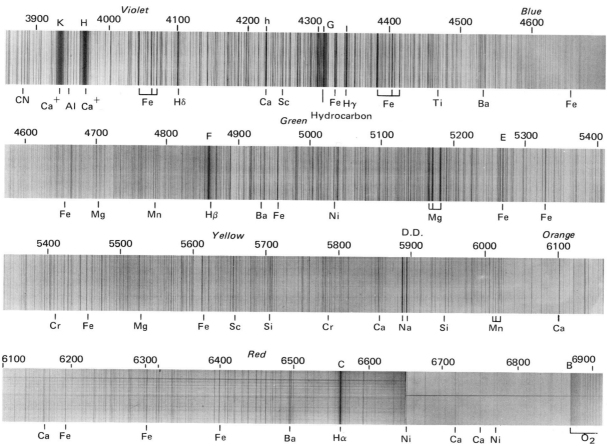

FIGURE 11.5
Solar spectrum from 3900 to 6900 angstroms. The lettered lines marked above were originally assigned by Fraunhofer. The chemical identification of a number of lines is shown below. B marks the head of a band, which extends to the right, due to molecular oxygen in the earth's atmosphere.

TABLE 11.2
Ten Most Abundant Elements in the Sun's Atmosphere

Atom	Atomic Number	Number (%)	Mass (%)
Hydrogen	1	90.7	69.1
Helium	2	9.1	27.7
Carbon	6	0.036	0.33
Nitrogen	7	0.009	0.10
Oxygen	8	0.072	0.88
Neon	10	0.003	0.04
Magnesium	12	0.002 ≈ 0.1	0.05 ≈ 2
Silicon	14	0.002	0.06
Sulfur	16	0.001	0.04
Iron	26	0.003	0.12

SOLAR ROTATION

The sun rotates eastward, like the planets (except Venus and Uranus). It does not rotate as a solid body. For the atmosphere, primarily the photosphere, the period of rotation progressively increases from 25 days at the solar equator to about 36 or 37 days at the poles. Astronomers think that this *differential rotation* results from the convection currents below the sun's surface. We see sunspots move across the disk because the sun is rotating, and we can measure their travel to find how long it takes the sun to complete one rotation on its axis.

Another method applicable to all solar latitudes (sunspots rarely appear beyond 40° on either side of the solar equator) employs the Doppler shift of the spectral lines from opposite limbs of the sun (see

Figure 11.6). Since the eastern limb of the sun rotates toward the earth and the western limb away from the earth, the measured difference in the Doppler wavelength displacement between the two limbs yields the relative velocity between the opposite limbs. The measured limb velocity relative to the sun's center is 2 kilometers per second at the solar equator. Dividing the distance traveled in one rotation—that is, the circumference—by the velocity gives the time of one complete rotation at the equator: 25 days.

FINE STRUCTURE OF THE PHOTOSPHERE

In ground-based photographs taken in the white light of the photosphere, a number of details are clearly visible. We have already noted that the photosphere darkens toward the limb, where the contrast is strong enough so that we can see bright patches called *faculae*. These are regions of transient solar activity. Photographs also reveal the sunspots. And with higher resolutions we can see that the entire disk is covered by small, bright features called granules (see Figure 11.1).

Remarkably clear pictures of the solar surface have been made by a telescope that was mounted in the *Skylab* station (see Section 5.5). At this altitude the finer details are not blurred by earth's atmospheric turbulence. Various high-resolution photographic studies reveal a potpourri of bright *solar granules* with dark in-

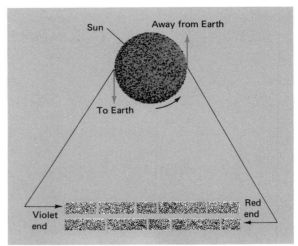

FIGURE 11.6
Doppler shifts of opposite limbs of the sun. The eastern limb rotates toward the earth, the western limb away from the earth. The Doppler shift between the two limbs is 4 kilometers per second, or 2 kilometers per second relative to the sun's center.

tergranular lanes; these give the surface its salt-and-pepper appearance in Figure 11.1. Sequences show the granules forming, disappearing, and reforming in cycles lasting several minutes. The granulation may be described as a pattern of quasi-polygonal cells with

TEMPERATURE OF THE SOLAR PHOTOSPHERE

We can find a representative temperature $T_\odot$ for the sun's photospheric layers from the radiation laws for blackbodies, since that is very nearly how the sun behaves. One of the three mathematical relations that specify the properties of blackbody radiation is the *Stefan-Boltzmann law*. It states that the flux of the radiation in ergs emitted per square centimeter each second in all wavelengths is proportional to the fourth power of the temperature. The formula is

$$F = (5.67 \times 10^{-5})T^4$$

where F is the radiation flux in ergs per square centimeter second and T is the temperature in degrees Kelvin. Dividing the luminosity of the sun (3.82×10^{33} ergs per second) by its surface area (6.09×10^{22} square centimeters) gives $F = 6.27 \times 10^{10}$ ergs per square centimeter second. Hence the surface temperature is

$$T_\odot = \left(\frac{6.27 \times 10^{10}}{5.67 \times 10^{-5}} \right)^{1/4} = 5770 \text{ K}$$

Because temperature varies somewhat over the solar disk and also at different depths in the photosphere, we can take this surface temperature as an average value for the entire solar disk.

characteristic diameters of a thousand kilometers and lifetimes of several minutes. At any given time the whole photosphere has roughly 4 million granules, each occupying about a million square kilometers of the surface. From the bright center of the granule to the darker intergranular region, the difference in brightness corresponds to a difference in temperature of about 300 K. Thus the photosphere is not uniform; it varies not only in height but also laterally across the face of the sun.

The granules in the photosphere are a form of convection, which is similar to, but not as permanent as, the hydrogen convection zone below the photosphere. As evidence of this convective exchange, spectral lines of the bright centers are Doppler shifted to the violet and those of the darker intergranular regions are shifted to the red. These shifts between the bright and dark regions indicate that the bright centers are the tops of hot, rising gas currents moving at a few tenths of a kilometer per second that radiate their excess energy and then form the sinking gas currents of the darker intergranular lanes.

In 1960 it was discovered that there was vertical oscillatory motion in and above the granulation. It has a well-defined average period of almost exactly 5 minutes and velocities of about 0.5 kilometer per second. Thus the layers above the convection zone are moving up and down with respect to the mean photosphere and low chromosphere. The mean excursion is on the order of 35 to 40 kilometers. The motion seems to be organized over areas as small as granule size, or about a few thousand kilometers, and has been reported to cover areas as large as 50,000 kilometers, with roughly two-thirds of the solar surface experiencing oscillations at any given moment. It appears that the 5-minute oscillation may be regarded as one extreme in the range of solar pulsation, with the 52-minute pulsation mentioned earlier as the other extreme.

Coexisting with the solar granules and the 5-minute oscillations of the solar photosphere is a completely different type of motion detected by Doppler studies of the full disk of the sun. These motions, shown in

Figure 11.7, are called *supergranulation cells* because of their resemblance to convective motions and the fact that they are typically an order of magnitude larger than granules (about 30,000 kilometers in diameter). Beyond the name, the granules and supergranules may have little in common. The supergranule cells have a very regular structure of rising gas at the center and a lateral flow toward the cell boundary with velocities of about 0.4 kilometer per second, in contrast to a much less regular motion in the granules. Also in contradistinction to the granules, the supergranules show no pattern of bright centers and dark boundaries, because the temperature differences are apparently not great enough. For this reason the

FIGURE 11.7
Supergranulation cells as seen in Doppler shift picture. Velocities of approach are indicated by lighter areas, and velocities of recession are shown by darker areas. Note the cell-like structures with lighter centers (approach) and darker boundaries (recession). These are known as supergranules and are typically 30,000 km across. Compare with Figure 11.14.

supergranules are seen only as Doppler shifts, as in Figure 11.7. Supergranules have a lifetime of about one day compared to the several-minute lifetimes of the granules. At what depth in the sun the supergranulation system begins is unclear. Also, we are not entirely sure how the motions in the supergranules affect the temperature and density structure of the chromosphere and corona.

11.3 Sunspots

WHAT ARE SUNSPOTS?

Sunspots are the most conspicuous features observed on the solar disk in white light. They were known centuries before the telescope was invented. Sunspots are but one of a number of transient phenomena, all of which are apparently confined to the solar atmosphere. A typical sunspot has a cellular structure with a dark center, the *umbra,* surrounded by a grayish filamentary region, the *penumbra*. The sunspot is in-

trinsically bright; the umbra is about one-fourth as bright as the photosphere and the penumbra about three-fourths as bright. The umbra looks dark because we see it against an even brighter photospheric background whose temperature is 1800 K higher. A large sunspot group is shown in Figure 11.1. Photospheric granules can be seen up to the edge of sunspots. However, the granule features in the umbra of a spot have low contrast, and it is difficult to determine their similarity to photospheric granules.

Sunspots develop in a matter of hours as small pores in the intergranular region of the photosphere. They grow rapidly, and they generally form in adjacent clusters, marking a *sunspot group* whose orientation is approximately parallel to the solar equator. Each end of the group is often dominated by a large spot surrounded by smaller spots. The largest groups may cover one-fifth of a solar diameter. Sometimes the sunspot group persists for several months, but the typical lifetime is about one week. The typical large spot in a sunspot group is about 10,000 kilometers across; exceptional ones are 50,000 kilometers in diameter, or about four times the diameter of the

FIGURE 11.8
Sunspot cycles for the period 1750 to 1980. The last minimum was in 1976 and the last maximum in late 1979.

Note that the 11-year cycle is very approximate, and the peak number is also highly variable. The period since

World War II has been a particularly active one for the sun as compared to earlier periods.

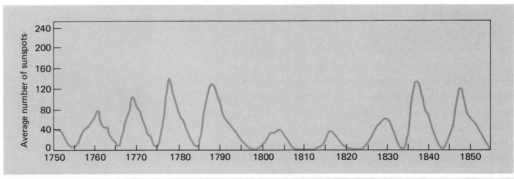

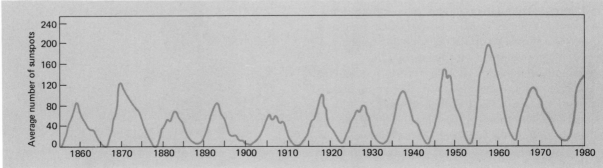

earth. In a week or so this large spot builds up to its maximum diameter, then its size slowly declines. Individual spots in a sunspot group undergo slow changes from day to day while they maintain their association.

SUNSPOT CYCLE

More than a century ago German amateur astronomer Heinrich Schwabe discovered, during 17 years of observations, that sunspots come and go in a *sunspot cycle* of about 11 years. The plotted sunspot number in Figure 11.8 takes into consideration both the number of sunspot groups and the number of individual spots. The heights of the successive maxima are unequal, and the interval between the peaks and troughs is not constant. The 11-year period is a very

rough average. The last minimum occurred in 1976; the last maximum occurred in late 1979. As the cycle progresses, the spots form closer to the equator in each succeeding year as shown in Figure 11.9. A few sunspots in the higher latitudes around 35°N or 35°S herald the beginning of a new cycle as the last spots of the preceding cycle disappear near the equator.

MAGNETIC FIELDS IN SUNSPOTS

Around sunspot groups we observe curved features (see Figure 11.13d). They are structures of gas whose shape is determined by curved magnetic lines of force in and around the spot group. Sunspots are apparently the centers of intense magnetic fields. Light from a radiating source, which is in a magnetic field, has a spectrum in which the absorption lines are split into three or more closely spaced, polarized components. This phenomenon is called the *Zeeman effect*, for the Dutch physicist who discovered it in 1896. It can be

FIGURE 11.9
Equatorial trend of sunspots during the sunspot cycle. As the cycle advances, spots appear progressively closer to the equator.

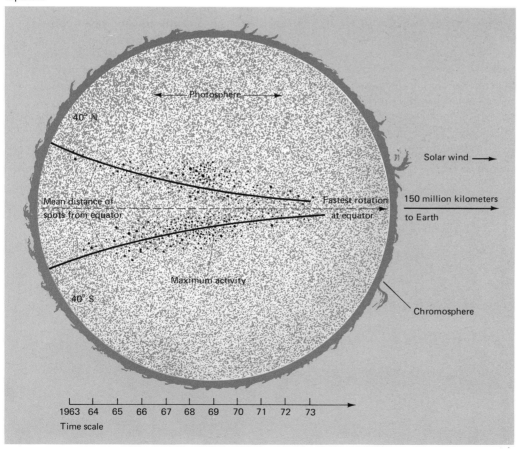

observed in the absorption lines of the sunspot spectrum. The strength and direction of the magnetic field in a sunspot can be determined from the degree and sense of the Zeeman splitting (see Figure 11.10).

The leading spots of a spot group (in the forward direction of the sun's rotation) are opposite in polarity from the following spots. Opposite polarities are like those of the north and south ends of a horseshoe magnet. The measured field strength in sunspots exceeds that of the earth's field by several thousand times. The unit of magnetic field strength is the *gauss*; the earth has an average field strength of 0.5 gauss, while the sunspot fields are several thousand gauss. In the opposite solar hemisphere the polarities of the leading and following spots are reversed, and those in both solar hemispheres reverse in the following sunspot cycle. As a whole, the sun may have a general magnetic field, but if so it is no more than a few times stronger than the earth's field, or several gauss altogether. The magnetic fields in the polar regions of the sun appear to reverse near the time of sunspot maximum. Although astronomers have explanations for sunspot behavior and the magnetic field phenomena that go with it, these explanations are not completely satisfying.

ORIGIN OF SUNSPOTS

Of major importance in understanding sunspots is the question of why they are cooler than their surroundings and how they are able to maintain that condition for so long a period of time. Since energy flows from hotter to cooler regions, a flood of photons from the high-temperature gas surrounding the sunspot should flow into the cooler spot and eliminate the difference in temperature. The reason this does not happen must be connected with the very strong magnetic fields that exist in the interiors of spots. The magnetic field can act to suppress some of the motions of the gas and hence cool the spot. But this still does not explain why the spot remains cooler. Estimates are that enough heat should flow into the spot within several days to bring it to normal photospheric temperatures. However, some large spots may persist for many weeks. At present we have no explanation of this aspect of sunspots other than to believe that it is connected with the existence of the magnetic field.

The most acceptable theory of the origin of sunspots suggests that distortions in the magnetic field below the photosphere are created by the sun's differential rotation. As the magnetic lines of force are pulled and stretched unequally at different latitudes by the flow of gas to which they are attached (maximum stretching is at the equator where rotation is swiftest), the magnetic lines eventually curl into a ropelike structure below the photospheric surface. If a kink forms, the magnetic rope of force may arch up into the photosphere, making a bipolar sunspot group, as shown in Figure 11.11.

As the bipolar group of sunspots breaks up and

FIGURE 11.10
Zeeman splitting in the spectrum of a sunspot. On the right is a white light photograph of a complex sunspot group taken 4 July 1974. The vertical black line indicates the position of spectrograph slit for the spectrum shown to the left. In the center of the spectrum is the FeI I λ5250 which is clearly split into three components by the sunspot magnetic field. The field strength of 4130 gauss is one of the largest ever measured.

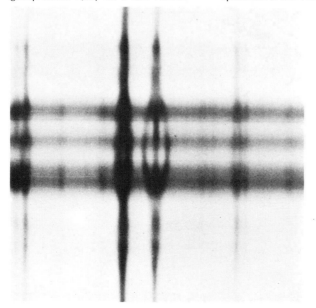
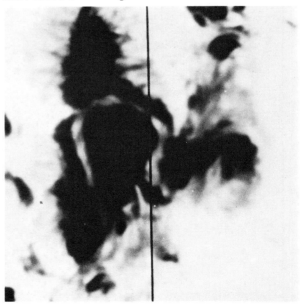

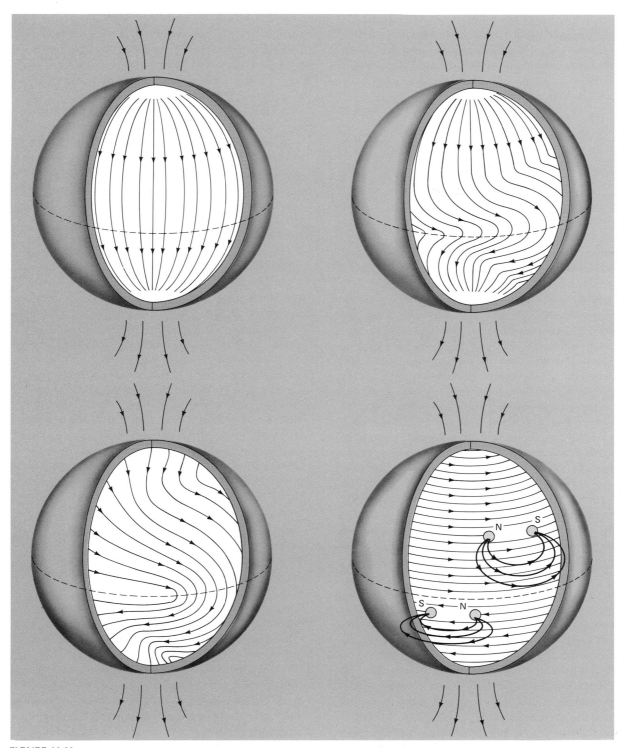

FIGURE 11.11

The origin of sunspots. The magnetic lines of force lying underneath the photosphere (upper left) are stretched out along the equator by the sun's differential rotation (upper right). After many rotations the field lines are stretched into an intense east-west field under the photosphere (lower left). Eventually a loop of magnetic field lines bursts through the photosphere (lower right) forming pairs of sunspots of opposite polarity (north-seeking, labeled N and south-seeking, labeled S). As shown, the leading spot (westernmost) in the northern hemisphere has an S-polarity and it has an N-polarity in the southern hemisphere. In the next cycle, the polarities of the leading spots in each hemisphere will reverse.

decays, the field of the following spots of the bipolar group migrate toward the polar regions and eventually neutralize the polar field, allowing a new one to form, its polarity reversed from the previous cycle. The period of complete reversal of the sunspot polarity is thus 22 years.

11.4
The Chromosphere

THE FLASH SPECTRUM

Most of the sun's radiation is from the photosphere and lies in the visible region, where a narrow optical window in earth's atmosphere lets it in (see Figure 4.4). During a total eclipse of the sun, when the moon has just covered the photosphere, a thin (about 2000-kilometer) pinkish fringe of light called the *chromosphere* appears beyond the moon's edge (see Figure 11.1). At the same time, the corona appears as a much fainter halo surrounding the sun and extending far out into space. Projecting from the chromosphere here and there are rosy arches and loops of gas called *prominences*, which may climb 100,000 kilometers or more into the corona. The chromosphere gets its reddish hue from the large amount of hydrogen gas emitting radiation in the hydrogen alpha line of the Balmer series.

For the few seconds during an eclipse when the chromosphere is exposed, the photosphere's normal absorption spectrum is no longer visible and we see a bright-line spectrum called the *flash spectrum* (shown in Figure 11.12). It is the spectrum of light originating in the chromosphere. Many of the emission lines match the wavelengths of the absorption lines, but among the exceptions is a bright orange line due to helium. Why do helium emission lines appear in the flash spectrum but not in the Fraunhofer absorption spectrum of the photosphere? The reason for this effect is the chromosphere's higher temperature—up to 30,000 K at the highest level—and lower density. Neutral helium can only be excited to emit radiation when the gas temperature is greater than 10,000 K. And the appearance of ionized helium lines requires temperatures in excess of 20,000 K. Thus the temperature must rise very rapidly from the top of the photosphere up through the chromosphere.

STRUCTURE OF THE CHROMOSPHERE

Many interesting chromospheric events can be monitored even when there is no eclipse by photographing the chromosphere in monochromatic light or in very narrow color bands. A single-color recording device, the *spectroheliograph*, can slowly scan the solar disk in the residual light of an absorption line, such as the hydrogen alpha line or the K line of singly ionized calcium visible in the solar spectrum in Figure 11.5. The finished picture, a *spectroheliogram*, is shown for hydrogen alpha in Figure 11.13b and for the calcium K line in Figure 11.13c. The reason for choosing the residual light in a strong absorption line is that most of the photons from the photosphere have been absorbed and the residual photons are those coming from the low chromosphere. Photons in the continuous spectrum between the lines come from the bottom of the photosphere, while residual photons in stronger and stronger absorption lines come from higher and higher up in the photosphere and the low chromosphere. Thus by choosing absorption lines of different strengths, the astronomer can photograph the solar atmosphere at various levels.

Bright enhanced patches in the chromosphere, called *plages,* surround the photospheric sunspot groups in spectroheliograms (see Figure 11.13d). Hotter and probably more dense than the normal chromosphere, with a magnetic field of a few hundred gauss, plages are much larger than the region of sunspots lying below them. Plages are nearly always found above regions in the photosphere in which a strong magnetic field exists, and they always appear before the spots form. Their usual life span is about forty to fifty days, during which several spot groups or none at all may form. They are typically ten times greater in area than that of the sunspot group. The plage areas do not look the same in the red hydrogen alpha line (Figure 11.13b) and in the violet calcium K line (Figure 11.13c), but they do mark the same general regions of the chromosphere. In a calcium K line spectroheliogram, one sees a *network* of dark cells with bright edges in addition to the larger plages. This chromospheric emission network contains cells that are approximately the size of the supergranules seen in the photosphere. It is strongly felt that the two phenomena are the same thing and that the plages are essentially network structures that are more closely packed and filled in.

In short spectroheliogram exposures the chromosphere is stippled with a myriad of tiny jetlike spikes

FIGURE 11.12

Flash spectrum of the chromosphere. The two most intense emission lines are those due to Hα (λ6562) and H (λ4861) in the Balmer series of hydrogen. Low-temperature lines due to sodium and magnesium are also present. However, note that there are also high-temperature lines of neutral helium, ionized helium and many times ionized iron.

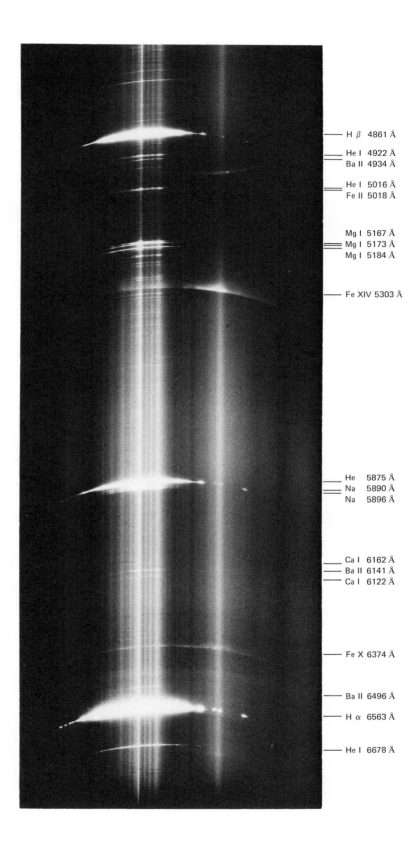

— H β   4861 Å

— He I 4922 Å
— Ba II 4934 Å

— He I 5016 Å
— Fe II 5018 Å

— Mg I 5167 Å
— Mg I 5173 Å
— Mg I 5184 Å

— Fe XIV 5303 Å

— He   5875 Å
— Na   5890 Å
— Na   5896 Å

— Ca I 6162 Å
— Ba II 6141 Å
— Ca I 6122 Å

— Fe X 6374 Å

— Ba II 6496 Å

— H α   6563 Å

— He I 6678 Å

of gas, or *spicules,* averaging about 1000 kilometers across (see Figure 11.14). Spicules rise at a rate of about 20 kilometers per second, attain typical heights of about 7000 kilometers, and then fade away or collapse in several minutes. At any instant 250,000 of them may cover a few percent of the sun's surface. They tend to cluster on the boundaries of the supergranules. Figure 11.14 shows a magnificent hydrogen alpha spectroheliogram with short dark lines, like blades of grass, that outline the supergranule cells.

FIGURE 11.13
Four views of the sun. The plage region (d) was photographed in the light of the hydrogen alpha line. Notice that the sunspot regions observable in the white-light view (a) are also present in the spectroheliograms in the hydrogen alpha line (b) and in the calcium K line (c).

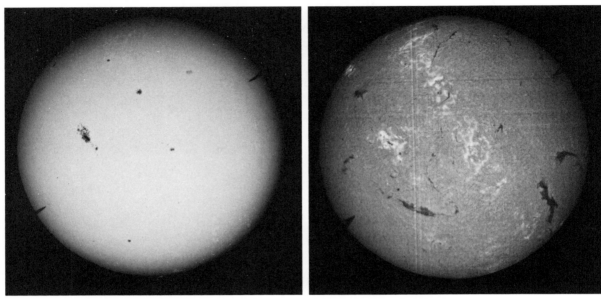

a. Direct view.

b. Hydrogen (alpha line) spectroheliogram.

c. Calcium (K line) spectroheliogram.

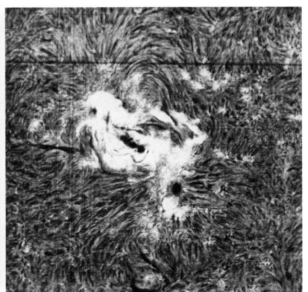

d. Plage region.

Skylab photos revealed that larger, longer-lived *macro-spicules* dominate the sun's polar regions. The photos also suggest that the chromosphere has granular features much larger than those observed in the photosphere.

SOLAR FLARES

Solar flares are perhaps the most complex of the sun's transient phenomena. A *solar flare* may suddenly erupt as an intensely bright area in a chromospheric plage (see Figure 11.15). Emitting radiation strongly throughout the electromagnetic spectrum; it rises to great brilliance in several minutes, then fades in half an hour to several hours. A flare outburst rises from the sudden release of energy in the chromosphere. The outburst of electromagnetic energy is apparently triggered by the collapse of the local magnetic field in the plage. The energy of the flare is stored in magnetic fields. If we think of magnetic lines of force as being like the rubber band of a toy airplane, then the storage of energy is analogous to winding the rubber band. Twisted magnetic structures are often seen in connection with strong magnetic fields. In a matter of seconds to minutes the energy stored in the magnetic field is converted into the kinetic energy of atomic particles. Free electrons are accelerated up to velocities of about half the speed of light. As these energetic electrons collide with the surrounding ambient gas, they share their kinetic energy and thereby heat the gas by a few thousand degrees Kelvin in the chromosphere to 10 million degrees Kelvin in the low corona. This heating phase may last from minutes to hours and is responsible for the X-ray, ultraviolet, and visible emission of the flare (see Figure 11.15). Some of the high-energy electrons pass out through the corona, where they excite successive layers to emit radio frequency radiation. Flares vary in size, brightness, and behavior, and they are most common when sunspots are most numerous.

A major flare is the most energetic solar disturbance, equaling a billion hydrogen bombs in power.

FIGURE 11.14
Hydrogen alpha spectroheliogram of chromospheric network, with spicules on the boundaries of the network cells. The dark blade-like features are probably spicules. The network cells are about 30,000 kilometers across. The network cells are probably the supergranule cells. The area shown is about 100,000 kilometers on a side. This level of the solar atmosphere is several thousand kilometers above the photosphere.

The ultraviolet and X-ray radiation arriving at the earth can disrupt the ionosphere and produce ionospheric storms, which sometimes persist for days. Also, over the few days following the flare outburst, subatomic particles (low-energy cosmic rays and enhanced solar wind particles) can spiral into the earth's polar regions, causing brilliant auroral displays and radio blackouts. All these events, which do not necessarily occur with every flare, partially distort the earth's magnetosphere,

> The great globe itself,
> yea all it inherit, shall dissolve
> And, like this insubstantial pageant folded
> Leave not a rack behind.
>
> Shakespeare

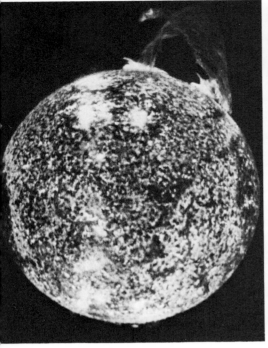
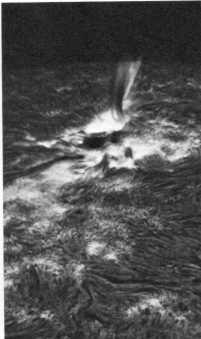
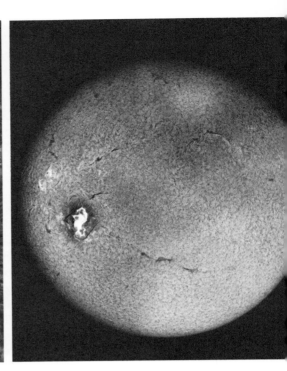

FIGURE 11.15

Solar flares photographed in hydrogen alpha light. On the right is a major flare of 7 August 1972 seen near peak brightness. Note the complexity of the region around the flare. On the left is a spectacular flare spanning more than 588,000 kilometers across the upper right edge of the solar disk. In the center is a close-up of a surge flare near the limb of the sun. The complexity and activity in the vicinity of a flare is quite evident in this picture.

generate geomagnetic storms, and induce electrical currents that interfere with worldwide radio and wire communication.

11.5
The Corona

APPEARANCE OF THE CORONA

The *corona* is the region of the solar atmosphere lying above the chromosphere. As seen during a total eclipse of the sun, it is the large halo of white, glowing gas extending out many solar radii (millions of kilometers) beyond the dark limb of the moon (see Figure 11.16). In white-light photographs one can see that the corona is irregular and structured. Beautiful long *streamers* extend outward from the sun in the equatorial regions. Their shape and motion are controlled by magnetic lines of force. Also in the corona are bright *coronal condensations.* Near sunspot max-

imum the corona is nearly circular, with streamers radiating out in all directions. However, near sunspot minimum it extends farther out in the equatorial region and terminates rather abruptly, with short thin *plumes* curving out of the polar areas.

Compared to our knowledge gained from rocket and satellite studies over the last thirty years, eclipse studies have yielded not more than a few hours of observation (an eclipse seldom lasts longer than a few minutes). In 1973 the launching of the *Skylab* orbiting observatory, with trained astronauts on board, gave us the opportunity to observe the corona in greater detail and to record its changing appearance in a virtually continuous fashion for up to eight months at a time. The more than 100,000 photographs of the sun, most in the X-ray and ultraviolet wavelengths, taken by the *Skylab* instruments have profoundly altered our understanding of the corona.

Many small, bright spots of unknown origin are visible in X-ray pictures of the solar corona (see Figure 11.17). And dark regions called *coronal holes,* which emit very little X-ray radiation, appear to extend

234   **11/The Sun: Our Bridge to the Stars**

downward into the upper chromosphere. Both the temperature and density seem to be lower in the coronal holes than in the normal corona. Because here the magnetic field lines stretch outward instead of looping back into the sun, it is suspected that these holes may be the source of the solar wind. The coronal hole is virtually invisible in radiation emitted from the chromosphere or photosphere. That is, we are unable to follow the phenomenon from its appearance in the corona to the cooler regions closer to the sun.

Approximately thirty emission lines have been seen in the coronal spectrum, part of which is shown in Figure 11.18. They originate in the peculiarly excited ions of familiar elements such as iron, nickel, chromium, and others, plus argon, from which nine to fifteen electrons have been stripped in the corona's extremely hot, tenuous gases. It takes a temperature of 1 to 2 million degrees Kelvin to produce such high ionization. Because some of the atomic transitions that cause them are very improbable ones, many of

FIGURE 11.16
The corona photographed in red light on 30 June 1973 in Kenya. The bright regions in the corona are primarily at low latitudes near the solar equator.

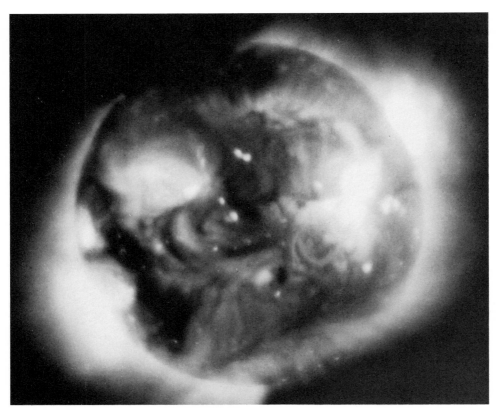

FIGURE 11.17
X-ray photograph of the solar corona
taken during a NASA *Aerobee* rocket
flight on 24 November 1970.

the coronal lines are called *forbidden lines*. Under the density and temperature conditions normal for laboratory experiments, the forbidden lines are absent, but in the corona's very low densities they can be relatively intense. Because of the high temperature, most of the coronal radiation should be in the ultraviolet and X-ray regions. This is substantiated in X-ray photographs revealing intense radiation whose source is well outside the photosphere, as is evident in Figure 11.17.

From the millimeter to the meter wavelengths, a wide spectral window in the earth's atmosphere lets in radio radiation. The sun, when quiet and undisturbed, normally emits thermal (blackbody) radiation which is characteristic of the hot corona, from the microwave region of about 1-centimeter wavelength to the long-wavelength region of about 20 meters. The layer in the solar atmosphere emitting electromagnetic energy at some radio wavelength depends on the density of free electrons. However, when the sun is disturbed, non-

thermal radio emission is very noticeable on the earth. It comes in short-lived *radio bursts* extending over many wavelengths, long-lived *noise storms* in meter wavelengths, and a slowly varying *thermal component* in centimeter wavelengths. The first and third effects occur in active regions surrounding sunspot groups; the second is due to interactions between magnetically trapped particles and moving plasma waves.

PROMINENCES AND FILAMENTS

A specially designed refracting telescope called a *coronagraph* is used to observe the corona from the ground. The coronagraph uses an occulting disk, which covers the image of the sun's disk, blocking light from the photosphere so that faint radiation from the chromosphere and corona can be photographed. By using a color filter that transmits a very narrow range of wavelength, we can photograph the solar corona in monochromatic light. Earth's atmosphere readily scatters photons from the photosphere, and those are millions of times more numerous than photons from the corona, so the coronagraph must operate from the highest mountaintop observatories to minimize this scattering.

With time-lapse photography we can see spectacular motions of towering masses of luminous gas, the *prominences*. Projected against the solar disk, these are the dark threadlike *filaments* seen in the photographs in Figures 11.13b and 11.1. Their forms vary from almost stationary, quiescent arches and graceful loops, to rapidly moving surges. The typical prominence is about 200,000 kilometers long and 5,000 kilometers thick and extends about 30,000 kilometers above the photosphere. It consists of cooler and denser gas than that in the corona around it. In the more active prominences gas may occasionally rise at more than 1000 kilometers per second, escaping from the sun. Frequently, however, matter appears to rain down from the corona in great luminous masses. Apparently, magnetic fields hold up these huge walls of gas against the sun's gravitational pull. We can see matter flowing along the body of a prominence, apparently following the curving and looping magnetic lines of force, as in the spectacular eruptive prominence shown in Figure 11.1.

The mean lifetime of large quiescent prominences is about two to three rotations of the sun. During sunspot maximum twenty filaments may appear on the disk; during sunspot minimum there are typically about four. Prominences always appear to be associated with a plage or sunspot group.

HEATING THE CORONA

There are several lines of evidence confirming the rise in temperature through the chromosphere into the corona. Emission lines of highly ionized atoms as mentioned is one. Second, coronal emission lines are highly spread out in wavelength due to Doppler shifts of the high-temperature gas particles moving toward and away from us. Finally, substantial emission of X-ray and radio radiation also indicates a high temperature. This temperature rise seems paradoxical,

FIGURE 11.18

Ultraviolet spectrum of the corona showing bright lines, photographed from an *Aerobee* rocket on 13 March 1959. I indicates a neutral atom; II, an atom that has lost one electron; III, an atom that has lost two electrons, and so forth. The horizontal scale is in angstroms.

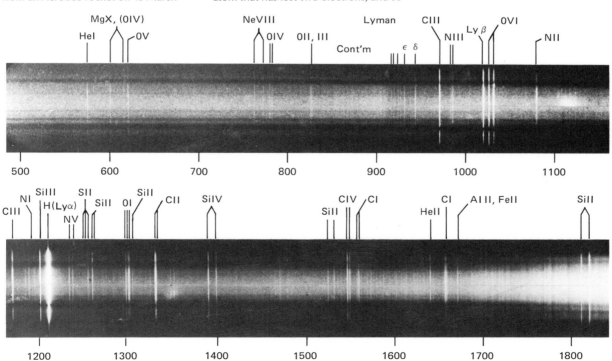

DEFLECTION OF STARLIGHT: TESTING RELATIVITY THEORY

A total eclipse of the sun gives us a chance to test relativistic against classical physics. Newtonian calculations say that a ray of light from a star, considered as a moving stream of photons (the photon is treated as having mass equal to its energy divided by the velocity of light squared) grazing the limb of the sun, will be deflected inward 0.875 arc second by the sun's gravitational field (see Figure 11.19). Relativity theory specifies that the space geometry is warped around the sun and a ray of light travels along the shortest path in this curved space. When we include this factor, starlight is deflected by 1.75 arc seconds, exactly twice the amount predicted by Newtonian theory.

How can we test these predictions? During a total solar eclipse, a photograph of the darkened sky around the eclipsed sun reveals nearby bright stars. Another photograph of the same area is made at night a few months earlier or later with the same telescope (the sun is in a different place in the sky then). The normal and displaced star positions are then compared. The amount of deflection decreases as distance from the sun's limb grows. Every check has favored the answer given by relativity theory, though the difficulties of measuring and the resulting uncertainties keep this from being a definitive test. This observational test has become almost a standard procedure at every total eclipse of the sun.

Bending of radio waves near the sun has also been checked in recent years by radio interferometry. The apparent position of a pointlike radio source about to be occulted or grazed by the sun can be simultaneously compared with that of a similar source nearby. Bending of the radio waves passing through the solar corona (caused by refraction by the corona) must be separated from that caused by the sun's gravitational field. The correction can be made because the refraction varies at different wavelengths but the gravitational deflection remains constant. The measured gravitational deflection agrees within 1.5 percent of the value predicted by Einstein's general theory.

When Mars was near conjunction with the sun from late November to mid December 1976, radio signals from the *Viking* orbiters grazing the sun on their way to the earth were bent and slightly delayed due to the warping of space around the sun. Mars was then about 321 million kilometers from the earth and radio signals took about 42 minutes for the round trip. The difference in the travel time of the signals when Mars was near the sun compared to that when Mars was well separated from the sun weeks later, 0.0002 second, was in exact agreement with that predicted by the general theory of relativity.

FIGURE 11.19
Gravitational bending of starlight grazing the sun's limb. The deflection angle α is 0.875 arc second according to Newtonian calculations but 1.75 arc seconds according to relativity theory.

The starlight, which is deflected when passing the sun, appears to an observer on the earth to be coming from a point farther from the edge of the sun than is actually the case.

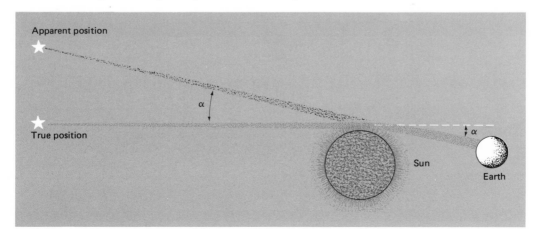

Apparent position

True position

α

α

Sun

Earth

but astronomers thought they had at least a reasonable explanation for it. However, evidence from *Orbiting Solar Observatory 8* experiments seems to cast doubt on whether we actually do understand it.

Astronomers think that most of the corona's heat comes from energy carried into the corona by mechanical waves starting in the turbulent convection zone below the photosphere. Supergranulation cells may be one way by which such waves are channeled into the sun's upper atmosphere. As the acoustic waves move through the chromosphere and into the corona, where the density of the gas is lower, their energy is absorbed by the gas through which they are moving, heating the coronal gas up to 1 or 2 million degrees Kelvin. This also creates a rapid flow of plasma (composed mostly of protons and electrons), which moves out from the base of the corona as the solar wind. Several radii outside the sun, the velocity becomes supersonic and continues to increase at least up to 1 astronomical unit. By then the solar wind is moving at about 450 kilometers per second, eight times the speed of sound in the gas. Because of the sun's rotation, magnetic field lines, which confine the solar wind particles, spiral outward like water from a rotating sprinkler. Perhaps 600,000 tons of plasma leave the sun every second, which amounts to about 10^{-15} of the sun's mass per year.

11.6 Solar Activity

ACTIVE REGIONS

During the sunspot cycle the general level of all activity in the solar atmosphere follows the number of sunspots. Thus the sunspot group seen in white-light photospheric pictures is just the most visible indicator of a large disturbed region in the solar atmosphere called an *active region*. Such regions can be up to several hundred thousand kilometers in extent. The various transient phenomena that are part of the active region are summarized in Table 11.3 and are illustrated in Figure 11.20. The common bond between these visible features is the magnetic field. It appears first, followed by the faculae in the photosphere and the plage in the chromosphere. This is

TABLE 11.3
Summary of Solar Active Regions

Atmospheric Region	Visible Activity	Description
Photosphere	Faculae	Observed in white light, denser and hotter than normal photosphere, granular and irregular, average life 15 days, can exist for 80–90 days.
	Sunspots	Observed in white light, cooler than photosphere, average life 6 days, strong magnetic fields, size about 10,000 kilometers.
Chromosphere	Plages	Lifetime about 40 days, hotter and denser than normal chromosphere, size about 50,000 kilometers.
	Flares	Brief brightening in plages, life about 20 minutes, size about 30,000 kilometers, enhanced particle emission in solar wind and solar cosmic rays.
Corona	Prominences (filaments)	Chromospheric material in corona, cooler than surrounding corona, lifetimes up to 60–90 days if quiescent, height 30,000 kilometers, length 200,000 kilometers, and thickness 5000 kilometers, exhibit motions associated with magnetic fields.
	Condensations	White-light features caused by increased electron density, enhanced emission in forbidden lines and ultraviolet lines, also radio emission.
	Radio bursts	Motions of fast electrons in upper corona.

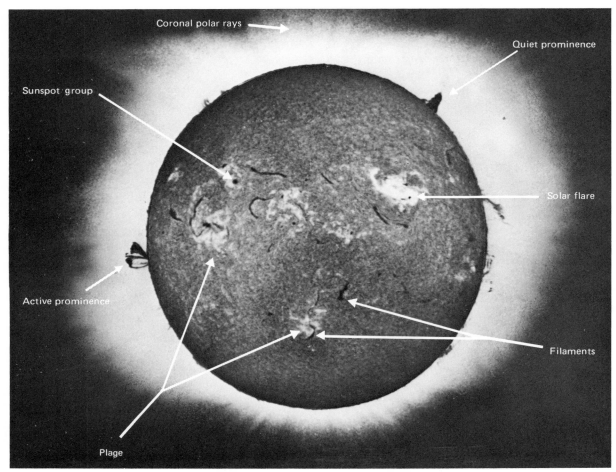

FIGURE 11.20
Composite spectroheliogram of the solar disk and the chromosphere, photographed in the light of the hydrogen alpha line, superimposed on the inner corona. Various transient solar features are marked.

followed by the sunspot group, flare activity, and prominences. The precise behavior is somewhat different for each active region; however, there is little doubt that the phenomena in different regions are related.

SOLAR ACTIVITY AND THE EARTH
As one might guess, transient activity on the sun does produce a variety of effects here on the earth. This transient behavior on the earth also follows a rough cycle in unison with the sun's cycle. An example of this relation is the increase in auroral activity in the earth's atmosphere during sunspot maximum and its decrease during minimum. Changes in solar activity also affect the earth's weather and climate, but how and over what time scales these effects occur we do not clearly understand. However, in recent years some interesting discoveries have been made about the constancy of the sun's activity and its relation to the earth.

IS SOLAR ACTIVITY CONSTANT?
As we stated at the beginning of this chapter, approximately 1.4 million ergs of radiant energy fall on each square centimeter of the earth every second. The theory of stellar evolution, discussed in Chapters 13

through 15, suggests that the sun's luminosity must have increased by perhaps 30 percent since it began its existence some 4.6 billion years ago. However, as recently as 1975, paleoclimate evidence concerning long-term temperature variations on the earth show that any change has been less than 3 percent in the last million years. One can argue that if the solar luminosity had been as much as 25 percent less than the present value, the oceans would have frozen. However, questions regarding the long-term constancy of the sun and the earth's probable response to any change are still unanswered.

Another question concerning constancy is the constancy of solar transient activity. Does the 22-year

FIGURE 11.21
The sun's variation in activity since the Bronze Age is shown in the top two panels. Below is shown the corresponding estimates of the temperature in Europe, the winter severity index from Paris and London, and Alpine glacier activity. It appears that long-term climatological activity rises and falls in response to long-term solar activity.

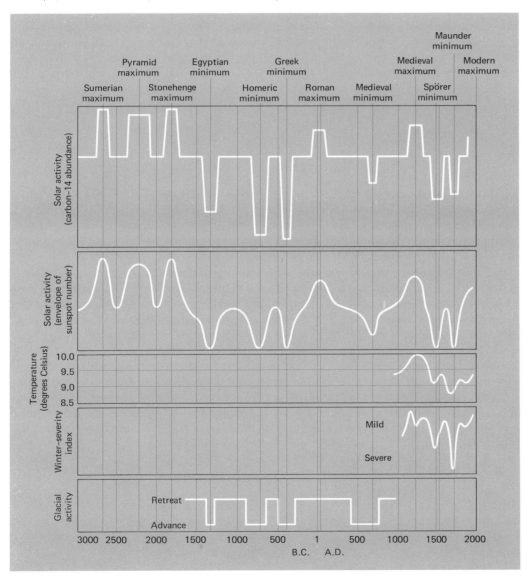

cycle of activity repeat itself, period after period, for millions of years? In 1893 E. Walter Maunder, British astronomer, found from historical records from Europe that very few sunspots were seen in the period from 1645 to 1715, now known as the Maunder minimum. Within the last several years Maunder's work has been confirmed and extended by the American astronomer John A. Eddy.

Eddy has shown that in addition to the absence of sunspots, very few auroras were observed in Europe, and during eclipses the corona was absent or very weak. No sunspots were reported in Asia during the Maunder minimum even though naked-eye sunspot-sighting reports exist there from as early as 28 B.C. Finally, measurements of the amount of carbon 14 ($^{14}C_6$), a radioactive isotope of carbon, in tree rings show that during the Maunder minimum there was an excess of this isotope in the earth's atmosphere. We believe that the very high energy subatomic particles, called *cosmic rays,* moving randomly through the Galaxy collide with the nucleus of nitrogen atoms ($^{14}N_7$) in our atmosphere, converting it to carbon 14. When the sun is very active, the interplanetary magnetic field is strong and galactic cosmic rays are deflected away from the earth. Thus high levels of carbon 14 in the earth's atmosphere correspond to low levels of solar activity.

Historical research by Eddy has shown a correlation between the carbon 14 abundance in tree rings, winter severity, galactic cosmic ray activity, and solar activity, as shown in Figure 11.21. Periods of colder climate appear to coincide with low levels of solar activity, as evidenced by unseasonably cold weather in Europe between the sixteenth and eighteenth centuries. There is evidence that at least a dozen similar periods of minimal solar activity lasting from 50 to 200 years have occurred since 3000 B.C. Thus after so much effort devoted to trying to understand the 22-year solar cycle, we have discovered that it may be only a modern transient feature itself.

SUMMARY

The sun is the only star whose surface we can study closely. It is our celestial laboratory. We observe the sun's atmosphere (photosphere, chromosphere, and corona), sunspots, plages, flares, and solar wind from space observations as well as from ground stations. Useful data on the sun's outer atmosphere trickle in during every total eclipse of the sun. Analyzing the solar spectrum builds knowledge of the sun's chemical composition, magnetic activities, sunspot behavior, differential rotation, and atmospheric temperatures, densities, and pressures. The various space projects have led to new insights about the sun's activity.

The sun makes its radiant energy by controlled thermonuclear fusion. In the superhot core hydrogen nuclei are fused into helium nuclei, releasing a large amount of energy, which is radiated into space. The little solar energy that the earth intercepts is enough to warm our planet and to sustain life. The mass of hydrogen still left in the sun's core is sufficient to keep it going at the present rate for another 5 billion years.

The eleven-year sunspot cycle is also an indicator for the earth of the amount of solar ultraviolet radiation, the frequency of auroral displays, the changes in the earth's ionosphere, and the variations in the earth's surface magnetism. The aftereffects of disturbances on the sun—flares, plages, and large sunspot groups—upon the terrestrial environment are being studied internationally. But we do not know yet just how the complicated sun-earth relationships work. The lack of sunspots from 1645 to 1715 suggests that the eleven-year sunspot cycle may not be a permanent feature of the sun.

REVIEW QUESTIONS

1. How do we know that sunspots are regions of intense magnetic activity?

2. What advantages does observing a total eclipse of the sun have over the normal means of observing the sun?

3. How do we know that the sun rotates pro-

gressively slower from its equator toward the poles?

4. An X-ray photon created within the sun's superhot core has no chance of getting through to the total surface. How is it possible that X-ray photographs of the sun are commonplace?

5. What is the general picture of the photosphere?

6. How do we obtain a single-color hydrogen or calcium scan of the chromosphere?

7. Describe the prominence activity in the solar atmosphere.

8. When viewed at the time of a total eclipse of the sun, the chromosphere has a pinkish glow. Why? What does its spectrum show at this time?

9. How do we account for a coronal temperature of 1 million degrees Kelvin when the sun's surface temperature is only 5800 degrees Kelvin?

10. How extensive is the corona? How is it related to the solar wind?

11. Why is helium prominent in the chromospheric spectrum and absent in the photospheric spectrum?

12. What gases have been identified in the coronal spectrum? What is so unusual about their physical condition?

13. Why are sunspots darker than surrounding regions?

14. About how many elements have been identified in the solar spectrum? What technique is used to identify them?

15. When will the next maximum of the sunspot cycle occur? When will the minimum that follows it occur?

16. What correlations have we found between unusual solar events and terrestrial disturbances?

17. Why are there dark lines in the spectrum of the photosphere and bright lines in the spectrum of the chromosphere and the corona?

18. Define the solar constant. How does its determination lead to the total output of the sun's radiant energy?

19. Why does the ultraviolet spectrum of the sun consist of bright lines?

20. What is the Maunder sunspot minimum? How did it affect the terrestrial environment?

12
The Nature of Stars

12.1 Introducing the Stars

Star maps have been included in this book to help you learn something about the organization of the sky. This might be a good time to spend a few evenings outside familiarizing yourself with the sky. Having spent a night looking at the stars, it is not hard to understand the fascination they have held for generations of human beings. But what are they? Only in the last hundred years or so have we been able to say with complete confidence that the stars are great spheres of glowing gas like the sun. However, there are still other questions to be answered about them.

As an example, how far away are the stars? How bright are they? How does brightness vary with wavelength? Is their brightness variable? Are they very hot? What are they made of? How much mass do they contain? How dense are they? Do they rotate, have magnetic fields, and have surface phenomena like the sun? Are they all the same, or is there a range in their properties? For decades many astronomers have been investigating these questions and others about the nature of the stars in our Galaxy. They have compiled a great body of knowledge about the distances, motions, sizes, brightnesses, and chemical compositions of stars—a stellar census of different population groups within the Galaxy. This chapter is devoted to surveying that data.

12.2 Finding Distances of Stars

TRIGONOMETRIC PARALLAX

Unless we know the distance of a star, it is impossible to evaluate many of its properties. The basic technique used is *trigonometric parallax*, the method we

◄ Winter constellation of Orion and the surrounding region. The three prominent stars in a line form Orion's belt. The Orion Nebula, the fuzzy object below the belt, is the scabbard of the mighty hunter.

discussed earlier to find the moon's distance (see Section 6.6). You can understand how it works by looking first with one eye and then with the other at a pencil held at arm's length: the pencil seems to shift its position against the background. The displacement of the pencil shrinks as the distance of the pencil from your eyes increases. Applied to stars (see Figure 12.1), the parallactic shift that we can see is very tiny; it was not until the years from 1837 to 1839 that astronomers were able to measure the parallax for three stars: Alpha Centauri, 61 Cygni, and Vega. Their distances are, respectively, 4.3 light years, 11 light years, and 26 light years.

The tiny apparent displacements of a star against the background stars in a photograph are compared, in principle, every six months, when the earth reaches opposite sides of its orbit. The comparison is best at this time because the parallactic shift is at its maximum at the opposite ends of the baseline whose length is equal to 2 astronomical units (the diameter of the earth's orbit). But even with a telescope of 15-meter focal length, the star closest to us, Alpha Centauri, has a parallactic displacement on a photographic plate of only 0.01 centimeter. Thus in practice, many photographs are secured at various times over a period of years.

To get the distance of a star from the sun in astronomical units, we insert the parallax angle p in the first formula of Figure 12.1 (derived by simple trigonometry). Because the distance in astronomical units may rise to very large figures, astronomers prefer two larger units of distance. The *light year* is the distance that light travels in 1 year (the velocity of light times the number of seconds in 1 year). A still larger unit is the *parsec*, the distance of a star whose parallax is 1 second of arc. The name parsec comes from the first three letters in the words *parallax* and *second*.

Table 12.1 gives the distances and other data on the known 25 nearest stars. About 35 stars are within 14 light years of the sun. This yields a density of stars in the solar neighborhood of about 0.1 star per cubic parsec, or about 0.003 star per cubic light year (a cube of space about 9.5 trillion kilometers on a side). Thus the parsec (equal to 3.26 light years) and the light year are units of distance characteristic of the separation between stars in the solar neighborhood. The limit for measuring parallax accurately is about 300 light years, or 100 parsecs. As many as 500,000 stars may be within this range, and 6,000 of these have been observed for

TABLE 12.1
The Twenty-five Nearest Stars

Designation	Distance (ly)	Proper Motion (arc seconds per year)	Radial Velocity (km/s)	Visual Apparent Magnitude[a]		Visual Luminosity[a] (sun = 1)		Spectral Type[a]		Mass (sun = 1)		Radius (sun = 1)	
				A	B	A	B	A	B	A	B	A	B
Sun				−26.81		1.0		G2 V		1.0		1.0	
Alpha Centauri[b]	4.3	3.68	−22	0.00	+1.4	1.3	0.36	G2 V	K IV	1.1	0.89	1.23	0.87
Barnard's star[c,m,n]	5.9	10.31	−108	+9.54		0.00044		M5 V		≈ 0.14		≈ 0.3	
Wolf 359[d,m,n]	7.6	4.71	+13	+13.66		0.00002		M8e V		≈ 0.1		≈ 0.1	
BD + 36° 2147[e,n]	8.1	4.78	−84	+7.47		0.0052		M2 V		0.35		≈ 0.4	
Sirius[f]	8.6	1.33	−8	−1.47	+8.7	23	0.0028	A1 V	wd	2.31	0.98	1.8	0.022
Luyten 726-8[m,n]	8.9	3.36	+30	+12.5	+13.0	0.00006	0.00004	M6e V	M6e V	0.044	0.035	≈ 0.1	≈ 0.1
Ross 154[d,m,n]	9.4	0.72	−4	+10.6		0.00037		M5e V		≈ 0.15		≈ 0.2	
Ross 248[d,m,n]	10.3	1.58	−81	+12.24		0.00010		M6e V		≈ 0.1		≈ 0.15	
Epsilon Eridani[h,n]	10.7	0.98	+16	+3.73		0.30		K2 V		0.98		0.5	
Luyten 789-6[d,m,n]	10.8	3.26	−60	+12.18		0.00012		M6e V		≈ 0.05		0.05	
Ross 128[m,n]	10.8	1.37	−13	+11.10		0.00033		M5 V		≈ 0.15		≈ 0.2	
61 Cygni[i,n]	11.2	5.22	−64	+5.19	+6.02	0.083	0.040	K5 V	K7 V	0.63	0.6	≈ 0.65	≈ 0.6
Epsilon Indi[m,n]	11.2	4.69	−40	+4.73		0.13		K5 V		≈ 0.6		≈ 0.5	
Procyon[j]	11.4	1.25	−3	+0.38	+10.7	7.6	0.0005	F5 IV–V	wd	1.77	0.63	1.7	0.01
Σ 2398[k,n]	11.5	2.28	+5	+8.90	+9.69	0.0028	0.0013	M4 V	M5 V	0.4	0.4	≈0.5	≈0.5
BD + 43° 44[l,m,n]	11.6	2.89	+17	+8.07	+11.04	0.0058	0.0004	M2e V	M4e V	≈ 0.3	≈ 0.3	≈ 0.5	≈ 0.4
CD − 36° 15693[m,n]	11.7	6.90	+10	+7.39		0.011		M2 V		≈0.4		0.6	
Tau Ceti[m]	11.9	1.92	−16	+3.50		0.44		G8 V		0.8		1.04	
BD + 5° 1668[m,n]	12.2	3.73	+26	+9.82		0.0014		M4 V		≈ 0.2		≈ 0.4	
CD − 39° 14192[m,n]	12.5	3.46	+21	+6.72		0.025		M1 V		≈0.4		≈ 0.7	
Kapteyn's star[m,n]	12.7	8.89	+245	+8.81		0.0040		M0 V		≈ 0.2		≈ 0.2	
Krüger 60[k,n]	12.8	0.86	−26	+9.77	+11.2	0.0017	0.00044	M3 V	M5e V	0.27	0.16	0.51	0.16
Ross 614[k,n]	13.1	0.99	+24	+11.13	+14.8	0.0004	0.00002	M5e V	?	0.14	0.08	≈ 0.3	?
BD − 12° 4523[m,n]	13.1	1.18	−13	+10.13		0.0013		M5 V		≈ 0.2		≈ 0.4	
van Maanen's star[m,n]	13.9	2.95	+6	+12.36		0.00017		wd		≈ 0.6		≈ 0.01	

[a] A is the brighter component and B is the fainter component of the star listed in the first column.

[b] AB system is visual binary, period 80 years; a third member is Proxima, 2.2° away; apparent magnitude, +10.68; luminosity 0.00007; spectral type, M5e V (flare star), mass 0.1.

[c] Appears to have two invisible planetary bodies circling it in periods of 12 years and 26 years; masses 0.8 and 1.1 of Jupiter's mass, respectively.

[d] Flare star; e signifies bright lines in spectrum.

[e] Unseen companion, period about 8 years.

[f] Visual binary, period 50 years. The faint member of the system is a white dwarf (wd).

[g] Visual binary, period may be about 50 years. One component a flare star, UV Ceti.

[h] A suspected invisible body, several times the mass of Jupiter, circles it in a period of 25 years.

[i] Visual binary, period about 720 years, has unseen component with mass 0.008.

[j] Visual binary, period 41 years; one of the components is a white dwarf.

[k] Visual binary.

[l] Visual binary, component A is a spectroscopic binary.

[m] Mass derived by authors from mass-luminosity relation.

[n] Radius derived by authors from temperature for spectral type and luminosity.

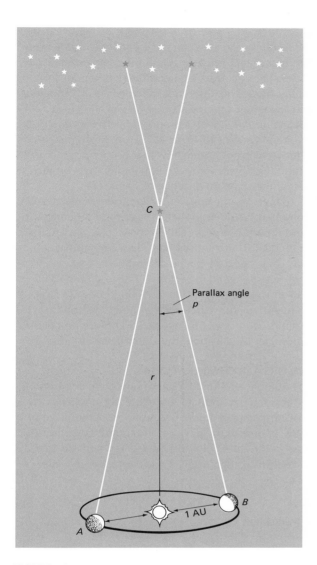

FIGURE 12.1
Parallax diagram. If we look at star C from points A and B on the earth's orbit six months apart, the star will seem to have moved slightly compared with the more distant stars in the background. Measurement of the parallax angle p, in arc seconds, combined with knowledge of the earth's mean distance from the sun (1 AU) yields the distance of the star r by trigonometry:

distance in astronomical units, $r = 206,265/p''$

distance in light years, $r = 3.26/p''$

distance in parsecs, $r = 1/p''$

1 ly $= 9.46 \times 10^{12}$ km; 1 pc $= 3.26$ ly

(See Appendix 1 for abbreviations of units of measure.)

trigonometric parallax. For stars whose parallaxes are too small to be measured, astronomers fortunately have other ways of measuring distances.

MORE REMOTE STARS

If we know an object's true brightness, we can use the relationship between apparent brightness and distance to find how far it is from us. By the inverse-square law of light (page 58), if one light source is twice as distant as another of the same intrinsic brightness, its apparent brightness is reduced by a factor of 4 (inversely as 2^2, the square of twice the distance); if three times as distant, the reduction is $3^2 = 9$ times; if four times as distant, the reduction is $4^2 = 16$ times, and so on. Hence, if we know the true brightness of a class of stars to which a star belongs, we can calculate its distance by the inverse-square law from its apparent brightness. This method depends on properly recognizing the subject as a specific kind of star whose luminosity is known, so that we have a standard of brightness.

Astronomers vary this technique to get a step-by-step scale of overlapping stellar distances. We begin with the nearby stars whose distances are known from trigonometric measurements of parallax and move toward the more distant stars, applying the appropriate technique.

12.3 Brightnesses of Stars

APPARENT MAGNITUDES

Before photography was invented the only way of knowing a star's apparent brightness, or the *apparent magnitude m*, was by visual estimates. Ancient Greek astronomers were the first to use a scale of brightness in terms of magnitude. They picked a scale to which the eye responds naturally—equal ratios of brightnesses corresponding to equal differences of magnitudes. Hipparchus and Ptolemy roughly graded the apparent brightnesses of the stars into six magnitude classes: from 1, the brightest, to 6, the faintest. Experimenting with telescopes of varied apertures, William Herschel concluded that a star of sixth magnitude was about a hundred times as bright as a star of first magnitude.

In 1856 British astronomer Norman Pogson quantified the system of magnitudes that we use today: the ratio of apparent brightness of first and sixth magnitude stars is taken to be 100:1. This ratio corresponds to any 5 magnitude difference. From this rule and a logarithmic formula, we can compare brightness ratios with magnitude differences, as in Table 12.2. As a means of developing a feeling for the great range in apparent magnitudes among celestial bodies, examine the values listed in Table 12.3. The numerically larger negative values of magnitude mean brighter objects; larger positive values mean fainter ones. From the table we see that the difference in visual apparent magnitude between Sirius and Venus at its brightest is $-1.5 - (-4.4) \approx 3$, which corresponds to a brightness ratio of about 16:1; hence Venus can appear 16 times brighter than Sirius in the night sky. However, in fact Sirius is intrinsically much brighter than Venus. Thus an important question to ask ourselves is whether the stars appear bright in our night skies because they are nearby or because they are intrinsically bright. The answer to this question involves what is termed a star's absolute magnitude.

ABSOLUTE MAGNITUDE AND DISTANCE MODULUS

To find the intrinsic brightness of a star, we must know its distance and apparent magnitude. We can calculate (using the inverse-square law of light) what the apparent magnitude an object would be if it were placed at the arbitrary distance of 10 parsecs (32.6 light years); the result is called the star's *absolute magnitude*. Comparing the stars' brightnesses at the same distance (10 parsecs) is, of course, a comparison of their intrinsic luminosities. The difference between a star's apparent and absolute magnitude, $(m - M)$, is called its *distance modulus*, and it is proportional to the ratio of the star's distance to 10 parsecs. From its numerical value we can determine how many times brighter or fainter the object appears compared with its brightness at the standard distance of 10 parsecs.

To measure the apparent magnitude of a star, astronomers have set up stellar magnitude sequences in various parts of the sky similar to the one shown in Figure 12.2. These are standards by which the magnitudes of other stars can be determined.

As we discussed in Section 5.2, the color of radiation affects our perception of its brightness. (At this point you might wish to review the discussion on radiation detectors on page 86.) So before proceeding further, let us talk about the color of stellar radiation.

COLORS AND TEMPERATURES

The numerical value of a star's magnitude depends on the spectral region that we are looking at when we measure the magnitude. Because photographic emulsions, photoelectric devices, and infrared detectors possess different color responses, standard practice is to use filters that admit a narrow range of wave-

TABLE 12.2
Brightness Ratios and Magnitude Differences

Brightness Ratio	Magnitude Difference
1:1	0.0
1.6:1	0.5
2.5:1	1.0
4:1	1.5
6.3:1	2.0
10:1	2.5
16:1	3.0
40:1	4.0
100:1	5.0
1,000:1	7.5
10,000:1	10.0
1,000,000:1	15.0
100,000,000:1	20.0

TABLE 12.3
Visual Magnitudes of Selected Objects

Object	Apparent Magnitude	Absolute Magnitude
Sun	−26.7	+4.8
Full moon	−12.5	
Venus (at brightest)	−4.4	
Sirius (brightest star)	−1.5	+1.4
Alpha Centauri	0.0	+4.4
Vega	0.0	+0.5
Antares	+0.9	−4.3
Andromeda galaxy	+3.5	−21.2
Faintest naked-eye star	+6.0	
Faintest star photographed by 5.1-m Hale telescope	+24.0	

William Herschel, destined to become one of the world's greatest observational astronomers, was born in Germany but lived most of his life in England. By the age of thirty-four he had established himself as a musician of note: teaching music; playing the organ at Bath; playing violin concerts; and composing military music, symphonies, and choral numbers. But his leisure hours were devoted to studying foreign languages, philosophy, and

mathematics. Then, in his thirty-fifth year, his desultory interest in astronomy was fired into action after reading Robert Smith's *Compleat System of Opticks* and Ferguson's *Astronomy.* And before long he was constructing his own reflectors.

By 1773 Herschel had built a 5.5-foot telescope, and he was now ready to begin his observations. Between 1774 and 1781 he recorded the observation of numerous individual objects. His study of the sky led him to several important discoveries. For example, in one project Herschel was searching for double stars in the expectation that a parallax shift of the brighter component might be detected relative to the fainter and presumably more distant component. Instead, Herschel discovered the orbital motion of one component around the other, the first tangible proof that gravity extended to the stars. Out of this work came the first *Catalogue of Double Stars.*

Another program Herschel initiated was the examination of every star in the standard star charts of the day. On the night of 13 March 1781 he made the historic discovery of a new planet in the constellation of Gemini, the planet Uranus. This discovery won him international fame and the royal patronage of King George III. After being knighted and awarded an annual stipend of 200 pounds by the king, Herschel could

now devote all his time to astronomy, unhindered by the necessity of earning a living as a professional musician.

One of Herschel's important discoveries was the sun's motion in space. From the proper motions of only thirteen stars he found that the sun is moving in space relative to its stellar neighbors toward a point on the sky in Hercules, not far from the bright star Vega.

His most ambitious undertaking was an attempt to determine the structure of the Milky Way system. This involved star gauging: making sample counts of the stars in the field of view of his telescope. By the time he finished nearly twenty years later in November 1802, he had counted over 90,000 stars in 2,400 sample areas. Along the way Herschel listed objects of interest: variable stars, pairs of stars, dark areas in the Milky Way that looked like holes, irregular bright nebulosities, clusters of stars, and several thousand small white nebulae. The latter three varieties of objects were incorporated in his 1802 *Catalogue of Star Clusters and Nebulae.* The Milky Way system, he concluded from his star counts, had the shape of a disk, like a grindstone, having a thickness of about one-sixth of its diameter. It was marked by many irregularities, and the sun was located near its center. Later studies confirmed Herschel's deduction that our Galaxy is disk-shaped.

---

lengths. This technique ensures that the results obtained by the different measuring devices with the same filter are comparable. Some common filter designations are ultraviolet U, blue B, yellow or visual V (which approximates the human eye's sensitivity), red R, near infrared I, and so forth. One magnitude scheme, known as the UBV photometric system, covers the spectral range from approximately 3000 angstroms to 6000 angstroms in three segments, as shown in Figure 12.3.

The difference in any two particular magnitudes measured for a star is called a *color index,* or simply a *color.* A frequently determined color index is the

difference between the blue and visual magnitudes, designated simply as (B-V). A blue star, for example, has a brighter blue magnitude B than its visual magnitude V. Since brighter means algebraically smaller values on the magnitude scale, (B-V) is negative in this case. The opposite is true for an orange or red star, where (B-V) is positive. The zero point of the (B-V) scale is arbitrarily assigned to stars whose spectrum is classified A0 V (see Section 12.4).

Do these stellar color indices tell us anything? Magnitudes for different spectral regions of the same object differ because the distribution of radiant energy varies with wavelength in accordance with Planck's radiation law for blackbodies (see Section 4.2). We can therefore use the color index to determine the star's temperature. [See Table 12.5 for the relationship of the (B-V) color index to the star's surface temperature.] For most stars it is an average of the high temperatures at the bottom of the star's photosphere and the lower temperatures found in the uppermost layers. For most of the nearby stars the surface temperature ranges from about 3000 K to 40,000 K.

BOLOMETRIC MAGNITUDES

A star's *luminosity* is the total amount of energy it radiates per second from its surface. But the magnitudes we have considered so far correspond to the amount of radiation in a specific spectral range that arrives at the earth's surface. A more fundamental measurement, the *bolometric magnitude*, represents all the star's radiation integrated over all wavelengths received outside the earth's atmosphere. Unless this magnitude can be observed directly from an instrumented satellite, a theoretical correction using the star's blackbody temperature must be applied to the ground-observed visual magnitude to compensate for the radiation that does not pass through the earth's atmosphere. The correction is largest where the peak of the energy curve lies outside the visual range, as it does for the hottest and coolest stars. Values for luminosity are given in Table 12.4 for the thirty stars with the greatest apparent brightnesses. As is evident

from the table, most stars are bright because they are intrinsically bright, not because they are nearby. Using Tables 12.1 and 12.4, we see that luminosity varies from about 10^{-5} times that of the sun to 10^5 times greater.

STELLAR RADII

Knowing a star's temperature and luminosity allows us to calculate another fact about a star—its radius. From the temperature we can determine the energy emitted per second from each square centimeter of the star's surface. The total amount of energy radiated per second by the entire star, which is its luminosity,

FIGURE 12.2
Photographic negative of a star field with several apparent magnitudes marked. An image's size and darkness are used to determine the apparent magnitude by comparison with known standards such as those labeled in this photograph.

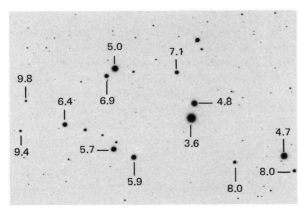

FIGURE 12.3
UBV photometric system. Color filters permit light to be transmitted only through the narrow spectral region shown. In the UBV system these spectral regions are centered around 3,500 angstroms for the ultraviolet, 4,400 angstroms for the blue, and 5,500 angstroms for the visual. The percentage of light transmitted through the filter drops off fairly rapidly on either edge of the spectral region. Notice that the (B-V) color index (the difference between the blue and the visual magnitudes) is positive for the 3000 K and 5000 K blackbody energy curves and negative for the 30,000 K energy curve. Thus the (B-V) color index can be calibrated as a measure of the temperature of the blackbody energy curve, which approximates the energy curve for a star. Such a calibration is given in Table 12.5.

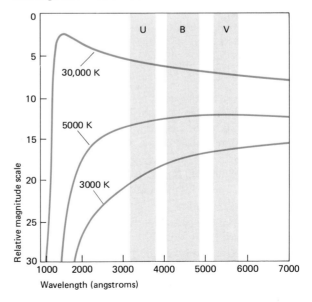

TABLE 12.4
Thirty Brightest Stars

Star	Constellation Designation	Distance (ly)	Proper Motion (arc second per year)	Radial Velocity (km/s)	Visual Apparent Magnitude A	B	Visual Luminosity (sun = 1) A	B	Spectral Type A	B
Sirius[a]	α CMa	8.6	1.33	−8 var	−1.47	+8.7	23	0.002	A1 V	wd
Canopus	α Car	80	0.02	+20	−0.72		800		F0 II	
Rigel Kent[b]	α Cen	4.3	3.68	−23 var	0.00	+1.4	1.3	0.36	G2 V	K1 V
Arcturus	α Boo	36	2.28	−5	−0.06		100		K2 III	
Vega	α Lyr	26	0.34	−14	+0.04		50		A0 V	
Capella[c]	α Aur	45	0.44	+30	+0.78	+0.88	75?	70?	G8 III?	F6 III?
Rigel[d]	β Ori	1600	0.00	+21 var	+0.14	+6.6	150,000?	100	B8 Ia	B5
Procyon[e]	α CMi	11.4	1.25	−3	+0.38	+10.7	7.6	0.0005	F5 IV–V	wd
Betelgeuse	α Ori	600?	0.03	+21 var	+0.41 var		13,000		M2 Iab	
Achernar	α Eri	120	0.10	+19	+0.51		600		B5 IV	
Hadar[f]	β Cen	400?	0.04	−12	+0.63	+4	9,000?	400	B1 II	?
Altair	α Aql	16.8	0.66	−26	+0.77		9.8		A7 IV,V	
Acrux[g]	α Cru	400?	0.04	+7 var	+1.39	+1.9	3,000?	1700	B0.5 IV	B1 V
Aldebaran[h]	α Tau	68	0.20	+54	+0.86 var	+13	150	0.002	K5 III	M2 V
Spica[i]	α Vir	230	0.05	+1	+0.91 var		1,700		B1 V	B3 V
Antares[j]	α Sco	420?	0.03	−3 var	+0.92 var	+5.1	8,000?	170?	M1 Ib	B4e V
Pollux	β Gem	35	0.62	+3	+1.16		30		K0 III	
Fomalhaut[k]	α PsA	23	0.37	+6	+1.19	+6.5	12	0.09	A3 V	K4 V
Deneb	α Cyg	2300	0.00	−5 var	+1.26		100,000?		A2 Ia	
β Cru[l]	β Cru	500?	0.05	+20 var	+1.28 var		5,000?		B0 III	
Regulus[m]	α Leo	84	0.25	+4	+1.36	+7.9	140	0.3	B7 V	K2 V
Adhara	ε CMa	700?	0.00	+27	+1.48	+8	8,000?	20?	B2 II	?
Shaula	λ Sco	300?	0.03	0	+1.60		1,600?		B1 V	
Castor[n]	α Gem	49	0.20	+4	+1.97	+3.0	28	11	A1 V	A5m
Ballatrix	γ Ori	500?	0.02	+18	+1.64		3,600?		B2 III	
El Nath	β Tau	200?	0.18	+8	+1.65		1,400?		B7 III	
β Car	β Car	86	0.18	−5	+1.67		110		A0 III	
γ Cru	γ Cru	200	0.27	+21	+1.69	+6.7	800		M3 II	
Alnilam	ε Ori	1600?	0.00	+26	+1.70 var		40,000?		B0 Ia	
Al Nair	α Gru	64	0.19	+12	+1.76 var?		57		B7 IV	

Note: Var means variable. Distances in light years of more remote stars are estimated from spectroscopic parallaxes.

[a] α CMa: The orbital period of the visual binary is 50 years; one component is a white dwarf.

[b] α Cen: The orbital period of the AB system is 80 years. Component C, Proxima Centauri, is 2.2° distant. It is a flare star (V645 Cen) of type M5e V; visual magnitude, +10.68, and visual luminosity, 0.00007.

[c] α Aur: The two bright stars form a spectroscopic binary with a period of 104 days. A distant comparison, 12′ from the bright star is a physical companion. It is itself a close binary of visual magnitudes +10.2 and +13.7 and corresponding luminosities 0.012 and 0.0005.

[d] β Ori: Component B is a spectroscopic binary with a period of 10 days and both stars are B9; it has been suspected of being a close visual binary.

[e] α CMi: The companion is a white dwarf. The orbital period is 41 years.

[f] β Cen: The separation of the two stars is 1″.

[g] α Cru: The separation of AB is now about 4.5″; component A is a spectroscopic binary with a period of 75.8 days.

[h] α Tau: Component B is 30″ from A.

[i] α Vir: The star is a spectroscopic binary with a period of 4.0 days. Shallow eclipses of about 0.1 magnitude have been observed.

[j] α Sco: The companion is about 3′j from A.

[k] α PsA: The companion, HR 8721, is 2° from the bright star.

[l] β Cru: The star is a variable of the β CMa type with a period of approximately 4 hours.

[m] α Leo: Component B is a double star whose companion has luminosity 0.0036.

[n] α Gem: Both A and B components are spectroscopic binaries with respective periods of 9.2 days and 2.9 days. A third component C, 72″ away, is YY Gem, an eclipsing binary with a period of 0.8 day.

depends not only on its temperature but also on its surface area. We can thus calculate how large the star must be to have its luminosity, which is known from its bolometric absolute magnitude (see below.) The radii of stars are found to vary from a hundredth of the sun's radius (or about the size of the earth) to about a thousand times that of the sun, which is about 5 astronomical units or five times the radius of the earth's orbit. Figure 12.4 compares the size of the red supergiant Betelgeuse to the size of the planets' orbits.

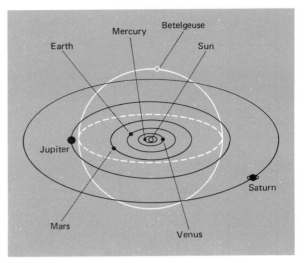

FIGURE 12.4
Comparison of the size of the red supergiant Betelgeuse in the constellation Orion with the orbits of the planets in the solar system. Betelgeuse is about as big as Jupiter's orbit—a thousand times the diameter of the sun.

12.4
Spectra of Stars

SPECTRAL CLASSES
Astronomers working to classify the spectra of the stars into an ordered sequence found themselves in a rich field of exploration. The first large study was a

FINDING A STAR'S RADIUS

Recall that the Stefan-Boltzmann radiation law (see Section 4.2) says that the radiation emitted from each square centimeter of the surface in one second of time is proportional to the fourth power of the temperature, or,

$$F = (5.67 \times 10^{-5})T^4$$

where F is the surface emission in ergs per square centimeter second. The luminosity L is equal to the surface area of the star multiplied by the unit surface emission; that is, $L = 4\pi R^2 F$, where R is the star's radius. Hence $L \propto R^2 T^4$, and $R \propto \sqrt{L/T^4}$. The angular diameters of some of the largest stars were measured directly many years ago with an optical interferometer attached to the Mount Wilson 2.5-meter reflecting telescope by the American physicist Albert Michelson.

We can express the star's radius in terms of the sun's, eliminating the constant of proportionality:

$$\frac{R_*}{R_\odot} = \sqrt{\frac{L_*}{L_\odot}} \left(\frac{T_\odot}{T_*}\right)^2$$

where * is the value for the star and ⊙ is that for the sun.

Example: Barnard's star (second closest to the sun) has a bolometric absolute magnitude of +10.3, compared to the sun's +4.8. From the magnitude difference of 5.6 between the two bodies, we find (from Table 12.2) that the luminosity of Barnard's star is 0.006 that of the sun. The temperature of Barnard's star is $T_* = 2900$ K; that of the sun is $T_\odot = 5800$ K. Then $L_*/L_\odot = 0.006$ and $T_\odot/T_* = 2$. Substituting the numerical values in the formula, we get

$$\frac{R_*}{R_\odot} = \left(\frac{5800}{2900}\right)^2 \sqrt{0.006} = 4(0.077) = 0.3$$

The diameter of Barnard's star is nearly one-third that of the sun.

photographic spectral survey of nearly a quarter million stars begun in 1884 at the Harvard College Observatory. Astronomers at Harvard completed their study forty years later.

The Harvard astronomers classified stellar spectra in a one-dimensional scheme. As the survey ended in 1924, when the relationship between atomic structure and emission of radiation was better understood, astronomers found that the great diversity in spectral appearance, shown in Figure 12.5, was due primarily to the stars' differing surface temperatures, not to differences in the abundance of their elements. Most stellar atmospheres have chemical elements in nearly the same proportions as the sun: an overwhelming amount of hydrogen, a little helium, and traces of the other elements (see Table 11.1).

Each of the seven *spectral classes*—O, B, A, F, G, K, M—in the classification scheme is divided into ten parts, or *spectral types*, from 0 to 9. The sun's type is G2, 0.2 beyond G0 toward the next class K. In Table 12.5 we have listed the spectral classes and described

their most distinguishing features, including the surface temperature and (B-V) color index for the hottest spectral type in each spectral class.

Four other classes take in a very small percentage of the stars. Class W (Wolf-Rayet) covers the very hot stars that have broad emission lines and an excessive abundance of carbon or nitrogen; this group precedes type O. Classes R, N, and S relate to cool red stars having prominent absorption bands of carbon (C_2), cyanogen (CN), or methylidine (CH)—for classes R and N—and zirconium oxide (ZrO) and lanthanum oxide (LaO)—for class S—along with the low-temperature lines of the neutral metals. These classes are alternatives to class M. Other subtle differences appear in the spectra of stars of the same temperature, depending on the sizes or luminosities of the stars.

Data, including the spectral type, are given in Table 12.4 for the thirty brightest stars. Spectral types for the nearest known stars are in Table 12.1. The significance of the Roman numerals in the spectral type column is discussed in the next paragraph.

TABLE 12.5
Spectral Classes of Stars

Spectral Class	Intrinsic Color	(B-V) Color Index[a]	Surface Temperature[a] (K)	Prominent Absorption Lines
O	Electric blue	−0.32	38,000	Ionized helium; multiply ionized oxygen, nitrogen, silicon; neutral helium; hydrogen weak.
B	Blue	−0.30	30,000	Neutral helium strongest; ionized helium weak or absent; hydrogen stronger; ionized oxygen, carbon, nitrogen, silicon.
A	Blue white	0.00	10,800	Hydrogen strongest; ionized calcium, magnesium, iron, and titanium strong.
F	Yellow white	+0.33	7,240	Hydrogen weakening; more ionized and neutral metals.
G	Yellow	+0.60	5,920	Hydrogen weaker; ionized calcium very strong; other ionized metals weaker; neutral metals stronger; CH present.
K	Orange	+0.81	5,240	Neutral atoms very strong; ionized calcium still strong; few ionized metals; hydrogen still weaker; CN usually strong.
M	Red	+1.41	3,920	Strong neutral atoms; hydrogen very weak; titanium oxide bands prominent.

[a] For the hottest spectral type in class, such as A0 in A.

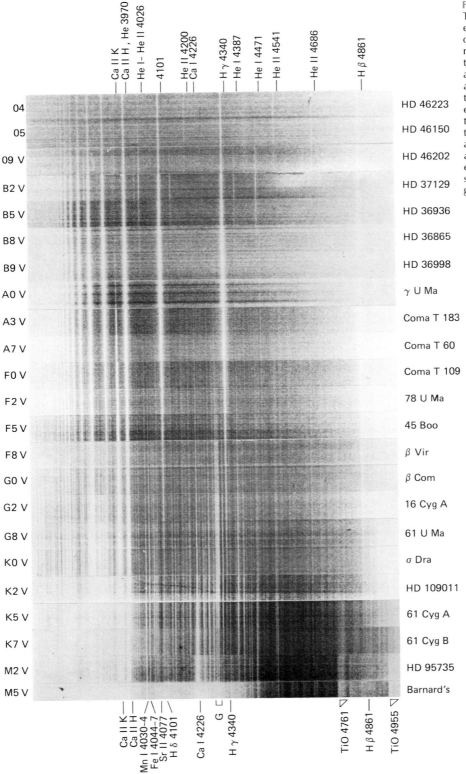

FIGURE 12.5
The spectral sequence, showing examples of each spectral class designated at the left. A number of prominent absorption lines of several elements and their wavelengths in angstroms are identified at the top and bottom. Note, for example, how the strength of the hydrogen lines increases toward the A spectral class and then diminishes. The rise and fall of line intensities of several elements throughout the spectral sequence is displayed graphically in Figure 12.9.

LUMINOSITY CLASSES

A two-dimensional extension of the Harvard spectral classification scheme was developed in the early 1940s; it subclassified stars of similar surface temperatures into *luminosity classes*. The scheme is based on fine but observable differences in the strength of certain absorption lines in a star's spectrum. The luminosity classes are designated by Roman numerals: I is for very luminous supergiants; II is for bright giants; III is for giants; IV is for subgiants; and V is for dwarf, or main sequence, stars. You can see in Figure 12.6 that though the spectra of the G8 supergiant, G8 giant, and G8 main sequence stars are much alike, some differences are visible in the indicated lines.

Luminosity differences in stellar spectra are much more subtle effects than temperature differences. Thus it is not always possible to determine a star's luminosity class even though its spectral type can be roughly estimated. Once a star's luminosity class is known, though, we can find its distance from its apparent and absolute magnitudes which give you (m-M) or its distance modulus.

The names for each of the luminosity classes are more than poetic. As we have already seen, two stars of the same spectral type or temperature must differ in radius if they are to have different luminosities. The more luminous the star, the larger will be its radius. Thus the supergiants are the largest, the bright giants next largest, and so forth. The significance of the name *main sequence* is that perhaps 85 to 90 percent of the stars are dwarf or main sequence stars.

In addition to stars brighter than the sun and other main sequence stars, there are stars that are less luminous than main sequence stars of the same spectral type. These are the subdwarfs and the white dwarfs. We will not be discussing the subdwarfs again, but we will have a great deal to say later about the white dwarfs, which are fascinating members of the world of stars.

MESSAGE IN THE SPECTRUM OF A STAR

Astronomers analyzing a star's spectrum can find out a great deal about the star's physical and chemical properties, particularly from the shape, width, and strength of absorption lines in the star's spectrum. You can see how intensity varies across part of a typical spectrogram in Figure 12.7. The absorption lines are the dips below the continuous background spectrum. The tiny wiggles are made by the photographic emulsion's graininess.

In actual practice, it is not easy to sort out all the processes that affect the shape of an absorption line. First, absorption lines are broadened by various atomic processes: electronic transitions between atomic energy levels which have a natural width; collisions between atoms and ions while either is absorbing a photon; and the Zeeman effect (discussed on page 227). As an example, the absorption lines in the spectra of supergiants and bright giants are narrow due to a lower pressure in their photospheres with consequently fewer collisions between atomic particles. From the Zeeman broadening of absorption lines, magnetic fields in a few stars are found to be as large as several thousand gauss compared to a few gauss for the sun.

Second, the motions in a star—such as random thermal motions, streams of gas analogous to solar

FIGURE 12.6
Differences in the appearances of certain indicated sensitive lines between a supergiant, giant, and main sequence star of the same spectral class. Note that the Fe I 4045 line is about the same in the spectra of all three stars, but the Sr II 4077 line grows progressively weaker from the supergiant star to the main sequence star.

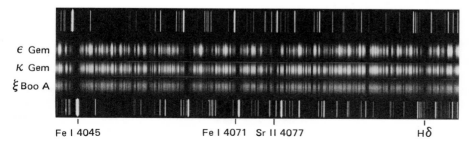

ε Gem — Supergiant
κ Gem — Giant
ξ Boo A — Main Sequence

Fe I 4045   Fe I 4071   Sr II 4077   Hδ

granules and supergranules, and rotation of the star—all broaden absorption lines through the Doppler effect. Rates of rotation for the O and B stars are several hundred kilometers per second compared to the few kilometers per second for stars like the sun, and this difference can be seen in the widths of the absorption lines in Figure 12.8.

Analyzing absorption lines to determine the chemical composition of a star's atmosphere is one of the most important fields of research in astronomy. The intensity of an absorption line (see page 257) depends on the excitation of neutral or ionized atoms of an element. This fraction depends on the star's temperature, the atmospheric gas pressure, and the abundance of the element. The astronomer generally knows something about the first two factors from the star's luminosity and spectral type. The uncertain factor is the abundance of the element. From judicious adjustment of the abundance, temperature, and pressure, the astronomer can calculate theoretical absorption lines that agree with the observed line intensities, element by element. Thus we find abundances for the elements in the star's photosphere and its atmospheric temperature and density.

Abundance studies for several hundred stars give values much like those for the sun (see Table 11.1). But some exceptions appear in the abundances both of all elements heavier than hydrogen and helium and of certain elements in some stars. The heavy-element abundances appear to vary from 0.01 percent to about 5 percent by weight among all stars.

DO STARS HAVE SURFACE ACTIVITY LIKE THE SUN'S?

Of great interest to astronomers is the question of whether other stars have chromospheres and coronas like the sun. Also, do all or only some stars have transient surface activity like the sun? A partial answer to these questions is suggested by the many main sequence red dwarf stars, called *flare stars*, which undergo sudden short flare-ups in brightness. A rise in brightness from 1 to 6 magnitudes builds up in a few seconds, then subsides within minutes. The flare is probably caused by a local hot disturbance on the star's surface, like a solar flare. Some flare-ups have been accompanied by a radio outburst and occasionally by X-ray emission much more intense than we find in the average solar flare. There are now what amounts to direct observations of stellar chromospheres and coronas for some stars of the later spectral types, and

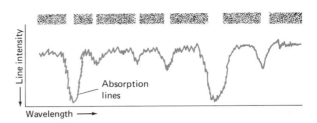

FIGURE 12.7
Microdensitometer tracing. The spectrogram (shown schematically above) is scanned with a narrow beam of light and recorded on a strip of moving paper. More light passes through where the absorption line is than elsewhere, so the recording pen shows a larger dip where the line is. The lower the dip, the stronger the absorption line. The tiny wiggles are not weak absorption lines; they are due to the photographic emulsion's graininess.

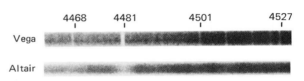

FIGURE 12.8
Differences in line widths between two stars. In the bottom strip the rapid rotation of the star is apparent by the washed-out appearance of the lines.

little doubt exists that at least a substantial fraction of the stars do have such atmospheric structure. Additionally, it is highly probable that other stars besides the sun have transient and periodic surface activity.

12.5 Variable Stars

CEPHEID VARIABLES

We now know of over twenty-six thousand stars that vary in brightness. To analyze their variability, astronomers usually employ photometric observations, spectrum analysis, and Doppler measurements. A plot of the observed change in magnitude in a specific time is a *light curve*; a plot of the Doppler line shift with time is a *velocity curve* (see Figure 12.10).

ABSORPTION LINE INTENSITIES

The differences observed in stellar spectra stem from the manner in which atoms of various elements absorb and reradiate energy under different temperature and density conditions in stellar atmospheres. Density influences spectra because the frequency with which gas particles collide increases with density. For most stars the atmospheric gas pressure, which is due principally to the free electrons, is a few thousandths or ten-thousandths of the air pressure at the earth's surface. In general decreased atmospheric pressure favors greater ionization of the atomic constituents; in-

creased pressure diminishes the effect.

By far the most important variable affecting stellar spectra, however, is temperature. In a hot gaseous medium there are many high-energy photons and more energetic and frequent collisions between atoms; as a result, more atoms are ionized and have their outermost bound electron excited to upper energy states. A cooler medium has insufficient numbers of the high-energy photons or violent collisions that are necessary for exciting electrons into upper atomic levels. Instead, low-energy photons and weak collisions are more

numerous. They control most of the excitation and ionization.

How the strengths of the absorption lines of the neutral and ionized atoms vary from one spectral type to another is shown in Figure 12.9. These curves are calculated from the theoretical equations that give the degree of excitation and ionization, element by element, for a specific temperature, atmospheric pressure, and excitation or ionization energy of the atom. The results agree closely with the observed absorption-line intensities throughout the range of stellar temperatures.

FIGURE 12.9
Relative strengths of absorption lines of various elements throughout the spectral sequence. The principal factor responsible for the variation in line intensities is the star's surface temperature. This determines the relative distribution of atoms of a particular element among

the different atomic levels and hence the strength of its lines. Higher temperature favors higher ionization. Corresponding spectral classes are shown on the horizontal scale above. The behavior shown here is described in Table 12.5.

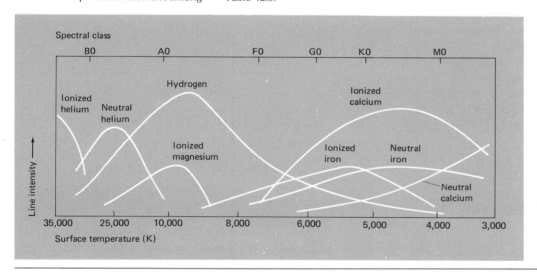

Many of the variable stars belong to a distinctive group, the *pulsating variables*, which we believe owe their variability to a regular expansion and contraction. The *classical cepheids*, numbering about seven hundred, form one important subdivision of this group. Named after the first star of their kind discovered,

Delta Cephei, they have the following characteristics: they are yellow supergiants of great brilliance and spectral types late F to early K; their brightness varies periodically over about 1 magnitude during an interval of 2 days to 40 days; their spectra show Doppler shifts synchronized with the periodic changes in brightness (see Figure 12.10). One of the most familiar cepheids is Polaris, the North Star, which varies by about one-tenth of a magnitude in a period just under 4 days.

The more luminous the cepheid, the longer the period of variation. This extremely important correlation between period and intrinsic brightness is the *period-luminosity relation*. It is represented graphically by a plot exhibiting the correlation between median absolute magnitude and the period (see Figure 12.11). (The median magnitude is halfway between the maximum and minimum magnitudes.) Calibrating the absolute magnitude scale is difficult because no cepheid is close enough to have its parallax measured; astronomers have fallen back on less direct methods to set the zero point of the absolute magnitude scale.

The classical cepheids are the most luminous cepheids. Sparsely sprinkled within the Galaxy's disk, they are a very small segment of the population of the spiral arms. Another less luminous group of cepheids, with periods generally between 10 days and 30 days, are the *W Virginis cepheids* (named after their prototype). They are found in the population of stars in the halo portion of the Galaxy and in several globular star clusters. Still another type of variable, present by the thousands in the Galaxy's central and halo regions and in globular clusters, constitutes a third class of cepheids. They are known as cluster variables or, more often, as *RR Lyrae variables* (also named for their prototype). These are bluish white giant stars fainter than the other cepheids but up to a hundred times brighter than the sun. Their periods of brightness variation average about a half day.

> And thus we die,
> Still searching, like poor old astronomers
> Who totter off to bed and go to sleep
> To dream of untriangulated stars.
>
> Edwin Arlington Robinson

The simple theory of stellar pulsation predicts that the period of oscillation is inversely proportional to the square root of the star's mean density. This is borne out by observational data for cepheids. A cepheid whose mean density is four times greater than

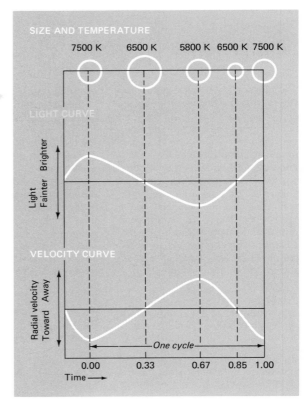

FIGURE 12.10
Typical cepheid variations during one light cycle. Stellar diameter changes are exaggerated in this drawing.

another will pulsate in half the time. If the sun were to pulsate, its period of oscillation would be about a half hour. This is very close to the period of the suspected pulsation of the sun (discussed on page 220).

LONG-PERIOD AND IRREGULAR VARIABLES

Nearly four thousand luminous red giants have cyclic brightness changes of several magnitudes with a period of about a year. These form another large group of variable stars, the *long-period variables*. Most celebrated of this group is the star Mira (Omicron Ceti), which varies between second magnitude (about as bright as Polaris) and ninth magnitude (sixteen times dimmer than the faintest naked-eye star) in a period of nearly a year. Infrared studies suggest that Mira-type stars are surrounded by extensive envelopes from which gas and dust are being expelled (see page 300). Variations in spectrum and velocity of the long-period variables are complex, with a poorly understood cause not identical to what makes cepheids pulsate.

Hundreds of other variable stars with widely differing luminosities and colors go through such irregular and often baffling changes that the cause of their behavior remains obscure. For example, one small group of yellow supergiants, the R Coronae Borealis variables, has a superabundance of carbon and a deficiency of hydrogen; the behavior of this group is completely unpredictable. After years of quiescence they suddenly dim by several magnitudes and fluctuate erratically before returning to their normal brightness months or years later. These stars have extended envelopes containing carbon grains, believed to be responsible for the star's change in brightness. More will be said about the reasons for a star's varying brightness in Chapters 14 and 15.

12.6
Binary Stars

Half or more of the stars may be stars that are orbiting about each other, whose fates are permanently linked by gravity. Among the sun's neighbors out to

15 light years, at least half are in multiple systems (see Table 12.1). One common subgroup of the multiple star systems are the double or *binary stars,* whose components may almost touch or may be separated by a large fraction of a light year. In a binary the stars orbit around the system's common center of mass (the *barycenter*) in an elliptical or circular orbit. The more massive component has a smaller orbit; the sizes of their orbits are inversely proportional to their masses. Visualize the system's relative motions by imagining how two unequal spheres of a dumbbell will move if you rotate it around its center of balance. This is the same kind of motion that we discussed between the earth and the moon.

Binary systems are classified according to the means of detection, which, in turn, depends on the separation between the components and the distance of the system from the earth. Three classes are recognized: visual, spectroscopic, and eclipsing binaries.

VISUAL BINARIES
Double stars that have large separations enable us to see both companions; they are the *visual binaries.* The secondary's orbital motion around its primary companion is often very apparent, as that of Krüger 60 in Figure 12.12. This apparent path is a projection of the true orbit on the plane of the sky. Yearly measurements of the visual binaries' apparent separations and motions around each other, as well as knowledge of their distances from the earth, provide the data needed to calculate their orbits. These vary in size from a few astronomical units to thousands of astronomical units, with periods of revolution from several years to many thousands of years. An example is shown in Figure 12.13.

In a small subgroup called *astrometric binaries* we cannot see the faint companion because of its closeness to the primary. What we do observe are tiny periodic wiggles in the visible star's apparent motion across the celestial sphere (called proper motion); from these we infer that an invisible companion exists. The variation arises from the visible star's orbital motion around the barycenter of the system combined with the system's motion relative to the sun. This action is pictured in Figure 12.14. As the visible star (in black) orbits around the barycenter, which moves upward, let us say, toward the right-hand side of the page (its proper motion), it makes a wavy motion along its path. In a few astrometric binaries the invisible companion seems to have a planetlike mass.

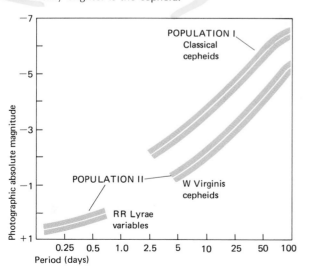

FIGURE 12.11
Period-luminosity relations for cepheids. The very important relationship between absolute magnitude and period helps us to determine the distances of single cepheids or of stellar systems that contain cepheids. Basically, the longer the period of light variation, the intrinsically brighter is the cepheid.

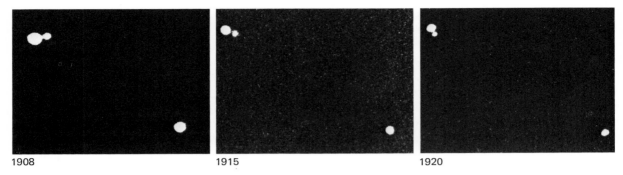

1908   1915   1920

FIGURE 12.12
Orbital motion of the visual binary
Krüger 60 during approximately one-
quarter of its 44-year period. This is one
of the nearby star systems listed in
Table 12.1.

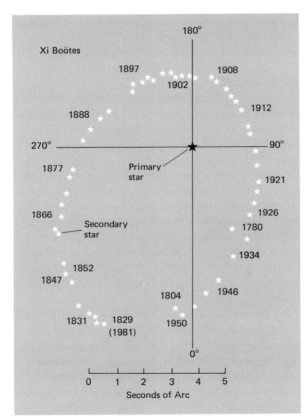

FIGURE 12.13
Orbit plotted for visual binary. Successive positions are shown for the fainter
secondary star in the visual binary
Xi Boötes as it orbited the brighter
primary from 1780 to 1950. The apparent elliptical orbit is actually the
projection of the true elliptical orbit
on the plane of the sky (the celestial
sphere).

SPECTROSCOPIC BINARIES

A system whose components cannot be resolved visually may nevertheless be revealed as a binary by the
periodic Doppler shifts in the spectral lines of the
composite spectrum due to one or both components.
These are the *spectroscopic binaries*, with orbits varying from a fraction of an astronomical unit to several
tens of astronomical units. The corresponding orbital
periods range from hours to several years. When the
absorption lines of both components are visible in the
composite spectrum, two sets of absorption lines are
periodically displaced from each other in opposite directions (see Figure 12.15). The maximum relative shift
occurs when the components' motion is along our line
of sight. There is no relative shift when the two stars
are moving across the line of sight a quarter of a
revolution later and the two sets of absorption lines
merge. Most often only the brighter component's
spectrum is visible on the spectrogram, and one set of
absorption lines shifts periodically. We can determine
whether the system is a binary system by analyzing the
velocity curve, a plot of the change in radial velocity
during the spectroscopic binary's period of revolution
(see Figure 12.16).

ECLIPSING BINARIES

Eclipses of the closely paired stars whose orbits we see
more or less edgewise appear as changes in brightness
at certain points in the system's orbital revolution.
These double star systems are *eclipsing binaries*. One
companion periodically passes between the other and
us, temporarily cutting off all or part of the eclipsed
star's light (see Figure 12.17).

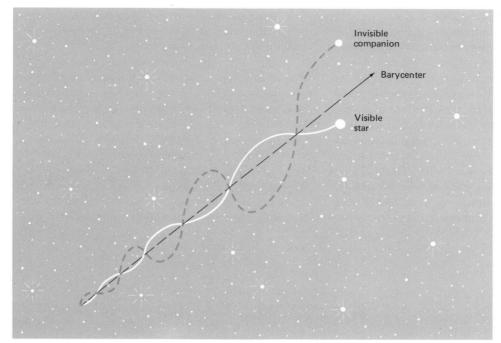

FIGURE 12.14
Path of astrometric binary across the
celestial sphere.

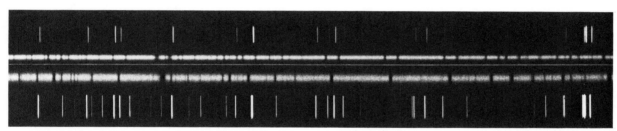

FIGURE 12.15
Composite spectrum of double-line
spectroscopic binary.

When the two eclipsing stars are equal in brightness and size (Figure 12.17a) the minima are V-shaped
and of equal depth, whether the eclipses are partial or
total. The more nearly total the eclipse, the deeper are
the minima. In Figure 12.17b most of the light comes
from the smaller, brighter component. We see the
deeper minimum when the larger, fainter star totally
eclipses the smaller star. Halfway around in the orbit,
the eclipse is annular when the brighter star passes in
front of the fainter one and the minimum is much
shallower. Both minima are flat-bottomed since that is
the time it takes the small star to cross the disk of the
large star. In Figure 12.17c, the stars are tidally distorted into ellipsoidal shape. The depths of the two
minima differ as in Figure 12.17b because the two
components are unequal in luminosity and size.

By analyzing the light curve's general shape and
measuring how long the eclipses diminish the light
and by how much, we can build a scale model of the
system. If we also observe the eclipsing binary as a
spectroscopic binary, we can combine data from the
light curve and the velocity curve to evaluate important stellar properties: radii, masses, densities, temperatures, and true orbital dimensions.

HOW MANY STARS ARE BINARIES

As we stated at the beginning of this section, observational evidence supports the contention that at least half of all stars are members of a binary system. However, we should qualify that by saying astronomers are not able at will to sample stars all across the Galaxy or in neighboring galaxies. Thus our statement is based on the extension of results found for nearby stars to a general truism about all stars. Also, if the two components of a binary system are of approximately the same mass, then their identication as a binary is fairly easy. But the greater the difference in the mass, the more difficult is the binary identification. As a point of illustration, the sun does not appear to have a companion star, although we can not entirely rule out the possibility that a very faint companion star exists far outside the edge of the solar system.

In the case of the sun, if there is no companion, does the existence of a planetary system take the place of a stellar companion? This is a very difficult question to answer. A study at Kitt Peak National Observatory of 123 nearby stars, which are similar in mass to the sun, found that, of the 123, 57 definitely have one companion, 11 have two companions, and 3 have three companions. Thus 58 percent of those studied definitely have companions. Using a variety of arguments, astronomers concluded that in the 123-star sample probably 67 percent have normal stellar com-

FIGURE 12.16
Spectroscopic binary system in which the spectral lines of both components appear. The black component is about 25 percent more massive than its companion. Thus both stars will be of approximately the same spectral type.

ELLIPTICAL ORBITS

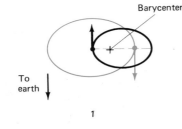
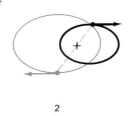
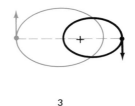
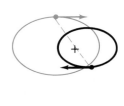

DOPPLER LINE SHIFTS IN THE COMPOSITE SPECTRUM

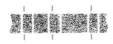

VELOCITY CURVES MEASURED FROM COMPOSITE SPECTRUM

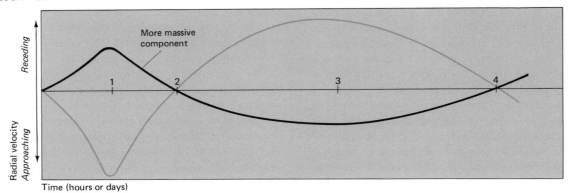

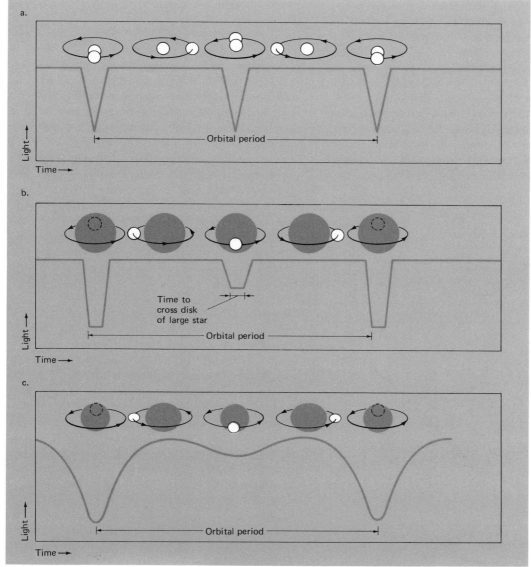

FIGURE 12.17
Representative eclipsing binary systems: a. partial eclipses; b. alternating total and annular eclipses; c. total and annular eclipses of tidally distorted companions.

panions, 15 percent black dwarf companions (a non-luminous stellar companion), and 20 percent planets. Or, in other words, there is at least inferential evidence for all sunlike stars having some kind of companion. It is probably too much to stretch this result to infer that all stars occur in some kind of companion relationship. But if in the future we find this to be true, it will not come as a surprise.

MASS-LUMINOSITY RELATION

We cannot directly obtain one very important quantity, the stellar mass, from isolated single stars, because their gravitational effects on other stars are insignificant. When stars are close enough, as double stars are, they orbit each other because of their mutual gravitational attraction. As a result, we can find the masses of these stars relative to the sun's mass.

In just this way an important correlation has been found to exist between the mass and luminosity of main sequence stars, and it is known as the *mass-luminosity relation* (see Figure 12.18). The relationship has also been derived theoretically from the fundamental laws governing matter and radiation in stellar interiors, to be discussed in the next chapter. Notice in Figure 12.18 that more massive stars shine more brightly: the luminosity is approximately proportional to the fourth power of the mass for stars brighter than the sun, and to slightly less than the third power of the mass for stars fainter than the sun. A star of two solar masses, for example, is sixteen times more luminous than the sun.

The correspondence between mass and luminosity in Figure 12.18 breaks down for the non-main sequence stars, which suggests that their internal structures differ from those of the main sequence stars. The very luminous supergiants, bright giants, and red giants lie somewhat above the curve and the white dwarfs considerably below it. Although luminosities vary greatly—from 0.001 of the sun's brightness to 50,000 times the sun's brightness—the range of masses is very modest: from several hundredths to about 50 or 60 times the solar mass. Below a minimum mass, it seems, a gaseous body does not radiate long enough at a constant luminosity to be a stable star. There is also apparently a maximum limit to the amount of material that can form a stable star, a limit set by the amount of material available and the motions inside a hot massive star.

MULTIPLE STAR SYSTEMS

Some gravitational linked systems have more than two stars. There may be a distant third star revolving around a close pair or, often, two close pairs, usually spectroscopic binaries, revolving around each other in a longer period. Other close combinations might have more than four stars. Take, for example, the second-magnitude star Castor in the constellation Gemini. With a small telescope we can easily pick out two visual components of Castor, which have a period of revolution of about 400 years. But when we observe the stars' spectra, we find that the visually brighter component is itself actually a binary with a period of 9.2 days. The Doppler shift from its orbital motion is shown in Figure 12.19. The fainter visible companion

FIGURE 12.18
Mass-luminosity relation for dwarf or luminosity class V stars. The plot shows that the more massive the star, the greater is its brightness. However, not all stars conform to this relation, such as the extremely luminous stars of luminosity classes I, II, and III.

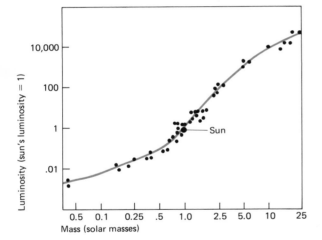

FIGURE 12.19
The spectrum of the spectroscopic binary α¹ Geminorum photographed at different times, showing the change in Doppler shift that results from its orbital motion. α¹ Geminorum is the brighter visual companion of the double star Castor in the constellation of Gemini. The bright-line comparison spectrum is that of titanium.

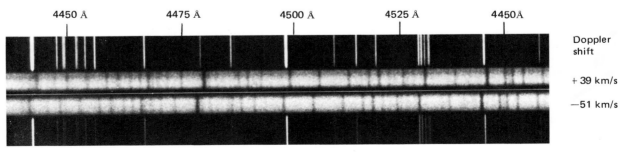

is also a spectroscopic binary; its period is 2.9 days. A ninth-magnitude star called Castor C, only 1.2 arc minutes away, takes many thousands of years to orbit the visual pair; this is an eclipsing binary with a period of 0.8 day. Thus all three classes of binaries are represented in Castor's sextuple system.

UNUSUAL BINARY SYSTEMS

Some binaries are unusual not because of peculiar characteristics of their orbital motion but because of the stars that compose the binary and because of their interaction with each other. As an example, unusual X-ray binaries have been recently discovered with X-ray telescopes carried by artificial satellites. They exhibit periodic variations in X-ray intensity. Their binary nature has been confirmed from the optical component, which displays orbital or light changes like those of the spectroscopic or eclipsing binaries. We will have more to say about these unusual binaries in Chapter 15.

SUMMARY

At first astronomers thought that the stars possessed almost infinite variety. However, over the first half of the twentieth century, astronomers were able to reduce the complexity presented by the stars by surveying the properties of the stars. The first step in this task was to obtain a means of determining the distances of the stars. The basis in what is an overlapping sequence of distance-determination techniques is trigonometric parallax. Knowing their distances, astronomers have been able for some stars to determine the brightnesses, colors and temperatures, radii, spectral features, variability, and masses. By doing this, we find that apparently, the brightest stars of our night skies are not necessarily the nearest stars; rather, they are intrinsically bright stars that can be seen at great distances across our Galaxy.

The stars possess a much greater range in their intrinsic brightnesses or luminosity than they do in their masses. Seeking correlations between various stellar characteristics, the astronomer has found relationships that help us understand how stars fit together. Important examples of such relationships are those between spectral types, and temperature and color; between masses and luminosities; and between the periods of pulsations and luminosities of cepheids. These relationships are important in helping us to understand how stars evolve into so many different varieties.

It appears that at least half of the stars are members of a binary or multiple star system. It is through their gravitational influence on each other in a binary system that we are able to determine their masses. Beyond this, the occurence of stars in multiple systems tells us about their origins and suggests that the formation of satellite bodies, like planets, may not be a rare event.

REVIEW QUESTIONS

1. In their study of the stars, what vital physical data do astronomers seek concerning the stars' properties?

2. Describe with the aid of a diagram the parallax method of determining the distance of a star.

3. What is one widely used method in finding the distances of stars too remote for trigonometric parallax determination?

4. Describe the general appearance of the Milky Way as it appears in your locality in the summer; in the winter.

5. Distinguish between the following magnitudes: (a) apparent magnitude; (b) absolute magnitude; (c) bolometric magnitude; (d) blue magnitude; (e) visual magnitude.

6. How is the color index or (B-V) magnitude of a star obtained? What is its significance?

7. What is the relationship between (a) spectral class and temperature; (b) spectral class and color index; (c) temperature and color index?

8. What is meant by distance modulus? What is its importance?

9. Why is the bolometric apparent magnitude more meaningful than the visual apparent magnitude?

10. What is the significance of the spectral sequence?

11. Why do the spectral line intensities of the various elements increase and decrease with temperature?

12. How can you distinguish between supergiants, giants, and main sequence stars of the same spectral type?

13. Why is the period-luminosity relation for cepheid variable stars so important to astronomers?

14. Differentiate between the classes of binary stars: visual, spectroscopic, eclipsing, astrometric, and X-ray.

15. How is the velocity curve of a spectroscopic binary obtained? The light curve of an eclipsing binary? What use is made of these curves?

16. How does the luminosity of a main sequence star vary with its mass?

17. Why can't we derive an accurate value of the mass of a single star?

18. Explain how it is possible for a double star to observationally be both a spectroscopic and eclipsing binary.

19. Discuss the probability of solarlike stars having planetary bodies circling them.

20. Discuss the differences in the prominent absorption lines between the various spectral classes.

13

The H-R Diagram and the Internal Structure of Stars

In the last chapter we presented some general physical characteristics of stars. The range of such aspects as radius, mass, mean density, surface temperature, and luminosity for as wide a sample of stars as possible is given in Table 13.1. For the majority of stars we see a tendency for them to cluster toward the middle of the range for a particular characteristic, such as mass, with only a few stars at either end of the range.

One characteristic that appears in Table 13.1 for which there has as yet been no discussion is the age of stars. If stars have definite ages, this implies they came into existence as stars at some definite time in the past, and they will cease to exist as stars at some definite point in the future. In this and the following two chapters we will consider questions such as how stars are created, what keeps them shining, and what is their ultimate fate.

Nothing is immutable in this universe. A star's existence is finite, and time brings change to it. A star evolves from a large cloud of gas and dust when it contracts under the force of gravity; unchecked, the gravitational force would squeeze the star down to almost nothing. Opposing forces, principally gas pressure, counterbalance this contraction, but the pressure tends to be undermined by the star's loss of energy as it radiates electromagnetic energy into space.

After a star is formed by accretion of interstellar matter, it begins to shine. Eventually it ends as a white dwarf, a neutron star, or possibly a black hole. In between, periods in which a star derives energy by contraction are separated by periods in which it resists gravity by nuclear burning,[1] which replaces the energy drained by its emission of radiation. That is the history of a star in its simplest form. However, before we can fill in the details, we must be able to interpret stellar data. We do this by using the Hertzsprung-Russell diagram, our single most important tool for studying stars. It allows us to organize stellar data and to look for relationships among different stars.

[1]By *nuclear burning* we mean those nuclear reactions in which light nuclei are fused together at very high temperatures to produce heavy nuclei plus energy; we do not mean ordinary combustion.

◀ Subatomic particle interactions in the 20-inch Bubble Chamber at Brookhaven National Laboratory. Shown is the annihilation of a proton and an antiproton.

TABLE 13.1
Range of Stellar Properties

Parameter	Approximate Limits[a]
Mass	$10^{-2} M_\odot$ to $10^2 M_\odot$
Radius	$10^{-2} R_\odot$ to $10^3 R_\odot$
Mean density	$10^{-7} \rho_\odot$ to $10^7 \rho_\odot$
Luminosity	$10^{-6} L_\odot$ to $10^6 L_\odot$
Surface temperature	$\approx 10^3$ K to 10^5 K
Heavy-element mass abundance	$10^{-2}\%$ to $10^1\%$
Age	Less than 10^4 years to 10^{10} years

[a]Solar units are used: $R_\odot = 6.96 \times 10^5$ km; $L_\odot = 3.83 \times 10^{33}$ erg/s; $M_\odot = 1.99 \times 10^{33}$ g; $\rho_\odot = 1.41$ g/cm³. The rare neutron stars (see Section 15.5) are not included.

13.1
The Hertzsprung-Russell Diagram

CORRELATING SPECTRAL CLASS AND LUMINOSITY
Between 1911 and 1913 the Danish astronomer Ejnar Hertzsprung and the American astronomer Henry Norris Russell independently stressed the importance of the diagram that we now call the *Hertzsprung-Russell,* or *H-R, diagram.* Plotting the spectral type (or, equivalently, the color index or surface temperature) horizontally across the diagram against the absolute magnitude, or luminosity, vertically for many different stars, we find that the plotted points are not scattered indiscriminately over the diagram. Instead, they lie in well-defined regions, as illustrated in Figure 13.1, which suggests a fairly continuous relationship between the stars in each region.

MAIN SEQUENCE, RED GIANTS, AND WHITE DWARFS
The most conspicuous region of the H-R diagram is the sequence of stars running from the bright hot stars in the upper left-hand corner to the faint cool stars in the lower right-hand corner. This sequence is called the *main sequence* and it contains most of the stars that could be plotted on the diagram. The sun is a G2

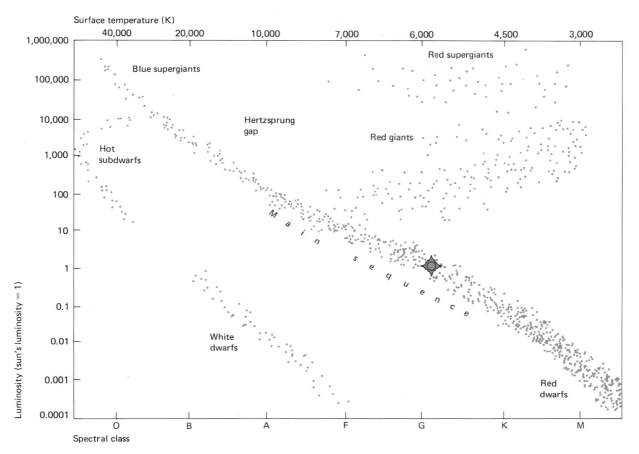

FIGURE 13.1

Schematic representation of the Hertz-sprung-Russell (H-R) diagram. When we plot a star's temperature (or spectral class) against its luminosity, we find that the plotted points for a large number of stars are not scattered at random but are confined to fairly well defined regions. The most prominent regions are these:

—*Blue supergiants:* bluest, most luminous, hottest; moderately large stars; low densities and large masses, very rare. Example: Rigel.

—*Red supergiants:* orange to red in color; the largest stars and among the brightest; large masses and extremely low densities; few in number. Example: Betelgeuse.

—*Red giants:* actually yellow, orange, and red; considerably larger and brighter than the sun; average to larger-than-average masses and low densities; fairly scarce. Example: Arcturus.

—*Main sequence stars:* blue, white, and yellow stars higher than the sun on the main sequence are somewhat larger, hotter, more massive, and less dense than the sun; plentiful in number. Example: Sirius. Orange and red stars below the sun on the main sequence are somewhat smaller, cooler, fainter, less massive, and denser than the sun; very plentiful in number. Example: Epsilon Eridani. An important subgroup are the red dwarfs, which are the coolest and reddest stars on the lower

end of the main sequence; considerably fainter and smaller than the sun; small masses and high densities; the most abundant stars. Example: Barnard's star.

—*Hot subdwarfs:* quiescent novas, central stars of planetary nebulas, and other subluminous hot blue stars; masses in the solar range and very high densities; fairly rare. Example: central star of the Ring nebula in Lyra.

—*White dwarfs:* mostly white and yellow; extremely faint and tiny by solar standards; enormously high densities; terminal evolutionary development; quite plentiful. Example: the companion of Sirius.

The stars shall fade away, the sun himself
Grow dim with age, and nature sunk in years.

Joseph Addison (1672–1719)

main sequence star and lies in roughly the middle of the H-R diagram among the yellow dwarfs. To distinguish the ends of the main sequence, the astronomer often refers to the blue stars as *early-type* stars and the red dwarfs as *late-type* stars.[2]

The second most prominent region of stars in the H-R diagram is that region broadly labeled as *red giants.* These are the luminous stars in spectral classes G, K, and M lying above the main sequence that angle up toward the upper right-hand corner of the diagram. The more luminous stars of luminosity classes I and II are called *red supergiants* if they lie on the cool side of the diagram in spectral classes G, K, and M, and *blue supergiants* if they are early-type stars of classes O and B.

The last region of some prominence, which spans the spectral classes B, A, and F, is that composed of the faint stars lying below the main sequence; these are called *white dwarfs.* (When we refer to a star as being *on* or *off* the main sequence, we refer to its position in the H-R diagram and not to its actual position in space.) As mentioned in the last chapter, the white dwarfs are an exotic group of stars of which we will have more to say later.

RADIUS ON THE H-R DIAGRAM

Stars of similar spectral type or surface temperature can be vastly different in size—remember that their luminosity is proportional to the radius squared and the fourth power of the surface temperature. For the three late-type stars of spectral class M presented in Table 13.2 (chosen from Figure 13.1), the radii are calculated from the formula on page 252. You can see how dwarfs, giants, and supergiants differ in size by comparing the numbers in the last column. These figures show (as you would expect) that, because of its larger surface, a more luminous star radiates more energy than a less luminous star of the same temperature. Figure 13.2 is an H-R diagram for the bright stars in Table 12.5. On it are plotted lines along which all stars have the same radius. Thus the radii of stars increases from the lower left-hand corner of the diagram toward the upper right-hand corner.

Although there is no precise correlation between

---

[2]At one point some time ago, it was thought that blue main sequence stars would evolve into red main sequence stars. Thus the designations *early-type* and *late-type* were thought to convey age. Astronomers now know this evolution theory is not true, but the labels remain.

TABLE 13.2
Intrinsic Brightness, Temperature, and Radius of Three Class M Stars

Object	Luminosity (sun = 1)	Temperature (K)	Approximate Radius (sun = 1)
Red dwarf	0.01	3,000	0.4
Red giant	100	3,000	40
Red supergiant	10,000	3,000	400

position in the H-R diagram and a star's mass, as was discussed in connection with the mass-luminosity relation, the more massive stars are, very roughly, toward the top of the diagram. Even though the masses of stars roughly increase as we move to larger-radii stars, the range of mass, from Table 13.1, is only a factor of about 5×10^3 between the least and the most massive stars. However, the radius varies by a factor of 10^5, and, since the volume is proportional to the cube of the radius, the volume varies by a factor of 10^{15}. For typical white dwarf stars, whose mass and radius are about 1 and 0.01 in units of the sun's values, the mean density is about 10^6 grams per cubic centimeter. On the other hand, the mass and radius of typical red giants are about 1 and 100 in solar units, so that the mean density is about 10^{-6} gram per cubic centimeter. Thus the mean density of the stars roughly decreases in the same direction as the radius increases.

BRIGHT STARS VERSUS NEARBY STARS

The H-R diagram for the bright stars in Figure 13.2 is striking in the sense that the majority of the stars are non-main sequence stars. Of the thirty stars plotted in the figure, 73 percent are stars above the main sequence, either subgiants, giants, bright giants, or supergiants, and only 27 percent are on the main sequence. More than half of the thirty stars are either giants or supergiants. This again points out that the apparently bright stars are bright because they are intrinsically bright and not because they are nearby.

In the H-R diagram for stars in the sun's immediate neighborhood (see Table 12.1) out to 15 light years, it is striking that no giants or supergiants appear (see Figure 13.3). The sun outranks all but three stars in size and luminosity. Three white dwarfs are set apart, and

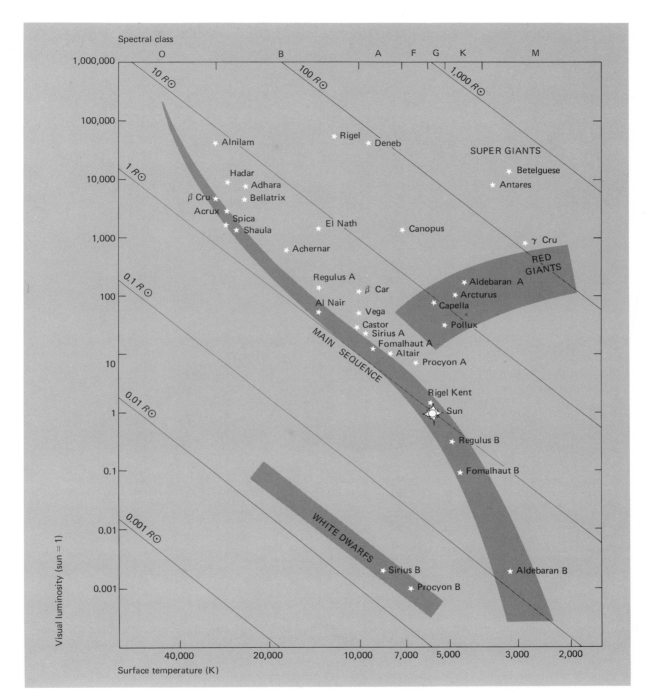

FIGURE 13.2
H-R diagram for the 30 brightest stars.
Note that the majority are intrinsically
more luminous stars than the sun. Two
companions are white dwarfs—Sirius B
and Procyon B. Compare this diagram
with the H-R diagram for the nearest
stars in Figure 13.3.

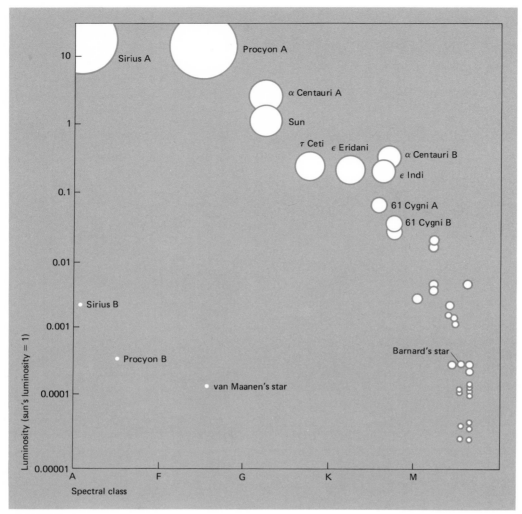

FIGURE 13.3
H-R diagram for stars within 15 light years of the sun. The approximate relative sizes are also shown. Letters A and B refer to the brighter and fainter components of a multiple system. Three white dwarfs appear at the lower left of the diagram; they are much smaller than indicated here.

many red dwarfs are at the tail end of the sequence. The relative number of stars of varying luminosity in a volume of space is known as the *luminosity function*. For every thousand small, red, class M stars, there are not more than a hundred class A-to-class K main sequence stars and one class B or class O giant or supergiant star. The evolutionary process that forms stars, as we will see in Chapter 14, appears to favor the formation of small, low-luminosity stars.

What we see when we scan the Galaxy are the intrinsically bright stars, not necessarily the most nu-merous type of star. We might anticipate the same result for other galaxies beyond the Milky Way. As an example, it would take about a thousand stars intrinsically as bright as Vega (A0 V), to equal the brightness of Rigel (B8 Ia), about fifty thousand as bright as the sun (G2 V), and almost 2.5 billion as bright as Wolf 359 (M8 V). Since the stars in the solar neighborhood do not suggest that the red dwarfs are that populous, then it is likely that it is the intrinsically bright stars that are responsible for the light emitted by a galaxy, while much of the mass of a galaxy is tied up in stars that do not contribute to its brightness.

INTERPRETING THE H-R DIAGRAM

The significance of the Hertzsprung-Russell diagram became clear several decades ago when astronomers realized that it is a panorama portraying how stars evolve. A star's position in the diagram is a function of its mass, radius, luminosity, composition, and evolutionary status. Some errors are present in the observational data, but the broadened regions of stars in the diagram are a composite of slightly different evolutionary positions determined chiefly by marginal differences in the chemical compositions of stellar populations. There is a physical reason why the points in the diagram are not scattered at random: the natural forces that settle the stars' destinies confine them to stable portions of the diagram while they are converting matter into radiant energy. The Hertzsprung gap (see Figure 13.1) is, for example, a semistable region through which the evolving stars quickly pass. (For the full story of stellar evolution, see Chapters 14 and 15.) Because the stars are in different stages of evolution, they serve as time machines in picturing the past and future of our sun.

13.2
Stars in Groups

OPEN CLUSTERS

In our Galaxy's disk are thousands of groups of stars that are bound together by their mutual gravitational attraction, the *open clusters,* sometimes called *galactic clusters.* Typical open clusters are shown in Figures 13.4 and 13.5, with data for a few selected open clusters listed in Table 13.3. The Galaxy may have about eighteen thousand of these clusters embedded in the spiral arms and disk of the Milky Way. They contain from a score to many hundreds of stars spread over several tens of light years and have a somewhat loose spherical shape. Although we have not explained how the ages of stars are determined, it is possible to make estimates of ages by arguments that will be given in Chapters 14 and 15. Using these, we find that most of the clusters' stars are moderately young with a large spread in color and luminosity. The clusters are from a

FIGURE 13.4
Open star cluster in the constellation of the Southern Cross. It has been titled the "Jewel Box" because a bright red star is situated in the middle of a cluster of white stars.

FIGURE 13.5
Open star cluster in the constellation of Cancer, known as M67. Note the difference in this cluster's appearance and that of the open cluster in Figure 13.4.

few million years old to as much as 10 billion years old (in a few exceptional cases). These old clusters are the more compact ones that have withstood the Galaxy's disruptive forces. In photographs of the youthful clusters the brilliant blue giants stand out. In the middle-

TABLE 13.3
Selected Open Clusters, O Associations, and Globular Clusters

Name	Distance (ly)	Diameter (ly)	Estimated Mass (sun = 1)	Estimated Age (yr)
Open Clusters				
Ursa Major	70	25	300	2×10^8
Hyades	140	15	300	6×10^8
Pleiades	430	15	350	5×10^7
NGC 752	1,200	15	—	1×10^9
M67	2,700	15	150	4×10^9
NGC 188	4,600	20	—	1×10^{10}
NGC 2362	4,900	10	—	5×10^6
M11	5,600	20	250	8×10^7
h Persei	7,300	50	1,000	1×10^7
χ Persei	7,800	50	900	1×10^7
O Associations				
I Orionis	1,600	—	3,000	—
IV Sagittarii	5,500	—	—	—
I Persei	6,200	—	180	—
Globular Clusters				
M4	9,100	60	60,000	—
ω Centauri	16,000	300	—	$\approx 2 \times 10^{10}$
M13	25,000	95	300,000	$\approx 1 \times 10^{10}$
M5	28,000	85	60,000	$\approx 1 \times 10^{10}$
M92	33,000	120	140,000	$\approx 1-2 \times 10^{10}$
M3	42,000	115	210,000	$\approx 1-2 \times 10^{10}$
M15	46,000	125	6,000,000	$\approx 1 \times 10^{10}$
NGC 7006	160,000	140	—	—

aged clusters most of the light comes from the yellow white main sequence stars; their bright blue stars have long since evolved into white dwarfs or neutron stars (see Section 15.5). The color-magnitude diagrams of two well-observed open clusters, the Hyades and the Pleiades, which are both in the constellation of Taurus, are shown in Figure 13.6. Both clusters clearly show a main sequence, with the Hyades possessing a few giant stars and the Pleiades none. We will have more to say about open clusters later and the role they play in our understanding of stellar evolution.

STARS IN ASSOCIATION

Sparsely populated *stellar associations* of highly luminous O and B stars are mixed in the concentrations of gas and dust in the spiral arms of the Galactic disk. They are up to several hundred light years in diameter; one estimate places their number at several hundred. Three associations are listed in Table 13.3. Evidence is strong that these stars have just recently formed out of the interstellar matter. The individual stars are separating rapidly from each other because the association has too little mass to bind them permanently by gravitation. These associations seem highly unstable, with a maximum life expectancy of only a few million years before they completely disperse into the Milky Way mainstream. We see these stars still together because they are so young. In Chapter 14 we will have more to say about the O associations as evidence for stellar evolution.

GLOBULAR CLUSTERS

Largest and most highly concentrated of the stellar groups are the *globular clusters*. (A photograph of a

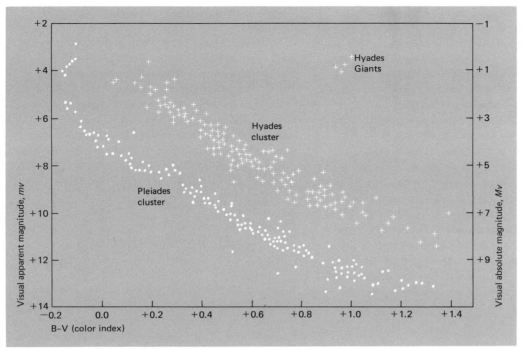

FIGURE 13.6
Color-magnitude diagram of two open clusters, the Pleiades shown as dots and the Hyades shown as plus signs. Notice that the Hyades cluster has a few giants off the main sequence; the Pleiades has none. The absolute magnitude scale on the right applies only to the Hyades cluster. Both clusters are visible to the naked eye and are located in the constellation of Taurus. It is evident from the apparent magnitudes that the Pleiades cluster is considerably more distant than the Hyades.

FIGURE 13.7
Globular star cluster in Hercules, M13. It is just visible to the naked eye.

fine globular cluster appears in Figure 13.7.) We have discovered about one hundred twenty of them surrounding the center of our Galaxy in a vast spheroidal distribution. Other clusters undoubtedly are hidden from us by dust in or near the Galactic plane. The largest may easily have half a million stars packed into a diameter of a hundred or so light years. The nearest clusters are about 7,000 to 8,000 light years from the sun, while the farthest are over 100,000 light years from us.

Red giants dominate photographs of globular clusters because they are more luminous, even though they may exist in fewer numbers than the main sequence stars. The *Orbiting Astronomical Observatory (OAO-2)* detected faint but very hot blue stars in several clusters. Their light does not register on photographs from the earth because our atmosphere absorbs ultraviolet light. The photographs look crowded, but the individual stars, though more closely spaced than in the solar neighborhood (up to a thousand times, or about 100 stars per cubic parsec), are far enough apart to avoid collisions. They swing in almost rectilinear orbits pendulumlike from one side through the center to the other side of the cluster. X rays have been detected coming from several globular clusters. In at least one instance powerful, irregular X-ray bursts lasting less than a second have been observed.

A color-apparent magnitude diagram of the globular cluster M3 is shown in Figure 13.8. The diagrams of various globular clusters differ somewhat. However, they differ quite dramatically from H-R diagrams for the majority of the open clusters. As in other globular clusters, many of the brighter original main sequence stars have evolved to populate the red giant branch and the horizontal branch. Because RR Lyrae stars are found in many of the globular clusters, their distances can be found from their median absolute magnitude ($M_v \approx +0.6$). In clusters too distant to resolve into stars, we can use the angular diameters and integrated light (with precautions) to estimate their distances.

Globular clusters are a distinctive stellar population. They are composed of the oldest stars; many of their spectra reveal exceptionally small amounts of metals compared with the stars in the disk of the Galaxy (some show abundances of metals ranging up to those found in younger stars); they are spherically distributed about the center of the Galaxy; and they move in far-ranging elliptical orbits about the Galactic center at all angles to the plane of the Galactic disk.

13.3 Stellar Populations

Among its billions of stars, our Galaxy contains a vast assortment of stars of varying size, mass, temperature, color, age, and chemical composition. At least four-fifths are main sequence stars; most of the rest are white dwarfs and other degenerate compact stars. There are rare giants and the still scarcer supergiants.

FIGURE 13.8
Color-apparent magnitude diagram of the globular cluster M3. RR Lyrae variables occupy the box in the horizontal branch.

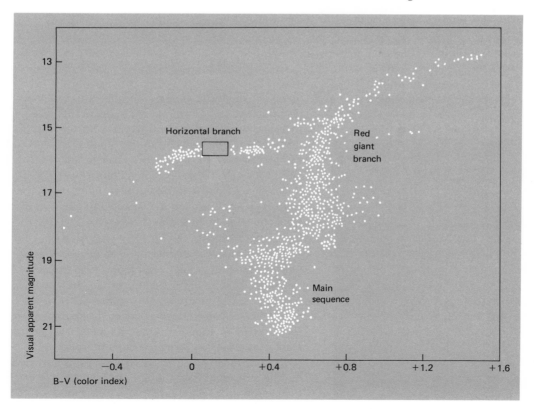

Less than half of the Galaxy's total population of stars are single stars; the rest are physically related doubles and multiples, clusters, and groups. Some unusual stars, such as the pulsating and eruptive ones, vary in brightness, which draws special attention to them. Even at great distances their variability is easy to recognize. But variable stars are few, probably less than one-tenth of a percent of the stars in the Galaxy.

Is there a pattern beneath the Galaxy's rich variety? A definite relationship exists between a star's physical characteristics and its location in the Galaxy. In many regions the relationship is not sharp, but generally we find two distinct groups, *Population I* and *Population II*, differentiated by their evolution, physical properties, chemical composition, velocities, and location.

Walter Baade (1893–1960) found the first clue to identifying these two stellar types in his research on the Andromeda galaxy in the early 1940s. Separating the never-before resolved central region of that galaxy into its stars, he found multitudes of red giants. These he named the Population II stars to distinguish them from what he called Population I stars, the highly luminous blue supergiants in the spiral arms (see Figure 13.9).

Soon it was clear that these population groups were like those in our Galaxy. Population I stars within our Galaxy are associated with the bright gaseous nebulae in the spiral arms. These are younger than the Population II stars, which are found in the globular clusters, the Galaxy's halo, and its central bulge. We now realize, because of accumulated observations and theories on stellar evolution, that there are gradations between the two populations. As we will discuss later, they have been subdivided into categories by their evolutionary history and their locations and motions in the Galaxy.

FIGURE 13.9
Population I stars resolved in the outlying portions of the Andromeda galaxy, M31, at top, and Population II stars resolved near the galaxy's nucleus, at bottom.

13.4 Physical Structure of Stars

The sun is our primary model for the stars. We have some clues that tell us something about the physical laws operating there and presumably in other main sequence stars.

First, the sun has neither expanded nor contracted appreciably during recorded history. Nor, probably, has its luminosity changed significantly in the last several hundred million years, for the fossil record, as we saw in Section 11.6, indicates a stable luminosity. (However, theoretical studies suggest the sun's luminosity has increased over the 4.6 billion years of its life.) The primitive organisms found in Precambrian rocks could have survived only if the earth's mean surface temperature was within 20°C of its present value. Such stability tells us that there must be a balancing of those forces, for long periods of time, that could alter the sun's size or brightness.

Second, we know that the temperature in the sun's photospheric layers is about 6000 K and that energy flows from places with high temperatures to cooler

> So may we read, and little find them cold;
> Not frosty lamps illuminating dead space,
> Not distant aliens, not senseless Powers.
> The fire is in them whereof we are born;
> The music of their motion may be ours.
>
> George Meredith

regions. The inside of the sun must therefore be hotter than the surface. If the sun's temperature climbs only a few tens of degrees with each kilometer of depth, the temperature at its center must be millions of degrees. If the sun emitted radiant energy simply because it is a hot body without any internal source of energy, it would cool enough for us to measure. But, again, geological evidence suggests that the sun is not cooling significantly. The sun—and other main sequence stars—must, then, have a source of energy deep in its interior that replaces the energy radiated away from its surface. We will see later how thermonuclear fusion provides this energy source.

HYDROSTATIC EQUILIBRIUM AND IDEAL GASES

For a star to exist without either expanding or contracting, all its layers from surface to center must have balanced forces, or *hydrostatic equilibrium*. At any distance from its center, the weight of the overlying layers due to gravity must be balanced by the upward pressure of gas in the layers closer to the center. Through most of the star's existence, normal gases conform to the simple ideal expressed in the *perfect gas law:* the pressure of a gas, which is the sum of the forces imparted by colliding gas particles, is proportional to its density and temperature. At the high temperatures inside stars, collisions will strip the electrons from atoms, leaving bare nuclei and free electrons—a *plasma.* Because most of the atom is just empty space, the electrons and nuclei may crowd closer, allowing gases of very high density to form. In the high-temperature and high-density regions near the star's center, the gas particles move faster, and so they collide more often and more violently. Hence the pressure exerted by the gas will be greater in the center and will decline outward. If the gas pressure exceeds gravity, then the material will expand; if gravity is greater, the material will contract. Finally,

in the hottest stars on the main sequence, high-energy photons in very intense radiation fields can add to the pressure, a phenomenon known as *radiation pressure,* which helps balance the weight of the outer layers.

THERMAL EQUILIBRIUM

During most of a star's existence the amount of energy generated inside the star equals the amount radiated away into space at its surface. As a result, the temperature and pressure at any given point inside the star do not change during these periods; the star maintains *thermal equilibrium.*

This equilibrium is self-regulating. Suppose that more energy is released in the center of the star than is radiated away at the surface. The temperature would therefore rise. Because the gas pressure depends directly on the temperature, this pressure will increase. The star will then expand, and heat energy will be transformed into *gravitational potential energy.* This change in turn cools the gas, decreasing the gas pressure, and hydrostatic equilibrium is achieved at a larger radius.

If, on the other hand, too little energy is produced, the star contracts, heats up, and the gas pressure increases. Hydrostatic equilibrium is now found at a smaller radius.

What would happen if the pressure of the gas did not depend on temperature in any way? Heating or cooling from changes in the amount of energy generated would not be checked by increasing or decreasing gas pressure and the subsequent changes in the star's size. This situation does exist and is called a *degenerate state.* It may apply to all the constituents of the gas or to only one of them, such as the free electron component, which would be called a *degenerate electron gas* (see page 279). We will see in Section 15.2 how degeneracy develops in stars.

DEGENERATE GASES

In Section 4.5 it was pointed out that there is a limit to the number of electrons that may occupy a particular orbit, or energy level, in the atom. This *exclusion principle,* originally proposed by the theoretical physicist Wolfgang Pauli (1900–1958), also applies to electrons that are not bound to a nucleus but merely confined to a fixed volume, such as the deep interior of a star. In such a volume there are only certain discrete energy states available to the electrons. When the density of a gas is quite low, such as in the air in a room, there are always enough unoccupied energy states available so that electrons may readily gain or lose energy in order to move from one energy state to another. However, when the density of a gas is quite large, such as in the interior of a star (see Table 15.1), the lowest energy states may all be filled, with only the highest still unoccupied. Thus electrons may not readily gain or lose energy, and the electrons in the gas will become highly incompressible. Thermal energy may not be extracted from the gas, so that the gas may cool down, since the individual electrons cannot give up energy by moving to a lower energy state.

In a degenerate gas the average pressure and kinetic energies are high enough to keep the material from being compressed by gravity. Also, because the kinetic energies are quite high and the rate of collision between the electrons and other particles is quite low, the degenerate electrons travel far at velocities that can approach the speed of light. And, unlike the pressure in a perfect gas, the pressure that degenerate electrons exert at temperatures they ordinarily encounter has little to do with their temperature. When electron degeneracy occurs within stellar matter, major changes in the thermal balance of energy can lead to mechanical instabilities.

TRANSPORTING ENERGY

Energy may be transported in one or more of three distinct ways, though most often one mode is more efficient under the existing physical conditions. One of these is *conduction,* which is the way heat is transferred along a metal rod. For the interior of main sequence stars obeying the perfect gas law, this is a very ineffective means of energy transfer. But in the cores of stars that have exhausted their nuclear fuels, it becomes a primary way of transporting energy. At this stage the core has been compressed by gravitational contraction to a very high density, as in the degenerate matter in white dwarfs and neutron stars.

A more frequent method for transporting energy is *convection,* in which gas circulates between hot and cool regions in the star, transferring thermal energy to the cool region (see Figure 13.10 and Section 11.1). Once a pattern of circulation is stabilized, convection can be very efficient in transporting energy. For layers in which temperature changes quite rapidly with depth, convection cells will develop as the principal means of carrying energy. In a *convective zone* the atomic constituents are well mixed by this continual stirring of the material.

Radiation is the third way of moving energy. Inside the star, photons from thermonuclear burning diffuse outward by almost random paths (see Figure 13.10). When a photon is emitted, it will either be absorbed by an atom or an ion or scattered by free electrons over distances ranging from small fractions of a centimeter deep in the star's interior to several kilometers in the photosphere. Although the direction in which the reemitted photon or the scattered photon moves is generally arbitrary, there will be a net drift of photons outward from the center to the surface of the star. Energy is transported by radiation through *radiative zones,* in which chemical elements go through very little mixing; any chemical inhomogeneity developing here will probably persist.

OPACITY

Radiation and matter continually interact by absorption, reemission, and scattering of photons, impeding the outward flow of radiant energy (see Figure 13.10). Matter's resistance to the flow of radiation through it is called *opacity.* If the resistance is high, the opacity is said to be large; if the resistance is low, the opacity is small. In regions of large opacity temperature drops rapidly outward and convection will take over as the primary way of transporting energy. When the density is low, radiation travels more freely through the star, because the open spacing between particles reduces the probability that the photon will interact with the gas particles. If matter did not hinder the flow of a star's radiant energy (if the gases were not opaque), the star would be transparent.

Energy is liberated in the deep interior in the form

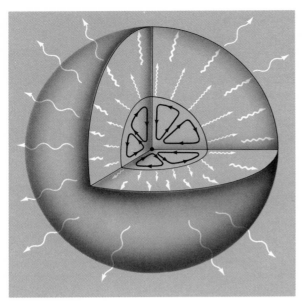

FIGURE 13.10
Schematic illustration of the interior of a star showing a convective zone in the center and a radiative zone as an envelope. Energy is carried out of the center by mass motions in the convective zone. Through the radiative zone, energy moves outward by successive emission and absorption processes. The greater the opacity of the material, the greater the number of absorptions and reemissions.

of a comparatively small number of high-energy gamma ray photons. As radiation works its way out of the star, the absorption and reemission of photons by overlying matter degrades the few high-energy gamma ray photons into millions of lower-energy photons by the time they reach the photosphere. Radiant energy takes hundreds of thousands of years to diffuse through the sun from its energy-generating core.

THERMONUCLEAR FUSION

Over a century ago astronomers knew that the energy already radiated by the sun could never have been supplied by ordinary combustion (e.g., burning wood or coal). Using even the most efficient chemical ways of producing energy, the sun would have burned out in less than forty thousand years.

Another way of producing energy is by converting gravitational potential energy into heat by contraction. In the nineteenth century this was thought to be the only source of the sun's energy. We now know that contraction does become a vital source of energy on which a star can draw at various stages in its existence. But at its current luminosity, our sun could not survive on gravitational contraction alone for more than about 15 million years.

For stars like the sun a source of energy must keep the luminosity approximately constant for billions (not just millions) of years. How? The answer is fusion of small-mass nuclei to form more massive nuclei. Sir Arthur Eddington (1882–1944) suggested in 1920 that fusion of hydrogen could form helium and that this could be the long-sought fuel. After it was found that the stars have vast quantities of hydrogen, nuclear physicist Hans Bethe proposed in 1939 a way in which four hydrogen nuclei, or protons, could be converted into a helium nucleus, releasing energy. If many hydrogen nuclei are converted, they will release enough power to keep stars shining for billions of years.

What controls this process, known as *thermonuclear fusion*? The temperature and density of the gas are critical in determining its rate. The higher the temperature (which measures the average kinetic energy of the colliding nuclei), the more readily two positively charged protons can overcome their electrostatic repulsion (the tendency of two like-charged particles, positive or negative, to repel each other). They may penetrate close enough for nuclear forces (whose range is very short) to fuse the two particles, while one proton undergoes a transformation to a neutron. High densities also make collision between the protons more likely, again improving the chances for fusion. Nuclear reactions will therefore be most numerous in the star's central region, where the temperature and density are highest, and reactions will gradually decline to zero somewhere out from the center where temperature and density are too low to sustain nuclear reactions. This distance from the center, then, defines the energy-generating *core* of the star. In a model of the sun's interior in Table 13.4, 95 percent of the energy released by thermonuclear burning takes place in the inner 20 percent of the radius, which holds 35 percent of the mass. The core is 20 percent of the radius, and the remaining 80 percent is the *envelope*.

THE P-P CHAIN AND CNO CYCLE

Hydrogen burning proceeds by two principal schemes: the *proton-proton chain* (p-p chain) and the *carbon-nitrogen-oxygen cycle* (CNO cycle) (see Figure 13.11).

In each process four protons are fused into one helium nucleus with a slight loss in mass, which is converted into energy (see page 283).

The rate at which energy is produced by the *p-p* chain in relation to that of the CNO cycle depends on the temperature. Up to about 16 million degrees the *p-p* chain dominates energy production. Beyond that temperature, however, the CNO cycle takes over as the most important thermonuclear reaction. The average rate of energy generation for the entire sun, which depends primarily on the *p-p* chain, is about 2 ergs per gram per second. For a star of ten solar masses the average rate of energy generation, supplied principally by the CNO cycle, is about a thousand times greater than that of the sun. It is interesting to note that our bodies radiate heat more efficiently than the sun does (about 10^4 ergs per gram per second), although obviously for an insignificant fraction of the time that the sun has been radiating energy.

The mass of the end product of these processes, ^{4}He, is 0.71 percent less than the combined masses of the four reacting protons (4^1H). What happened to the missing mass? Just after this century began Albert Einstein pointed out that there is an *equivalence between mass and energy*. Mass is just one more manifestation of energy, and what is conserved in any type of

STANLEY ARTHUR EDDINGTON (1882–1944)

Eddington was a brilliant scholar. His first position after graduating from Cambridge University (1906) was chief assistant at the Royal Observatory in Greenwich, where he excelled in practical astronomy. In 1913 he was appointed Plumian Professor of Astronomy at Cambridge and a year later made director of the observatory.

In the years following, his intuitive insight, bold imagination, and mastery of mathematics led him to important discoveries over a wide range of problems: the motions and distributions of the stars and other classes of objects; the internal structures of the stars and their sources of energy; the physical nature of white dwarfs; the dynamics of pulsating stars; and the physics of interstellar matter.

Eddington's pioneering work in astrophysics began in 1916. He was the first to model the interior of a star, pointing out that the condition for stellar equilibrium involved three forces: gravitation, gas pressure, and radiation pressure. He recognized the importance of ionization in stellar interiors. He boldly assumed, and this was later accepted, that due to the high ionization of the internal gases, the perfect gas condition prevailed within the interiors of the stars, except for the white dwarfs. He demonstrated that energy could be transported by radiation as well as by convection, and that the centers of stars must be at very high temperatures—in the millions of degrees. An extremely important result that emerged from his research was his theoretical formulation of the mass-luminosity relation, verified from stellar data. Eddington suspected that the chief source of stellar energy was subatomic and that the overwhelming abundance of hydrogen played a dominant role in supplying this energy. Later, in 1938, physicist Hans Bethe introduced the theory of the CNO cycle of hydrogen fusion into helium, which clarified the picture of stellar energy generation and substantiated Eddington's theory.

In 1919 Eddington organized a solar eclipse expedition to Brazil to photograph the stars in the neighborhood of the eclipsed sun to observe the bending of starlight as predicted by relativity theory. Although difficult to measure, the observed amount of deflection was in rough agreement with Einstein's predicted value—the first observational test of relativity theory. In 1924 Eddington sought further confirmation of the theory, based on his calculated high density and small size for Sirius's white dwarf companion. He wrote to Director Adams of the Mount Wilson Observatory that a measurable red shift should be observable in the spectrum of Sirius's companion. Mount Wilson astronomers measured a red shift that tallied fairly well with the predicted value.

Eddington's complete mastery of relativity theory led him on to further studies in cosmology. He helped to clarify the pulsation mechanism in the cepheid variables. He was one of the first to develop a theory of absorption lines in stellar atmospheres. He extended the theory to the absorptive properties of interstellar matter, suggesting that the intensities of the interstellar absorption lines could be used to estimate the distance of a star, a technique that is still used today.

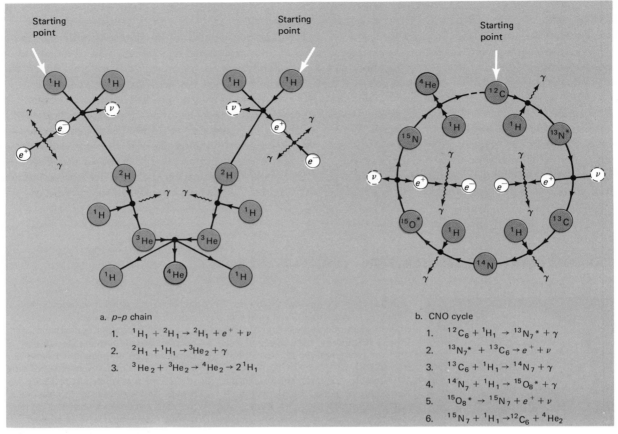

FIGURE 13.11

a. *p–p chain*

1. $^{1}H_{1} + {}^{2}H_{1} \rightarrow {}^{2}H_{1} + e^{+} + \nu$

2. $^{2}H_{1} + {}^{1}H_{1} \rightarrow {}^{3}He_{2} + \gamma$

3. $^{3}He_{2} + {}^{3}He_{2} \rightarrow {}^{4}He_{2} \rightarrow 2\,{}^{1}H_{1}$

b. CNO cycle

1. $^{12}C_{6} + {}^{1}H_{1} \rightarrow {}^{13}N_{7}{}^{*} + \gamma$

2. $^{13}N_{7}{}^{*} + {}^{13}C_{6} \rightarrow e^{+} + \nu$

3. $^{13}C_{6} + {}^{1}H_{1} \rightarrow {}^{14}N_{7} + \gamma$

4. $^{14}N_{7} + {}^{1}H_{1} \rightarrow {}^{15}O_{8}{}^{*} + \gamma$

5. $^{15}O_{8}{}^{*} \rightarrow {}^{15}N_{7} + e^{+} + \nu$

6. $^{15}N_{7} + {}^{1}H_{1} \rightarrow {}^{12}C_{6} + {}^{4}He_{2}$

The two principal fusion processes for hydrogen burning. The proton-proton chain predominates in the sun and less massive stars. The carbon-nitrogen-oxygen cycle takes about 7 million years to complete and is predominant in stars hotter than the sun.

Key to symbols: ^{1}H = hydrogen 1 nucleus (proton); ^{2}H = hydrogen 2 nucleus (deuteron); ^{3}He = helium 3 nucleus; ^{4}He = helium 4 nucleus; e^{-} = electron; e^{+} = positron; ν = neutrino; γ = gamma ray photon; ^{12}C = carbon 12 nucleus; $^{13}N^{*}$ = unstable nitrogen 13 nucleus; ^{13}C = carbon 13 nucleus; ^{14}N = nitrogen 14 nucleus; $^{15}O^{*}$ = unstable oxygen 15 nucleus; ^{15}N = nitrogen 15 nucleus.

interaction between particles of matter is the total energy, including the energy equivalent of the mass. The equivalence is symbolized in Einstein's equation $E = mc^{2}$, where E is the energy, m is the mass, and c is the velocity of light.

In hydrogen burning, 1 gram of hydrogen is converted into 0.9929 gram of helium plus 6.4×10^{18} ergs of energy—exactly 0.71 percent of the original 1 gram of hydrogen times c^{2}. In what form does the energy appear? In the various steps several gamma ray photons are created and degraded by absorption and reemission into many photons having the same total energy. Also, some of the material particles created have large kinetic energies, which they will soon redis-

tribute to other particles by collisions. Thus both radiant energy and heat energy come from the mass that is lost in these thermonuclear reactions.

Translated into practical units, for every passing second the sun converts 600 million metric tons of hydrogen into 596 million metric tons of helium, and 4 million metric tons of mass into energy. This energy will diffuse to the surface, where it will supply the 3.83×10^{33} ergs of energy radiated away into space each second. In its core the sun has enough hydrogen to keep it shining for about 10 billion years. So far in 4.6 billion years it has used up about half of its core's hydrogen supply and lost about 0.043 percent of its mass.

STEPS IN HYDROGEN BURNING

The first step in the *p-p chain* (see Figure 13.11) is fusion of two colliding protons (^{1}H) to form a *deuteron* (^{2}H), which is the nucleus of the hydrogen isotope deuterium, emitting a positron (e^+) and a neutrino (ν). This reaction happens, on an average, once every several billion years for each pair of protons.

The *positron* is a positively charged particle with the mass and other characteristics of an electron; it is the *antiparticle* for the electron. The collision of a positron with an electron destroys them as matter and creates two gamma ray photons. The *neutrino*, on the other hand, is a massless, chargeless particle traveling at the speed of light and is most unlikely to interact with matter. It immediately escapes from the star, carrying away about 2 percent of the energy released in the *p-p* chain of reactions.

The second step in the *p-p* chain is the collision within a few seconds of another proton with the deuteron to fuse and form the light isotope of helium, emitting a gamma ray photon. Finally, two ^{3}He nuclei collide every few million years and fuse to form the heavy isotope of helium (^{4}He), accompanied by the return of two protons. (We should point out that there are other branches of these reactions leading to the same end.)

All told, six protons have taken part in producing two ^{3}He nuclei, from which one ^{4}He nucleus is produced and two protons are returned to the reservoir of fusionable matter. The time for the whole thermonuclear reaction is determined, of course, by the first step, and it is only the enormous quantity of ionized hydrogen in the cores of stars that makes this process a significant source of energy.

The other hydrogen-burning reaction, the *CNO cycle,* has six steps (see Figure 13.11) occurring at rates between 80 seconds and 300 million years but leading to the same result as the *p-p* chain: conversion of four protons to produce one helium nucleus and to liberate energy. The cycle begins with ^{12}C and closes with the return of ^{12}C, so that carbon is only a catalyst that makes the reaction go. Along with the ^{4}He nucleus at the end of the chain of reactions, three gamma ray photons will be liberated along with two neutrinos (ν), which will escape from the star, carrying away some energy, and two positrons (e^+). The two positrons (e^+) will immediately combine with two electrons (e^-) and in the process will be converted into four additional gamma ray photons.

THE STARS: MATHEMATICAL MODELS

Each of the physical processes described above involves several physical quantities, among them mass, temperature, density, pressure, and luminosity. We can use a symbol to represent the numerical value of each quantity and combine these symbols into mathematical equations embodying their relationships. These equations, called *equations of stellar structure,* describe how mass, pressure, temperature, and luminosity vary outward from the center of the star. Within the equations are additional quantities, such as density, chemical composition, opacity, and rate at which energy is generated. To construct models of stars, we take their observed properties—such as mass, radius, luminosity, surface temperature, and suspected chemical composition—as constraints in solving the equations of stellar structure at a discrete number of points along the radius. The solution is a *mathematical model* or a *stellar model.* The electronic computer makes it possible to develop these models in reasonable lengths of time for several hundred points. Without the high-speed digital computer much of the progress we have made in

the last twenty years in this field would not have been possible.

An example of such a mathematical model for the present sun is given in Table 13.4 and is shown schematically in Figure 13.12. As the star evolves, it alters the structure of its layers from center to surface. In the layers responsible for nuclear burning, the chemical composition is changing as well.

We can calculate a sequence of mathematical models simulating the restructuring that the real star presumably undergoes as it ages during its span of millions of human generations. For each model in the sequence the surface temperature and luminosity represent a fixed point in the H-R diagram at a given time. How do we test the validity of our model? We see how well this time sequence of points, or *evolutionary path,* for one star helps us to predict the distribution of real stars in the H-R diagram. Both open and globular clusters are important observational keys in checking our results. The color-magnitude diagrams of clusters, such as Figures 13.6 and 13.8, give the best evidence on stellar aging, because a cluster is a group of stars with a range of masses that began their exis-

TABLE 13.4
Mathematical Model for the Sun

Fraction of the Radius	Radius (10^3 km)	Temperature (10^6 K)	Density (g/cm³)	Fraction of Central Pressure[a]	Fraction of Mass	Fraction of Luminosity
0.0	0	15.5	160	1.00	0.0	0.0
0.04	28	15.0	141	0.84	0.008	0.08
0.1	70	13.0	89	0.46	0.07	0.42
0.2	139	9.5	41	0.15	0.35	0.94
0.3	209	6.7	13.3	0.035	0.64	0.998
0.4	278	4.8	3.6	0.007	0.85	1.00
0.5	348	3.4	1.0	0.0014	0.94	1.00
0.6	418	2.2	0.35	0.0003	0.982	1.00
0.7	487	1.2	0.08	0.00004	0.994	1.00
0.8	557	0.7	0.018	0.000005	0.999	1.00
0.9	627	0.31	0.002	0.0000003	1.000	1.00
1.0	696	0.006	3×10^{-7}	4×10^{-13}	1.000	1.00

[a] $P_c = 3.4 \times 10^{17}$ dyne/cm².

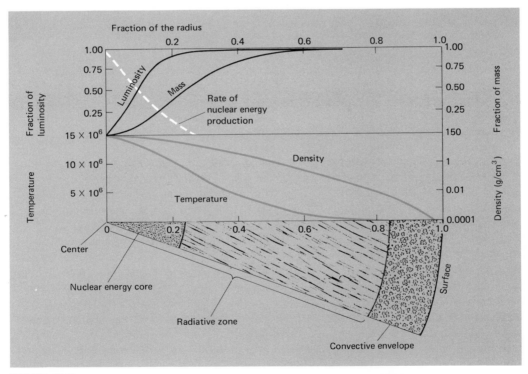

FIGURE 13.12

Plot of a mathematical model of the solar interior along a radius from the center to the surface, based on Table 13.4. The wedge through the sun at the bottom shows the primary zones of interest: energy-generating core (also a radiative zone), radiative zone, and convective envelope. Temperature and density variations along the radius are shown in the bottom panel; the fraction of the mass and luminosity to a given point along the radius are shown in the top panel.

tence at about the same time and in the same place. They also formed from the same material and so at first were reasonably similar chemically. From the distribution of cluster stars in the H-R diagram, we can then deduce details of how individual stars age.

13.5
Main Sequence Stars

RESULTS FROM STELLAR STRUCTURE AND EVOLUTION STUDIES

What is the fundamental problem in stellar structure and evolution? Stated in its simplest form, it is this: If we know the star's physical characteristics (mass, radius, luminosity, temperature, and chemical composition), what do the laws of physics tell us about its internal structure and evolution? If our knowledge of the physical processes going on inside the star is correct, then we should be able to predict the star's position in the H-R diagram at different stages during its existence. Then we can explain the existence of the various regions of the diagram and trace the evolutionary cycle that carries stars from one region to another (see Figure 13.13). Also, the fraction of stars in any region of the H-R diagram should equal the fraction of a star's existence spent in that region of the diagram. If the red supergiant phase is a short part of the star's life, then we should see relatively few red supergiants—and we do. If the main sequence is a long phase in the star's existence, then many stars should be main sequence stars. We do actually observe the main sequence to be the most densely populated region in the H-R diagram.

Let us begin the study of the structure and evolution of the stars with the most common—the main sequence stars. In this sequence the hot O and B stars are the most massive and the cool red dwarfs of class M the least massive. The main sequence for Population II stars differs just slightly from that for Population I stars, a difference traced to the smaller abundance of elements heavier than helium in the Population II stars (see Figure 13.13).

MEANING OF THE MAIN SEQUENCE

Stars spend most of their lives on or near the main sequence for two main reasons: the large yield of energy per gram from hydrogen fusion compared with other sources of nuclear energy, and the vast amount of available hydrogen. We would therefore expect to find most stars on the main sequence. The H-R diagram for the stars in the sun's neighborhood (see Figure 13.3) and the color-magnitude diagrams for open clusters (see Figure 13.6) show that this is the case.

The approximate time a star spends on the main sequence is proportional to its mass divided by its luminosity as shown on page 287. This proportionality comes from Einstein's mass-energy equivalence. The more massive a star, the greater is its emission of radiant energy per gram of matter and the shorter is its time on the main sequence. See the numerical estimates for luminosity per gram and duration on the main sequence in Table 13.5, where we assume that the sun will be a main sequence star for about 10 billion years.

> There is one glory of the sun, and another glory of the moon, and another glory of the stars, for one star differeth from another in glory.
>
> New Testament, I. Corinthians XV.

Now let us consider the shorter hydrogen-burning phase for more massive stars. After contraction onto the main sequence, the central temperature must be higher than for lower-mass stars because higher gas pressure is needed in the center of the star to balance gravity. The temperature difference between the center of the star and its photosphere will be larger if the star is more massive. For a 15 $M_\odot$ star (fifteen times the mass of the sun), the central temperature is around 35 million degrees Kelvin compared with 7 million for a star of 0.25 $M_\odot$. More thermal energy flows from the interior of massive stars, making them more luminous than less massive stars. The more massive a star, the more rapidly it must burn hydrogen to supply the energy loss from its surface. The CNO cycle in the massive stars also depends more strongly on temperature and consumes the hydrogen faster than does the p-p chain that operates in stars of low mass. The dividing line between the two is about 2 $M_\odot$.

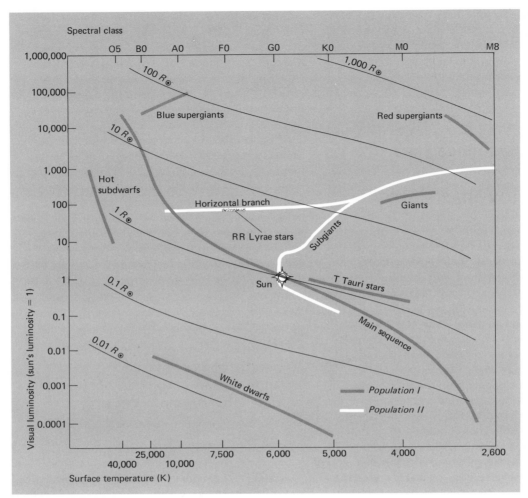

FIGURE 13.13
H-R diagram showing the average distribution of Population I and Population II stars. The diagram is a composite of many thousands of stars, giving us a snapshot of stellar populations today. The black lines indicate stellar radii in units of the sun's radius. The astronomer's task is to explain the significance of the regions in the diagram and to trace the evolutionary process that causes stars to move from one region to another during their lifetimes.

SOLAR NEUTRINO EXPERIMENT

How certain is the astronomer about the thermonuclear processes going on inside stars? So far the only experiment designed to directly test the theory involves the neutrino. Neutrinos pass freely out of the sun into space because the probability that they will interact with matter is extremely low. They carry away, at the speed of light, a small fraction of the energy generated in the sun. By detecting solar neutrinos, astronomers can have first-hand information on the average temperature in the hydrogen-burning core, ar-

rived at earlier partly by theoretical studies and partly by experiments on colliding nuclei conducted with large particle accelerators.

One scheme for detecting solar neutrinos uses a huge tank filled with 400,000 liters of dry-cleaning fluid (C_2Cl_4), located deep in a South Dakota gold mine to shield the chlorine atoms in the solvent from cosmic ray particles; nothing is allowed to reach them but neutrinos. When the nucleus of a chlorine atom (^{37}Cl) absorbs a neutrino, it is transformed into a radioactive argon (^{37}Ar) nucleus (with a half-life of

TABLE 13.5
Approximate Time Stars Are on the Main Sequence

Spectral Class	Surface Temperature (K)	Mass (sun = 1)	Luminosity (sun = 1)	Luminosity/ Mass (erg/s · g)	Time on Main Sequence (yr)
O7	35,000	25	80,000	6140	3×10^6
B0	30,000	15	10,000	1280	15×10^6
A0	10,800	3	60	38.6	500×10^6
F0	7,240	1.5	6	7.68	2.5×10^9
G0	5,920	1.0	1	1.92	10×10^9
K0	5,240	0.8	0.6	1.44	13×10^9
M0	3,920	0.4	0.02	0.09	200×10^9

about 35 days). The argon nucleus is recovered and its decay is monitored. We find that the observed flux of solar neutrinos is 2.3 argon atoms per week, which is about two to three times smaller than the rate predicted from standard mathematical models of the sun. By manipulating the solar model, which is subject to some uncertainties anyway, the discrepancy can be reduced but not eliminated. Other explanations for the discrepancy have been proposed but none has received widespread endorsement. Plans are being made to measure the flux of lower-energy neutrinos, which are not counted in the experiment above.

Recently there has been another explanation which arises from two studies of the historical records of measurements of the sun's diameter. Both studies seem to find that the sun has been shrinking for the last 100 years or so. The question is by how much. The contraction, if real, amounts to about 0.01 to 0.1 percent per century. Further speculation suggests that the sun may undergo a long-term cycle of expansion and contraction. During contraction, the sun derives heat

MAIN SEQUENCE LIFETIMES

From Einstein's relation for the equivalence between mass and energy, we can determine a relation for the length of time a star will spend on the main sequence. The luminosity (L) multiplied by the time on the main sequence (t) is the total energy radiated away (E), which must be equal to the fraction of the star's mass ($f \times M$) that will be converted to helium by hydrogen fusion multiplied by c^2, or

$$E = L \times t = f \times M \times c^2$$

Hence

$$t \propto M/L$$

which says that the time on the main sequence is proportional to the star's mass divided by its luminosity. Also, the mass-luminosity relation tells us that the luminosity is roughly proportional to the mass to about the 3.5 power or

$$L \propto M^{3.5}$$

Hence, substituting in the proportionality for the time gives us,

$$t \propto M/M^{3.5}$$

$$t \propto 1/M^{2.5}$$

energy from gravitational potential energy and lowers its rate of hydrogen burning. This would account for the emission of fewer neutrinos. However, these results are a long way from being thoroughly verified.

The solar neutrino problem is serious, for it casts doubt on our knowledge of the details of the structure and/or energy generation in main sequence stars. Thus we are forced to look more carefully at the details of these processes in the sun if the experiment is completely correct. However, it is unlikely that it forecasts the failure of the present theory of stellar evolution. The details will change in time due to the solar neutrino problem and others, but the general outline seems likely to endure.

MAIN SEQUENCE STARS OF THE NIGHT SKY

How does the time a star spends on the main sequence affect the stars in our night skies? A star like Spica, in the constellation Virgo, is a B1 star, not older than about 15 or 20 million years (see Table 13.5). Stars like Sirius in Canis Major and Vega in Lyra are early A stars and are less than 500 million years old. The main sequence life of nearby Barnard's star, an M5 red dwarf in the constellation Ophiuchus, is greater than 200 billion years, and it will still be fusing hydrogen in its core long after the sun has ceased to shine. During the sun's 4.6 billion years on the main sequence, many generations of massive O and B stars have come into existence and left the main sequence phase, but all the small-mass M stars born in that period have hardly begun their stay on the main sequence.

LIMITS OF THE MAIN SEQUENCE

How much matter is necessary to form a star? How much mass may a star have? To find out we must first answer another question: How do we define what a star is? If we say it is an object held together by gravity that is or has been self-luminous, then the smallest mass for a star that gets its energy from thermonuclear burning is about $0.1\ M_\odot$. This limit is set by the mass needed to bring the temperature at the center high enough to start hydrogen burning. Yet objects even smaller than $0.01\ M_\odot$ can survive a very long time on just the energy from gravitational contraction. These "stars" would, however, be so faint that they could easily escape detection.

At the upper end of the main sequence the mass of a star is probably determined more by the amount of interstellar matter available when it was formed than by its internal structure. Stars of about $60\ M_\odot$ and greater have such a delicate balance between gravitational and pressure forces, however, that equilibrium could be prevented by any irregularities of motion inside the star. Limits on the masses of main sequence stars determined from binaries are consistent with these figures.

As hydrogen burning progresses on the main sequence, the central core of the star is slowly depleting hydrogen and converting it to helium. Because the gas pressure depends on the density, or number of particles, converting four hydrogen nuclei to one helium nucleus must reduce the gas pressure. Hydrogen burning will therefore be accompanied by a very slight contraction of the energy-generating core and a heating up of the gas. The star brightens slightly because of the added heating of the center; this increase in the temperature difference between the center and surface causes a greater outflow of radiation, and the outer portion of the star expands, increasing the radius. This is part of the reason for some of the width of the main sequence evident in H-R diagrams. For example, estimates from mathematical models for the sun suggest that it has increased its luminosity by 20 to 30 percent during its 4.6-billion-year existence as a main sequence star.

In Chapters 14 and 15 current ideas on the birth and death of stars will be discussed. These are the phases in the star's life that bring it to the main sequence and lead it away later in its life.

SUMMARY

Stars range enormously in size, density, luminosity, and temperature but much less in mass and chemical composition. By analyzing these stellar properties, astronomers can decipher significant relationships that help in understanding stars.

A very important relationship is that between the stars' luminosities and their temperatures or spectral types, known as the Hertzsprung-Russell (H-R) diagram. Not only does this diagram sort out different luminosity classes of stars (dwarfs, giants, and supergiants), but, most importantly, it is our guide in the study of stellar evolution.

Scattered within our Milky Way Galaxy are groups of stars called open clusters, stellar associations, and globular clusters. The first group may contain up to several hundred stars of Population I, bound by gravity and spread over a distance of 10 to 50 light years; they are observed mostly in the spiral disk of our Galaxy. Population I stars are the younger stars of the Galaxy. The second group is a loose, unbound system of several dozen or so very young, highly luminous Population I stars located in the spiral arms of the Galaxy. The distribution of the globular clusters is very different. Over a hundred globular clusters are scattered all around the center of the Galaxy, out to considerable distances. The typical globular cluster has a spherical shape and may contain up to a million Population II stars within a diameter of about 100 light years. Population II stars are the older ones.

The light coming from a star started its journey within the star's central region as gamma rays up to a million years ago, and it has been degraded to visible light by the time it leaves the surface of the star. Applying the laws of physics, astronomers can derive, with the aid of fast electronic computers, mathematical models that simulate the internal structure and physical conditions that prevail within the star as it ages. For stars up to about two solar masses, the proton-proton (p-p) chain is the dominant hydrogen fusion process; for larger stellar masses the carbon-nitrogen-oxygen (CNO) cycle is the dominant hydrogen fusion process, and the rate of energy production is much greater in this case. Computer calculation leads to a theoretical H-R diagram that shows the star's changing surface temperature and luminosity during its life cycle. This is an evolutionary track that can be compared to the actual distribution of the stars in the H-R diagram. The latter is a snapshot of many stars at different stages in their evolution. The best test of our predictions comes from a comparison of the predicted and observed temperatures and luminosities of star clusters. Such a comparison indicates that we are basically on the right track in working out the details of how stars age.

One worrisome difficulty in the theory of stellar structure and evolution is the rate at which neutrinos emerge from the sun. An experiment in a South Dakota gold mine yields a solar neutrino flux that is about one-third the predicted flux from part of the proton-proton chain. The discrepancy has not yet been satisfactorily explained, but we believe that the general picture we have of what goes on inside the sun is fairly secure.

REVIEW QUESTIONS

1. What is the distinction between Population I and Population II stars?

2. Explain how a sunbeam striking the ground may have begun its long journey as a gamma ray photon inside the sun.

3. How do the two balancing forces that keep a star in hydrostatic equilibrium react on each other?

4. Describe the three modes of energy transport inside a star. Under what conditions does each of them operate best?

5. Explain why the perfect gas law still prevails in normal stellar interiors even at high densities, since it is known to fail when the density of an ordinary gas reaches a density of one-tenth that of water.

6. How is it possible to obtain an enormous amount of energy from the conversion of the lighter elements (such as hydrogen) into the heavier elements?

7. How many protons are fused in a stellar interior to produce helium plus energy? At what temperatures does the proton-proton chain reaction operate best? The carbon-nitrogen-oxygen cycle?

8. Why do the more massive stars shine more brightly than the less massive stars? Why are their life cycles shorter than those of their less massive counterparts?

9. Why do stars spend most of their lives on the main sequence?

10. What is meant by thermal equilibrium?

11. Indicate on the main sequence of the H-R diagram where stars of different masses lie.

12. Discuss the range of the physical properties of the stars: their dimensions, masses, densities, luminosities, temperatures, and chemical composition.

13. Why do the H-R diagrams of clusters test our theory of stellar structure and evolution?

14. Do the H-R diagrams of clusters validate our theory?

15. Why does a star spend most of its life in the hydrogen-burning stage?

16. Discuss the solar neutrino problem and any effect it may have upon our theory of stellar structure.

17. Discuss the differences in appearance and location between open clusters, associations, and globular clusters.

18. Are the following stars older or younger than the sun (see Table 12.4): Sirius, Rigel Kent, Spica, Fomalhaut, Regulus, Shaula, and Castor?

19. Discuss the definition of a star. What are the limits in masses of stars? Why are there no stars of 200 $M_\odot$ or 300 $M_\odot$?

20. Discuss the use of stellar models or mathematical models in understanding the evolution of stars.

14
Interstellar Matter: The Birthplace of Stars

That nova was a moderate star like our good sun; it stored
    no doubt a little more than it spent
Of heat and energy until increasing tension came to the
    trigger point
Of a new chemistry; then what was already flaming found a
    new manner of flaming ten-thousand fold
More brightly for a brief time; what was a pin-point fleck
    on a sensitive plate at the great telescope's
Eyepiece now shouts down the steep night to the naked eye,
    a nine-day super-star.

Robinson Jeffers

Before we can proceed with the discussion of the evolution of stars begun in the last chapter, we need to know something about the material and the environment out of which stars will form. Therefore, the first three sections of this chapter will be devoted to the nature of the interstellar medium, some of the physical processes going on in it, and how it has changed during the history of the Galaxy. With these points as a basis, the current theory for the birth of stars should be more understandable. However, we give a note of caution: The interstellar medium is important not only as the birthplace of stars but also as a vital element in the structure of the Galaxy. And, as such, a theory of galaxies cannot ignore its contributions.

14.1
Gas, Dust, and Molecules Among the Stars

Early in this century astronomers believed that interstellar space was fairly transparent and that any dimming of starlight could be ignored. Then in the 1930s astronomers discovered that distant stars are dimmed and reddened more than the nearer ones by an *interstellar medium* that absorbs and scatters starlight. The interstellar matter is a mixture of

◀ Lagoon nebula, M8, in the constellation of Sagittarius, which lies in the richest part of the Milky Way.

atomic and molecular gases, mostly hydrogen, along with small solid particles called grains or dust, concentrated mostly in the plane of the Galaxy.

If you exhale your breath once and let it expand into an evacuated cubical enclosure 1 kilometer on a side, the density of your breath will exceed the density in most parts of the interstellar medium. Although this suggests that interstellar space as nearly a vacuum, there is a significant amount of matter lying between the stars because of the vast volume of space. Altogether, significant amounts of interstellar matter can be found in about 10 percent of the Galaxy's volume. Studies show that the interstellar medium is not uniform but is very clumpy. These clumps are referred to as interstellar clouds, even though they have a variety of sizes, compositions, and structures.

INTERSTELLAR GAS

The gaseous component of interstellar matter is almost entirely confined to a fairly thin disk in the plane of the Galaxy. About 90 percent of it is hydrogen, of which perhaps half is in molecular form. Atomic hydrogen is there in both neutral and ionized forms. However, molecular hydrogen and ionized hydrogen are found in only a tiny fraction of interstellar space compared to the distribution of hydrogen atoms. Ten percent of the gas is helium, primarily in atomic form, with traces of other elements. The average density of the interstellar gas is about 1 atom per cubic centimeter. It appears to vary from less than 1 atom per cubic centimeter outside the clouds to at least 10^4 atoms per cubic centimeter inside them. In the dark neutral clouds of hydrogen the density may reach

10^6 atoms per cubic centimeter. Because hydrogen is the main ingredient in the *interstellar gas,* we frequently designate a region in which hydrogen is predominantly ionized as an *H II* region and a neutral region as an *H I* region.

Starlight passing through cool interstellar gas is selectively absorbed producing a few absorption lines superimposed on the normal spectra of stars (see Figure 14.1). These *interstellar lines* can be differentiated from the spectral lines of the O and B stars because they are usually narrow and have different Doppler shifts from the stellar absorption lines. They are more difficult to identify in stars of later spectral classes, which have many absorption lines. Frequently we see several sets of Doppler-shifted interstellar lines meaning that the starlight has passed through separate intervening clouds moving at different speeds along the line of sight (see Figure 14.2).

In the visible region of the spectrum astronomers have identified lines due to sodium, calcium, iron, potassium, and titanium, and the molecules cyanogen (CN) and neutral and ionized methylidine (CH). In the ultraviolet part of the spectrum molecular hydrogen, carbon, nitrogen, phosphorus, sulfur, chlorine, manganese, and carbon monoxide lines have been identified in observations from the *Copernicus* satel-

lite (see Figure 7.5). Since 1964 radio frequency lines of hydrogen, helium, and carbon have been observed, due to electron transitions between closely spaced energy levels near the series limit. For example, a free electron may be captured into the 110th level of the hydrogen atom from which it may cascade down to the 109th level emitting a photon with a radio wavelength of 6 centimeters.

THE 21-CENTIMETER LINE

New paths for exploring the interstellar medium and its structure opened in 1951 when a radio spectral line at *21 centimeters* (1420 megahertz) emitted by cool neutral hydrogen was discovered. Why is such a spectral line emitted? Seven years before, the Dutch astronomer H. C. van de Hulst had predicted it might be present because of the way electrons and protons spin.

Electrons revolve about the proton, but electrons and protons also spin like tiny rotating tops (see Figure 14.3). About once in 11 million years (on the average) the electron, if spinning in the same sense as the proton to which it is bound, and if not disturbed by collisions with other atomic particles, will spontaneously change its spin to an opposite rotation. This change

FIGURE 14.1
Absorption or scattering of starlight by interstellar gas (atoms, ions, and molecules). View of observer along the line of sight and off to one side.

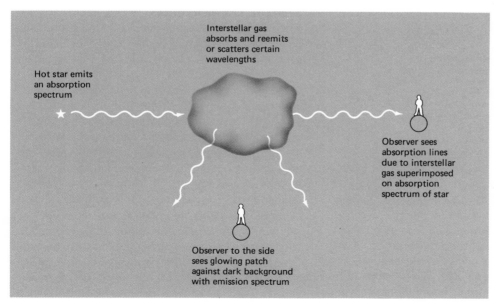

Interstellar gas absorbs and reemits or scatters certain wavelengths

Hot star emits an absorption spectrum

Observer sees absorption lines due to interstellar gas superimposed on absorption spectrum of star

Observer to the side sees glowing patch against dark background with emission spectrum

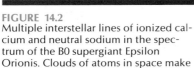

K line of calcium II   H line of calcium II   D lines of sodium I

FIGURE 14.2
Multiple interstellar lines of ionized calcium and neutral sodium in the spectrum of the B0 supergiant Epsilon Orionis. Clouds of atoms in space make their presence known by their effect on transmitted light. They absorb small amounts of energy from the starlight as it passes through them, producing interstellar absorption lines in the spectra of the most distant stars. Epsilon Orionis is about 1600 light years away (see Table 12.4). The strength of such lines depends on the number of absorbing atoms lying along the line of sight. The five different sets of Doppler-shifted lines of singly ionized calcium (H and K lines) and neutral sodium (D lines) give the following velocities for the five absorbing clouds through which the starlight has passed: +3.9, +11.3, +17.6, +24.8, and +27.6 kilometers per second, respectively.

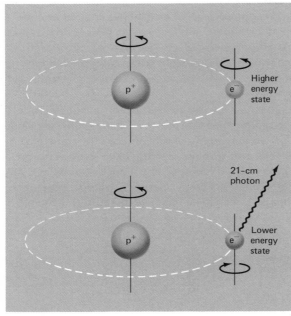

FIGURE 14.3
In the upper neutral hydrogen atom the spins of the proton and the electron are parallel, while in the lower atom they are anti-parallel. When they are parallel, the electron is in a higher energy state than when they are anti-parallel. The transition from parallel (upper) to anti-parallel (bottom) produces a 21-centimeter photon as shown.

drops the atom into a lower energy state, creating a photon whose wavelength is 21 centimeters. Actually, within an interstellar cloud the electron may reverse its spin much sooner, as often as once every 400 years after a collision with a passing atom. This random collision may transfer kinetic energy to the electron, causing it to flip over and align its spin with that of the spinning proton. Even though the time lag is inordinately long for producing a 21-centimeter photon, a ready supply of 21-centimeter radiation is always available because of the enormous number of hydrogen atoms along a line of sight through the Galaxy.

The emission of 21-centimeter photons not only confirms the importance of hydrogen as the primary constituent of the interstellar medium but also provides radio astronomers with a valuable tool for studying its structure. Because of its long wavelength, a 21-centimeter photon can travel greater distances through interstellar space than photons of visible light. In general, electromagnetic waves are more likely to interact with matter the closer the characteristic size of the matter is to the wavelength of the wave. Thus electromagnetic waves with visible wavelengths more readily interact with atoms, molecules, or very small solid particles than do those of radio wavelengths.

INTERSTELLAR MOLECULES
Since 1963 radio astronomers have found a surprising number of *interstellar molecules,* including many organic ones (those containing carbon), by searching for their spectral fingerprints in the centimeter and millimeter region of the electromagnetic spectrum. From approximately a hundred fifty radio spectral lines, some fifty molecules (see Table 14.1)—containing mostly combinations of hydrogen, carbon, nitrogen, and oxygen—have been identified; the number is increasing yearly. Some of the familiar inorganic compounds are ammonia (NH_3), water (H_2O), and several containing sulfur—including sulfur monoxide (SO), sulfur dioxide (SO_2), hydrogen sulfide (H_2S), nitrogen sulfide (NS), and a metallic sulfide (SiS)—and one metallic oxide, silicon monoxide (SiO), has also been discovered. Some of the interesting organic molecules are carbon monoxide (CO), cyanogen (CN), formaldehyde (H_2CO), methyl alcohol (CH_3OH), and ethyl

PLATE 13
Spectroh...
angstrom...
solar erup...
disk in th...
angstrom...
At the top...
which ex...
sun's surf...
image at...
eruption,...
emission...
lower cor...
increases...

PLATE 14
X-ray vie...
Large brig...
loops or l...
and Engi...

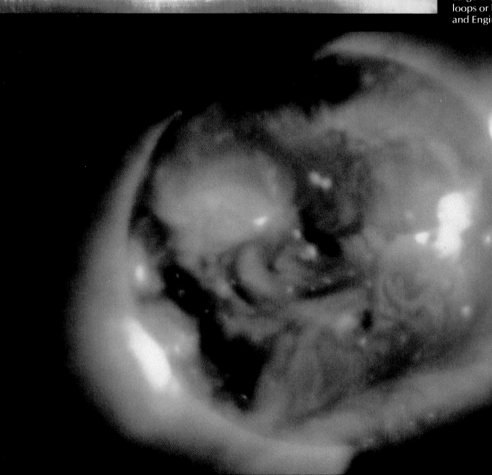

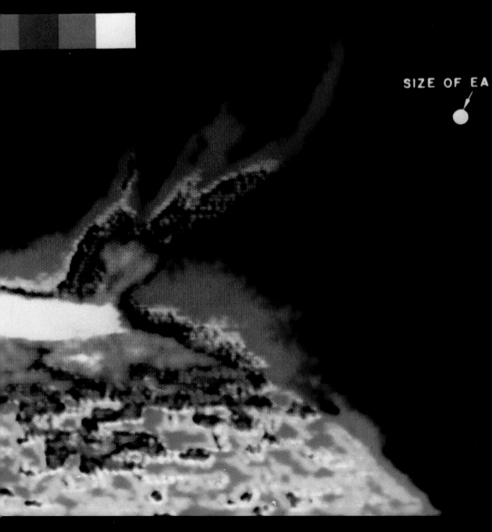

e seen in this
during the
eme Ultravio-
liograph. It
n from 150 to

650 angstroms. This photograph reveals
for the first time that helium erupting from
the sun can stay together to altitudes up to
800,000 kilometers. (NASA)

PLATE 16

Great Nebula in Orion (M42) photographed with the Hale 5.1-meter reflector. The Orion Nebula is the birthplace of new stars some 1500 light years from earth. Its overall diameter is about 20 light years. The nebula consists mostly of glowing hydrogen gas with some helium and lesser amounts of heavier atoms, all stimulated to fluorescence by the ultraviolet light of its recently created stars. This phenomenon is similar to the mechanism causing certain mineral substances to glow when irradiated with ultraviolet ("black") light. In the nebula stellar ultraviolet light striking the atoms is absorbed and energy is reradiated in the visible colors. (Copyright by the California Institute of Technology and Carnegie Institution of Washington. Reproduced by permission from the Hale Observatories.)

PLATE 17

This unusual gaseous nebula in Serpens (M16) is associated with an open star cluster. The opaque nodules or dark globules in the nebula are believed to be gravitationally collapsed condensations of gas and dust that will begin to shine as embryo stars when their internal temperatures become high enough to cause them to radiate heat and light. (Copyright by the California Institute of Technology and Carnegie Institution of Washington. Reproduced by permission from the Hale Observatories.)

PLATE 18

North America Nebula in Cygnus (NGC 7000) photographed with the 1.2-meter telescope of the Hale Observatories. The pinkish glow arises from hydrogen gas shining by the same mechanism that causes gas in an emission nebula to radiate: the ultraviolet light of nearby hot stars, which is absorbed by the gaseous medium, is reradiated in the visible region of the spectrum. (Copyright by the California Institute of Technology and Carnegie Institution of Washington. Reproduced by permission from the Hale Observatories.)

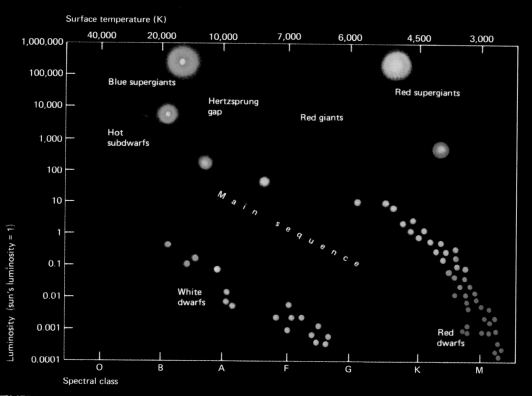

Surface temperature (K)

| 40,000 | 20,000 | 10,000 | 7,000 | 6,000 | 4,500 | 3,000 |

Luminosity (sun's luminosity = 1)

1,000,000
100,000
10,000
1,000
100
10
1
0.1
0.01
0.001
0.0001

Blue supergiants

Hertzsprung gap

Red supergiants

Hot subdwarfs

Red giants

M a i n   s e q u e n c e

White dwarfs

Red dwarfs

| O | B | A | F | G | K | M |

Spectral class

PLATE 19

A Hertzsprung-Russell diagram showing the integrated or composite color for stars on the main sequence, in the red giant region, and the white dwarfs. The differing diameters of the stars are illustrated for each of the regions in the diagram.

PLATE 20

Pleiades and associated nebulosity in Taurus (M45) photographed with the 1.2-meter telescope on Mount Palomar, Hale Observatories. The nebulosity around the brighter stars of the Pleiades star cluster contains dusty material that scatters star-light. The phenomenon is analogous to the color glow observed around a street lamp on a foggy night, which results from the scattering of the lamplight by moisture particles. The nebulosity is bluer than stars because the shorter waves are scattered more strongly than the longer waves. (Copyright by the California Institute of Technology and Carnegie Institution of Washington. Reproduced by permission from the Hale Observatories.)

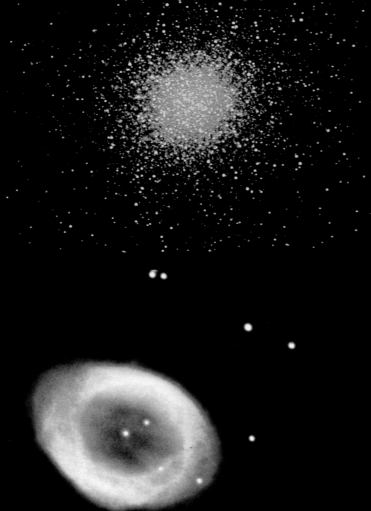

PLATE 21
Globular cluster M13 in the constellation of Hercules. (NASA)

PLATE 22
Ring Nebula in Lyra (M57) photographed with the 5.1-meter Hale reflector. The central star is a hot dwarf whose strong ultraviolet light has stimulated its ejected "smoke rings" into luminescence. (Copyright by the California Institute of Technology and Carnegie Institution of Washington. Reproduced by permission from the Hale Observatories.)

PLATE 23
Crab Nebula in Taurus (M1) photographed with the 5.1-meter Hale reflector. Its distance is about 5000 light years and its diameter, which is still growing, is 6 light years. (Copyright by the California Institute of Technology and Carnegie Institution of Washington. Reproduced by permission from the Hale Observatories.)

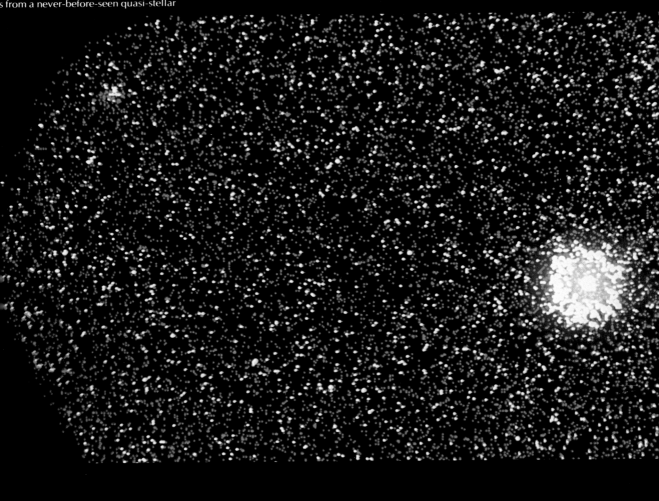

TABLE 14.1
Molecules Detected in the Interstellar Medium

Number of Atoms in the Molecule

2	3	4	5	6	7	8	9	11
H_2^a	H_2O	NH_3	CH_4	CH_3OH	CH_3NH_2	$CHOOCH_3$	CH_3CH_2OH	HC_9N
CH^b	C_2H	H_2C_2	CH_2NH	CH_3CN	CH_3C_2H	CH_3C_2CN	CH_3OCH_3	
CH^{+b}	HCN	H_2CO	NH_2CN	NH_2CHO	CH_3CHO		HC_7N	
OH	HNC	$HNCO$	$HCOOH$		CH_2CHCN			
CN^b	HCO	H_2CS	HC_4		HC_5N			
CO^a	HCO^+	C_3N	HC_3N					
CS	HN_2^+							
SiO	H_2S							
NS	OCS							
SO	SO_2							
SiS								

[a]Identified by ultraviolet lines.
[b]Identified by visible lines.

alcohol (C_2H_5OH). The latter has nine atoms in the molecule and along with the heaviest, cyano-octatetra-yne (HC_9N), is a fairly complex molecule to be produced in the interstellar medium. Two of the organic molecules, methylamine (CH_3NH_2) and formic acid ($HCOOH$), can react to form one of the amino acids known as glycine so essential to life.

Compared with the ubiquitous hydrogen, a small amount of other molecules occurs—less than one part in a thousand of hydrogen. A few of the spectral lines are observed in absorption instead of emission, whenever enough molecules happen to lie in front of a Galactic or extragalactic source emitting continuous radiation. The hydroxyl radical (OH), water (H_2O), and carbon monoxide (CO) have been detected in several galaxies.

A large number of molecules are found in a dense cloud, Sagittarius B2, near the Galactic center in Sagittarius, and in the Orion Nebula. Other sources appear to be randomly distributed in localized regions of the interstellar clouds. Some are concentrated in tiny regions comparable in size to the solar system. How they were formed is not well understood. Two-atom collisions can produce the diatomic molecules, but it is very difficult to imagine a sequence of successive collisions producing polyatomic molecules with nine atoms. But before we discuss possible formation mechanisms, we should discuss the interstellar dust.

INTERSTELLAR DUST

Interstellar dust consists of microscopic solid grains whose composition and properties are not like the dust on earth. Photographs of regions along the Milky Way are laced with dark patches that are large clouds containing dust. The dimming of starlight is due almost entirely to these interstellar grains, as the gaseous component of the interstellar medium is fairly transparent to starlight. In fact, the interstellar gas is billions of times more transparent to visible light than is sea-level air on the earth for the same path length. Clinging close to the Galactic plane, interstellar dust completely shuts off our optical view of the Galactic center and keeps us from seeing extragalactic objects near the Galactic plane. In edge-on spiral galaxies this dust is the dark lane passing centrally across the galaxy (see Figure 17.3).

Interstellar absorption is greatest for ultraviolet light, less for visible light, and least in the infrared wavelengths. For visible light it can be as large as about 0.7 magnitude per thousand light years (the average is about half this value) near the Galactic plane. Thus for a star at the center of the Galaxy, about 30,000 light years away only about one photon out of a hundred billion reaches us. In the X–ray, infrared, and radio spectral regions, however, we can observe all the way to the Galactic center. Because blue light is affected twice as much as red light, the color of a distant star whose light has been selectively scattered looks not only dimmer but redder than it really is (see Figure 14.4). If we do not suitably correct the observed magnitude of a distant star, its distance calculated from the distance modulus is too great. For example, if a star's visual light is dimmed by 1 magnitude, the calculated distance is about 60 percent too great.

About 1 percent of the mass of interstellar clouds is due to the interstellar dust and 99 percent to the interstellar gas. The average density of the dust is about one grain per 10^{13} cubic centimeters, or one grain in a cube 200 meters on a side. Compare this to the typical gas density of one atom per cubic centimeter. Density of the dust grains can be much larger in small localized regions such as the interstellar clouds. However, the density of gas will also be larger. It appears that the ratio of dust to gas is constant in most of the interstellar medium to within a factor of about two.

What is the size and composition of the dust grains? The scattering of photons is strongly dependent upon the size of the scattering particle. The strong scattering of visible light by the interstellar grains suggests that a grain is comparable in size to the wavelength of light, 4×10^{-5} to 7×10^{-5} centimeter. Best estimates set the size as about 2×10^{-5} centimeter, with some evidence for an even smaller grain, which is responsible for scattering ultraviolet light. At this size the typical grain contains about 100 million atoms. At the temperature and density of the interstellar medium, hydrogen and helium are gases. In order for the dust grains to make up 1 percent of the interstellar medium, the grains are probably composed primarily of the elements heavier than hydrogen and helium.

From all the reddening, scattering, polarizing, and spectral absorption of starlight, astronomers conclude that the dust grains in the clouds are probably assorted graphite, iron particles, silicon carbide, silicates, and frozen gases (ices). The silicate grains possess marked spectral signatures in the infrared region of the electromagnetic spectrum around 10 and 20 microns. Another spectral feature appears at 0.2 micron, which is in the ultraviolet, probably due to either graphite or some other carbon-rich compound. Finally, an absorption feature at 3.1 microns in the infrared has been identified as due to water ice.

Starlight that passes through an interstellar cloud is usually slightly polarized by the dust grains of the cloud. For polarization to occur, as shown in Figure 14.5, grains are probably elongated or rodlike and spin rapidly about their short axis. The grains appear to have a reasonably uniform alignment over major distances in interstellar space due to the weak magnetic field that pervades the disk of the Galaxy. The magnetic field of the Galaxy has a strength of about a millionth of that of the earth.

Possibly most of the interstellar dust comes from the material that condenses after being blown out of

FIGURE 14.4
Scattering and reddening by interstellar dust.

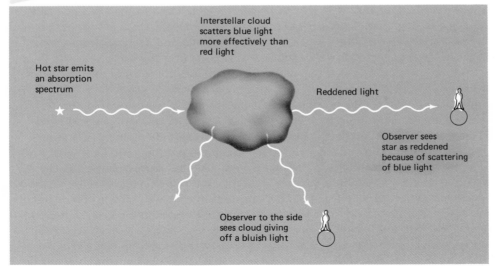

Interstellar cloud scatters blue light more effectively than red light

Hot star emits an absorption spectrum

Reddened light

Observer sees star as reddened because of scattering of blue light

Observer to the side sees cloud giving off a bluish light

stellar atmospheres. There are several phases in a star's life, starting from birth and ending with its death, when it will lose matter. Among the brightest infrared sources are the glowing dust shells around some stars, called *circumstellar shells* (see Figure 14.6). Apparently the grains intercept the short-wavelength radiation of the central star, heat up, and reradiate the energy as long-wavelength infrared photons. The spectral energy distribution of the infrared radiation is what would be expected from dust grains composed of various silicates, silicate carbides, and graphite. This chemical makeup is about the same as that of the material that formed the solar nebula, which was discussed in Section 6.1.

Interstellar dust grains may have something to do with forming interstellar molecules. It is thought that hydrogen atoms can accrete on the cold surfaces of the grains, forming molecules that may escape from the surface by absorbing a photon of starlight (or by some other means). In the interstellar gas are ions produced when cosmic ray particles (high-energy subatomic particles) collide with the interstellar atoms and molecules and strip one or more electrons from them. The ions thus formed can trap atoms to build more complex molecules out of the simpler ones. Apparently the enveloping dust clouds prevent ultraviolet starlight or other energetic photons from reaching the interstellar molecules and dissociating them. In some imperfectly understood way, the molecules are being replenished faster than they are being destroyed. Some of the molecules found are also those present in the organic compounds associated with living matter on the earth, although they were probably produced by nonbiological means. These fascinating discoveries have opened up exciting possibilities for the evolution of organic matter.

14.2
Interstellar Clouds and Nebulae

INTERSTELLAR CLOUDS
In photographs of the Milky Way our view of the starry background is partly or wholly blocked by *dark nebulae* (see Figure 14.4). They contain denser concentrations of interstellar dust than occur generally in the Galactic plane. One long region is a deep, chain-

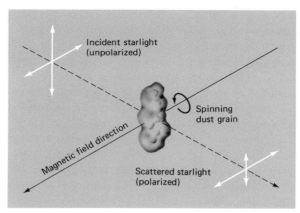

FIGURE 14.5
Polarization of starlight by an interstellar grain. The grains spin with their long axes perpendicular to the magnetic field. Transmitted light is polarized parallel to the magnetic field.

FIGURE 14.6
Circumstellar shell. Radiation emitted by the central star is absorbed by dust grains in the circumstellar shell warming them and causing them to radiate energy in the infrared spectral region. The thicker the dust shell, the greater the amount of energy emitted by the central star that will be converted to infrared radiation.

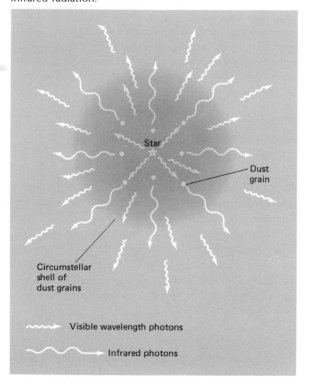

like complex with dozens of isolated and connected dark *interstellar clouds* stretching about halfway around the middle of the Galactic band from the constellation Cygnus to Crux. This obscuring strip forms the Great Rift dividing the Milky Way into two branches, as shown in Figure 16.1. In many regions the dark nebulosity separates into tangled absorbing lanes cutting across bright gaseous nebulae.

Ground-based observations have shown that there is a wide range in the properties of interstellar clouds. However, much of our knowledge about them has come from ultraviolet studies with the *Copernicus* and *IUE* satellites. The clouds can be divided into *diffuse clouds,* which are thin enough for us to observe stars behind them, and *dark clouds,* which are so opaque that stars behind them cannot be seen. Both types of irregularly shaped clouds are from 0.1 light years to 50 light years in diameter. Their temperatures go from about 100 K for the diffuse clouds down to 10 to 20 K for the dark clouds. Interstellar clouds may take up about 4 percent of space in the Galactic plane, with typical masses of several solar masses up to 10^4 solar masses for diffuse clouds and up to 10^5 times the mass of the sun for dark clouds. Their densities, which may vary from 100 particles per cubic centimeter for diffuse clouds to more than one million particles per cubic centimeter for dark clouds, are low. But dark clouds are remarkably opaque because of the accumulated effect of light extinction as starlight passes through an enormous length of absorbing material. Typical separations between clouds appear to be on the order of 100 light years.

RADIO EMISSION FROM INTERSTELLAR CARBON MONOXIDE

In the dark interstellar clouds hydrogen is primarily in the form of molecules rather than atoms, so that the clouds are not sources of 21-centimeter radiation. However, with the discovery of a strong emission feature at 2.6 millimeters due to carbon monoxide (CO), radio astronomers have had a new probe for investigating dark clouds. Dark clouds are the primary locations for the interstellar molecules, and the CO molecule is much more abundant in them than in the general interstellar medium. CO has proven to be an invaluable aid to the study of dark interstellar clouds, as shown in Figure 14.7.

As an example, maps of dark clouds made from the radiation from interstellar CO reveal a number of hot

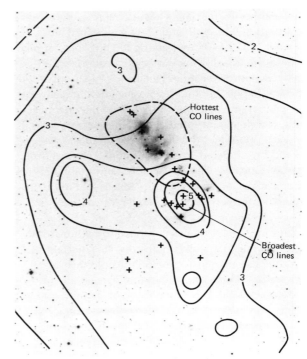

FIGURE 14.7
CO contour lines superimposed on an interstellar cloud complex.

spots in the clouds. These are regions of relatively higher temperatures, 20 to 50 K compared to the 10 to 20 K for the rest of the cloud. Infrared maps of these hot spots often show one or more bright infrared sources within them. Thus the radio emission and the infrared emission tell us that something is happening in small localized regions of the dark clouds that is heating the region. We believe that this is probably an indication of star formation.

INTERSTELLAR MASERS

In 1965 radio astronomers accidentally found microwave emission from the hydroxyl radical (OH) coming from interstellar clouds. The character of the emission was peculiar and the region from which it came was very small. Astronomers found that these small regions were bright sources of infrared radiation also, but they emit virtually no visible light.

The radio emission was much stronger than could be accounted for by random thermal collisions. Clearly there had to be a pumping mechanism that was selectively exciting the OH molecule into the energy state whose de-excitation produced the observed emission. In *masers* imbedded in a molecular cloud, it is thought that infrared radiation from nearby stars excites OH molecules into a quasi-stable level from which they can be stimulated to fall into a lower energy level after interaction with a stellar photon of the right wavelength. The resulting emitted radiation stimulates other molecules in the cloud to radiate in the same fashion, producing a cascading avalanche of emission. Thus the radiation in a normally weak line is greatly amplified. The word *maser* used to describe the phenomena is an acronym for *microwave amplification by stimulated emission of radiation*. The essential requirement for maser amplification is that there must be a larger number of molecules in the upper level of the transition than in the lower.

Astronomers know of some 300 OH masers operating in interstellar clouds, in the atmospheres of the red giant variables like Mira, and in the atmospheres of red supergiants like Betelgeuse. Several dozen water (H_2O) masers emitting at 1.35 centimeters have been discovered in molecular clouds, and at least eighty in the atmospheres of red giant and supergiant variable stars. In addition to the OH and H_2O masers, silicon monoxide (SiO) masers and a methyl alcohol (CH_3OH) maser have been found.

EMISSION NEBULAE

Far outside hot O and B stars, their ultraviolet radiation ionizes hydrogen gas. These regions of ionized hydrogen are the H II regions we have mentioned (see Figure 14.8). Almost all hydrogen atoms have their electrons in the ground state since the interstellar medium is so cold. Neutral hydrogen atoms thus absorb the ultraviolet photons, whose wavelengths are less than 912 angstroms, emitted by these hot stars. These photons are energetic enough to remove the electron and thus ionize hydrogen. In their wanderings free electrons can be recaptured by other hydrogen nuclei. As the captured electrons cascade down into various energy levels, the energy given up by the electron appears as photons that produce the many emission lines in the various hydrogen series throughout the electromagnetic spectrum, among them the well-known Balmer series. The radiation observed coming from bright gaseous nebulae is mostly the integrated light of emission lines created in this way. Stars of spectral type O5 emit enough ultraviolet photons to ionize hydrogen out to distances of 300

light years from the star. For cooler spectral types the surrounding H II region is smaller. An A0 star creates an ionized region about it that is less than 1 light year in radius. Table 14.2 lists some of the properties of a few selected emission nebulae.

Some atoms, chiefly ionized oxygen, nitrogen, and neon, can also be stimulated to emit by infrequent collisions with free electrons that may excite the atoms into rare, low-lying metastable levels. These are long-lived atomic levels with lifetimes up to several hours instead of the ordinary hundred-millionth of a second. In the rarefied medium such excited atoms will long endure in these excited metastable states because collisions that could depopulate them are infrequent. Eventually they will spontaneously drop to lower levels, emitting a photon. These transitions produce the so-called *forbidden lines* prominent in the red, green, and ultraviolet portions of the nebular bright-line spectrum. (They are forbidden in the sense that these photons are emitted only under conditions approaching a vacuum; the phenomenon is similar to that observed in the bright-line spectrum of the solar corona.) An interstellar cloud of gas with a bright-line spectrum (see Figure 14.9) is an *emission nebula*. See Plate 16 for an excellent example of an emission nebula.

14.3
Mass Loss by Stars

There are many stars that, for one reason or another, will lose mass at one or several times during their lives. Such mass loss is important to the star for its effect upon the star's evolution. It is also important since it alters both the dynamics and the chemical composition of the interstellar medium.

The evidence for the loss of matter by stars comes from direct telescopic observations and from indirect studies of the spectra of stars. For some stars matter is ejected in one gigantic explosion, such as for the supernova and to a lesser extent the nova. The planetary nebula is another example of a single expulsion of material from the star. However, the planetary nebula, compared to the supernova, is very gentle. There are also stars for which the evidence points to an almost continuous loss of matter for substantial periods of the star's life.

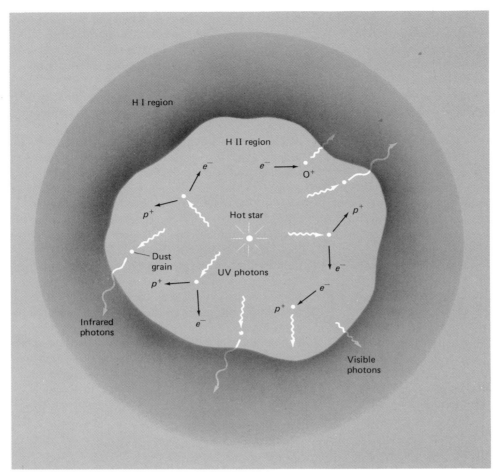

FIGURE 14.8

H II region about a hot star. Ultraviolet photons are absorbed by neutral hydrogen atoms, ionizing the atom and creating free electrons that excite the surrounding gas. The dust near the hot star also absorbs some radiation, is heated, and emits infrared radiation. The wavelength of most of the infrared radiation emitted lies between 30 and 100 microns.

STELLAR WINDS

The continuous loss of matter is called a *stellar wind,* and it is thought to be roughly analogous to the loss of mass by the sun in the solar wind. As discussed earlier, the solar corona is a tenuous, spherical halo of very hot gas, 1 to 2 million degrees Kelvin. These high temperatures may arise from the dissipation of energy carried into the corona by mechanical waves, much like sound waves. The pressure of the hot corona exceeds that of the interstellar medium surrounding the sun, and thus the solar corona continuously expands, being replaced by material from the photosphere. Such a stellar wind is said to be thermally driven.

Do other sunlike stars also have thermally driven stellar winds? There is no reason to believe that the sun is unique in this respect. However, the amount of mass lost by the sun in this way is so small that it would take literally trillions of years for it to lose a significant fraction of its total mass. It is therefore unlikely that we can detect such a low rate of mass loss in other stars. Even though there are numbers of stars like the sun, it is also unlikely that stellar winds by these stars appreciably alter the balance of the interstellar medium.

Red giants and supergiants are stars with very large radii and very low mean densities. Atomic and

TABLE 14.2
Properties of Some Selected Emission Nebulae

Name	Messier Number	Distance (ly)	Diameter (ly)	Mass (sun = 1)	Density (particles/cm³)	Temperature (K)	Spectral Type of Exciting Stars
Orion	M42	1600	16	300	600	9000	O
Trifid	M20	3200	12	150	100	8200	O
Lagoon	M8	4000	30	1000	80	7500	O
Eagle	M16	5600	20	500	90	8000	O
Omega	M17	5200	30	1500	120	8700	B

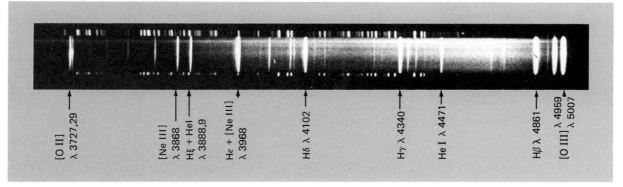

FIGURE 14.9
Bright-line spectrum of the Orion Nebula. Bracketed entries refer to forbidden lines. The Roman numerals indicate atomic state: I, neutral; II, singly ionized; III, doubly ionized. The continuous background spectrum arises from the scattering of starlight by dust particles.

molecular particles in the outer layers of their atmospheres are not held as tightly by the gravity of the star as is the case for main sequence stars. A more substantial stellar wind is occurring in the red giants and supergiants, but with smaller velocities than the solar wind. Astronomers have actually photographed a faint halo of scattered light from enormous envelopes of gas surrounding the red supergiant Betelgeuse (see Figure 14.10). These large, gently expanding stellar winds have also been detected by radio astronomers using the Doppler shifts of OH and H_2O masers in the atmospheres of red giants and supergiants.

The recognition that some hot stars of spectral classes O and B continuously lose mass dates from about 1929. Surveys using the *Copernicus* ultraviolet satellite have now shown that all stars brighter than bolometric magnitude −6.0 are losing appreciable amounts of mass by stellar winds. The velocity of stellar winds in hot stars is as much as a hundred times as great as that in the red giants and supergiants. The loss of matter by the very luminous stars appears to be significant enough to effect the evolution of the star. However, there is still much we do not know about the mechanism driving the stellar winds of the luminous stars.

PLANETARY NEBULAE

In contrast to the continuous loss of matter is the expulsion at one time of the outer layers of a star. A mild and slow ejection of the surface layers appears to form the *planetary nebula*.[1] A small, hot subluminous central star is surrounded by a slowly expanding,

[1]So named by the eighteenth-century astronomer William Herschel, who noted its resemblance to the disk of a planet. However, it certainly has nothing to do with the planets of our solar system.

HENRY NORRIS RUSSELL (1877–1957)

Henry Norris Russell, director of the Princeton observatory, was one of America's most beloved and distin-

guished astronomers of this century. His grasp of all phases of astronomy was truly awesome.

In 1912 Russell developed a method of calculating the orbital elements of an eclipsing binary, a method that is still used today. He showed how to derive the relative dimensions of the two components and their densities. Beginning in 1903, in collaboration with British astronomer Hinks of Cambridge, he undertook a program of deriving stellar parallaxes photographically. The work, which was completed in 1910, was instrumental in his discovery of a relationship between the absolute magnitudes and spectral types of the stars. A plot of this relationship showed the existence of two types of red stars: one highly luminous, the other quite faint. By 1913 he had refined this correlation, now known as the Hertzsprung-Russell diagram, and was using the terms *giants*

and *dwarfs* to distinguish between the two groups.

In the early 1920s Russell applied the newly developed quantum theory to the determination of stellar abundances of the elements. From the analysis of the intensity profiles of the solar absorption lines, he derived the relative abundance of some fifty different elements in the solar atmosphere. He also applied this technique to a number of stars. The research revealed the very high abundance of hydrogen in the sun and stars, a result of great importance in our understanding of the role of hydrogen in astrophysical processes. Russell also made important contributions to the theory of stellar structure. He showed that the physical properties of a star can be found solely from its mass and chemical composition (the Vogt-Russell theorem).

---

frequently convoluted nebulous shell of ionized gas moving outward at speeds of about 30 kilometers per second.

The nebulous envelope has a bright-line spectrum of rarefied common gases, including the forbidden lines of ionized oxygen, neon, sulfur, and ordinary lines like the more prevalent ones of hydrogen and helium. The ultraviolet light of the hot central star stimulates these gases to fluoresce. The surface temperature of the central star appears to be around 100,000 K, so that it provides a large number of ultraviolet photons. The bolometric luminosity of the star is typically a thousand times that of the sun's, while the radius is only a few tenths that of the sun. Such conditions provide the nebula surrounding the star sufficient energy to give the gas a kinetic temperature of about 10,000 K, a density of several thousand particles per cubic centimeter, a diameter of several tenths of a light year, and a mass of a few tenths that of the sun. Infrared observations indicate that much dust accompanies the gases. The dust heated by the absorption of ultraviolet and visible photons radiates in the infrared region of the spectrum between about 2 and 75 microns.

The planetary nebula's precursor may be a red giant star of moderate mass in the Galactic disk. Toward the end of its nuclear burning, the distended envelope is ejected because of some thermal instability, and the hot core becomes the central star. Many stars may pass through the planetary nebula stage, though we have identified only about a thousand. This transition is brief—lasting a few tens of thousands of years—reducing our chance to see it. Astronomers estimate that there are somewhere between 20,000 and 50,000 planetaries in our Galaxy and that a few form each year. This rate of formation is consistent with the rate at which moderate-mass stars are predicted to advance to later stages in their evolution. Assuming one or two planetaries form per year that lose a couple of tenths of a solar mass of material, then during the life of the Galaxy, about 15×10^9 years, on the order of 10^9 to 10^{10} solar masses of stellar matter are returned to the interstellar medium. This is a considerable sum of material. Faint planetary nebulae have also been observed in some of the nearby galaxies, including the Large and Small Magellanic Clouds, the Andromeda galaxy, and the two galactic companions of Andromeda.

FIGURE 14.10
Halo of gas around the red supergiant
Betelgeuse due to its stellar wind is
seen in this computer enhanced image.

FIGURE 14.11
Expanding nebulosity around Nova Persei 1901, photographed with the 5.1-
meter Hale reflector in 1949.

NOVAE

Most spectacular of the eruptive stars are the novae
and supernovae. The sudden appearance of a *nova* is
signaled by a rapid rise in brightness amounting to
tens of thousands of times its original brightness in a
few hours. A slow decline in light then may persist for
a year before the star settles down to its former obscurity.

We think we understand the cause of this sudden
outburst of light. A subluminous hot star of approximately solar mass undergoes a violent but superficial
explosion because of some temporary instability. For
evidence that matter has been expelled from the star,
we look to the nova's spectrum. There we find a large
Doppler shift of the absorption lines toward the blue,
indicating a velocity of approach. Soon after the outburst, very broad emission lines appear, indicating that
a transparent gaseous shell has been ejected at a high
speed, usually a few hundred to two or three thousand
kilometers per second. The front part of the shell has
the largest blue shift, indicating it is approaching us,
while the back part has the largest red shift, indicating
it is receding. All other parts of the shell have velocities lying between these two extremes. The expansion

can often be confirmed much later by direct photographs of a growing envelope, as shown in Figure
14.11. In several recent novae radio radiation from
their thermal energy has been detected in the expanding envelopes of ionized gas. About thirty novae may
occur in our Galaxy each year; a few even go through
recurrent outbursts.

An extremely bright nova suddenly appeared in
Cygnus in August 1975. It reached a peak brightness
greater than the Pole Star. This nova gave astronomers
a rare chance to study its spectral behavior in great
depth throughout the electromagnetic spectrum.

There is strong spectroscopic and photometric
evidence that many—and perhaps all—novae are the
hotter member in a close binary. In the proposed
model an expanding red giant star and a hot white
dwarf coexist at relatively close quarters. During its
evolution the red giant has swollen and it overfills its
Roche lobe (Figure 14.12). The lobe's boundary is the
limit within which the expanded star can retain material when a nearby star exerts a strong pull on it. Now
the overexpanded red giant encroaches on the *Lagrangian point*. This is the position at which the gravitational attraction between the surfaces of the two stars

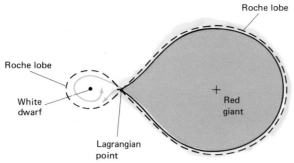

FIGURE 14.12
Production of a nova. The transfer of mass from an expanded red giant toward its small, compact white dwarf companion eventually results in an unstable condition that leads to a nova outburst.

is balanced, making it possible for matter to be attracted equally to both stars. Hydrogen-rich gas streams from the cool giant onto its small, hot companion at a rate of about ten to a hundred-millionth of the mass of the sun per year. Eventually enough matter collects on or about the companion to unsettle the equilibrium of the hot white dwarf and force its layers to expand rapidly and to violently eject material as a visible nova. The amount of matter ejected in the nova outburst is not known. However, estimates place it somewhere on the order of a tenth to a hundred-thousandth of the mass of the sun. After a relatively quiet period that may last months or centuries, another buildup comes with similar consequences. In recent years several X-ray novae have been identified by orbiting spacecraft; unfortunately, little is definitively known about them as yet.

SUPERNOVAE

An explosive event so catastrophic that it stands out even when the rest of an external galaxy cannot be seen—that is a *supernova*. These exploding stars sud-denly attain a luminosity up to several billion times the sun's brightness. Nearly four hundred supernovae have been discovered in other galaxies. As many as five supernova outbursts may occur in our Galaxy each century, according to present estimates. The majority in our Galaxy probably escape detection because of the heavy obscuration by interstellar dust in the plane of the Galaxy. Elsewhere, their occurrence varies from about several times a century in the brightest and largest spiral galaxies to one every few centuries in the faintest spirals.

Of the hundred or so supernova remnants that radio astronomers have found, at least eight are X-ray objects ranging in age from 300 years to 100,000 years, and thirteen have been identified optically. Table 14.3 lists some examples of supernova remnants in the Milky Way. Identifying old supernovae optically is difficult because the ejected material expands and thins out so rapidly that it becomes very difficult if not impossible to see. Presumably many supernova remnants exist throughout the Galaxy. The rapidly expanding remnants of the expelled clouds of gas from some supernovae, colliding with pockets of interstellar material, may be excited into emitting electromagnetic radiation. This seems to be what happened to the Loop Nebula in Cygnus, shown in Figure 14.13.

The enormous Gum Nebula (Figure 14.14) extends more than 50° across the southern sky. It is about 2300 light years in diameter and its closest edge is only 320 light years from the earth. This nebula may represent a wide region of ionized hydrogen gas that was excited when a moving pulse of ultraviolet and X-ray emission left the exploding supernova Vela X about 10,000 years ago.

At least two kinds of supernovae are recognized. The major difference between them is in their spectra and maximum luminosity. The basic physical processes are pretty much the same. Type I supernovae have been observed in all types of galaxies, but they occur most often in the disk of the spiral galaxies. Their maximum luminosity is about 4 billion times that

When men are calling names and making faces
   And all the world's a jangle and ajar,
I meditate on interstellar spaces
   And smoke a mild seegar.

Bert L. Taylor

TABLE 14.3
Some Supernova Remnants

Object	Age (yr)	Distance (ly)	Diameter (ly)
Cassiopeia A	300	10,000	15
Kepler's SN	370	20,000	13
Brahe's SN	400	9,800	20
Crab Nebula[a]	925	6,500	8
Lupus	975	4,000	35
Puppis A	4,000	7,200	55
Vela X[a]	10,000	1,600	130
Cygnus Loop[b]	20,000	2,500	120
IC 443[c]	60,000	5,000	70

[a]Pulsar inside remnant.
[b]Suspected neutron star at center.
[c]Pulsar near remnant.

FIGURE 14.13
Loop Nebula in Cygnus, photographed in red light with the 1.2-meter Schmidt telescope, Hale Observatories. The photograph reveals the filamentary remnants of a supernova that exploded about 20,000 years ago and has since reached a diameter of 120 light years. X-ray emission has been detected coming from a point source at the center of the loop, possibly from the surviving neutron core of the supernova.

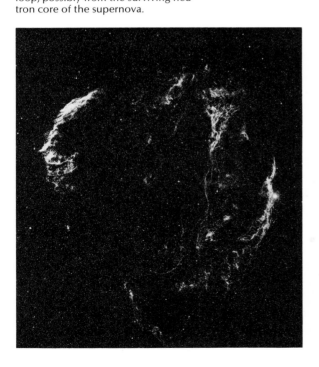

of the sun. A photograph of one very bright supernova is shown in Figure 14.15. Type II supernovae reach a maximum brilliance of up to 600 million times the sun's and exhibit a variety of light-curve shapes and spectra. They appear most often in the arms of spiral galaxies. They rarely appear in elliptical galaxies.

Both types of supernovae have very complex and variable spectra, which are not yet fully understood. They give evidence, however, of very high expansion velocities, on the order of 15,000 to 20,000 kilometers per second.

For Type I supernovae the rapid decline in brightness following maximum luminosity is caused by a cooling of the still-opaque expanding shell. Continuation of the expansion causes the shell to eventually become transparent. The decline in brightness becomes less with the passage of time.

How much matter is blown off to return to the interstellar medium? The amounts of mass ejected are not known, but best estimates at present suggest that the Type I supernovae are Population II stars and lose about half a solar mass of material. The Type II supernovae, on the other hand, are Population I stars and may eject more than five solar masses (in some cases 50 $M_\odot$). The total energy output for both types of supernovae is somewhere between 10^{49} and 10^{51} ergs per second.

CRAB NEBULA

In A.D. 1054, old Chinese chronicles say, a "guest star" suddenly appeared in the constellation Taurus; for many months it was visible to the naked eye. This date corresponds very closely to the estimated birth date of the Crab Nebula shown in Figure 14.16, the remnant of the most celebrated and most studied supernova. Its birth has been estimated from the rate at which the nebula is expanding. By combining the proper motion of the nebula with the velocity from the Doppler shift of its spectral lines, astronomers have estimated that the nebula is about 6500 light years away.

In Figure 14.16 you can see the Crab Nebula's outer filamentary structure and the inner smoothed-out amorphous region. Two kinds of spectra have been observed in the Crab Nebula. In the spectrum of the expanding filamentary network, emission lines characteristic of planetary nebulae are present; an underlying continuous spectrum is also present, arising from the amorphous region. The emission lines are produced by (1) the capture of free electrons into

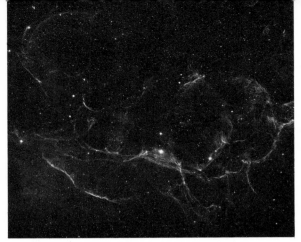

FIGURE 14.14
Gum Nebula photographed in ultraviolet light. A huge, expanding
network of ionized hydrogen surrounds
the Vela X supernova remnant near the
center, where the pulsing neutron core
or pulsar is located. The pulsar emits
both X-ray and radio pulses.

FIGURE 14.15
Supernova in galaxy IC 4182, Type I, in
Virgo. *Top:* 10 September 1937—peak
brightness; exposure 20 minutes. It out-
shines the entire galaxy, which is not
recorded during the exposure time.
Middle: 24 November 1938—about 400
days after maximum brightness; exposure 45 minutes. The galaxy is faintly
recorded. *Bottom:* 19 January 1942—
about 1600 days after maximum brightness; exposure 85 minutes. The supernova now is too faint to be detected
in the galaxy, which is clearly recorded
in the longer exposure.

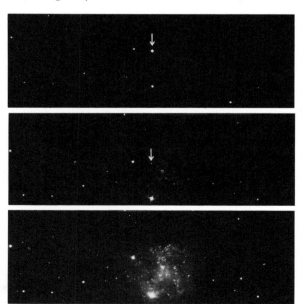

highly excited states of ionized atoms (mostly hydrogen and helium) followed by their spontaneous
cascade down into the lower energy levels and
(2) collisions of free electrons with the ions of oxygen, neon, argon, and sulfur, which excite the outermost electron of the ion into the energy states
responsible for forbidden lines. The continuous
emission is spread in wavelength throughout the
electromagnetic spectrum. It arises (1) from the
capture of free electrons mentioned above, producing the continuum beyond the series limit such as
in the Balmer series of hydrogen; (2) from close
noncapture encounters between ions and free electrons, causing the electrons to lose kinetic energy
and emit it as photons, mostly in the radio spectrum
(known as free-free emission); and (3) from highly
polarized synchrotron radiation. This kind of non-
blackbody radiation is emitted by fast-moving electrons spiraling along the magnetic lines of force
permeating the nebula (see page 181). These electrons
are probably ejected from the rapidly rotating remains of the star whose explosion caused the supernova outburst. The electrons are accelerated to high
velocities by expanding magnetic fields in the nebula.

Where did the energy sustaining the Crab Nebula
for so many centuries come from? It was a mystery until the discovery of the *pulsar,* which we will discuss in
the next chapter.

14.4
Birth of Stars

How are stars born? Where can we observe newly
forming stars? The observational evidence points
to the interstellar gas and dust clouds along the
Galaxy's spiral arms as the present birthplaces of stars.
As some stars—like those responsible for the planetary
nebula, the nova, and the supernova—are approaching
the end of their life, they return some of their mass to
the interstellar medium. New generations of stars are
thus forming from the ashes of previous generations.

STELLAR BEGINNINGS
Apparently, the particles composing interstellar matter
do not experience a net gravitational attraction from
their neighbors, pulling them together. If they did,
within several hundred million years all the matter in

the interstellar medium would collapse and fragment into stars. Then all the matter would have been used up early in the Galaxy's history and no more stars could form.

The very existence of interstellar matter and its organization into clouds of up to several hundred thousand solar masses argue that the gas pressure is sufficient to balance the effects of gravity. The first step in making new stars is to unbalance the forces, so that the material in the cloud can break into smaller clouds and eventually collapse to form stars. A promising way of getting this process going is the traveling compressional wave, or density wave, which we presume is responsible for the Galaxy's spiral-arm structure (see Section 16.4). As the wave moves past the interstellar cloud, it compresses the cool molecular cloud, driving the particles closer together. Their mutual gravitational attraction is now greater than the gas pressure. If the compressed cloud has no other force that can halt contraction, the collapse continues until the matter heats up, raising the gas pressure sufficiently to resist further contraction. Another possible mechanism for unbalancing the forces acting on a molecular cloud is a supernova outburst. Expanding matter expelled from a supernova explosion impinging upon an interstellar cloud can compress it by factors of ten or more and trigger gravitational collapse. The discovery that some young stellar associations may each be located inside the expanding shell of an old supernova remnant certainly supports this as one possible mechanism. Finally, the collapse of the cloud could begin if the cloud could be cooled, so that the gas pressure would go down. There are several possible ways of cooling the cloud, such as radiating away its energy by dust grains.

As the fragmenting molecular cloud breaks into smaller units, the fragments attract surrounding matter and grow in mass. The rate at which stars are created out of the fragments of an interstellar cloud and the number of stars of different masses formed probably depend on several factors: total mass, density, temperature, magnetic fields, and the amount of internal motion stirring the material. The mechanism forming the stars favors the small-mass stars (see pp. 270–271), since we observe many more small-mass stars than large-mass stars. Finally, it appears that only a small fraction, 1 or 2 percent, of the matter in dark clouds actually forms stars.

Some dark clouds in our Milky Way, such as the Coalsack in the constellation Crux, have a clumpy in-

FIGURE 14.16
Crab Nebula, photographed in the red spectral region. Its visible diameter is 8 light years. The filamentary network and associated amorphous region are clearly visible. The arrow points to the pulsar inside the nebula, which is thought to be the star responsible for producing the nebula.

ternal structure, which may mean that fragmentation is starting and the fragments are collapsing. Also, many small dark blobs have been photographed against bright star-filled regions and luminous H II nebulosities of the Milky Way. They are called *globules* (see Figure 14.17). These may be an early stage in the coalescence of matter on the way to forming stars. Their diameters are thousands of times that of the solar system, and they contain several tens of solar masses of material.

The matter of the collapsing fragments converts its gravitational potential energy into thermal energy, some of which is lost as infrared radiation. At some point a significant amount of energy is also taken up as molecular hydrogen dissociates to form atomic hydrogen; later, energy is needed to ionize all chemical species. Because this energy is not available as thermal energy, the collapsing fragment is prevented from even approaching hydrostatic equilibrium. In a very short time (hundreds to tens of thousands of years), it collapses from a small fraction of a light year (several million solar radii) in diameter to a few thousand solar radii. Eventually the central regions of the forming star become opaque, slowing the outward flow of radiation. Then the temperature rises, which increases the gas pressure. The central regions slow from a free-fall collapse to a gradual contraction as they approach a balance between the gas pressure, which is pushing outward, and the weight due to gravity, which is pushing inward. Now the embryo star is on the coolest fringes of the H-R diagram toward the beginning of the evolutionary tracks; see the right-hand side of Figure 14.18.

PROTOSTARS

Once the forming star has stabilized somewhat, it is in the red giant region, though not yet called a red giant; it is a *protostar*. The temperature of the surface is about 4000 K, and energy is transported entirely by convection, which now extends from center to surface. In this slower contraction phase, when the protostar decreases its luminosity but keeps about the same surface temperature, most of the accretion has been completed (see Figure 14.18, right).

During evolution toward the protostar stage, the gases in the stellar envelope or nebula have been growing hotter, emitting more visible light and less infrared radiation. Because it is cooler, however, the surrounding dust cloud or stellar nebula out of which the star is forming absorbs visible photons, then heats up, and reemits the energy in the infrared. Thus the stellar nebula conceals the developing star until most of the surrounding gas and dust is attracted to the protostar or blown away by it. Astronomers apparently have witnessed several times how, within a few years, the dust rearranges itself to reveal, if not a developing star, the place where one or more stars probably will be. You can see how quickly such objects can appear in Figure 14.19. If they do contain collapsing protostars, such regions may be only a few hundred thousand years old. Infrared observations have revealed a number of young cocoon stars imbedded in H II regions like the Orion Nebula. The envelopes around these stars contain large quantities of silicate grains inherited from the molecular cloud.

We have supporting evidence that stars begin in the gas-dust clouds in the coexistence of T Tauri variable stars and veiling clouds out of which they seem to have originated. The T Tauri variables have observable characteristics that we might expect of pre-main sequence evolution; in particular, they lie above the main sequence in the H-R diagram (see Figure 14.20). They have gas and dust envelopes around them, have much lithium, vary erratically in brightness, have large quantities of infrared radiation associated with them, and apparently are losing significant amounts of matter. This loss may represent excess material in the surrounding stellar nebula

FIGURE 14.17
Unusual gaseous nebula in Serpens, photographed in red light with the 5.1-meter Hale reflector. The tiny dark spots (globules) projected against the bright background are believed to be condensations of gas and dust that may some day begin to radiate as stars.

being pushed out of the system by a stellar wind, as we saw in the solar nebula after the sun and planets formed.

Sometimes the very luminous O and B stars, which are definitely quite young, are intermingled with T Tauri stars. Because the O and B stars are massive and are already on the main sequence, it is presumed that the T Tauri stars are in the $0.2\,M_\odot$-to-$2\,M_\odot$ range, contracting toward the main sequence. We do not know definitely, however, if all stars in the cloud form simultaneously. The formation period may be as long as 10 million years. T Tauri stars then probably range in age from 100,000 years to 10 million years. The Orion Nebula is a typical stellar breeding place for both the more slowly evolving T Tauri stars and the rapidly evolving blue supergiants.

As a protostar of low mass nears the main sequence, its energy, derived from gravitational contraction, is transported mainly by convection. In stars

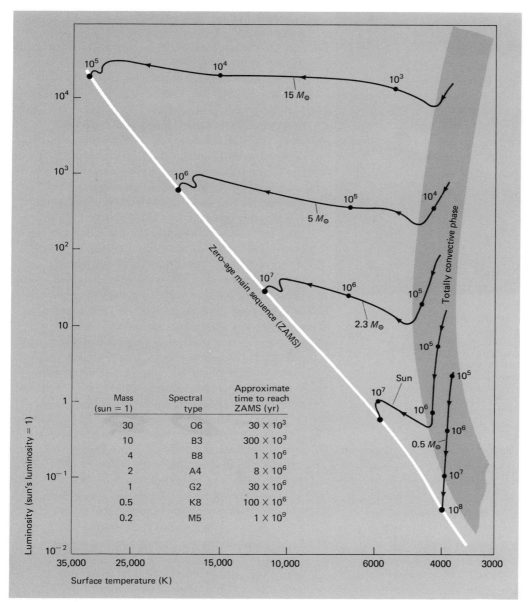

FIGURE 14.18
Theoretical pre-main sequence evolutionary tracks for star models ranging from 0.5 $M_\odot$ to 15 $M_\odot$. The points along each path are labeled with the approximate time (in years) elapsed during contraction to the *zero-age main sequence* (when hydrogen burning supplies all of a star's luminosity).

of greater mass (including the sun), convection dies out except in the layers just below the photosphere, and it is supplanted by a large radiative core in which energy is moved by radiation. Gravitational contraction has now raised the temperature in the radiative core to several million degrees, hot enough to destroy quickly the light nuclei of deuterium, lithium, beryllium, and boron (initially present in small quantities) as they react with protons, producing small amounts of helium. In the first stage of the star's thermonuclear existence, these reactions are an extremely modest step, from which the star derives very little energy.

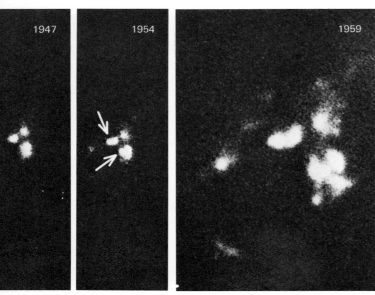

FIGURE 14.19
Herbig-Haro objects photographed with the 0.91-meter and 3-meter reflectors of Lick Observatory. These are probably newly emerging stars, or at least reflection nebulae near emerging stars. Notice the change in the shapes of these condensations (indicated by the arrows). The 1947 and 1954 pictures were taken in blue light; the 1959 picture is in red light.

We can use the sun as an example. By the time the young sun's central temperature had risen by gravitational contraction to several million degrees, the *p-p* chain was ignited and hydrogen burning began to supply the luminosity, at first in small amounts. Several million years later, the sun arrived in the H-R diagram at the *zero-age main sequence*—where hydrogen burning supplies 100 percent of the luminosity—and contraction virtually ceased. The zero-age main sequence, (see Figure 14.18), is the position in the H-R diagram occupied by protostars of different masses that have ceased to contract and have found a stable configuration by burning hydrogen. The protostar is now a full-fledged star. It is convenient to date the star's age from the zero-age main sequence, since the time spent contracting out of the interstellar cloud to the main sequence is only a small fraction of its life span. About 4.6 billion years ago, after a protostar stage of about 30 million years, the sun settled down on the zero-age main sequence for a long uninterrupted time of stability. This stability will continue for another 5 billion years.

EVOLVING TOWARD THE MAIN SEQUENCE

How long does it take to reach the hydrogen-burning phase from the protostar stage? Approximate times are shown for different masses in Figure 14.18. Stars exceeding the sun's mass evolve quite rapidly along paths above and to the left of that for the sun in the H-R diagram (see Figure 14.18). For stars less massive than the sun, evolution is longer and follows tracks below and to the right of the sun's evolutionary path.

According to some astronomers' calculations, protostars with masses of less than about 0.1 $M_\odot$ never become hot enough at their centers to begin fusion of hydrogen. They pass the lower end of the main sequence in a wholly convective state and continue contracting toward extremely high densities. They apparently bypass normal stellar evolution and proceed slowly to their fate of becoming a degenerate red dwarf. With less than a hundredth the mass of the sun, the object may well end up as a Jovian-like planet.

When a large, dark interstellar cloud begins to fragment in selected regions into a cluster of protostars of differing masses, the evolving stars will reach the main sequence at different times (see Figure 14.18). The more massive stars begin burning hydrogen first, and in beadlike progression the others arrive along the zero-age main sequence from the top down. Stars of lower mass will lie progressively farther above and to the right of the zero-age main sequence at some time after contraction starts. Open cluster NGC 2264 shows this progression quite well in the H-R diagram (see Figure 14.20). Many stars in this cluster have gas and dust envelopes. The less massive ones will eventually be spread along the lower portions of the main sequence according to their masses.

By the time several O stars arrive on the zero-age main sequence, they will produce enough ultraviolet radiation to evaporate the dust grains and ionize

They cannot scare me with their empty spaces
Between stars—on stars void of human races.

Robert Frost

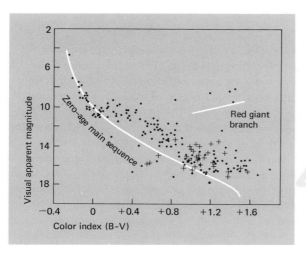

the gas in their vicinity. An H II region will then form. Finding H II regions in dark cloud complexes definitely shows that stars are forming, as shown in Figure 14.21.

There is little doubt among astronomers that rotation is a crucial factor during the collapse of interstellar clouds and the contraction of the resulting fragments into main sequence stars. It probably determines whether the results will be multiple star systems, binaries, stars with planets, or just single stars. At present, the attempts to include the effects of rotation have been crude and not very satisfactory. The role of rotation throughout the universe is still not well understood.

As stated earlier, the star's life on the main sequence is a quiescent phase. After it exhausts its supply of hydrogen fuel, the star will undergo some rapid changes, which lead eventually to the death of the star.

FIGURE 14.21
Birthplace of stars, the Omega Nebula, M17. The bright H II region M17 shows an asymmetrical pattern of development. Note that to the left the expand-ing H II region seems to advance unopposed. However to the right the H II region is adjacent to the dark molecular cloud shown by the radio contour lines for the CO molecule. The numbers on the contour lines indicate the CO temperature in degrees Kelvin.

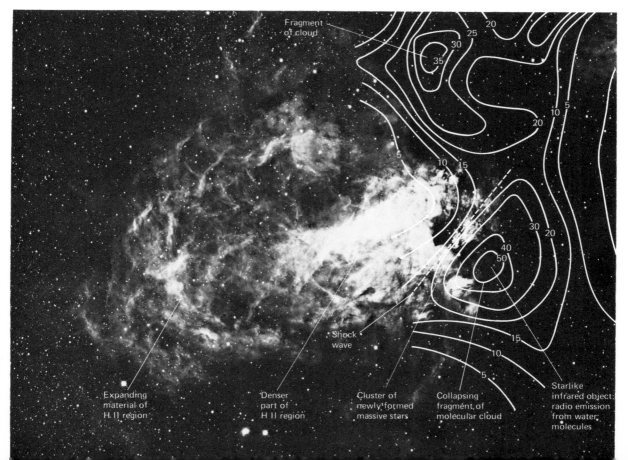

SUMMARY

The interstellar medium is the nonstellar matter in the space between the stars. It is a mixture of gas, mostly atomic and molecular hydrogen, and some dust. Nearly all of it is confined close to the central plane of the Galaxy. Concentrations of this material occur as bright and dark nebulae in the Galaxy's spiral arms. Within the interstellar medium there are clouds of cold gas and dust, which contain complex molecules combining hydrogen, carbon, nitrogen, and oxygen, including some of the organic compounds found in biological matter.

The gas component of the interstellar medium is fairly transparent to visible light. The dust component, which comes mostly from grains formed in the outer atmosphere of some stars, amounts to only about 1 percent of the interstellar medium's mass. It is very effective in absorbing visible light by scattering, reddening, and dimming starlight, which prevents us from observing the central portion of our Galaxy. Deep penetration of the Galaxy is achieved by observing in the longer infrared and radio wavelengths and in the very short X-ray and gamma-ray wavelengths.

Of great importance to astronomy is the 21-centimeter radiation of neutral hydrogen, which is prevalent throughout the Galaxy. It is produced by the atom radiating a photon when the spinning electron suddenly changes its spin to the opposite direction of the spinning proton. The penetrating power of this radiation has enabled us to probe the structure of the Galaxy more thoroughly than by any other means.

Stars are continuously losing mass through stellar winds and mass exchange in binary systems, and intermittently through explosive mechanisms near the end of their life cycles. The explosive outbursts may be mild, as in the case of the planetary nebulae, or violent, as in the case of the novae and supernovae. The material expelled in a supernova explosion enriches the interstellar medium, from which new generations of stars will form.

Although there is strong observational evidence that stars form within the dust-laden gas clouds of interstellar space, the mechanics of star formation are not well understood. Denser clouds of interstellar matter contract gravitationally, heat, and within a relatively short time, depending on their masses, form into protostars that become main sequence stars.

REVIEW QUESTIONS

1. Why is light of distant stars dimmed and reddened going through the interstellar medium?

2. How can we determine the composition of the interstellar dust grains?

3. What is the composition of the interstellar compounds? How do we identify which among them are the organic molecules?

4. Since the gas component of the interstellar medium is transparent to starlight, how is its existence revealed?

5. What triggers the outburst of a nova?

6. What is a planetary nebula? How is it formed?

7. What data suggest that stars are created out of a concentrated interstellar mix of gas and dust?

8. Describe in a general way the evolutionary history of the sun from its birth inside an interstellar cloud to its full-fledged status as a main sequence star.

9. What sort of unstable conditions are the recently created stars likely to experience on their way toward the main sequence? Have we discovered any such stars?

10. How is the 21-centimeter radiation of neutral hydrogen produced?

11. In what spectral regions besides the 21-centimeter wavelength can we detect hydrogen?

12. Describe the nova model astronomers have put together and how it leads to a sudden outburst.

13. What is the difference between a Type I and Type II supernova and in what kind of galaxies does each appear?

14. Why will the sun never become a supernova?

15. Describe the light and spectral changes undergone when a star explodes as a supernova.

16. What possible mechanism might compress a gas-and-dust cloud sufficiently for it to fragment into globules that will evolve into protostars?

In the preceding two chapters we have discussed the hydrogen-burning, or main sequence, phase of stars and the birth of stars. After stars have exhausted their hydrogen fuel while on the main sequence, they are unable to support the weight of their outer layers with their internal gas pressure. Thus they will renew their battle against gravitational compression by contracting, heating, and igniting new sources of nuclear fuel in order to establish thermal equilibrium. This chapter discusses these advanced stages of stellar evolution, which ultimately lead to the death of the star.

15.1
The End of Main Sequence Life: Hydrogen Exhaustion in the Core

It can take a star 10 billion years to evolve from the zero-age main sequence to hydrogen exhaustion (the sun, for example), or as little as several million years (stars of 15 $M_\odot$), as shown in the H-R diagram in Figure 15.1. During hydrogen burning the core had been contracting very slowly, however, as the star fought to maintain hydrostatic equilibrium. This contraction converted gravitational potential energy to thermal energy, raising the core temperature and the star's overall luminosity as the star evolved off the zero-age main sequence toward the hydrogen exhaustion point.

The main sequence consists of a broad avenue of stars of different ages evolving from the zero-age main sequence to their respective hydrogen exhaustion points (for example, see Figure 15.3). Contrary to the changes in pre–main sequence contraction, the star's overall radius has not been decreasing but increasing as the outer layers absorb some of the outflowing energy. Enough energy is absorbed to lift these layers against the gravitational pull of the underlying layers. The H-R diagram records the increases in luminosity and radius at the surface of the star; it does not show

◀ Artist's conception of an accretion disk around a black hole.

the core's slow contraction. For a star like the sun, the radius roughly doubles during the main sequence phase as about 12 percent of its hydrogen is depleted.

As hydrogen burning in the core ends, the hydrostatic equilibrium shifts in favor of gravity, and the core contracts more rapidly now. The contraction releases substantial amounts of gravitational potential energy, further heating the deep interior. The layers just outside the former energy-generating core are now hotter, too, and hydrogen burning migrates to a relatively thin shell surrounding an inactive helium core. Although the core continues to contract and heat, it is the burning of hydrogen in the surrounding shell that primarily supplies luminosity at the star's surface. The evolutionary path now carries the star into the red giant region in the H-R diagram.

The *red giant branch* in Figure 15.1 is the portion of the evolutionary path that extends steeply upward at the extreme right. When the contracting helium core reaches about 100 million degrees Kelvin (see Table 15.1), the second major thermonuclear reaction, *helium burning*, begins. This *triple alpha process*, or fusion of three helium ^{4}He nuclei, forms a carbon ^{12}C nucleus and two gamma ray photons. Stars burn helium in their cores for about 5 to 20 percent of the time spent burning hydrogen on the main sequence.

When helium begins to burn, rapid gravitational contraction in the stellar core ends. And just as hydrogen burning on the main sequence slowly contracted the core, helium burning will also be accompanied by a slow contraction of the energy-generating core. On the main sequence the core was 15 percent to 20 percent of the star's inner radius, but when helium starts to burn, the core is more like 0.1 percent of the radius. The star's structure is thus changing drastically from what it was on the main sequence.

When will contraction raise the temperature in the core to 100 million degrees so that helium may begin burning? This event is determined by the star's main sequence mass. We can summarize the approach to helium burning and subsequent evolution by dividing the stars into two groups, those whose masses are less than about 2 $M_\odot$ and those whose masses are greater than 2 $M_\odot$. This is a somewhat gross generalization but more than adequate for an introduction to stellar evolution. As a means of distinguishing these two groups, we will refer to those less than 2 $M_\odot$ as *low-mass stars* and those greater than 2 $M_\odot$ as *high-mass stars*.

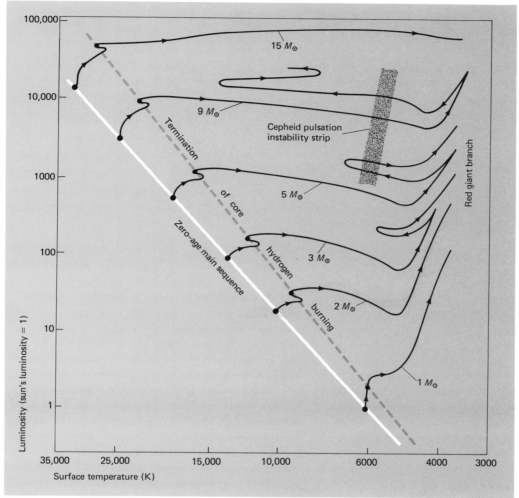

FIGURE 15.1
Theoretical evolutionary tracks in the H-R diagram for representative Population I stars from 1 $M_\odot$ to 15 $M_\odot$. Stars evolving away from the zero-age main sequence eventually leave the main sequence at the point (dashed line), at which time they have consumed their core hydrogen.

15.2
Red Giant Evolution for Low-Mass Stars

HELIUM FLASH

Let us consider first the post–main sequence evolution of the low-mass stars. As hydrogen is exhausted, stars that have slightly greater (or less) mass than the sun evolve rapidly for a while at nearly constant luminosity and decreasing surface temperature from the hydrogen-exhaustion point in Figure 15.1. These stars will then brighten appreciably before helium burning begins, and they will take an evolutionary path upward into the red giant region. This phase takes about a billion years for the sun. For stars less massive than the sun the time is even longer.

During the core's gravitational contraction the densities become so large that the perfect gas law no longer applies to the free electrons. They have become a degenerate gas (see page 279). By the time a star like the sun reaches the red giant region, the helium core

TABLE 15.1
Thermonuclear Burning and Electron Degeneracy

Thermonuclear Process	Minimum Star Mass Needed (sun = 1)	Ignition Temperature (K)	Approximate Density (g/cm³)	Electrons Degenerate at Densities Greater Than
Hydrogen burning H → He	0.1	4×10^6	10^1–10^2	$\approx 10^3$
Helium burning He → C, O	0.4	100×10^6	10^3–10^6	$\approx 10^5$
Carbon burning C → Ne, Na, Mg, O	4.0	600×10^6	10^5–10^8	$\approx 10^7$
Oxygen, neon, and silicon burning Ne → O, Mg O → Si, S, P Si → Ni → Fe	8.0	1×10^9 to 3×10^9	10^7	$\approx 10^9$

(containing about 30 percent of the star's mass) will have been compressed to a volume about twice that of the earth and a density of about 10^6 grams per cubic centimeter. Surrounding the core is a hydrogen-burning shell several thousand kilometers thick. All of this is enclosed in a highly distended, hydrogen-rich envelope about 7 million kilometers thick, whose density is about like that in our best earth-made vacuum chambers. The star's radius is ten to a hundred times its zero-age main sequence radius, and the star is a hundred to a thousand times more luminous. In this part of its history the sun will look quite different from its appearance today (see Figure 15.2).

For low-mass stars the degenerate electron core causes helium to ignite very differently from the way it would in a perfect gas. In a perfect gas any rise in temperature would also increase the pressure and the contraction then underway would be halted. The increase in pressure would expand the core slightly, cooling it and reducing the thermonuclear burning rate until thermal equilibrium is restored.

But in a degenerate electron gas, matter behaves more like a solid than a gas. The core does not readily expand with an increase in temperature, and if it cannot expand, it cannot cool. The rate of helium burning will increase with the temperature. As more energy is generated, the temperature continues to rise. In turn increasing the rate at which energy is generated, this pushes the temperature still higher, and a runaway condition follows. In a few hours temperatures leap to hundreds of millions of degrees and generate as much energy as 100 billion stars of solar luminosity—about as much as a whole galaxy emits. This explosive flash of energy is called a *helium flash*. Huge though the energy flux is, the helium flash is not enough to blow the star apart. Instead, all of the energy goes into removing electron degeneracy, and this so alters the structure that the star may now burn helium in a perfect gas.

The helium flash we have described is generated in our computer models of stars—we have not seen and never will see the flash in a real star.

I seem to have stood a long time and watched the stars pass.
They shall also perish I believe.
Here today, gone tomorrow, desperate wee galaxies
Scattering themselves and shining their substance away
Like a passionate thought. It is very well ordered.

Robinson Jeffers

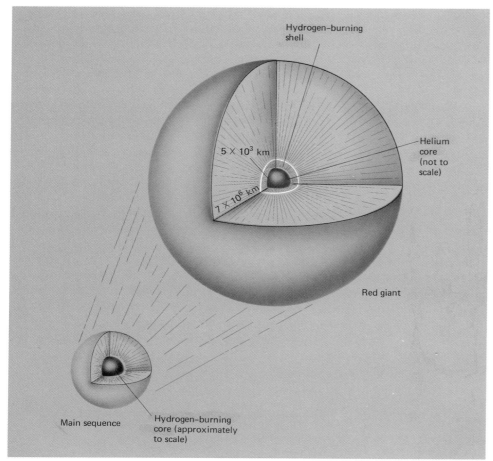

FIGURE 15.2
Helium core and hydrogen-burning shell in low-mass stars prior to helium flash. As star radius (7×10^6 km) and helium core radius (5×10^3 km) show, the drawing is not to scale. Hydrogen-burning core while low-mass star is a main sequence star is approximately to scale.

OPEN CLUSTER H-R DIAGRAMS

Astronomers find observational evidence confirming much of this evolutionary sequence in the H-R diagrams for open clusters. As the individual stars in a cluster evolve, their rate of aging is determined by their masses. The more massive stars, because they are more luminous, move through the evolutionary stages more rapidly than stars of lower mass. From the relative aging of the cluster's members, we can tell the age of a cluster. The critical time at which a star reaches the hydrogen-exhaustion position depends on the mass of the star, which in turn is related to its luminosity and hence its age. With the model star calculations we can place an age corresponding to the observed *turnoff point* of the least massive star to reach hydrogen exhaustion along the right side of the composite H-R diagram, as shown in Figure 15.3.

The youngest cluster is NGC 2362, whose age we estimate as less than 2 million years. None of its blue supergiants have yet crossed the Hertzsprung gap. In the next youngest cluster, h and χ Persei, some of the massive blue supergiants have evolved across the diagram and are now in the red giant region. Next youngest is the Pleiades, followed, in order, by M41, M11, the Hyades, NGC 752, M67, and NGC 188.

The oldest open clusters shown in Figure 15.3 are M67 and NGC 188. The most obvious difference between these two is that all subgiant and giant stars in M67 are more luminous than comparable ones in NGC 188. Stars leaving the main sequence in M67 are more massive and therefore younger than those in NGC 188. The subgiant branches in M67 and NGC 188 strongly resemble the evolutionary tracks of stars of about 1.25 $M_\odot$ and 1 $M_\odot$. Estimated ages are 5 billion years for M67 and 11 billion years for NGC 188. The sun, at about 4.6 billion years, is still on the main sequence, and the 1 $M_\odot$ stars in NGC 188 that are 11 billion years old are in the red giant region. This makes NGC 188 not as old as our Galaxy, but it is certainly a very old collection of stars.

HORIZONTAL-BRANCH STARS

One conspicuous difference between an H-R diagram for open clusters and one for globular clusters is the *horizontal branch* in globular cluster H-R diagrams (see Figure 15.3). This region of stars runs almost horizontally across the diagram from the red giant branch to the very blue stars at a visual absolute magnitude of about zero.

The helium flash in the solar-mass stars reduces the stars' luminosity over a very short time—about ten thousand years. The drop in core temperature, after the flash, lowers the energy output of the hydrogen-burning shell, reducing the star's luminosity and, at the same time, the internal pressure that kept the outer envelope expanded. As the star's envelope shrinks, the surface temperature must increase because of its smaller radiating surface. The star moves down and to the left of its position in the red giant region toward the horizontal branch, where it carries on the helium-burning phase (see Figure 15.4). From stellar models we are led to believe that stars on the horizontal branch have a mass as small as 0.5 $M_\odot$ to 1 $M_\odot$.

The star settles into a relatively stable helium-burning phase on the horizontal branch. However, during this period the star can also become unstable temporarily and pulsate as an RR Lyrae variable (see Section 12.5). According to model star calculations, a range of masses of stars must populate the horizontal branch, but we would expect a very limited range to move onto the horizontal branch at any one time. A well-developed horizontal branch, how-

FIGURE 15.3
Composite H-R diagram for several open clusters and one globular cluster. A cluster's age can be estimated from the location of the turnoff point which is the point of the most massive star still on the main sequence.

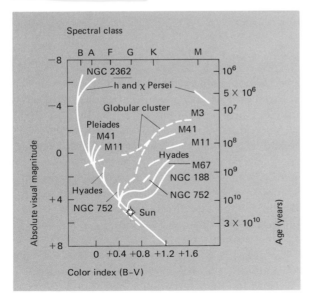

FIGURE 15.4
Theoretical evolutionary tracks for representative Population II stars that have low mass and a low abundance of heavy elements. Globular clusters include such stars. Dashed lines represent periods of rapid evolution.

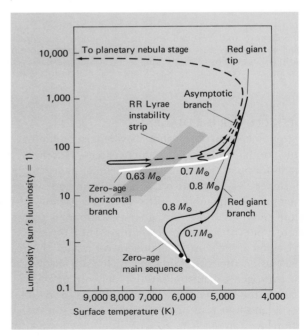

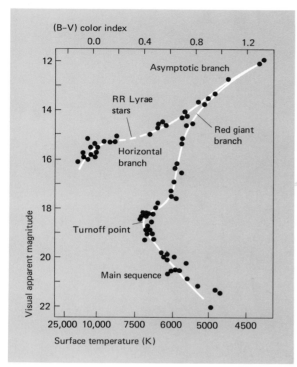

FIGURE 15.5
H-R diagram for stars in the globular cluster M92. Solid lines indicate the average theoretical positions of various stages in the evolution of model stars of different masses.

ever, could be accounted for by significant loss of mass either before or during the helium flash. Evidence from observation does not convincingly suggest that sufficient mass loss is occurring. Although a comparison of Figure 15.5 (H-R diagram of globular cluster M92) and a theoretical diagram made up from tracks of model stars similar to those in Figure 15.4 shows reasonable agreement, we still have much to learn about horizontal-branch stars in globular clusters.

As helium quietly burns at the star's center, the battle between gravity and gas pressure resumes. Depletion of helium causes the core region to contract slowly and to heat, and the star moves upward from the horizontal branch, following the contorted path in the H-R diagram in Figure 15.4. Eventually helium is exhausted in the center of the star as hydrogen had been, but because the central regions have been heated by contraction, helium burning can continue in a shell surrounding the core, now rich

in carbon and oxygen (a fourth helium nucleus may fuse with the carbon nucleus, forming oxygen). Hydrogen burning continues in a shell outside the helium shell, so that now two energy-generating shells in the star contribute to its luminosity. Most important, however, compression does not heat the core enough to start carbon burning in the low-mass stars, as shown in Table 15.1. The end of nuclear burning signals the beginning of these stars' final phases.

After some contortions above the horizontal branch, the star evolves along the path called the *asymptotic branch,* where it moves into the red giant region for a second time. The envelope is again greatly distended, and the outer layers are held so loosely by the gravitational attraction of the inner layers that the star may expel large amounts of material in the form of a stellar wind. Current observational evidence suggests that the more luminous red giants and supergiants are indeed losing significant amounts of matter to the interstellar medium (as discussed in Section 14.3).

15.3 Death of Solarlike Stars

STARS IN PLANETARY NEBULAE AND EJECTION OF MATTER

A most obvious loss of mass is the star's ejection of almost all the hydrogen-rich envelope to form a planetary nebula (see Section 14.3). While it evolves along the asymptotic branch, burning helium and hydrogen in two separate shells, the star undergoes oscillations of increasing amplitude in the envelope. These oscillations, known as *thermal pulses,* are extremely short, separated by intervals of a hundred thousand years or less. Eventually the outer layers separate from the core and expand outward at about 30 kilometers per second, forming a planetary nebula. The path in the H-R diagram, along which the star is evolving very rapidly, is the dashed line at the top of Figure 15.4. Whether the sun will eject a planetary nebula is somewhat uncertain. Most, if not all, of the intermediate-disk Population I stars with main sequence masses between about 1 $M_\odot$ and 4 $M_\odot$, as well as some low-mass Population II stars, may eject their outer envelopes during their evolution after being a red giant.

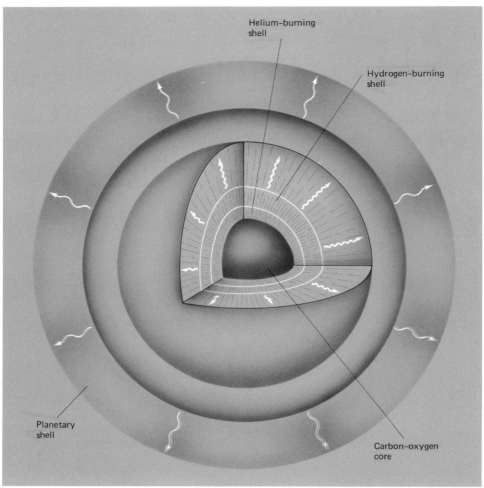

FIGURE 15.6
Low-mass star undergoing planetary
shell ejection. Outer layers are expelled
as a shell at velocities of about 30 kilometers per second.

The ejected shell holds most of the original outer layers that missed thermonuclear burning because they were never hot enough (see Figure 15.6). Left behind is a very hot and very dense carbon-oxygen core. It is surrounded by a thin helium-burning shell and a small layer of matter that has not evolved chemically. Because the star's energy output depends on the burning rate of helium, its luminosity remains constant as it rapidly evolves horizontally across the H-R diagram to the high-temperature side (see Figure 15.7). Near the red giant branch the luminosity and the temperature of the star's original surface are what appear in the H-R diagram. By the time the star reaches the blue side of the diagram, however, the luminosity and the temperature of the now-exposed hot core are what are plotted.

Calculations suggest that a final helium flash swings the star around in a flat loop lasting a hundred to a thousand years, depending on the core mass (see Figure 15.7). Afterward, the star increases its surface temperature before it begins to decline in luminosity toward the white dwarf state (see Figure 15.7). Then burning of the helium shell ceases. By the time the central star has moved into the white dwarf region, the expanding nebula has dissipated into the interstellar medium.

WHITE DWARFS

When the star has exhausted its nuclear fuel supply, gravitational contraction has produced very high densities, and the electrons are once again degenerate. Electron degeneracy in the carbon-oxygen core prevents further contraction and the star cannot ignite the carbon fuel. The only source of energy the star now has is its store of thermal energy, which supplies the luminosity at the surface, cooling the interior of the star. It is now a *white dwarf*.

Mathematical modeling of white dwarfs leads to a surprising result: the larger the mass, the smaller will be its radius. For example, a white dwarf of 0.4 $M_\odot$ has a radius equal to about 1.5 percent of the sun's radius, or about 10,000 kilometers, but one of 0.8 $M_\odot$ has about 1 percent of the sun's radius, or about 7,000 kilometers, which is about the size of the earth. A theoretical limit, the *Chandrasekhar limit,* is reached for a white dwarf of mass 1.4 $M_\odot$, which would have a zero radius. We can infer that a star whose initial mass is less than 1.4 $M_\odot$ can evolve into a stable white dwarf. For an initial mass exceeding 1.4 $M_\odot$ the star must lose mass sometime during its life if it is to become a white dwarf. We think it most probably does so during its red giant or post–red giant evolution.

No clear correlation connects a white dwarf's radius, surface temperature, or spectral appearance like those we saw during its predecessor's main sequence phase. A star of solar mass enters the white dwarf stage as a small, hot blue object. As energy is radiated away, thermal energy is depleted and the temperature declines. As the white dwarf cools, it changes from blue to white to yellow and eventually to red. Since the star can no longer contract, the evolutionary tracks of white dwarfs roughly follow lines of constant radius. The tracks are parallel to the main sequence but well over to the left and below on the diagram (see Figure 15.7).

Most white dwarfs that have been discovered are relatively close to the sun. Their very low luminosity makes them too faint to be seen at greater distances even in large telescopes. Of the twenty nearest stars, two are white dwarfs: the companions of Sirius and Procyon (see page 323). The total number of known white dwarfs is several thousand, of which some four hundred have been studied spectroscopically. A conservative estimate places their total number in the Galaxy in the billions. Most of them are presumably descendants of old disk population stars that have

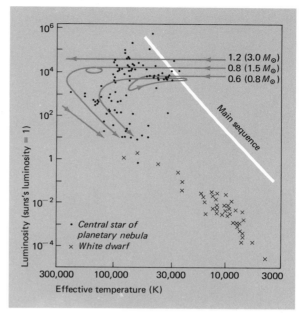

FIGURE 15.7
Theoretical evolutionary tracks of three red giants that have evolved beyond the asymptotic branch in the red giant region. Numbers in parentheses give the mass before the envelope was ejected. Evolutionary tracks are shown for the central star into the white dwarf region of the H-R diagram.

evolved into white dwarfs during the Galaxy's 15 billion or so year life span.

After billions of years a white dwarf's thermal energy will be exhausted. As the star cools, its rate of cooling slows, making its approach to the final nonluminous state as a *black dwarf* quite long. The Galaxy is probably not old enough for very many white dwarfs to have cooled to black dwarfs. This approach to obscurity, though, is definitely a one-way track from which nothing can save the white dwarf.

Our knowledge of the evolution of low-mass stars is not complete enough that we can state precisely what evolutionary path a particular star, such as the sun, will follow. What has been described in the preceding paragraphs is a general evolution for all low-mass stars. If the sun does proceed along that general evolutionary path, then Figure 15.8 traces out its entire evolution from its accretion in the interstellar medium to its final state as a white dwarf.

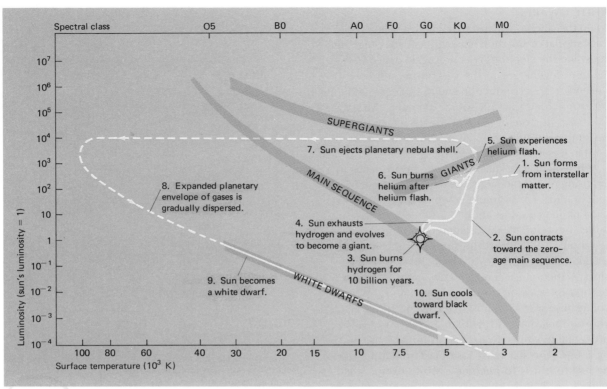

Spectral class

- 1. Sun forms from interstellar matter.
- 2. Sun contracts toward the zero-age main sequence.
- 3. Sun burns hydrogen for 10 billion years.
- 4. Sun exhausts hydrogen and evolves to become a giant.
- 5. Sun experiences helium flash.
- 6. Sun burns helium after helium flash.
- 7. Sun ejects planetary nebula shell.
- 8. Expanded planetary envelope of gases is gradually dispersed.
- 9. Sun becomes a white dwarf.
- 10. Sun cools toward black dwarf.

SUPERGIANTS

MAIN SEQUENCE

GIANTS

WHITE DWARFS

Luminosity (sun's luminosity = 1)

Surface temperature (10^3 K)

FIGURE 15.8
An H-R diagram of the evolutionary history of the sun from birth to death, showing a number of critical stages in its life.

15.4
After the Main Sequence for Massive Stars

Just as for low-mass stars, the day comes in the lives of high-mass stars (greater than about 2 $M_\odot$) when they also exhaust their supply of hydrogen fuel in their cores. The evolutionary path in the H-R diagram followed by the high-mass stars is described in this section and the next two. It is a very different evolution in many ways from that for the low-mass stars that we have just considered.

HELIUM BURNING IN HIGH-MASS STARS
With hydrogen gone in the massive main sequence star, the core shrinks drastically, driving up the density

and temperature and exaggerating the differences between the star's central regions and its outer layers.

For stars above 2 $M_\odot$ in Figure 15.1, evolution away from the main sequence is quite rapid, almost horizontal, across the H-R diagram toward the red giant region. These stars expand their radii by about ten times—a hundredfold increase in their surface area. Still, the temperature is lowered enough to keep the luminosity roughly constant in this phase of evolution. The region across which the stars are evolving in the H-R diagram is the *Hertzsprung gap* (see Figure 13.1). Very few stars have been found in this gap, so the transition must be brief. Calculations for models of different masses also indicate that this change is rapid—occurring in about 50 million years to less than a million years.

Just when will contraction raise the temperature in the core to 100 million degrees so that helium may begin burning for the massive stars? Again this is decided by the star's mass. For stars between 2 $M_\odot$

SIRIUS B—A WHITE DWARF

In the constellation of Canis Major, the Great Dog, we find the brightest star in the sky, Sirius (Alpha Canis Majoris). Sirius, also called the Dog Star, is one of the three members of the winter triangle of very bright stars, the other two being Procyon (Alpha Canis Minoris) and Betelgeuse (Alpha Orionis). Sirius B, the faint companion of Sirius, was one of the first white dwarf stars discovered. Procyon also has a white dwarf companion. Table 15.2 contrasts some of the physical properties of Sirius B with the sun, our stellar yardstick, and the earth. Sirius B is about ten thousand times less luminous than Sirius and is very difficult to photograph, as shown in Figure 15.9.

TABLE 15.2
Some Physical Properties of Sirius B, the Sun, and the Earth

Quantity	Earth	Sirius B	Sun
Mass (sun = 1.99×10^{33} g)	$3 \times 10^{-6}\ M_\odot$	$1.05\ M_\odot$	$1.00\ M_\odot$
Radius (sun = 6.96×10^5 km)	$0.009\ R_\odot$	$0.008\ R_\odot$	$1.00\ R_\odot$
Luminosity (sun = 3.83×10^{33} erg/s)	$\approx 0.0\ L_\odot$	$0.03\ L_\odot$	$1.00\ L_\odot$
Surface temperature (K)	287	27,000	5,770
Gravitational red shift (km/s)	0.0	89 ± 16	0.6
Mean density (g/cm³)	5.5	2.8×10^6	1.41
Central density (g/cm³)	9.6	3.3×10^7	1.6×10^2
Central temperature (K)	4,200	2.2×10^7	1.6×10^7

FIGURE 15.9
White dwarf companion of Sirius. In the short exposure to the left, the white dwarf is the small dot below the over-exposed image of Sirius.

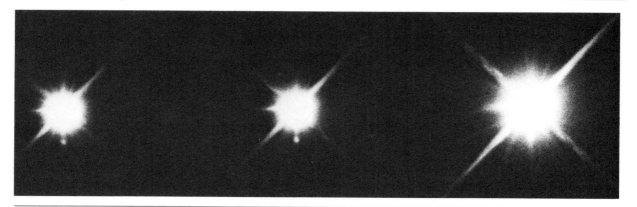

and 9 $M_\odot$ or so, the time comes when the star is in the red giant region. Before helium burning begins, the luminosity is supplied primarily by a hydrogen-burning shell surrounding the helium-rich core. Hydrogen keeps burning in the shell even after helium burning begins, and hydrogen burning still contributes much of the star's luminosity. When helium starts burning, the star's evolution upward along the red giant branch in the H-R diagram stops. Then the star, still burning helium in the core and hydrogen in a surrounding shell, contracts its radius and moves to the high-temperature side of the red giant branch and then back to the low-temperature side a second time (see Figure 15.1).

The path these stars follow in the H-R diagram apparently leads some of them across the *cepheid instability strip* around 6500 K (see Figure 15.1). A cepheid pulsates only in its outermost layers. Small rhythmic expansions and contractions of these layers cause the cyclic variations in the cepheid's light and

In his general theory of relativity Einstein showed how an object in a gravitational field would contract, gain mass, and slow down in its clock time. Implied in the last item, time dilation is a gravitational field's effect on atoms emitting photons. Relativity predicts how much the wavelength of a photon emitted by an atom is lengthened, or shifted to the red, when the atom is in a strong gravitational field. The relative change in wavelength ($\Delta\lambda/\lambda$) is proportional to the mass of the attracting body divided by its radius (see Figure 15.10).

This *gravitational red shift* has practical astronomical interest, because it occurs when a photon of light escapes from a star. If the star's gravitational field is sufficiently intense, we can measure the change in wavelength. We cannot easily observe this effect in the sun, but for a white dwarf of solar mass and small size (where the mass divided by the radius is large), the gravitational red shift is detectable. It has been observed in several white dwarfs in binary systems. Also, we can differentiate between the gravitational red shift of the spectral lines and the Doppler shift that arises from the system's orbital motion and radial velocity. The measured red shifts agree satisfactorily with those predicted by theory. And the gravitational red shift has been verified with even greater accuracy in a physics laboratory experiment.

FIGURE 15.10
Photon escaping from the strong gravitational field of a white dwarf with mass M_2 and radius R_2. The wavelength of the escaping photon is longer than an identical photon escaping from a main sequence star of mass M_1 and radius R_1, if $M_1 = M_2$ and $R_1 > R_2$. Hence the absorption lines in the white dwarf's spectrum should exhibit a small red shift.

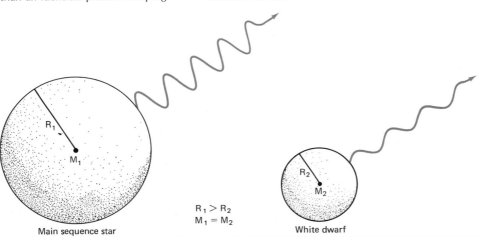

$R_1 > R_2$
$M_1 = M_2$

Main sequence star   White dwarf

velocity (described in Chapter 12). Pulsation will occur only if there is an outer zone of ionized hydrogen separated from an inner zone of partially ionized helium. At a certain critical depth below the surface, the zones will act as a heat engine and drive the pulsation. A cepheid is apt to pulse for a relatively brief time late in its evolution. Once pulsation begins, it will continue for a considerable time until evolution carries the star toward higher temperatures in the H-R diagram, when the ionization zones occur at such shallow depths that there is too little energy to drive the pulsation. The classical cepheids appear to be Population I stars in the $3 M_\odot$-to-$9 M_\odot$ range in their helium-burning phase.

Stars more massive than about $9 M_\odot$ will have begun and completed helium burning in their helium-rich core after hydrogen exhaustion and before they have reached the red giant branch. They will still burn helium while they are red supergiants, but they will do it in a shell surrounding a core that is

now exhausted of helium and that is in turn surrounded by a shell burning hydrogen (see Figure
15.11). The helium burning increases the number of
carbon nuclei at the expense of helium nuclei.

Now other nuclear reactions can take place. Carbon and helium nuclei will fuse, forming an oxygen
nucleus (^{16}O). The parts of the star burning helium
will become rich in carbon and oxygen. A star more
massive than about 25 $M_\odot$ will even initiate the next
step in its thermonuclear evolution before evolving
into a red giant: the step is *carbon burning,* in which
two carbon nuclei fuse to produce primarily neon
(^{20}Ne), or sodium (^{23}Na), or magnesium (^{24}Mg),
above 600 million degrees.

ADVANCED THERMONUCLEAR BURNING IN MASSIVE STARS

By the time helium is exhausted in the core of the
3 $M_\odot$-to-9 $M_\odot$ stars, they will have returned to the red
giant branch for a second time (see Figure 15.1). Now
they will begin to burn helium in a thick shell, which
provides most of the surface luminosity, and to burn
hydrogen in a thin shell farther out from the core. The
core is slowly contracting and heating. The increased
density of the already dense core material squeezes
the electrons and ions even closer together, causing
degeneracy for the electrons. Once the central temperature in the core (rich in carbon and oxygen) has
reached about 600 million degrees Kelvin, carbon will
undergo thermonuclear fusion in carbon burning. Carbon burning in an electron degenerate gas can start
explosively, just as helium burning does. Thus the core
can have a *carbon flash,* which removes the degeneracy and allows the carbon burning to proceed in thermal equilibrium. Studies of model stars suggest, however, that carbon burning in some stars may start so
explosively as to blow the star completely apart, as
shown in Table 15.3, where we summarize evolution
of all masses of stars. We are not yet absolutely certain
that this explosion actually occurs and is one type of
supernova outburst, but it seems probable.

Thermonuclear burning is a viable energy source
for a shorter time as synthesis proceeds toward heavier
elements. Hydrogen burning may last a few hundred
million years; helium burning only up to 20 percent as
long, or tens of millions of years, and carbon burning
may last no longer than a few thousand years. In part,
the reason that carbon burning is so short is that the
energy loss due to escaping neutrinos can actually exceed the energy carried away by photons.

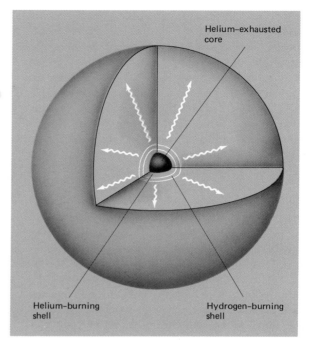

FIGURE 15.11
Red giant phase for high-mass star of
about 9 $M_\odot$ or greater. Inert core,
helium and hydrogen burning shells.

For stars with main sequence masses greater than
about 9 $M_\odot$, carbon burning begins in a gaseous core
that still obeys the perfect gas law. There is no carbon
flash. The more massive stars will become hot enough
for their cores to initiate *oxygen, neon, and silicon
burning* at 1 to 2 billion degrees (see Table 15.1). Nuclear burning for these massive stars ends with the nuclear reactions that produce the iron nucleus. Because
the iron nucleus is the nucleus most resistant to any
type of structural alteration, thermonuclear burning
beyond iron does not release energy but takes up thermal energy from the surroundings. From here on the
star cannot go through the alternate contracting, heating, and nuclear-burning sequence that has sustained
the star against gravity through most of its existence.

The destiny of the massive star (greater than about
15 $M_\odot$) is not at all clear (see Table 15.3), but several
important predictions can be made from mathematical models. For a massive Population I star, most
of the time in which it burns helium in the core is
spent as a blue supergiant near the main sequence.
A star of 25 $M_\odot$ or greater also burns carbon in its
core as a blue supergiant. During all evolutionary

phases beyond this, most massive stars are probably red supergiants. The star has evolved through nuclear burning into a very small, extremely dense core with shells like those of an onion, which have different chemical compositions because of the nuclear burning. Various thermonuclear reactions can take place simultaneously: silicon burning at the center and neon, oxygen, carbon, helium, and hydrogen burning in successive shells outward (see Figure 15.12). Surrounding the core is a highly distended hydrogen-rich envelope. The star is very large, very bright, and quite red. The 25 $M_\odot$ star has spent about 5 to 10 million years in hydrogen burning, 0.5 to 1.0 million years in helium burning, 500 to 1000 years in carbon burning, 6 to 12 months in oxygen burning, and a mere day or so in silicon burning. Again, the energy carried away by neutrinos shortens the burning phases from carbon on. There may well be periods during which energy is transported by convection, which will mix the newly processed elements over much of the radius. If the star also loses mass significantly in this phase, some of the heavy elements synthesized in the core may be returned to the interstellar medium.

A striking example of mixing and element synthesizing was found about a quarter of a century ago with the discovery of spectral lines of technetium, element number 43 in the periodic table, in the spectra of some M, S, and N types of red giants. All technetium's isotopes are unstable, with the longest half-life of 212,000 years belonging to $^{99}Tc_{43}$. In order for us to see it in the spectrum of a star, it must be brought to the atmospheric layers from the interior before it decays. This suggests that the time scale for mixing in the star is less than a few hundred thousand years.

15.5 Supernovae, Pulsars, and Neutron Stars

SUPERNOVA OUTBURSTS

In massive stars the final phases of nuclear burning of oxygen and silicon can be violently explosive, model star studies tell us, if they occur in degenerate or nearly degenerate matter. The star may undergo a bomblike detonation or, after exhausting all its nuclear fuels, suffer a final catastrophic collapse and explode. Supernova outbursts (see Section 14.3) probably are these violent events, predicted by our computer studies of model stars. Although it may seem at first that whatever remains of a star after

TABLE 15.3
Summary of Stellar Evolution by Mass of the Star

Range of Mass While Main Sequence Star	Thermonuclear-Burning Sequence	Evolution After Red Giant Stage	Final State of Star
Less than 0.1 $M_\odot$	None	None	Black dwarf
0.1 $M_\odot$ to 0.5 $M_\odot$	Hydrogen	Red giant	White dwarf
0.5 $M_\odot$ to 1.4 $M_\odot$	Hydrogen Helium	Red giant Horizontal branch Small mass loss or planetary nebula (?)	White dwarf
1.4 $M_\odot$ to 8 $M_\odot$	Hydrogen Helium Carbon	Red giant Horizontal branch (?) Large mass loss Pulsation Explosive supernova (?)	White dwarf or neutron star (?)
9 $M_\odot$ to 60 $M_\odot$	Hydrogen Helium Carbon Oxygen Neon Silicon (?)	Red giant Large mass loss Explosive or implosive supernova (?)	Neutron star and/or black hole

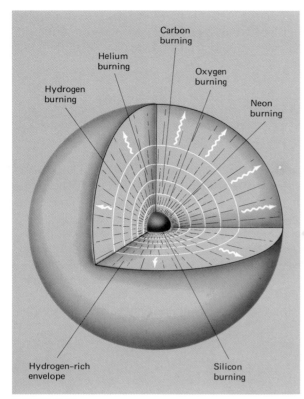

FIGURE 15.12
Successive burning shells about a silicon-burning core (not to scale) in Population I star with a mass greater than about 15 $M_\odot$.

Labels on figure: Hydrogen burning, Helium burning, Carbon burning, Oxygen burning, Neon burning, Silicon burning, Hydrogen-rich envelope

a supernova outburst might be less than 1.4 $M_\odot$ and a white dwarf, this is not necessarily what happens. Even as far back as the 1930s a few astronomers felt that the remaining star was likely to be too massive and much too dense to be a white dwarf. We now believe that these massive stars end in one or the other of two of the strangest objects in all of the cosmos: the remaining star will be either a neutron star or a black hole.

What causes supernova outbursts? Astronomers do not agree on their exact cause. There is no case in which a star that underwent a supernova outburst was identified and studied before the outburst. It appears that most precursors of Type I supernovae may be low-mass stars in binary systems. Perhaps one star is a white dwarf that accretes hydrogen-rich material from an expanded red giant companion. As material flows onto its surface, the white dwarf eventually grows so unstable that when its mass exceeds 1.4 $M_\odot$, it detonates as a supernova.

A Type II supernova, however, is believed to start with the gravitational collapse of a massive star having a degenerate iron core. For stars greater than about 8 $M_\odot$, when core thermonuclear reactions stop with the last fuel, the star is in a catastrophic state. If the core is greater than 1.4 $M_\odot$, a free-fall collapse drives the core densities well beyond that of the white dwarf, toward nuclear densities (10^{14} grams per cubic centimeter). Protons and electrons are forced to combine to form neutrons. As the core heats up to a very high temperature, a large number of neutrinos are released in a variety of nuclear processes. Under the conditions in and about the collapsing core, the neutrinos do not immediately escape from the star but diffuse their way out. This enormous burst of escaping neutrinos hastens the collapse of the core, while also exerting possibly enough pressure to blow off the outer layers of the star. Nuclear reactions in the material surrounding the core also contribute to the excess pressure that blows the envelope off the star. Which of these mechanisms or what combination of them is responsible for blowing off the outer layers is not known. The star explodes, and its gravitational potential energy is released as radiant energy and as kinetic energy of ejected matter, leaving a rapidly spinning compressed remnant of the former star. The collapse will halt at approximately the density of the atomic nucleus, if the collapsing mass is less than about 3 $M_\odot$ with a radius of the order of 1 to 10 kilometers. This object is a *neutron star*. If the collapsing mass is much greater, the end is presumably a *black hole* with a radius of a few kilometers.

> A leaf of grass is no less than the journey work of the stars.
>
> Walt Whitman

As the outer layers of the star are blown off in the catastrophic explosion, a great flux of high-energy radiation and high-energy particles is released along with the avalanche of neutrinos. This outburst may be so violent that it liberates as much energy as the sun expends in a hundred million years. As the rapidly expanding gas plows into the interstellar medium, a shock wave develops and a hot expanding shell forms, radiating strongly throughout the electromagnetic spectrum. The broad emission lines in the supernova's visible spectrum are evidence for this outflow of gases at speeds of several thousand kilometers per second. Collision with interstellar matter eventually slows the shell, which cools as it radiates away its thermal energy. By the time a large amount of interstellar matter has been swept up, the shell has slowed to the speed of the interstellar matter and we can no longer recognize the shell optically. But fragments may persist long afterward and can be identified by radio astronomers from their radio wavelength emission.

Calculations indicate that large numbers of free neutrons would be very likely to exist during a supernova outburst. When a star blows up, the neutrons can rapidly react with the different nuclei to create the heavy elements beyond iron in the periodic table. The expanding cloud disperses into the interstellar medium and increases its heavy-element composition.

Thus successive generations of stars formed from interstellar clouds, enriched with supernova debris, will have more of the heavy elements than the older stars do.

PULSARS

In 1967 a strange object in the constellation Vulpecula was discovered emitting pulses of radio radiation with a precise period of 1.337 seconds between the bursts. The object was later named a *pulsar,* and it is strongly believed to be a neutron star. Within a few weeks after the discovery, three more pulsars were found. Over three hundred have since been discovered. Their periods range from 0.033 second to 3.75 seconds with the majority having periods from 0.5 to 1.0 seconds. Of all these, only two have been identified with an optical source: the Crab Nebula pulsar and Vela supernova remnant pulsar.

The Crab Nebula was one of the first discrete radio

KARL G. JANSKY (1905–1950)

After graduating from the University of Wisconsin, Jansky joined the Bell Telephone Laboratories in New Jersey in 1928. His work dealt with the problems of short-wave radio telephony. In 1931 he was assigned the task of tracking

down the crackling static noises that plagued overseas telephone reception. At the Holmdel station he constructed a large directional antenna system tuned to 14.5 meters wavelength. It was mounted on a wooden frame that could be rotated by four wheels on a circular track 50 feet in diameter. To pick up the signals arriving at the antenna he employed a very sensitive receiver coupled with an automatic recorder.

Jansky recorded two well-known kinds of atmospheric static: crashes from local thunderstorms and noise from distant thunderstorms reflected from the ionosphere. From his records he later singled out a weak third kind of static that could hardly be distinguished from the internal receiver noise. In headphones it sounded like a steady hissing noise. At first Jansky thought that the interference came from the sun but after a year of careful measurement he concluded that the radio waves came from a specific region on the sky every 23 hours and 56 minutes. Suspecting that the radiation was coming from

an astronomical source, he attempted to trace its origin. He knew from his study of astronomy that the period of the earth's rotation relative to the stars was four minutes less than the 24-hour period relative to the sun. This was the clue that the radio noise originated in space beyond the solar system. He found that its direction coincided with the constellation of Sagittarius toward the center of the Milky Way.

At the age of 26, Jansky had made a historic discovery—that celestial bodies could emit radio waves as well as light waves. His results, published first in the *Proceedings of the Institute of Radio Engineers* and then in *Popular Astronomy* in 1933, received little attention, however. Not until the end of World War II was his achievement widely recognized.

Jansky's serendipitous discovery gave birth to a new branch of astronomy—radio astronomy. In Jansky's honor, radio astronomers named the unit of radio flux the *Jansky* (10^{-26} watts per square meter per Hz).

sources discovered. In 1968 radio astronomers found that a source within the nebula was emitting very short bursts or pulses of radio radiation 30 times a second. The source was soon pinpointed: it was one of the two central stars in the nebula (see Figure 14.16). The star was observed to flash optically at the same frequency as the radio pulses—30 times a second (see Figures 15.13 and 15.14). Wispy structures moving back and forth near the center of the nebula have been observed traveling up to several tenths the speed of light. From their association with the X-ray–emitting regions, it seems they may be carriers transmitting energy from the pulsar to the nebula.

The Crab Nebula was also the first discrete X-ray source identified; this happened during a 1964 rocket flight directed by Naval Research Laboratory scientists. In 1969 they found pulses of X-ray radiation whose rate matched the bursts of optical and radio radiation. Finally, in 1972, astronomers discovered that the Crab Nebula was also emitting gamma radiation. The rate at which the Crab Nebula is emitting radiant energy in all wavelengths is on the order of 10^{38} ergs per second, which is a hundred thousand times greater than the 10^{33} ergs per second emitted by the sun. Thus the

Crab Nebula is comparable in luminosity to the most luminous supergiant, although the way in which the emitted energy is distributed in wavelength is entirely different.

The Vela pulsar (see Table 14.3) also emits electromagnetic radiation in pulses spanning the wavelength range from gamma rays to radio waves. The pulse rate is 11 times per second. The Vela pulsar, as well as the Crab and other pulsars, is not at the geometrical center of the surrounding nebula. If the supernova outburst was not symmetrical, the neutron star may have been forced out in one direction and the expanding nebula in the opposite direction, an example of Newton's third law. If enough time has elapsed since the supernova outburst, then the neutron star may no longer even lie inside the nebula, and this might explain why no pulsar has been found for most supernova remnants.

The amount of energy emitted in the radio wavelengths by one pulsar, whose period is 0.2530646 second, is plotted in Figure 15.15. Although the interval between pulses is very constant (better than one part in a billion), the pulse amplitude or amount of energy, the shape of the pulse, and its width in time vary

FIGURE 15.13
Optical and radio light curves of the Crab Nebula. The radio curve shows a split down the middle of the main pulse. Its ragged appearance results mainly from scintillation effects in the interstellar medium.

FIGURE 15.14
Crab pulsar at maximum and minimum light. This body flashes on and off about thirty times per second. Rapid atmospheric scintillation is responsible for the difference in the appearance of nonvarying stellar images.

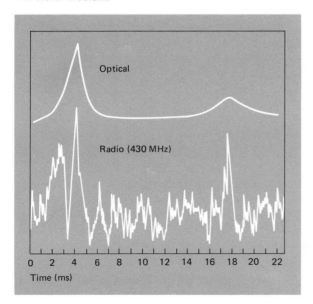

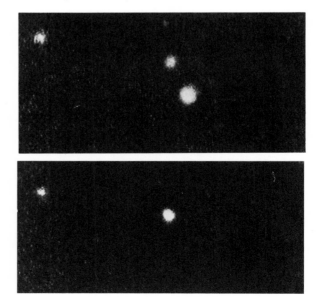

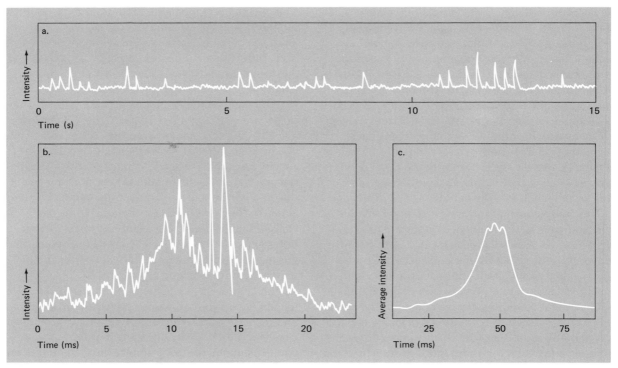

FIGURE 15.15
Pulsar recordings. a. The pulses occur at precise intervals of 0.2530646 seconds but their amplitudes vary. b. Fine structural details of a single pulse recorded with the 305-meter telescope at Arecibo, Puerto Rico, at a frequency of 195 megahertz. c. Pulse envelope, averaged over many pulses.

considerably from one pulse to another. The pulses themselves last from several thousandths of a second to about a tenth of a second. The amount of energy in the pulses usually decreases toward shorter wavelengths, an indication that the source of radiation is nonthermal. The amount of radio energy emitted during one burst is from 1 million times to 10 billion times greater per square centimeter each second than the energy the sun emits from a similar area per second.

The first pulsar in a binary system was discovered in 1975. The second star in the binary system unfortunately is not visible, although the guess is that it also is a compact star. During the pulsar's one-third–day orbit, its burst period of 0.059 second varies by 78 microseconds because of the Doppler effect as it alternately approaches and recedes from the earth due to its orbital motion. Because the variable velocity of the pulsar can be measured accurately, we therefore have a velocity curve for it from which we can calculate the orbital parameters of the binary system, like those of a spectroscopic binary, and

from these we estimate the pulsar's mass. The pulsar is a neutron star with a mass slightly larger than that of the sun, with the companion having a mass about one and a half times that of the sun.

When a source of radiation is very small, scintillation (twinkling) becomes noticeable. The trace of the radio-frequency radiation in Figure 15.13 clearly reveals such scintillation as the radiation from the pulsar passes through interstellar clouds and the interplanetary plasma. (This scintillation is like the twinkling of starlight passing through the earth's atmosphere.) The sharp peaks in the pulsar's radiation indicate that the energy is coming from a very small region. For a pulse duration of 100 microseconds (0.0001 second), the diameter is about 30 kilometers, which is the distance that light travels in that time.

Electrons in interstellar space affect the velocities of radio waves: the longer the wavelength, the slower is the wave's velocity. Radio waves coming from a sharp pulse are therefore dispersed along the way into a train of waves. The shorter wavelengths arrive on the earth before the longer ones. The difference in arrival time between the extreme ends of the radio spectrum may be as much as 60 seconds. From this, we can estimate the distance to the pulsar (the path length) from the delay time. Calculations of this kind tell us that most pulsars are Galactic objects relatively close to us, at an average distance of about 3000 light years. Most of them are in or near the plane of the Galaxy, as shown in Figure 15.16.

THEORETICAL MODEL OF A NEUTRON STAR

Astronomers now seem agreed that pulsars are rapidly rotating neutron stars, the possible remnants of supernovae. It is hypothesized that the stellar remnant, rapidly contracting after a supernova outburst, is a rapidly spinning, highly magnetic, superdense *neutron star*. Only a body 10 kilometers to 30 kilometers in diameter with a density approaching nuclear densities (approximately 10^{14} grams per cubic centimeter) could survive the disruptive force of such rapid rotation. The magnetic field in neutron stars may run as high as 10^{12} gauss. By comparison, the earth's magnetic field intensity is half a gauss and the largest magnetic field produced in a laboratory is about 300,000 gauss.

In the model the neutron star's high rotational speed provides the reservoir of energy needed to maintain the continuous flow of charged particles streaming from the star's magnetic poles. The radial magnetic field lines in the polar region provide an escape route for the particles. The flow of charged particles emit coherent electromagnetic waves in a highly directed cone of radiation that spins with the neutron star. When this searchlight beam sweeps across our line of sight, we see a pulse of radiation from the pulsar every fraction of a second or so. The emission of intense electromagnetic radiation is typically confined to 1 to 5 percent of the rotation period. A diagram of this proposed model is shown in Figure 15.17.

The model predicts that the magnetic field will slow the pulse rate by dissipating the neutron star's rotational energy. This prediction has been confirmed by observations for over ninety pulsars that show a decrease in the pulse rate between 14×10^{-5} and 36×10^{-9} second each day. Pulsars with the shortest periods are thus apparently the youngest. It also follows that the Crab pulsar, with the shortest period yet discovered (0.033 second), must be the youngest pulsar with an estimated age of a thousand years, or the age of the Crab Nebula. It is slowing by about 13×10^{-6} second each year, corresponding to an energy loss of about 10^{38} ergs per second, or about a hundred thousand times more than the sun loses by radiation each second. Ap-

FIGURE 15.16
Galactic Distribution of pulsars. The coordinate system is such that the equator is the plane of the Milky Way and the point 0 is the direction toward the Galactic center. Note that pulsars are located fairly close to the Galactic plane. Those at high Galactic latitudes are primarily nearby and lie above or below the sun in the plane of the Galaxy.

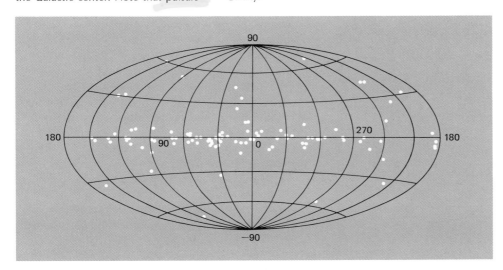

proximately the same amount of energy in synchrotron radiation comes from the nebula and from the kinetic energy of the nebula's expansion. The average age of the remaining pulsars is about 2 million years; the oldest pulsars are about 10 million years old. Our failure to detect optical pulsations in all but the Crab and Vela pulsars may mean that the light flashes are a transient phenomenon, occurring only during a pulsar's early history.

Astronomers picture the interior of a neutron star as a degenerate gas of neutrons mixed with a smattering of protons and other subatomic particles. The pressure of the degenerate neutrons provides the force balancing gravity. This superdense material (approximately 10^{14} to 10^{16} grams per cubic centimeter) is packed into a diameter of several kilometers at perhaps 10 billion degrees Kelvin. Over-

lying the fluidlike interior is a rigid crust of heavy nuclei several kilometers thick. The outer skin is a lattice of iron atoms only a few centimeters thick.

15.6 Black Holes

THEORY OF BLACK HOLES

In Chapter 13 we pointed out that throughout a star's existence there is a continuing struggle between gravity and various countering forces to prevent the contraction of the star. Different thermonuclear reactions maintain the kinetic energy of random thermal motion of the atomic constituents. And it is this random thermal motion that provides the kinetic pressure to balance the weight of overlying layers during much of the star's life. However, once the nuclear fuels have been used up, kinetic pressure can no longer be the mainstay against the forces of gravity. For stars whose masses are less than 1.4 $M_\odot$, contraction leads, as we have seen, to a degenerate electron gas, and the degenerate electrons can provide indefinitely a pressure that balances gravity. These stars are the white dwarfs, and they are stable and will suffer no further contraction. This is the most common form of stellar demise.

For high-mass stars, if mass loss during the stars' lives does not reduce them to less than 1.4 $M_\odot$ by the time they exhaust their nuclear fuels, then they cannot become stable white dwarfs. Such stars may go through a supernova outburst. And if the core remaining after the outburst is less than 2 $M_\odot$ to 3 $M_\odot$, gravitational collapse cannot be halted by degenerate electrons, but it can be halted by the nuclear forces in the degenerate neutron gas that forms. The object is now a neutron star. The pressure provided by the degenerate neutrons is sufficient to maintain the balance against gravity, and the neutron star is a stable object at a radius considerably smaller than that of the white dwarf. This is the second form for a dying star.

However, what happens if the remaining core mass after the supernova outburst is larger than 3 $M_\odot$? Is there a third form for the dying star? The pressure of the degenerate neutrons cannot halt the collapse and as far as we know there are no forces that can halt it.

FIGURE 15.17
Model of a pulsar. The rotating neutron star is typically about 1 $M_\odot$, with a radius of about 10 kilometers. The charged particles are accelerated by the magnetic field of the neutron star (up to about 10^{12} gauss in intensity) and flow out along the magnetic axis, producing radio radiation. The magnetic axis of the rapidly rotating neutron star must be properly oriented for us to catch the flash of radiation when one of the rotating beams sweeps past our line of sight. Otherwise we will not detect any pulses.

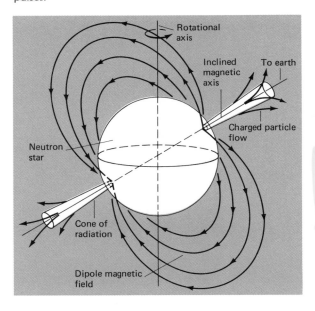

There is thus nothing to balance gravity and the collapse continues. As the star becomes more compact and its size continues to decrease, the intensity of its gravitational field increases dramatically. In most situations in our normal experience, the gravitational force is the weakest of all forces in nature. But in this case, gravity overpowers everything, and this contracting, superdense mass of matter causes the space-time geometry in its vicinity to warp about it. Eventually this warping becomes so great that space-time folds in over itself as the star passes through the *event horizon* (see Figure 15.18), the point of no return, and the star disappears into a *black hole*. The enormous forces of gravity have so modified the local space-time geometry of the black hole that there are no paths by which photons or particles of matter may escape the black hole. There are only paths into it, none away from it, and thus the name black hole.

The distance from the black hole's center to the boundary of the event horizon marks the region of curved space within which the collapsed body becomes invisible to an external observer, for neither light nor matter can escape from the powerful gravitational field of the black hole, and all communication is lost with the outside world. This critical distance is the *Schwarzschild radius*, named after the German astronomer who first explored the space geometry around a point mass, using Einstein's theory of relativity. The numerical value of the Schwarzschild radius is given in the equation on page 334 along with some representative values. The calculated Schwarzschild radius for the sun is about 3 kilometers, though the sun is not the kind of star that we expect would experience such a collapse.

A distant observer will see the first stages of collapse as the contracting matter approaches the Schwarzschild radius, followed by a slowing down of the action. The observer cannot witness the final stage of collapse because time slows down in a gravitational field. The greater the intensity of the gravitational field, the more time slows down. In other words, time stands still at the event horizon for the outside observer. An observer inside the Schwarzschild radius would see matter crushed to a stupendously high density in a relatively short period of time. This means that when we observe light from a collapsing star we see two effects. The first is that the star grows dimmer because of the warping of space-time over itself such that it is

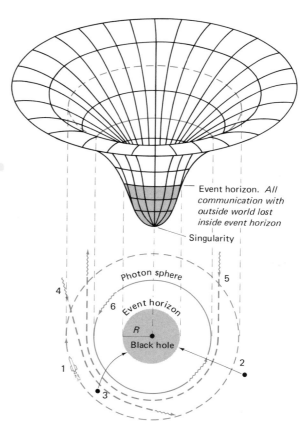

Event horizon. *All communication with outside world lost inside event horizon*

Singularity

FIGURE 15.18
Space geometry of a black hole and possible trajectories. *R* is the Schwarzschild radius. The increased warping of space as we come closer to the black hole results in several possible trajectories for objects in its vicinity. The trajectory will depend on three variables—the object's speed, its direction, and its distance.

1. An object (spaceship) moving at one-half the speed of light in the last stable orbit.

2. An object approaching head-on is sucked into the black hole.

3. An object on nonradial trajectory falls into the black hole on a curved path.

4. A photon passing outside the photon sphere is deflected at a large angle.

5. A photon outside the photon sphere returns in a direction opposite to its original path.

6. At the boundary of the photon sphere a photon moves continuously in a circular orbit.

harder for photons to find exit paths to the outside. Second, the atom can be thought of as being a small clock, and as photons are emitted, their frequencies decrease as the collapse proceeds. That is, the entire spectrum of emitted radiation is shifted further and further to the red. As we watch the collapse, light from the star grows fainter and redder.

Let us examine the space geometry near a black hole (see Figure 15.18). The Schwarzschild solution was built on the simplest idea: a spherical, nonrotating, gravitationally collapsed body. The nearer we come to the collapsed body, the more space is warped. An un- wary spaceship pilot caught in the gravitational clutches of the curled-up field might be sucked into the black hole if the ship had too little kinetic energy to escape (trajectory 3 in Figure 15.18). At a critical distance corresponding to a definite fraction of the speed of light, a spaceship or any other object might just manage to stay in a stable orbit around the black hole (trajectory 1). With proper acceleration in the right direction, it could burst free and leave the vicinity for good.

Closer in is a critical distance at which a photon would circle continuously around the black hole

SCHWARZSCHILD RADII FOR BLACK HOLES

The *Schwarzschild radius* is the radius at which the space-time geometry about a nonrotating collapsing mass folds over itself so that contact with the outside world is lost; it is the radius of the *event horizon*. The numerical value of the Schwarzschild radius R can be found from the relation

$$R = \frac{2\,GM}{c^2} \approx (1.5 \times 10^{-28})M$$

where G is the gravitational constant (6.67×10^{-8} cubic centimeters per gram second squared), c is the velocity of light ($c^2 = 8.99 \times 10^{20}$ square centimeters per second squared), and M is the mass of the collapsing body in grams. If we express the mass in units of the sun's mass, then R in kilometers is given by

$$R(\text{km}) \approx 3.0M$$

The density of the body when it has collapsed to the size of the Schwarzschild radius is given by

$$\rho = \frac{M}{4/3(\pi R^3)} = \frac{3c^6}{32\pi G^3 M^2} \approx \frac{7.3 \times 10^{82}}{M^2}$$

or it is inversely proportional to the square of the mass. Table 15.4 lists the Schwarzschild radius for various objects, even though there is no evidence that any of them (except perhaps the last) will ever form a black hole.

TABLE 15.4
Some Schwarzschild Radii

Object	Mass	Schwarzschild Radius	Density (g/cm³)
Hydrogen atom	2×10^{-24} g	3×10^{-44} Å	10^{130}
Human being	7×10^4 g	1×10^{-15} Å	10^{73}
Earth	6×10^{27} g	0.9 cm	2×10^{27}
Sun	2×10^{33} g	3 km	2×10^{16}
Galaxy	$10^{11}\ M_\odot$	0.03 ly	2×10^{-6}
Cluster of galaxies	$10^{14}\ M_\odot$	60 ly	5×10^{-13}
Closed universe	$10^{22}\ M_\odot$	16×10^9 ly	6×10^{-30}

(trajectory 6). This is the boundary of the *photon sphere.* A little farther out a photon would curve in the warped space surrounding the black hole (trajectory 4) or could even leave in the direction opposite that from which it came (trajectory 5). Between the photon sphere boundary and the event horizon, photons would move in unstable orbits that would spiral them into the black hole. The photon's spiral path is still the shortest distance between points in the warped geometry of space. Once past the event horizon, the photons are trapped inside the black hole and cannot escape. At the very center is a *singularity,* a point where the mass of the collapsed body is concentrated into zero volume and infinite density! No one knows if nature abhors a singularity, but theoreticians feel uncomfortable whenever a mathematical formulation of a physical phenomenon leads to a singularity. Different objects falling into a black hole lose all their identity except mass, angular momentum, and electric charge. This meager set of descriptive properties for the black hole is in stark contrast with the extensive set of descriptive properties for the object when it was an active star, such as temperature, pressure, density, energy-generation rates, and so on, as they vary with the radial distance from the center.

Even though an object may become a black hole, it is not isolated gravitationally from the rest of the universe. One place to look for black holes is in binary systems, where we might detect a black hole (which should have a mass greater than 2 $M_\odot$ or 3 $M_\odot$) from the orbital motion of the binary system's visible component. Another possibility is that gas streaming from the visible component of the binary onto its black hole companion might emit detectable X-ray radiation. The X-ray source Cygnus X-1 may be a black hole companion of a nearby blue supergiant star physically associated with it, a possibility we will discuss in the next section.

ROTATING BLACK HOLES

Since all bodies in the universe rotate, it seems logical to investigate the more complex possibilities of rotating black holes. A solution based on the general theory of relativity has been found for the rotating black hole. This solution reveals that two kinds of surfaces exist: a one-way event horizon as in the nonrotating black hole and an outer surface called the *static limit* (the surface of infinite red shift). Both surfaces merge at the poles of rotation, as shown in

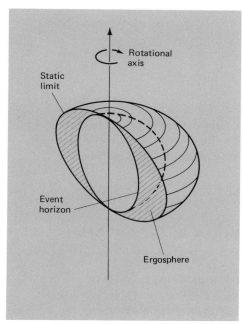

FIGURE 15.19
Cross-sectional view of rotating black hole. Between the static limit or surface of infinite redshift and the event horizon lies the atmosphere.

Figure 15.19. The space-time region between the two surfaces is called the *ergosphere;* it vanishes if there is no rotation. Within the ergosphere no object can remain at rest. Should an astronaut venture into the ergosphere, he can exit from it so long as he does not cross the event horizon.

Two interesting phenomena appear in the case of rotating black holes. Due to the interaction between the gravitational field and the rotation, the space-time environment is dragged around the spinning black hole. As a consequence an observer stationed near a black hole in a nonrotating (inertial) frame of reference would notice distant stars revolving around him. The other phenomenon is that energy can be extracted from objects spiraling into rotating black holes. Since more energy is removed than is put in, the rotational energy decreases—that is, the spin slows down. When rotation stops, all energy possible has been extracted from the black hole. In theory one could use such a hole to efficiently extract copious amounts of energy. The singularity in a nonrotating black hole is not visible to an outside observer; it is said to be clothed by the event horizon. It is as if nature has imposed a kind of cosmic censorship. A black hole that rotates very rapidly and yields energy has no event horizon; thus the singularity is exposed as a naked singularity and cosmic censorship is lifted.

15.7
Evolution of Binary Systems

MASS TRANSFER IN BINARY SYSTEMS

In Chapter 14 we saw that many stars lose matter to their surroundings on either a slow time scale, such as in stellar winds, or a rapid time scale, such as a supernova. Since at least half of all stars appear to be members of binary or multiple star systems, it is important to inquire whether the various mass loss processes can actually result in the exchange of mass between the components of a binary system. If this does occur, then how does that process affect the evolution of the stars involved? Finally, what signs would we look for to verify that mass transfer is indeed occurring?

In a binary system gravitational theory predicts that there exists for each star a critical limiting surface known as the Roche limit. The significance of this surface is that if either star expands outside the Roche limit, then its outer surface layers, under the gravitational attraction of the companion, can flow away onto the companion star. If the two stars in the binary system are relatively close, on the order of sev-eral tens of their radii apart, the Roche surface is close enough to each star (see Figure 15.20) that the course of normal evolution for either star can cause it to overflow its Roche lobe. Computations for close binary systems show that significant amounts of mass can be transferred from one star to the other several times during the evolution of the system. This mass transfer process can in the extreme case completely change the normal evolution of both stars. Let us briefly outline the process of evolution in a close binary before discussing some examples that are of current interest.

For a binary system in which the two components have different masses, the larger-mass star will have the shorter main sequence evolution time. After it exhausts its hydrogen in the core, the envelope of the star will expand as the star restructures itself to become a red giant. In that process the star may fill its Roche lobe and begin the process of transferring its outer layers to its companion. Matter will probably form a gaseous disk about the companion in the plane of the orbit before accreting onto the companion's surface. Theoretical studies of the evolution show that the mass transfer may include some or all of the envelope of the massive star and will occur in a relatively short period of time. What may remain of the original primary is the evolved core and possibly some of the original envelope. The original primary can now appear as an overluminous main sequence star for its new mass, while the original secondary is just a massive main sequence star with possibly some atmo-

FIGURE 15.20
Roche limit surfaces for three possible close binary stars. In the system to the left, neither star fills its lobe. After the larger star expands, it may fill its lobe as shown in the center figure and begin to transfer mass to the companion. In the system at the right, both stars fill their lobes and may exchange mass as shown.

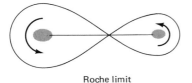

Roche limit

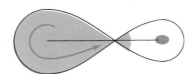

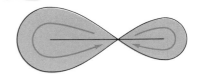

spheric chemical anomalies. If the mass of the core of the original primary is less than 1.4 $M_\odot$, the object could also be a white dwarf, depending upon how complete the transfer process was.

After the mass transfer the original secondary can have become more massive than the original primary, and it will proceed to finish its main sequence evolution in a relatively short time. When it exhausts its core hydrogen, the now-massive secondary will expand to become a red giant filling its Roche lobe. This can lead to a reverse transfer of matter back to the original primary, which may be an extremely compact object such as a white dwarf. Matter coming from the normal star is heated as it falls through the intense gravitational field of the compact companion. This process is extremely efficient in producing X rays. In fact, the inward-falling matter can produce more energy in the form of X rays than it could in thermonuclear processes. In some cases it is possible to envision several exchanges of matter between the two components of the binary and in each exchange the course of evolution being greatly altered. Another possibility is that during the exchange process matter is completely lost from the system, so that two white dwarfs are left at the end.

There are obviously a number of different endpoints of the evolutionary process that can occur in binary systems involving mass transfer. The mass transfer may, for example, precipitate the gravitational collapse of one or even both components, leading to a supernova outburst. As we have discussed in the preceding pages, the end result of the supernova outburst may be the formation of either a neutron star or a black hole.

X-RAY BINARIES

Among the most fascinating stellar objects are the X-ray binary stars, whose orbital periods range from a fraction of a day to many days. The usual X-ray binary combines a visible hot blue star from which matter is streaming to the neighborhood of a nearby, gravitationally collapsed body such as a white dwarf, neutron star, or even a black hole. Let us examine the remarkable behavior of such systems (see Tables 15.5 and 16.4).

The model of a typical eclipsing and pulsating X-ray binary, SMC X-1, which is 190,000 light years away, is shown in Figure 15.21. It was discovered in the Small Magellanic Cloud by the *Uhuru* satellite in 1971. The rapidly spinning neutron star emits X rays from a hot spot on its surface. The pulse rate is 0.716 second, which corresponds to the neutron star's rotation period. When the neutron star moves behind its blue supergiant companion, the X-ray emission disappears for about fourteen hours. After the eclipse, as the neutron star approaches the earth, the pulse rate is slightly higher due to the Doppler effect. In the other half of the orbit the pulse rate decreases slightly prior to eclipse. Shown in the figure is matter captured by the neutron star from the strong stellar wind of the supergiant coming from the tidal bulge produced by the strong gravitational pull of the neutron star. As the inward-falling material collides with the neutron star, it causes the hot spot on the surface from which X rays are emitted. The total amount of energy radiated is 3×10^{38} ergs per second, the highest of any known neutron star.

Another example of an eclipsing X-ray binary is Hercules X-1, in which the X-ray source is eclipsed by

TABLE 15.5
Some X-ray Binary Systems

Binary System	Mass of X-ray Source (sun = 1)	Mass of Companion (sun = 1)	Period (d)	Spectral Type of Companion
SMC X-1	0.5–1.8	13–22	3.9	B0 Ib
Vela XR-1	1.0–3.4	19–32	8.9	B0.5 Ib
Cen X-3	0.7–4.4	14–20	2.1	B0 Ib–III
Her X-1	0.4–2.2	1.4–2.8	1.7	late A
3U 1700-37	≤ 0.6	≥ 10	3.4	O7f
Cyg X-1	9–15	≥ 30	5.6	O9 Ib

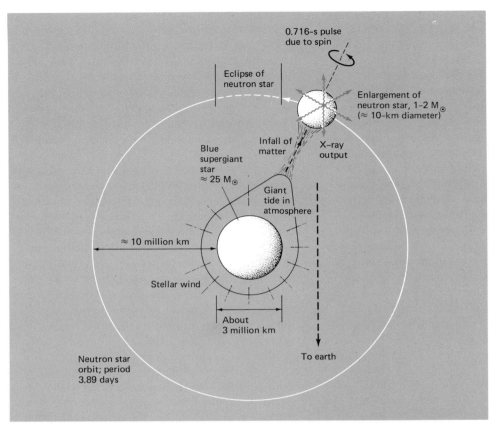

FIGURE 15.21

SMC X-1 eclipsing X-ray source. A rapidly rotating neutron star orbits a massive blue supergiant star. The neutron star has a hotspot on its surface from which X rays are emitted. When the hotspot faces the earth, we can detect X-ray emission. Periodically, the neutron star is eclipsed by its companion as shown and emission from the neutron star ceases.

a blue giant companion, HZ Hercules, for about 6 hours every 1.7 days. The binary system is about 12,000 light years from us. Analyzing the orbital data gives upper mass limits of 2.8 $M_\odot$ for the visible component and 2.2 $M_\odot$ for the X-ray component. The low mass of the X-ray source suggests that it is a neutron star, a view reinforced by the short X-ray and visible light pulsations of 1.24 seconds, presumably due to the neutron star's rapid rotation. As the hot gas flows from the giant, it rains down upon the neutron star and emits X rays. This radiation, according to one interpretation, heats the side of the giant facing the companion and is reflected as blue visible light in unison with the X-ray pulsations. When the X-ray source is behind the giant companion, the light from the giant star is fainter and less bluish. The picture is complicated by X-ray

emission that goes through an on-off cycle of about 35 days. It turns on for about 12 days and turns off for about 23 days. The system also undergoes ultraviolet flickering that lasts from seconds to minutes. No satisfactory theory explains this strange behavior other than to say that some kind of abnormal gas flow exists in the system.

Still other examples abound, each with a unique behavior. Some X-ray binaries are more or less stable in their behavior, while others suddenly turn on briefly, then turn off for weeks, months, or even years, resembling recurrent novae in their actions. The latter are highly variable in their output, with pulsing periods ranging from a fraction of a second to many seconds when turned on. It is suspected that there may be a sporadic transfer of gas from one binary component to a compact degenerate companion giving rise to an outburst between long-duration quiet periods.

Soft X-ray emission, punctuated by occasional flare outbursts, has been observed coming from certain yellow subgiant short-period binaries (whose periods are 2.8 to 6.4 days). The prototype of these systems is RS Canum Venaticorum. It is theorized that the emission originates in a hot corona, at a temperature of 10 million degrees, whose energy is derived from flares arising from massive starspots (similar to sunspots) on the surface of the cooler companion.

Weak, soft X-ray emission has been detected in dwarf novae, which are also binary systems, simultaneously with optical and ultraviolet outbursts. Two examples are SS Cygni and U Geminorum, well known to variable star observers. These objects appear to consist of an overextended late-type star and a close white dwarf or degenerate companion that undergoes frequent outbursts at intervals of several weeks. Material from the late-type companion is spewed onto the surface of the compact companion (see Figure 14.12) by way of an accretion disk. The gravitational potential

FIGURE 15.22
Proposed model for Cygnus X-1. A black hole in a binary system with a very massive blue supergiant star pulls matter escaping the supergiant into an accretion disk as shown in the lower figure. The X rays probably come from as close as several hundred kilometers from the black hole.

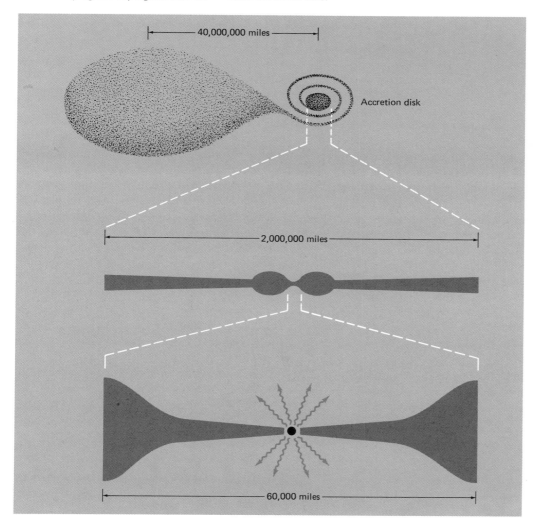

energy of the inward-falling gas is converted into thermal energy. X rays result when the gas reaches temperatures in the range of 10 to 100 million degrees.

Finally there are the suspected black hole candidates: Cygnus X-1 (period of 5.6 days) and Circinus X-1 (period of 16 days). Both are eclipsing binaries containing a visible blue supergiant component and an X-ray source associated with a body of many solar masses that is high enough to be considered a black hole rather than a neutron star. These objects exhibit variable X-ray emission believed to originate in a compressed disk surrounding the black hole. Consider Cygnus X-1, whose estimated distance is 10,000 light years. A proposed model of the system is shown in Figure 15.22. The black hole mass, derived from orbital

parameters, is definitely over 5 $M_\odot$ and probably lies in the range of 9 to 15 $M_\odot$, while that of the supergiant is about 30 $M_\odot$ or greater. Gas is apparently being pulled from the overextended blue supergiant by the black hole's powerful gravitational field. Most of the stellar wind escapes from the system, except that headed toward the black hole. As matter streams toward the black hole, a large part of it is captured and swirls around it to form a thin, gravitationally compressed disk. Friction within the spiraling gas heats it to very high temperatures so that it emits variable X-ray radiation. Like water draining out of a bathtub, the tightening vortex of the X-ray–emitting gas is finally sucked into the black hole; as it crosses the event horizon, it disappears from view.

SUMMARY

After exhaustion of the hydrogen in the central core, the stars restructure themselves by contracting and heating in the central regions as they expand their outer layers to become red giants. During this period both hydrogen in the surrounding shell and helium within the core burn at a much higher temperature. As further nuclear burning of hydrogen into helium and then into carbon and the heavier elements proceeds, the stars track back and forth near the red giant region or horizontal branch of the H-R diagram. While pursuing this course they may temporarily emerge as pulsating stars. The stars continue to undergo successive internal adjustments as they synthesize the heavier elements from their dwindling stock of nuclear fuels.

Eventually a stage is reached where stars like the

sun and stars of lesser mass can no longer generate energy through their internal thermonuclear processes. They are transformed into small, dense white dwarfs. Their internal heat slowly shines away and they end their lives as black dwarfs—stellar corpses.

The more massive stars may terminate their existence by exploding as supernovae and gravitationally collapsing into tiny neutron stars with densities a million billion times that of water. The most massive stars may undergo such a complete gravitational collapse near the end of their lives that they contract into minute, superdense objects called black holes, from which no light can escape. The vast majority of the stars, however, have relatively quiet lives from birth to death.

REVIEW QUESTIONS

1. What is meant by the hydrogen exhaustion point above the zero-age main sequence line? What is so significant about this position?

2. Describe the probable post–main sequence evolution of the sun after it leaves the main sequence until it arrives on the horizontal branch of the H-R diagram.

3. What is the probable terminal fate of the most massive stars?

4. Describe the sun's predicted evolution after it has exhausted its available nuclear fuel.

5. What produces a helium flash in a star? At what stage in the star's evolution does it take place?

6. Indicate on the main sequence of the H-R diagram where stars of different masses lie.

7. What is meant by electron degeneracy? By neutron degeneracy? When and where do they occur in stars?

8. What determines whether a star becomes a white dwarf, a neutron star, or a black hole?

9. Describe the internal structure of a star when evolution moves it to the red giant stage for the first time; for the second time.

10. In how many ways can stars lose mass? At what points in their evolution will mass loss occur?

11. Why does a massive star possess a relatively short life expectancy compared to a solar-mass star?

12. How do astronomers account for the difference between Type I and Type II supernovae?

13. What is the difference between a pulsar and a neutron star?

14. Why has the Crab Nebula been such an important object to astronomers?

15. Describe the internal structure of a neutron star.

16. Why do pulsars gradually slow down in their rotation?

17. What is a black hole and why is it so named?

18. What is meant by the event horizon of a black hole?

19. Describe the current model of an X-ray binary.

20. How does a rotating black hole differ from a nonrotating black hole?

Although the title of this chapter indicates that we will be studying the Milky Way, the preceding chapters have also been about the Milky Way. Stars are the principal components of the Milky Way Galaxy, and they in all their variety, along with the interstellar gas and dust, determine its general appearance. In this chapter we bring together our knowledge of individual stars, of the interstellar medium, and of mechanics to study the Galactic system as a whole. Let us begin with a brief survey of the way we view the Galaxy from the earth along with the history of galactic research.

16.1 Island Universes

THE MILKY WAY

If we pick a clear, moonless night, away from city lights, and look up, there arching above us is a misty, irregular, beltlike cloud of light—the Milky Way. In the temperate latitudes you see different parts of this band of stars that are inclined to the horizon at different angles, depending on the time of night and season of the year (see star maps in Appendix 3). About 30 percent of the Milky Way in the temperate latitude lies beneath your horizon; to see all of it you must go to the equatorial zone. Visualizing the parts of the Milky Way Galaxy in the sky is easier if you are familiar with the constellations.

From midnorthern latitudes in the late summer months, the Milky Way shines its richest; south toward the great star cloud in Sagittarius lies the center of our Galaxy. In winter we see to the south the dimmer and more sparsely populated portion of the Milky Way in Orion and up toward Taurus, for that direction leads away from the Galaxy's center. See the mosaic of the entire Milky Way in Figure 16.1.

It is not obvious from the way the heavens look that the sun is immersed in the disk-shaped system of stars that we call the Galaxy, although our edgewise view explains why it has a bandlike appearance. As

◄ A detail from *The Origin of the Milky Way* by Tintoretto (c. 1578).

long ago as 1750 Thomas Wright claimed that the Milky Way system had a roughly disklike form. By counting stars, William Herschel in 1784 found the first scientific evidence that verified Wright's speculation. Not until 1917 was it discovered by Harlow Shapley that the sun is not at the center of the Galaxy. He did this by studying the distribution of globular clusters, which he correctly assumed are centered on the center of the stellar system (see Figure 3.6).

NEBULAE—INTERNAL OR EXTERNAL?

More than two centuries ago Thomas Wright and Emanuel Swedenborg imaginatively suggested that the small nebulous patches of light observed through telescopes might be other stellar systems like our own. In 1755 Immanuel Kant pursued this idea further. He proposed that these objects might be distant Milky Way systems (island universes) distributed at random angles of inclination to the line of sight. He reasoned that these systems of stars look misty because they are so far from us. (A modern photograph of several distant stellar systems is shown in Figure 16.2.)

"Lo," quoth he, "cast up thine eye,
See yonder, lo! the galaxie,
The which men clepe the Milky Way
For it is white; and some parfay
Callen it Watling streete."

Geoffrey Chaucer

Lord Rosse made telescopic drawings midway through the nineteenth century, with a 1.8-meter reflector, then the world's largest. He saw a spiral pattern in several nebulae. But the first substantial evidence that the spiral nebulae were stellar systems came only in 1915 when Vesto Slipher (1875–1969) of the Lowell Observatory recorded the absorption line spectra of a number of spiral nebulae. Their spectra were dark lines on a continuous background—the kind that would be expected from the composite light from vast numbers of stars. The majority had lines whose wavelengths were displaced to the red (or redshifted), with recessional velocities up to 1100 kilometers per second. In at least two instances Slipher

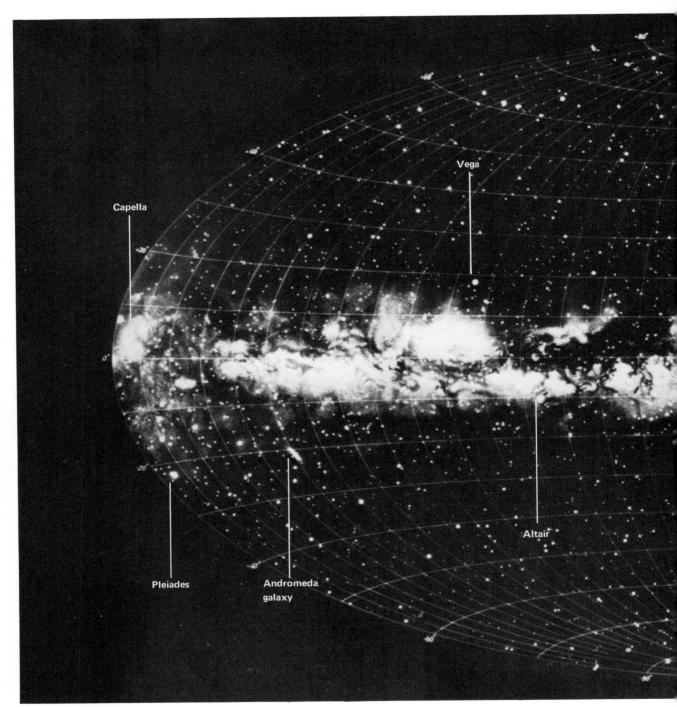

FIGURE 16.1
Panoramic view of the Milky Way
Galaxy. The coordinates shown are
Galactic longitude and latitude.

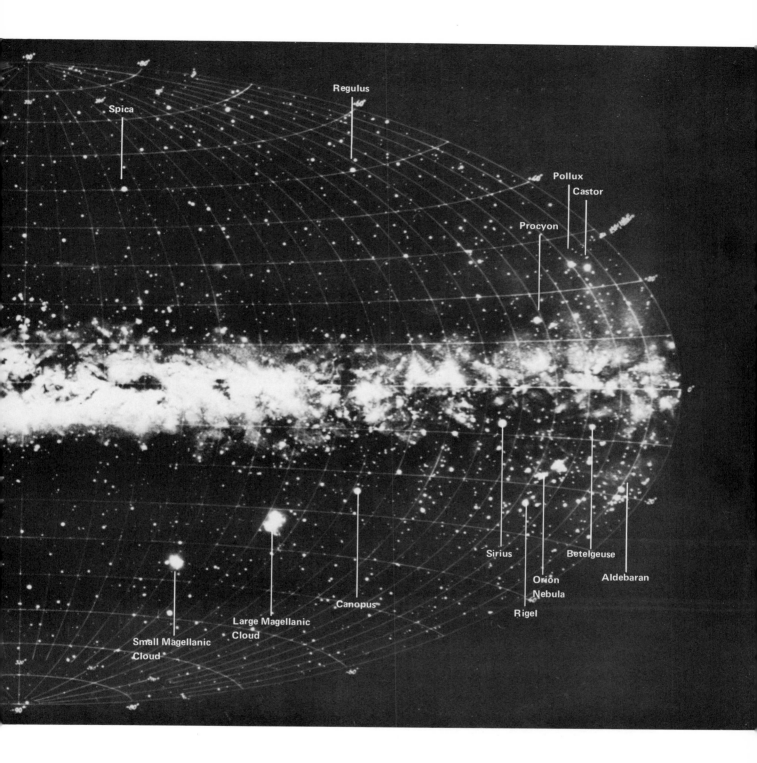

thought that the spectral lines were slightly tilted, which indicated to him that the nebulae were rotating.

If the spiral nebulae were really huge systems of stars, why could they not be resolved into separate stars with the largest telescopes early in the twentieth century? A dozen bright novae and supernovae had been found in the spiral nebulae when no other stars could be seen, but no one recognized that here was an indication that these objects were extremely distant. Were the spiral nebulae well outside our own Galaxy, as Heber Curtis of the Lick Observatory advocated, or were they really associated with the Galaxy, as Harlow Shapley of the Mount Wilson Observatory argued? By 1924 the answer was definite. Edwin Hubble of the Mount Wilson Observatory succeeded in resolving into discrete stars the irregular nebulous structure of NCG 6822 in Capricornus and the peripheral portions of two large spirals, M31 in Andromeda (Figure 16.3) and M33 in Triangulum (see Figure 17.11). (We show NGC 4622 in Centaurus—Figure 16.4—here because its similarity to our Galaxy will help us to visualize its structure.) He identified some of their stars as cepheid variables. Their distances, derived from the cepheid period-luminosity relation, placed these objects far outside the Galaxy.

NAMING THE GALAXIES
Astronomers had observed, described, and cataloged small nebulous patches and stellar knots even before

FIGURE 16.2
Four galaxies in Leo—three spirals and one elliptical.

Nature is a network of happenings that do not unroll like a red carpet into time, but are intertwined between every part of the world; and we are among those parts. In this nexus, we cannot reach certainty because it is not there to be reached; it goes with the wrong model, and the certain answers ironically are the wrong answers. Certainty is a demand that is made by philosophers who contemplate the world from outside; and scientific knowledge is knowledge for action, not contemplation. There is no God's eye view of nature, in relativity or in any science: only a man's eye view.

J. Bronowski

the advent of photography. The French comet hunter Charles Messier (1730–1817) was the first to assemble a catalog. Completed in 1781, it listed 103 galaxies, star clusters, and gaseous nebulae. To avoid mistaking them with the comets, which he was looking for, Messier carefully described these objects. A more extensive catalog, including some 5000 star clusters, gaseous nebulae, and galaxies, was published in 1864 by John Herschel, whose father William had recorded about half the entries in an 1802 catalog. Between 1888 and 1908 Danish astronomer John Dreyer (1852–1926) compiled the New General Catalog (NGC), the most comprehensive of the older catalogs still used, and two supplemental Index Catalogs (IC).[1]

From the brightest objects down to the twelfth apparent magnitude, these catalogs list about 13,000 star clusters, planetary nebulae, diffuse nebulae, and galaxies. A recent collection lists about 200,000 extended optical sources—galaxies, star clusters, nebulae of all descriptions, and quasi-stellar objects. Even this number is dwarfed by the total number, running into the hundreds of millions, that can be photographed with modern telescopes.

[1]The catalog numbering system is simple. M13 is Messier's thirteenth catalog entry for the great globular cluster in Hercules, which is NGC 6205, or entry number 6205 in Dreyer's New General Catalog.

FIGURE 16.3
The Andromeda galaxy, M31, and its two elliptical companions. They are a little over 2 million light years away.

The stars scattered in the photograph are foreground stars of our Galaxy.

Besides catalog designations, a number of galaxies are identified by a proper name, such as the Andromeda Galaxy or the Large and Small Magellanic Clouds. Somewhat less confusion exists in identifying galaxies as compared to star names.

16.2 Stellar Motions

Most of the changes we see in the night sky result from the earth's motions, but the stars, too, are moving. Do they move chaotically or is there some pattern to their motion? There is, as we will see, a slight individual random motion (detected only by careful telescopic observation) superimposed on a large systematic movement that all stars share in as they revolve around the Galaxy's center.

MEASURING STELLAR MOTIONS
Astronomers cannot observe directly the actual motions of the stars relative to the sun, but they can detect their motions projected on the sky by comparing two photographs of the same star field taken years apart. Then from the photographs they can measure the minute change in position between the two observing periods. Reducing this measurement to the annual amount of angular change in position in arc seconds, they derive what astronomers call the star's *proper motion*.

FIGURE 16.4
Spiral galaxy in the constellation Centaurus. This galaxy is a member of the Centaurus cluster of galaxies. Note its remarkable smooth and thin spiral arms including either a coalescence of the arms on the right side or a forking of the arms. The distance of the galaxy is about 200 million light years. This photograph was taken with a 4-meter telescope.

FIGURE 16.5
Proper motion of Barnard's star between 1937 and 1962. During these twenty-five years the star moved a little over 1/15 degree, mostly toward the north.

The distances involved are so immense that the observed proper motions are exceedingly small. In 1718 Edmund Halley discovered the proper motions of several bright stars by comparing their positions with those listed in star catalogs that Hipparchus and Ptolemy compiled some seventeen hundred years earlier. Stars close to the sun naturally have larger proper motions. Barnard's star has the largest known proper motion, 10.3 arc seconds. Figure 16.5 shows its proper motion during a quarter of a century. Some 3300 stars have proper motions of more than 0.5 arc second, which is only about 1/4000 of the sun's angular diameter. Proper motions have now been measured for more than a quarter of a million stars.

The constellations we see today looked very much the same in ancient days, but eventually the familiar star configurations will change due to the proper motion of the constellation stars. This is illustrated in Figure 16.6 for the Big Dipper, which will change drastically in appearance, as shown, over the next hundred thousand years.

A star's proper motion combined with its distance gives us its speed at right angles to our line of sight. This is called the *tangential velocity*. We can also find its line-of-sight motion, or *radial velocity*. This is found from the Doppler shift of the star's spectral lines. Radial velocity can be measured to an accuracy of a few kilometers per second or less. Most of the stellar radial velocities that we observe in our region of the Galaxy are under 80 kilometers per second. Together, a star's tangential velocity and its radial velocity define its speed and direction, its *space velocity*, relative to the sun.

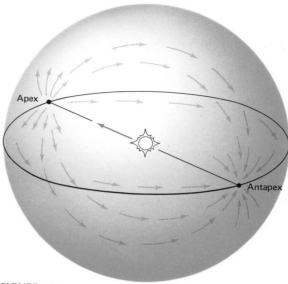

FIGURE 16.7
Effect of the sun's motion on nearby stars. The stars appear to scatter outward from the point on the sky (apex) toward which the sun is headed. The nearby stars appear to converge toward the opposite point (antapex).

FIGURE 16.6
Changes in the appearance of the Big Dipper due to proper motions. The five inner stars are parts of an open star cluster whose members are moving in parallel tracks in space.

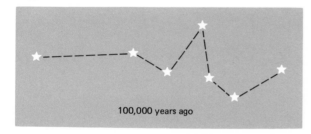

100,000 years ago

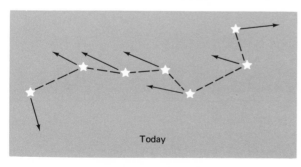

Today

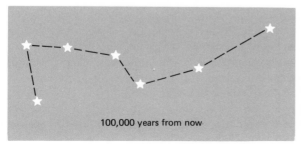

100,000 years from now

WHERE ARE THE STARS GOING?

Even without knowing how far away they are, we can make some generalizations about the stars' movements from their proper motions and radial velocities. Statistically analyzing proper motions, we find that nearby stars generally diverge outward from a point on the sky toward which the sun appears to be headed. Simultaneously, on the opposite side of the sky, nearby stars are converging toward the point on the sky from which the sun appears to be receding. The *solar motion* (see Figure 16.7) is taking the sun toward a point on the celestial sphere called the *apex*; it lies in Hercules within 10° of the bright star Vega. The opposite point on the sky away from which the sun is receding is called the *antapex*; it is in the constellation Columba. Radial velocities of the sun's closest neighbors further tell us that the sun's motion relative to them is fairly slow: 15.4 kilometers per second (3.25 astronomical units per year).

From data on stars at great distances from the sun, we know that the sun and its neighbors are moving in nearly parallel circular orbits at about 250 kilometers per second around the center of the Galaxy, or about one orbit in 225 million years. This solar neighborhood of stars is about 30,000 light years from the nucleus,

By the year A.D. 30,000 Alpha Centauri will be not 4.3 light years from the sun, as it now is, but 3 light years (see Figure 16.8). During the 28 millennia required to reach this, its minimum distance, its motion in space will place it in the constellation Southern Cross (Crux) around A.D. 14,000. Precession of the equinoxes will then have made Vega the North Star and the constellation Crux with its altered configuration will be visible farther north. This very bright interloper, outshining the four stars of the Southern Cross, will move through the constellation on its way toward Hydra, reaching there around A.D. 30,000. Alpha Centauri will then get farther and farther from the sun as it next passes into the constellation of Cancer in A.D. 131,000. By that time it will have traveled slightly more than a quarter of the way around the sky.

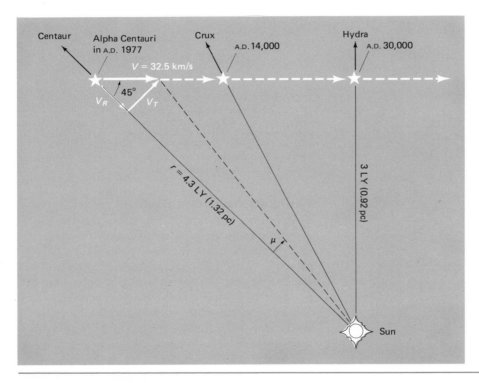

FIGURE 16.8
Future space motion of Alpha Centauri Components of space velocity are r = distance in light years, μ = proper motion; V_R = radial velocity; V_T = tangential velocity = $15.4\,\mu \cdot r$ (km/s); V = space velocity = $\sqrt{V_R^2 + V_T^2}$

which is about two-thirds of the Galaxy's radius out from its center. The nearby stars have enough slight variations in their individual motions, caused by their mutual gravitational attractions, to make it seem as though they move randomly within their local frame of reference. It is with respect to its stellar neighbors that we see the sun's approach toward Vega.

GALACTIC COORDINATES

Astronomers have devised a coordinate system (see Figure 16.1) based on the Galaxy to help them describe the sky from the perspective of the Galaxy (see Appendix 3 on *Astronomical Coordinate Systems*). Figure 16.9 shows an idealized sketch of the Galaxy, both in a face-on view to the disk and in a cross-sectional view through the disk. On the right hand side of the figure are shown the Galactic coordinates, which are known as Galactic longitude and Galactic latitude. *Galactic longitude* is the angular distance measured in the central plane of the disk starting from the Galactic center and measuring along the Milky Way through the constellations shown. The direction 90° to the

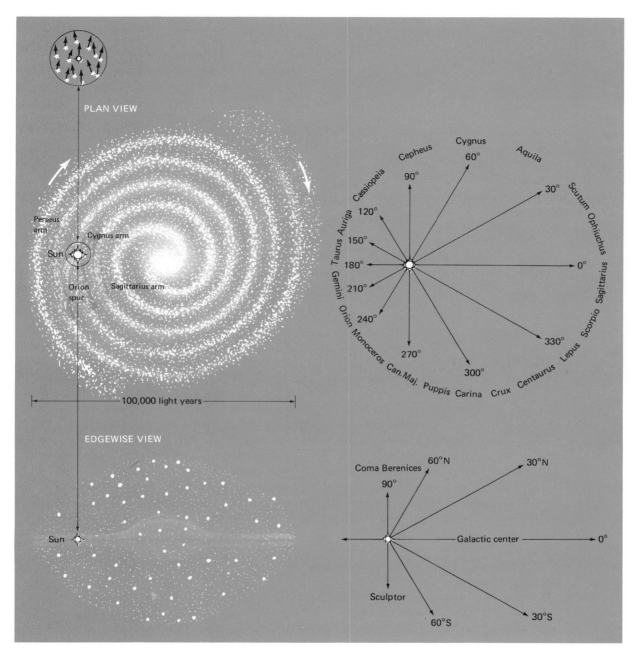

PLAN VIEW

Perseus arm

Cygnus arm

Sun

Orion spur

Sagittarius arm

|— 100,000 light years —|

EDGEWISE VIEW

Sun

Cepheus   Cygnus   Aquila
Cassiopeia   60°
Auriga   90°   30°   Scutum
Taurus   120°   Ophiuchus
    150°   0°   Sagittarius
Gemini   180°   Scorpio
    210°   330°   Lepus
Orion   240°   270°   300°   Centaurus
Monoceros Can.Maj. Puppis Carina Crux

Coma Berenices   60°N   30°N
    90°
    ←————— Galactic center ——————→ 0°
Sculptor
    60°S   30°S

FIGURE 16.9

Our Galaxy and its structure. Shown here are a plan view of an idealized model of the Galaxy with a corresponding diagram showing Galactic longitudes and an edgewise view and corresponding diagram of Galactic latitudes. Our Galaxy is a huge system of stars having the shape of a disk with a central bulge out of which wind several spiral arms interspersed with clouds of gas and dust. Surrounding it is a spheroidal halo of stars and some 120 scattered globular clusters. The entire Galaxy is rotating, and the stars are moving in orbit around the Galaxy's center. The sun is close to one of the outer arms. It is in slow random motion with respect to its neighbors. The sun and its neighbors as a whole, however, are orbiting around the Galactic center at 250 kilometers per second toward Cygnus-Cepheus on a line 90° from the direction toward the Galaxy's center in Sagittarius. The sun completes its journey around the center in approximately 225 million years.

Galactic center in Sagittarius is toward the constellations of Cepheus and Cygnus; 180° is toward Taurus, Auriga, and Perseus, not far from the Pleiades open cluster; 270° is toward Canis Major and Puppis. *Galactic latitude* is the angular distance above and below the plane of the Milky Way. The North Galactic Pole at 90° N galactic latitude lies in the constellation of Coma Berenices and is the same hemisphere as the North Celestial Pole. The South Galactic Pole, 90° S, lies in the constellation of Sculptor.

16.3
Rotation of the Galaxy

If the Galaxy rotates, as was stated in the previous section, how did astronomers show that it does and how did they measure its velocity of rotation? To answer this question, we begin again with the nearby stars. Their motion relative to the sun as a frame of reference is their space velocity. However, another local frame of reference would be to assume that all the nearby stars partake of a general rotation of the Galaxy, which for the solar neighborhood is the same for all nearby stars, and a small random motion relative to each other. This is similar to breaking the motion of a flock of birds into the motion of the flock as a whole and the motion of individual birds relative to the flock. What was described in the preceding section was the motion of the stars of the solar neighborhood relative to the neighborhood. Now how do we measure the motion of the solar neighborhood, which is the rotation of the Galaxy?

One way is to use a frame of reference outside the Galaxy, such as distant galaxies. A systematic study of the Doppler shifts of distant galaxies reveals the fact that the solar neighborhood is moving toward the galaxies in the direction beyond the stars of Cygnus and away from those beyond the stars of Canis Major. As stated, the velocity of the solar neighborhood,

which is about 30,000 light years from the center of the Galaxy, is about 250 kilometers per second, giving a revolution period of about 225 million years.

But use of distant galaxies as a frame of reference was not the first method used by astronomers. The first observational clues for rotation were found in 1926 and 1927 as astronomers studied the motions of distant stars. They uncovered a pattern of movement among the stars indicating approximately circular, quasi-Keplerian motion in the outer parts of the Galaxy. Within 15,000 light years of the Galactic center, the denser, inner portion of the Galaxy rotates more like a solid wheel. Here the orbital velocity increases outward from the center. Beyond 20,000 light years, however, the orbital velocity begins to decrease and becomes more Kepler-like farther from the Galactic center.

What astronomers observe for stars in the Galactic plane is the radial component of the difference in orbital velocity between the sun and a star. The observed radial motions of stars on our side of the Galaxy agree with the theoretical stellar pattern deduced from the differential Galactic rotation shown in Figure 16.10. Stars in the direction of Galactic longitude 45° and 225° appear to be receding from the sun; stars in the direction of 135° and 315° appear to be approaching the sun; those stars in the 0°, 90°, 180°, and 270° directions show no differential radial motion. Stars whose average distance from the sun is 3000 light years exhibit a maximum differential radial motion of about 15 kilometers per second to 20 kilometers per second. Stars whose average distance from the sun is greater than that show a larger effect because these

FIGURE 16.10
Differential radial motions of nearer stars resulting from Galactic rotation.

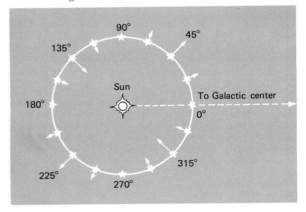

are either closer to the Galactic center where the rotation is faster or farther from the Galactic center where the rotation is slower. The individual stars well above or below the Galactic plane—the halo population and the globular clusters—move around the Galaxy's center at all angles of inclination in highly eccentric elliptical orbits. Their motions are like those of the far-ranging comets orbiting the sun.

16.4 Spiral Structure

DIFFICULTY OBSERVING SPIRAL STRUCTURE

Nineteenth-century astronomers had found that the nebulous patches they saw were not uniformly distributed in the heavens: few spiral and elliptical nebulae were observed near the Galactic plane. At that time no one was sure if these objects were part of our Galaxy, nor whether the absence of these bodies in the Galactic plane was real or an odd preference by nature. But after Hubble proved that these nebulous objects were other systems of stars, there was no reason to doubt that the galaxies were uniformly distributed over our sky. The natural inference was that the apparently unequal distribution of the galaxies was caused by obscuring material between the stars, as Hubble proceeded to demonstrate.

Plotting the positions of the galaxies on a map of the sky, Hubble found an irregular region centered on the Galactic plane where galaxies were absent, the *zone of avoidance* (see Figure 16.11). As was discussed in Chapter 14, it is caused by interstellar dust concentrated in the plane of our Galaxy (as one can see in Figure 16.1). The large vacant place bounded in part by the lines at the extreme right of the diagram coincides with the direction toward the center of the Galaxy in Sagittarius, which is completely hidden in the visible

EDWIN P. HUBBLE (1889–1953)

Hubble attended high school in Chicago, where he excelled as a student and as an athlete. In 1906 he received a B.S. degree in mathematics and astronomy from the University of Chicago. At the university he earned such an excellent reputation as a boxer that a sports promoter wanted him to train for a fight with Jack Johnson, the world's heavyweight champion. He was awarded a Rhodes scholarship for Oxford, England, where he studied jurisprudence.

Hubble returned to the United States in 1913 to practice law in Louisville. A year later he went back to his alma mater to begin studies for an astronomical career at the Yerkes Observatory. No one, he stated, should go into astronomy without a genuine call, and the only way to test a call is by having another calling to be called away from.

In 1919 Hubble joined the staff of the Mount Wilson Observatory. Toward the end of 1924 Hubble made his first great discovery. With the new 100-inch reflector, he was able to sort out a number of bright cepheids in some large spirals. Employing the cepheid period-luminosity relation, he demonstrated that these spirals were other Milky Way systems. This put an end to the raging controversy about whether these objects belonged to our Milky Way system or were beyond it.

In 1925 Hubble established a classification system of the galaxies into three groups: the spiral, the elliptical, and the irregular galaxies (see Figure 17.1). Four years later he made his greatest discovery. After the tedious process of determining the distances of a number of galaxies and observing their Doppler red shifts, he found a proportional relationship between distance and radial velocity, now known as the Hubble law of recession. Although at first Hubble rejected the notion that this relationship was evidence of an expanding universe, he came around to this view after theoretical cosmologists pointed out that this was the only logical explanation.

An individual of extraordinary talents—scholar, athlete, lawyer, and astronomer—Hubble is best remembered as the founder of the observational cosmology and explorer of the deep cosmos.

wavelengths. Light from an external object is dimmed by about 15 percent in the direction vertical to the Galactic plane. It is made even dimmer in the direction away from the vertical until light is nearly or totally extinguished when moving directly in line with the plane of the Galaxy.

In spite of the obscuring dust, optical astronomers in the early 1950s identified sections of the spiral arms by tracing where the O and B associations and their emission nebulosities are most prominent. The Orion arm, as shown in Figure 16.12, its inner edge skirting the sun, is closest to us. It arcs over an angle 3000 light years long around our position. Near the sun it bulges outward in a short extension called the Orion spur. A second arm, Perseus, extends in the same general direction nearly 4000 light years beyond the Orion arm, closer to the edge of the Galaxy. A segment of a third arm lying inside the Orion arm, the Sagittarius arm, has been observed toward the Galactic center several thousand light years from the sun. The radio view of the spiral structure not only encompasses these three arms but includes many more hydrogen lanes.

RADIO MAPPING OF THE GALAXY

Radio astronomers analyzing profiles and Doppler shifts of the 21-centimeter line in many directions through the Milky Way have mapped the distribution of neutral hydrogen in detail (see Figure 16.13). Our knowledge of our Galaxy's spiral arm structure, originally discovered by optical means, has thus been greatly extended by radio mapping. More arms or segments of arms have been discovered in the 21-centimeter surveys of the Galaxy (see Figure 16.13). Radio observations have also revealed hydrogen streamers or jets projecting out of the Galactic disk.

By way of illustrating radio-mapping techniques, let us consider a small segment of the arm structure as revealed by the 21-centimeter-line profiles along Galactic longitude 85°; see Figure 16.14. What are the

FIGURE 16.11
Apparent distribution of the galaxies. This type of projection covers the entire sky. The horizontal line running through the center of the chart coincides with the Galaxy's equatorial plane. Numbers along this line correspond to Galactic longitudes measured from the direction to the Galactic center in Sagittarius; numbers along the periphery represent Galactic latitudes measured from the Galactic plane. The zone of avoidance is the empty, irregularly bounded region running laterally across the center. It outlines the obscuration centered on the Galactic plane. Vacant areas in the diagram were too far south to be seen in California, where the observations were made.

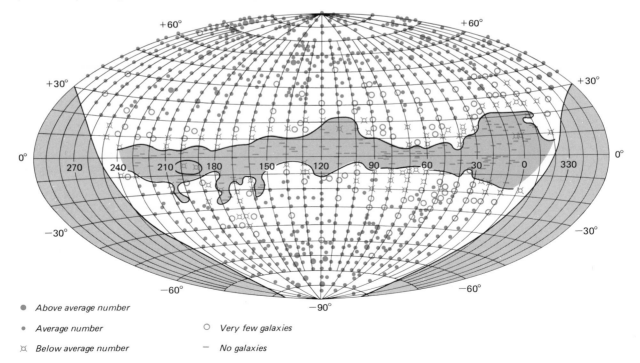

● Above average number

• Average number

¤ Below average number

○ Very few galaxies

— No galaxies

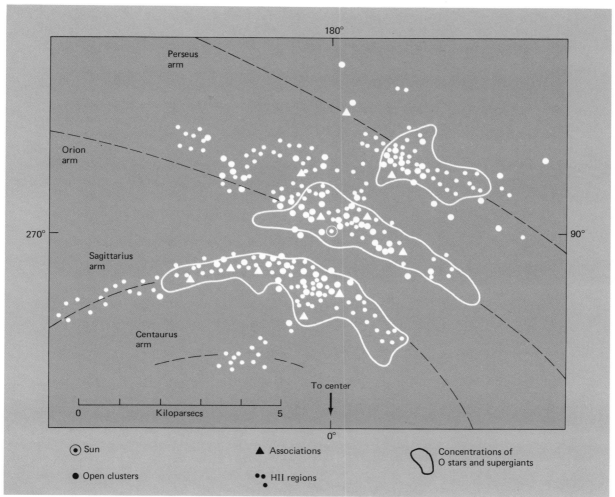

FIGURE 16.12
Spiral arm features in the vicinity of the sun. Looking down on the plane of the Galaxy, we see that the features outlin- ing its spiral structure are the O associ- ations, H II regions, and open clusters. The sun's position is shown on the in- side of an arm, with an arm visible inside and outside the sun's arm.

distances from the sun to the three observed maxima in part b of the figure (where the hydrogen intensity is greatest)? To determine this, we can use the Galactic rotation data first derived by optical astronomers to establish a model of the arm structure with an approx- imate scale of distances. From this model we can cal- culate what the radial velocities should be where the spiral arms cross our line of sight at Galactic longitude 85°, assuming the hydrogen gas moves in a circular orbit around the center of the Galaxy.

The arm portion at A in part a of Figure 16.14 is or- biting at approximately the same distance from the Galactic center as the sun. The difference in the mo- tions between the sun and region A along the line of sight produces a slight negative Doppler shift of the 21-centimeter line corresponding to a distance of about 1,600 light years. Since the arm portion at B is farther from the Galactic center than is the sun, its or- bital motion along the line of sight produces a large negative Doppler shift corresponding to a distance of about 12,000 light years. In the region around C, which is farthest from the Galactic center, the orbital motion is slower. Its projected motion along the line of sight produces an even larger negative Doppler shift corre- sponding to a distance of about 25,000 light years. If our model is incorrect, we can slightly relocate the

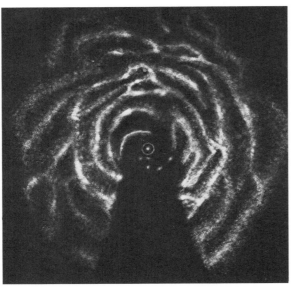

FIGURE 16.13
Radio picture of the spiral structure of our Galaxy derived from the 21-centimeter observations made by Dutch and Australian radio astronomers. The sun is represented by the small white dot and enclosed circle above the Galaxy's center.

FIGURE 16.14
Mapping Galactic rotation with 21-centimeter radio emission. The observed Doppler shift peaks in Part B labeled A, B, and C correspond to the positions of the spiral arms in Part A labeled A, B, and C. Their distances are about 1,600, 12,000, and 25,000 light years, respectively.

portions of *A*, *B*, and *C* in part a of Figure 16.14 to conform more closely to the observational data, which are shown in part b.

The actual analysis of the line profiles and the mapping of the spiral arms is far more complex than the simplified approach we have used here. Application of the proper technique at different Galactic longitudes makes it possible to map the distribution of neutral hydrogen throughout the Galaxy, shown in Figure 16.13. The absence of structure in the sector centered at Galactic longitude 0° is not real. Since the radial motion with respect to the sun in this direction is zero or close to zero, it is not possible to differentiate one spiral arm from another in this direction.

WHAT MAINTAINS THE SPIRAL STRUCTURE?

Gravity is probably dominant in shaping and preserving the structure of the spiral arms in normal spiral galaxies. The arms trail behind in rotation. The more rapid motion of the spiral arms close to the nucleus should wind them so tightly after several galactic revolutions that the spiral pattern in the Milky Way Galaxy would soon disappear. However, from the large number of spiral galaxies that exist, we can see that the spiral-arm pattern is fairly stable and is probably not a transient feature of our Galaxy.

We have two promising theories that explain why the arms are stable. One says that density waves or waves of compression move through the gaseous and stellar matter of the disk as a result of gravitational variations in the disk. They spiral outward from the center at a constant rotational rate somewhat slower than the inner portions and faster than the outer portions of the disk. The waves do *not* consist of moving matter but pass through it. As the compressional waves move through the interstellar medium, they pile

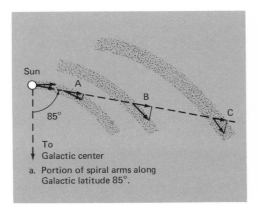

a. Portion of spiral arms along Galactic latitude 85°.

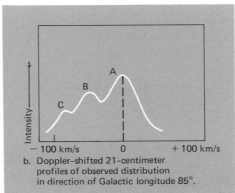

b. Doppler-shifted 21-centimeter profiles of observed distribution in direction of Galactic longitude 85°.

up gas and dust into a spiral-arm pattern dense enough to initiate star formation. Between the spiral arms, interstellar matter is not dense enough to contract under its own gravitation to form stars.

To help visualize a density wave, think of the following analogy. Imagine a slowly moving road crew painting white divider strips on a freeway where traffic moves in one direction. The traffic piles up wherever the crew is working before it can proceed normally. From a plane the congestion would seem to be moving slowly forward as the crew plods along.

The extensive dust lanes frequently observed along inner edges of the spiral arms seem to confirm that these compressional waves are real. The newly formed blue supergiants and H II regions, like brilliant beacons, illuminate and indicate the spiral arms' present locations. In time the wave moves on, and the H II regions and the short-lived massive blue stars soon disappear. Long-lived stars of small mass, like the sun, are left to mix with and become part of the disk population of stars. Thus there is a continual replenishing of the arms.

A second theory for the persistent spiral-arm structure is based on a repeating process in which clusters of stars are created from shock waves generated by supernova outbursts within the differentially rotating Galactic disk. Star formation is supposedly triggered by the rapidly expanding shock wave fronts from the supernova explosions that compress interstellar clouds (see Section 14.4). Out of the stars formed, the bright, young, massive ones soon live out their lives and explode as supernovae, whose shock waves trigger the formation of a new generation of stars. Thus the process is repeated from one generation to the next, while the differential rotation of the Galaxy stretches new star groups into the recognized spiral features. The spiral structure can be preserved by having a single supernova outburst take place once per century within a stellar assocation, which is a reasonable value. The results for one differentially rotating model are shown in Figure 16.15 for the galaxy M101. The calculated spiral segments, marked by crosses, are closely aligned with M101's spiral arms and possess the same curvature.

FIGURE 16.15
Plot of the model spiral-arm structures, represented by the crosses, superimposed on the spiral galaxy M101.

GENERAL FEATURES OF THE GALAXY

Let us review the large features of our Galaxy before we look at some of its finer details. From a variety of different measurements of stars and the interstellar medium, astronomers have obtained the information summarized in Table 16.1. These figures tell us that our stellar system is a galaxy of major size, constructed along the same lines as our large neighbor in the constellation Andromeda some 2 million light years away.

The Milky Way is a large, flattened, disk-shaped system of approximately 200 billion stars. The stars are most numerous in the thicker central portion and decline through the disk of the Galaxy and even more rapidly away from the disk into the halo. Within the Galaxy stars occur singly, in multiple star systems, or in clusters, such as the open clusters of the disk and the globular clusters in the halo. The oldest stars are in the halo and the youngest can be found in the spiral arms winding their way out from the central bulge through the Galactic disk. The sun is a yellow middle-aged star that is part of the disk population of stars.

About 95 percent of the Galaxy's mass (below) may be tied up in its stars. The remainder is gas and

TABLE 16.1
The Galaxy

Galactic Property	Numerical Value
Diameter of disk	100,000 ly
Thickness of disk	3,000 ly
Thickness of central bulge	10,000 ly
Diameter of central bulge	13,000 ly
Diameter of halo	Greater than 100,000 ly
Mass of Galaxy	$1.7 \times 10^{11}\ M_\odot$
Mass of Galactic halo	Less than 10% of total mass
Approximate number of stars	2×10^{11}
Distance of sun from center	30,000 ly
Velocity of sun around center	250 km/s
Period of sun's revolution	225×10^6 yr
Direction toward Galactic center	Constellation Sagittarius
Mean density of matter	10^{-23} g/cm³
Average star density	$\approx 0.075\ M_\odot/\mathrm{pc}^3$
Population I	$\approx 0.06\ M_\odot/\mathrm{pc}^3$
Population II	$\approx 0.015\ M_\odot/\mathrm{pc}^3$
Gas density	$\approx 0.018\ M_\odot/\mathrm{pc}^3$
Dust density	$\approx 0.002\ M_\odot/\mathrm{pc}^3$

MASS OF OUR GALAXY

Once again we put Kepler's modified third law to work; we can use it to obtain an approximate mass for our Galaxy. Assuming that the entire mass is concentrated at the center, we write

$$P_G{}^2 = \frac{4\pi R_G{}^3}{G(M_G + M_\odot)}$$

where P_G is the period of orbital revolution at the sun's distance, R_G; M_G is the Galaxy's mass; and $M_\odot$ is the sun's mass. This example is similar to the one on planetary motion on page 41. Because the sun's circular orbital velocity is $V_\odot = 2\pi R_G/P_G$, we can restate the preceding equation after some transformation; we also neglect $M_\odot$, which is insignificant compared to M_G:

$$M_G = \frac{V_\odot{}^2 R_G}{G}$$

Substituting $V_\odot = 2.5 \times 10^7$ centimeters per second, $R_G = 2.84 \times 10^{22}$ centimeters, and $G = 6.67 \times 10^{-8}$ cubic centimeter per gram per second squared in the equation, we find $M_G = 2.66 \times 10^{44}$ grams $= 1.34 \times 10^{11}\ M_\odot$. This is somewhat less than but quite close to the more accurate value given in Table 16.1.

cold dust grains strewn about more abundantly along the arms than in the regions between the spiral arms. In many places the interstellar clouds remain dark; in other places they are rendered visible by the radiation from the recently formed hot young stars imbedded within the clouds. Their bright patchy appearance forms the spiral framework of the Galaxy. Table 16.2 summarizes the Galaxy's features by its stellar constituency. Because the stars are thickest toward the center of the Galaxy, the gravitational attraction is great enough to make the central regions rotate like a solid body out to about a third of the radius. At the sun's distance, about two-thirds of the radius out from the center, the stars are moving in almost circular Keplerian orbits. The Galactic halo as a whole rotates more slowly around the center.

DISK OF THE GALAXY

As stated, the disk of the Galaxy is a large grindstone-shaped distribution of late Population I and early Population II stars. Threading through the disk are the spiral arms, in which the density of interstellar matter

is larger than between the arms. Because of this, active star formation is occurring in the arms, producing the very bright, young, blue supergiants and H II regions. With a greater brilliance than the interarm region, the spiral arms dominate the visual appearance of the Galaxy. However, there is a reasonably uniform distribution of stars by numbers throughout the disk. Evidence to date also suggests that the arms do not constitute high concentrations of mass in the disk. Apparently the total mass, including all the stars and interstellar medium, declines out from the center of the Galaxy and away from the central plane of the disk in a fairly uniform fashion. If this is true for our Galaxy, we suspect it is also true for other spiral galaxies in general. Therefore, we should remember, when looking at pictures of galaxies, that the way in which their light is distributed does not necessarily tell us how the mass of the system is distributed.

CENTER OF THE GALAXY

Sagittarius A is a small, very bright source of radio frequency radiation at the Galactic center that is not

TABLE 16.2
Stellar Populations in the Galaxy

	Extreme Population II	Intermediate Population II	Intermediate Population I	Extreme Population I
Location	Galactic halo	Galactic bulge and substratum	Disk	Spiral arms
Age in years	10–15 billion	1.5–10 billion	0.2–10 billion	1–100 million
Star types	Globular clusters; RR Lyrae variables with periods greater than 0.4 days; subdwarfs	Planetary nebulae, novae; RR Lyrae with periods less than 0.4 days; long-period variables	Stars of solar type; open clusters; A stars; giants	O–B supergiants; open clusters; associations; T Tauri stars; classical cepheids
Concentration toward Galactic center	Strong	Mild to considerable	Light	Weak
Concentration toward Galactic plane	Weak	Moderate	Strong	Very strong
Galactic orbits	High eccentricity	Moderate eccentricity	Low eccentricity	Circular
Approximate heavy-element ratio to hydrogen	Less than 1%	Up to 1%	Up to 2%	Up to 4%

visible optically. The most powerful source of non-thermal radiation in the Galaxy, Sagittarius A shows on our map of radio radiation toward the center of the Galaxy (see Figure 16.16). A source at the center of the Galaxy is also a strong emitter of X-ray, ultraviolet, and infrared radiation and appears to be a small fraction of a light year in diameter with an estimated mass equal to about one million solar masses. Less than one de-

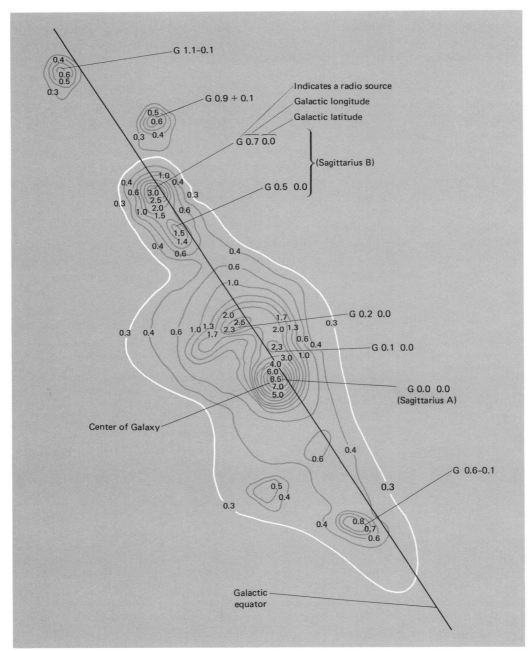

FIGURE 16.16
Radio map in the vicinity of the Galactic center. The map was prepared by Downs, Maxwell, and Meeks of the Lincoln Laboratory, using the 36.6-meter Haystack antenna at a wavelength of 3.75 centimeters.

gree away from Sagittarius A is the small radio source Sagittarius B. Interstellar molecules have been discovered there; it is also a strong source of infrared radiation.

At 12,000 light years from the Galaxy's core is a rotating, broken arm of hydrogen gas expanding outward at an average velocity of 50 kilometers per second. At 8,000 light years from the opposite side of the core is a second rotating arm expanding at approximately 135 kilometers per second. Both arms may be parts of complete rings. A rapidly rotating disk of nonexpanding gas extends out about 2,600 light years from the Galactic center. Within this disk a partial ring of molecular clouds appears to be moving outward at 100 kilometers per second; it may have been ejected from the Galactic nucleus as the result of a short-lived explosion earlier in the Galaxy's history. Gaseous material also appears to be moving inward from the direction of the Galactic halo into the disk. In addition, eruptive movements of gas have been observed streaming outward at an angle, about 23,000 light years from the Galactic center. They form a turned-up and turned-down extension on opposite sides of the Galaxy.

As the sun moves with the stars of the solar neighborhood toward the constellations of Cepheus and Cygnus, material from the interstellar medium flows into the solar system at a speed of about 15 kilometers per second. As the gas in the Galactic plane nears the solar system, it is attracted by the sun and sweeps through the solar system, forming an extended tail downwind from the sun.

HALO OF THE GALAXY

The halo is apparently a remnant of the Galaxy's very early formation and development. Stars and globular clusters in it are clearly the oldest in the Galaxy (see Chapter 15) and show how very different conditions in the Galaxy were at the time the clusters were formed. The halo stars are very deficient in elements heavier than helium, compared with the youngest stars in the Galactic plane. Stars seem to have less of the heavy elements as we move away from the Galactic disk. Stars in the bulge, once also thought deficient in heavy elements, apparently are not, even though they are quite old. Stars elsewhere, then, formed under conditions that were different from those in the halo. Orbits of the halo stars are highly elongated, carrying them in toward the nucleus and above or below the Galaxy's plane. These motions probably were taken over from

the hydrogen gas out of which they formed early in the Galaxy's history.

Most evidence suggests that the halo's mass is only a few percent of that in the disk. But there are some astronomers who believe the evidence suggests that the mass of the halo must be comparable to the disk's in order for the Galaxy to maintain its stability over its life span. If that were true, most of this mass would be tied up in stars of very low luminosity and small mass. Such a finding would alter our idea of the distribution of mass in all spiral galaxies, and it would greatly alter our understanding of spiral systems.

> Geographers . . . crowd into the edges of their maps parts of the world which they do not know about, adding notes in the margin to the effect that beyond this lies nothing but sandy deserts full of wild beasts, and unapproachable bogs.
>
> Plutarch

STELLAR EVOLUTION IN OUR GALAXY: A BRIEF HISTORY

All classes of objects moving around the Galactic center usually occupy the volume of space they inhabited at the time of their formation. For that reason globular clusters, obviously formed some distance from the Galactic plane, moving in highly elongated elliptic orbits distributed almost spherically around the Galactic center. On the other hand, the disk population objects move in more nearly circular orbits confined to the Galactic plane.

Early in our Galaxy's history the gas was denser, favoring star formation. Stars then must have formed more prolifically. During this formative period numerous, very short-lived, supermassive stars might have formed out of the fragmented gas in the protogalaxy while it was contracting. We have seen that stars tend to contract under self-gravitation toward a small, dense sphere of matter, the loss of energy from their surfaces having robbed them of the ability to withstand compression by gravity. Their store of nuclear fuels lengthens this process greatly. By contracting, the star heats its interior, initiating nuclear reactions.

Most matter in the central parts of the star is converted in stages from hydrogen to helium to carbon and oxygen, and even to iron in some instances. Other nuclear reactions, less important for producing energy, also go on during the nuclear-burning phases. These reactions synthesize various elements in the periodic table rather slowly. During the supernova outburst all the elements not already synthesized are produced very rapidly. During a supernova outburst at the end of a massive star's life, chemically evolved matter would have been thrown back into space to mix with the interstellar medium.

Out of this evolved gaseous mixture came the present Population II stars, such as the globular cluster stars. They in turn synthesized in their interiors more heavy elements, some of which were expelled into interstellar space by slow loss of mass due to stellar winds or eruptions. In this further-enriched gaseous environment, the old Population I stars evolved. And finally the young Population I stars appeared inside the gas-dust residue still in the outer disk portions of the Galaxy.

From our laboratory knowledge of the nuclear reactions in these synthesizing schemes, it seems that we can predict the relative abundance of the chemical elements in successive generations of stars. The composition we observe by spectroscopic analysis of the sun, stars, and nebulae and by chemical analysis of the meteorites (see Table 16.3) agrees quite well with what we predict if we assume that the Galaxy started with hydrogen and some helium in a roughly spherical form and that 15 billion years of star formation, evolution, and mass loss have passed during which the Galaxy has become a more flattened system.

The cyclic production and distribution of elements diagrammed in Figure 16.17 shows how chemical evolution will proceed in the Galaxy. The cycle will stop when the bulk of the matter in the Galaxy is tied up in dead stars so that new stars can no longer be formed. Today about 75 percent of the matter in the Galaxy is in active stars, about 15 percent in dead stars, and about 10 percent in the interstellar medium.

Nearly the whole of that part of our everyday world made up of elements heavier than hydrogen and helium was synthesized in the cores of stars billions of years ago. They were spewed back into space in supernova outbursts and stellar winds, mixed with the interstellar matter, and captured as the solar system gathered matter. The matter in our bodies is but one link in our Galaxy's cycle of chemical evolution.

16.6 Ultraviolet Rays, X Rays, and Gamma Rays in Our Galaxy

ULTRAVIOLET STUDIES

Although yet in its infancy, astronomical studies in the invisible portion of the electromagnetic spectrum promises to be an extremely valuable tool in the study of our Galaxy. It was originally thought that all ultraviolet light from 912 angstroms, where the Lyman continuum begins, to about 20 angstroms would be absorbed by interstellar hydrogen. Hence it was supposed that starlight in the extreme ultraviolet region would not penetrate any distance through the interstellar medium and therefore would not be observed. Apparently there are gaps between the interstellar clouds or regions of very low hydrogen density in some directions where the extreme ultraviolet photons can penetrate the plane of the Galaxy. In many directions extreme ultraviolet observations are possible out to several hundred light years. Thus the hottest

TABLE 16.3
The Most Abundant Elements in Stars

Atomic Number	Element	Relative Number of Atoms
1	Hydrogen	662,500
2	Helium	46,040
8	Oxygen	450
6	Carbon	245
7	Nitrogen	78
10	Neon	71.6
12	Magnesium	22
14	Silicon	21
26	Iron	17
16	Sulfur	10
18	Argon	2.4
13	Aluminum	1.8
20	Calcium	1.5
11	Sodium	1.25
28	Nickel	1
	Remaining Elements	≈ 0.04

Source: Derived from A. G. W. Cameron.

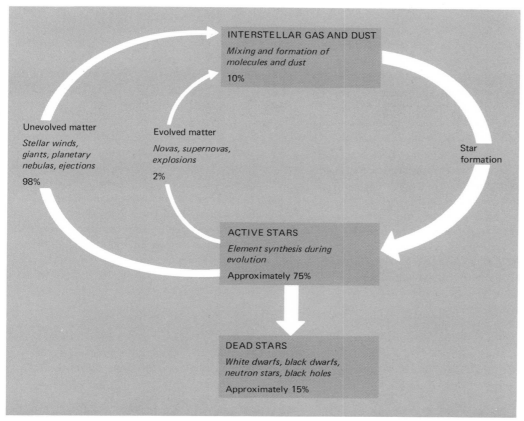

INTERSTELLAR GAS AND DUST

*Mixing and formation of
molecules and dust*

10%

Unevolved matter

*Stellar winds,
giants, planetary
nebulas, ejections*

98%

Evolved matter

*Novas, supernovas,
explosions*

2%

Star
formation

ACTIVE STARS

*Element synthesis during
evolution*

Approximately 75%

DEAD STARS

*White dwarfs, black dwarfs,
neutron stars, black holes*

Approximately 15%

FIGURE 16.17
Chemical evolution cycle. Matter cycles in the Galaxy between the interstellar medium and stars and back to the interstellar medium. Some matter is taken out of the cycle by dying stars. Percentages in the boxes give the present division of matter between the interstellar medium, active stars, and dead stars; percentages on the left of the drawing give the division between chemically evolved matter and unevolved matter that has been expelled by various processes and is returned to the interstellar medium.

white dwarfs and subdwarfs, and the central stars of planetary nebulae, which have effective temperatures of 50,000 K or more, can be seen in the extreme ultraviolet in the solar neighborhood and possibly beyond.

As an example, an ultraviolet telescope flown aboard the U.S. Apollo and U.S.S.R. Soyuz mission in July 1975 detected the first extrasolar extreme ultraviolet object. It is a very blue white dwarf star in Coma Berenices with an effective temperature of 107,000 K and a radius of 5,000 kilometers, and is part of a binary system whose other member is a faint red dwarf. A second white dwarf in the extreme ultraviolet was found in Cetus, with an effective temperature of 60,000 K and a radius of 17,000 kilometers. In another example, the closest known star, Proxima Centauri, the third component in the Alpha Centauri system and a

red dwarf flare star, was found to be emitting extreme ultraviolet photons during an outburst. The ultraviolet measurements showed that the flare temperature was about twenty times the star's surface temperature of 3000 K. A fourth object, the recurrent binary nova SS Cygni, which experiences outbursts every few weeks, was discovered to be emitting extreme ultraviolet photons during one of its outbursts. For 26 other stars, mostly white dwarfs, no extreme ultraviolet photon counts were recorded. This result sets an upper limit for the temperatures of these stars and a lower limit on the number of hydrogen atoms per cubic centimeter along the line of sight. It is hoped that the *International Ultraviolet Explorer* satellite will greatly expand these first steps in ultraviolet exploration of our Galaxy.

X-RAY STUDIES

The great penetrating power of X rays makes it possible to observe X-ray objects through the interstellar haze of our Galaxy and even beyond in intergalactic space. Unfortunately reliable distances of X-ray sources are not easy to estimate unless they have optical counterparts or visible binary companions whose distances can be derived by one or more well-established methods. In some instances the relative strength of the soft (12-120A) and hard (1.2-12A) X rays from the same source serves as an indicator of distance. Intensity measurements from a distant X-ray source will show a larger ratio of hard-to-soft X rays than a nearby source because of the higher penetrating power of the hard X rays. This situation is somewhat analogous to estimating the distance of a star by the amount of interstellar reddening of its light, since red light has twice the penetrating power of blue light.

The majority of X-ray sources are distributed in or near the plane of the Milky Way, with a marked concentration in the central bulge (see Figure 16.18). Generally, objects at the high Galactic latitudes are extragalactic. However, some faint objects (smallest dots) as well as some bright objects (large dots) at low Galactic latitudes may also be extragalactic. Identifications of these sources with optical objects can settle the issue.

The rich variety of presently known X-ray sources, numbering over a thousand, is bewildering, and more sources are constantly being discovered. They include neutron stars or black hole binary systems (see Section 15.7), many having blue supergiant companions; dwarf binary novae; globular clusters; supernova remnants; active galaxies (Seyferts, BL Lacertae objects, and quasars; to be discussed in Chapter 17); clusters of galaxies (see Chapter 17). One characteristic that nearly all X-ray sources possess is their variability, which may or may not be superimposed on a steady flux. Some idea of their diversity and distinguishing characteristics for the Galactic sources can be obtained from Table 16.4. The explanations given to account for X-ray emission are far from settled.

X-RAY BURSTERS

In 1975 a class of X-ray sources that emit X rays in powerful bursts was discovered. Several dozen of these *bursters*, as they are called, are presently known

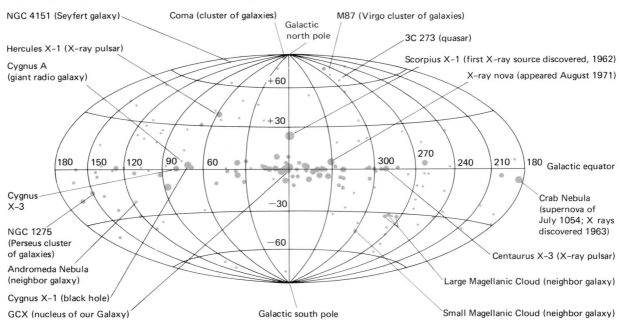

FIGURE 16.18
Sky distribution of X-ray sources. In general, those objects near the Galactic equator (in the plane of the Milky Way) belong to our Galaxy; those well removed from it are beyond our Galaxy. The larger the dot, the more intense is the X-ray radiation. Uniform, diffuse X-ray radiation also is observed over the entire sky.

and their number is increasing. Their X-ray energy output rises steeply from a steady value to a peak about a million times the sun's luminosity, followed by a slower decline. Two kinds of bursters have been identified: type I, whose bursts recur at intervals of hours or longer and which are in the majority, and type II, whose bursts repeat on time scales of seconds to minutes, the so-called *rapid bursters*. Some bursters have steady optical sources associated with them. In one instance (MXB 1735-44) optical bursts were detected in 1978 coinciding more or less with the X-ray bursts. This is the first of such observed events, but it is expected that more, involving other bursters, undoubtedly will follow.

TABLE 16.4
X-ray Phenomena in Our Galaxy

X-ray Source	Example	Observed Event(s)	Possible Cause
Dwarf nova binary	SS Cygni	Frequent outbursts every four weeks, consisting of soft X-ray pulses	Gas transferred from nearby companion onto degenerate subdwarf or white dwarf component, which ejects it when a critical stage is reached
Short-period binary	RS Canum Venaticorum	Soft X-ray outbursts at temperatures of 10 million degrees in yellow subgiant	Injection of energy into corona of cooler companion from large, active region on its surface
Neutron star binary	SMC X-1	Pulsed X rays from neutron star component of close binary	X rays originate mainly from hot spot on rapidly spinning neutron star
Black hole binary	Cygnus X-1	Variable, hard X rays in vicinity of black hole companion	Matter funneled into accretion disk surrounding black hole is heated by friction and other effects to very high temperatures, resulting in X-ray emission
Recurrent transient binary	AO114+63[a]	Novalike characteristics; large sudden increase in X-ray, optical, and radio emission, lasting hours to days, followed by long unpredictable turnoff	Transfer of mass from nuclear-burning component to compact component on a long time scale
Globular cluster center	NGC 6624	X-ray bursts several hours apart but variable with time; bursts stop when steady X-ray source is high	Uneven flow of gas swirling into massive black hole at center of cluster
Semiperiodic burster	MXB 1735-44[b]	Time scale between bursts in hours or longer; rise time of burst about 1 second with few seconds decline	Short-time burst implies neutron star is involved; gas flows at spasmodic intervals onto accumulated material on surface of neutron star, resulting eventually in a nuclear detonation
Rapid burster	MXB 1730-355	X-ray bursts at intervals of seconds to minutes; the greater the burst size, the longer the interval between bursts	Gas flows from a nuclear-burning component onto a neutron star companion at an uneven rate; following a temporary interruption, gas explodes
Supernova remnant	H 1207-52[c]	Soft X rays with long-period variability	Shock excitation of hot supernova expanding shell colliding with pockets of gas and dust in the interstellar medium

[a]British *Ariels* satellite discovery at right ascension 01^h14^m, declination $+63°$.
[b]MIT X-ray burster (MXB) investigated at MIT from *Small Astronomy Satellite* (*SAS-3*) data at right ascension 17^h35^m, declination $-44°$.
[c]*High Energy Astronomy Observatory* (*HEAO-1*) discovery of right ascension 12^h07^m, declination $-52°$.

The first X-ray burster discovered appeared, to the astonishment of astronomers, in the center of the globular cluster NGC 6624, about 25,000 light years distant. It was thought that, because of the great age of globular cluster stars, there should be no activities, such as X-ray emission (usually associated with the evolution of massive stars), going on in them. The cluster possesses a small, bright optical core with high stellar density. The bursts are superimposed on a steady X-ray flux. When first observed, the bursts came several hours apart. Several months later the interval between bursts had shortened to about half the earlier value. The X-ray bursts took place only when the steady X-ray emission was low; when the steady emission was high, the bursts stopped. A number of bursters have subsequently been located in other globular clusters.

> Ten years of radio astronomy have taught humanity more about the creation and organization of the universe than thousands of years of religion and philosophy.
>
> P. C. W. Davies

Let us generalize what is known so far about X-ray bursters. The majority of bursters are confined to the Galactic bulge within 30° on either side of the Galactic center where the concentration is highest. This distribution indicates that their distances are of the order of 30,000 light years. If so, the energy output in a single burst is about a hundred thousand times greater than that of the sun. In general, the more powerful the burst, the longer is the time interval between bursts, the shorter is the duration of the burst, and the less intense is the steady X-ray flux prior to the burst. At most, the type I bursters produce 5 to 10 bursts per day. On the other hand, one type II rapid burster experienced 5000 bursts in one day. Typically the burst period lasts several weeks, then stops and resumes at intervals of about 6 or 7 months.

What produces the bursts? The short duration and rapidity of the bursts implies a very small source,

possibly no larger than a neutron star of 10 to 20 kilometers diameter. The supposition is that a binary system is involved in which one member is a degenerate compact star as discussed in Section 15.7. Several-bursters are identified with faint blue stars. As gas is funneled from the companion into an accretion disk surrounding the compact object, be it a neutron star or a black hole, gravitational potential energy is converted into the thermal energy that is the source of the steady X-ray flux. This kind of mass-to-energy conversion is far more efficient than nuclear fusion and can provide a very high output of X-ray photons. It is thought that the bursts arise from instabilities, gravitationally induced or otherwise, resulting from an uneven flow of inward-falling gas. When the gas stream is temporarily interrupted and then suddenly released, an X-ray burst follows simultaneously with a much weaker optical burst.

For the rapid burster inside a globular cluster, it is speculated that a black hole of from 30 to 100 solar masses may be hidden at the cluster's center. One scenario envisions inflowing gas being heated by X-ray radiation emanating from gas closer to the black hole. When a critical temperature is reached at a certain distance from the X-ray source, heat pushes the gas outward to form a gaseous shell around the black hole. Later the shell is suddenly cooled at a certain density; it then collapses inward and produces an X-ray burst. The process is repeated with each newly formed shell.

Since the majority of rapid bursters are not located in globular clusters, it is suggested that at least some of them may originate in massive black holes formed from the collapsed cores of old globular clusters no longer recognized as clusters. Or the burster may be a remnant of an ancient population of massive stars formed early in the life of the Galaxy. Opinion presently leans toward the binary system hypothesis. However, it may be some time before a satisfactory theory is developed to adequately account for the burst phenomena and its implications for the structure and evolution of the Galaxy.

GAMMA RAY STUDIES

One component of an observed diffuse gamma radiation is isotropic, that is, it has the same intensity in all directions; it is presumably extragalactic in origin. The other component is apparently of Galactic origin since it is coming from the plane of the Milky Way. The isotropic component has an energy spectrum quite different from that of the Galactic component. Most of the Galactic radiation is thought to come from gamma ray photons (at least those with energy over 100 million electron volts) produced by collisions of cosmic ray particles (see Section 16.7) with atoms of the interstellar gas.

Although gamma rays come from all parts of the Galaxy, more come from two very bright regions: the nuclear bulge, from which they are most intense, and a region which is spread out in the form of a ring or belt 15,000 to 20,000 light years from the center of the Galaxy (see Figure 16.19). The ring portion seems to mark an unusually active region of the Galaxy. The evidence for this comes from the number of supernova remnants and pulsars of supernova origin, which reach a maximum in this zone.

Theory predicts that gamma ray spectral line emission should be observable from a number of sources: long-lived radioisotopes synthesized in stellar interiors and in supernova outbursts; excitation and fragmentation of interstellar gas and dust by low-energy cosmic ray particles; electron-positron annihilation in the intense magnetic fields of neutron stars; and the accretion of matter onto neutron stars and black holes. In 1978 the first evidence was found for the existence of gamma ray spectral lines.

In 1967 *Vela* satellites, launched by the United States to monitor violations of the treaty banning atmospheric nuclear tests, detected gamma ray bursts coming from outer space. Since then several such events have been recorded each year by different satellites. In a number of instances there is an observed correlation between gamma ray and X-ray bursts and occasional radio outbursts. This indicates that the

FIGURE 16.19
Radial distribution of gamma ray intensities in the Galaxy. The emission reaches maximum strength at the Galactic center. The broad hump in the middle of the graph is due to radiation within the circular ring surrounding the center of the Galaxy at a distance of 15,000 to 20,000 light years. We view this emission strongest in a tangential direction along the edge of the ring at Galactic longitudes 35° and 325°. The two minor humps on either side may come from discrete sources within the Galaxy.

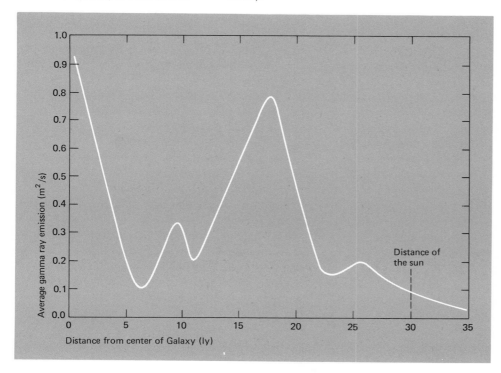

bursts in different parts of the electromagnetic spectrum originate in the same event. The bursts are not all alike. But, in general, each burst lasts a second or so, which implies that the source of the radiation is small. If Galactic in origin, the energy radiated in a single burst is around 10^{39} ergs per second, a quarter of a million times the sun's output. It is estimated that at least a dozen of them are less than 3000 light years from us and that the total number in our Galaxy may be about 1000.

16.7 Cosmic Rays

The earth is constantly bombarded by high-speed atomic nuclei from all directions in space, giving us a fine opportunity to sample elementary matter from outside the solar system. First detected in our upper atmosphere early in this century, these energetic charged particles were misidentified as high-energy photons and named *cosmic rays*, and we still call them that. The sky is as bright with cosmic rays as with starlight. The very energetic cosmic rays arriving from distant parts of our Galaxy carry the imprint of their sources and therefore provide us with an important source of astrophysical information. Cosmic rays also tell us something about the interaction of charged particles with matter, magnetic fields, and photons within the interstellar and intergalactic mediums.

Cosmic ray detectors aboard high-altitude balloons, rockets, and space vehicles have shown that cosmic ray particles are about 83 percent protons, about 16 percent helium nuclei, about 1 percent electrons, and much less than 1 percent heavier nuclei. Their energies range from 10^8 to 10^{20} electron volts. (The kinetic energy of a 10^{20}-electron-volt particle is equal to that of a tennis ball moving at 100 kilometers per hour.)

From cosmochemical evidence we think that cosmic ray flux has remained substantially unaltered since the solar system began. The intensity of cosmic rays from outside the solar system varies with the eleven-year sunspot cycle. During maximum sunspot activity and flare events, the solar wind's magnetic field strength grows. Cosmic ray particles with energies below about 10^{11} electron volts are then deflected, and only those with higher energies can penetrate the solar wind's magnetic fields. The result is that the extrasolar cosmic ray flux decreases when the sun is most active.

The abundance of elements in the cosmic ray particles, with some exceptions, closely matches that of matter within the solar system. Exceptions are the light nuclei—lithium, beryllium, and boron (atomic numbers 3, 4, 5)—and the nuclei from phosphorus to chromium (atomic numbers 15 to 24). These are rare in the solar system. Their abundance in cosmic rays may be caused by fragmentation of heavier nuclei colliding with nuclei of particles of the interstellar medium. During their 20 million years or so of interstellar travel before they arrive on the earth, cosmic ray particles pass through about a spoonful of the thinly spread interstellar matter. As they pass through the earth's atmosphere, their chance for a collision increases and their collisions with the atmospheric gases create decay products such as electrons, neutrinos, and gamma rays, which are observed at or near ground level. Only about one in a million cosmic ray protons manages to make it to the ground.

All nuclei in the periodic table of elements (see page 66) are present in cosmic rays. Because of this we can learn a great deal about the evolution of matter in the universe and the origin of the elements. The chemical composition of cosmic rays suggests a thermonuclear origin in some explosive event. Supernova outbursts have long been thought to be a prime source of the heavier cosmic ray particles from our Galaxy. Other potential candidates for the extragalactic cosmic rays (the very high energy ones) may be quasars, exploding galaxies—any source of violent activity. How are cosmic ray particles accelerated? We suppose they are ejected at high speeds after a catastrophic outburst and accelerated by interaction with interstellar magnetic fields. Our Galaxy confines all but the most energetic particles (energies below about 10^{16} electron volts) within it. Thus the majority of the cosmic ray particles originate in our Galaxy.

Cosmic rays reach their highest intensity within the Galactic nuclear bulge and in the circular belt 15,000 to 20,000 light years from the Galactic center. These are also the same regions where the gamma ray flux is highest, as mentioned previously (see Figure 16.19), and where the supernova remnants and pulsars are most numerous. The common distribution of supernova remnants and pulsars, gamma radiation, and cosmic rays lends further support to the contention that supernova explosions are a prime source of cosmic rays in our Galaxy.

SUMMARY

As the twentieth century opened, exploration of the Milky Way disclosed a structure that was nowhere near as simple as astronomers had imagined. In the 1920s the Galaxy's real shape began to unfold. Astronomers learned that this small part of the cosmos, the Milky Way Galaxy, is a large, flat, rotating spiral galaxy that, like thousands of other galaxies, has gas, dust, and billions of stars. The sun, well outside the center of the Galaxy, is one of those billions. The Galaxy has a scattering of about a hundred globular clusters and an extended halo of stars around its bulging center. Like the planets in the solar system, stars closer to the center revolve around the Galactic center faster than those farther away. The sun's period of revolution is on the order of two hundred million years. In and near the central bulge, where the stars are most dense, the motion is more like a rotating solid body; the rotation is faster farther from the center.

Astronomers classify the stars in the Galaxy into two broad population groups: the younger Population I stars and the older Population II stars. Their numbers, distances, motions, locations in the Galaxy, and physical properties have been studied. The stars of the globular clusters characterize the properties of the Population II stars, while the stellar associations of O and B stars are typical Population I stars.

Nonstellar matter in the space between the stars is the interstellar medium—a mixture of gas, mostly hydrogen, and dust. Concentrations of this material appear as bright and dark nebulae in the Galaxy's spiral arms. Within the interstellar medium are clouds of cold gas and dust that obscure our view of the whole Galaxy.

Some fragmentary spiral aspects of our Galaxy were originally recognized in optical studies of the distribution of high-luminosity stars and their associated bright nebulosity. Radio analysis of the 21-centimeter radiation emitted by neutral hydrogen has built up more detailed information about the spiral structure of the Galaxy. Optical and radio investigations verify that the Galaxy is a large rotating spiral. The sun is about two-thirds of the Galactic radius outside the center in the disk portion of the Galaxy. The oldest, metal-poor Population II stars tend to be concentrated in the globular clusters and in the Galactic halo; the youngest, metal-rich Population I stars are in the outer spiral arms; mixed population groups exist throughout the disk portion of the Galaxy.

High-energy astrophysics, which utilizes the extreme ultraviolet, X-ray, and gamma ray radiations of celestial bodies, has become one of the most active fields of astronomical research. Its scope includes the hot stars near the end point of their life cycles (central stars of planetary nebulae, white dwarfs, neutron stars, supernovae, and black holes); compact binaries; X-ray and gamma ray transients and bursters. The wealth of information presently accumulating in high-energy astrophysics is leading to a deeper understanding of the various physical processes that govern the activity of the many kinds of bodies that populate our Galaxy.

REVIEW QUESTIONS

1. How have radio astronomers succeeded in tracing the spiral structure of the Galaxy from their observations of the 21-centimeter line of neutral hydrogen?

2. How did we discover the rotation of the Galaxy?

3. In an edgewise sketch of the Galaxy, locate the interstellar medium, the Population I and Population II stars, the halo population, and the globular clusters.

4. Sketch the approximate structure of the Milky Way system and show the location of the sun.

5. What information is needed to find the space velocity of a star?

6. Describe the motion of the sun with respect to its neighbors; with respect to the center of the Galaxy.

7. What is meant by the "zone of avoidance"? How was it determined?

8. What maintains the spiral structure of the Galaxy?

9. List several kinds of Population I and Population II stars and their properties.

10. What are cosmic rays? What may be their origin?

11. Give a brief account of stellar evolution in our Galaxy.

12. Which is the most penetrating electromagnetic radiation in the Galaxy and why?

13. List some major discoveries made by astronomers researching the short wavelength radiations in our Galaxy.

14. What is a recurrent transient? a rapid burster? Give an example of each.

15. How were gamma ray bursts first discovered?

16. Discuss the differences in the space distributions between the open clusters, the globular clusters, and the stellar associations.

17. Trace the sequence of steps in the chemical evolution of our Galaxy assuming it started out possessing only hydrogen and a trace of helium.

18. Describe the nucleus of our Galaxy. If the center of the Galaxy contained a massive black hole, what should the astronomer look for in order to verify its existence?

19. Discuss some of Edwin Hubble's contributions to the study of galaxies.

20. What does NGC stand for? What is its significance in astronomical work?

17
The Galaxies

In the preceding chapters we have dealt primarily with our Milky Way Galaxy and its contents. Although it is home to us, our Galaxy is an extremely small entity in the universe. It is only one of billions of galaxies, which are the building blocks out of which the universe is made. Galaxies occur in pairs, small groups, great clusters, and clusters of clusters. They constitute the bulk of matter that is emitting visible light. And, as far as we know, the visible galaxies contain the bulk of the matter in the universe. So let us begin our discussion of the diversity of these fascinating cosmic building blocks.

CLASSIFICATION OF GALAXIES: THE HUBBLE SEQUENCE

From his extensive collection of photographs, Hubble chose about six hundred well-defined bright galaxies on which to base a classification scheme for galaxies.

◀ The Whirlpool galaxy, M51, and its close companion in Canes Venatici, about 32 million light years away. They appear to show the effects of a close encounter in which matter is exchanged between them.

He arranged them in an orderly progression now called the *Hubble sequence*. Shaped like a tuning fork, his sequence ran from essentially spherical configurations of stars, through very flat spiral systems, to irregular systems of stars, as Figure 17.1 shows.

Today astronomers classify galaxies according to several criteria: (1) degree of flattening; (2) relative size of the nucleus compared to the disk; (3) how tightly the spiral arms are wound; (4) type of stars; and (5) relative amount of interstellar matter (gas and dust). Many thousands of galaxies are close enough so that we can examine their structure in detail and classify them. With these characteristics in mind, let us look at the three major types of galaxies: elliptical, spiral, and irregular galaxies.

The *elliptical galaxies* (*E*) are ellipsoidal or spherical-shaped without spiral structure or any evidence of a disk or plane. They have a smooth, symmetrical look, as shown in Figure 17.2, and they vary from dwarfs to giants. In a given volume of space they are far more numerous than spiral galaxies. The absence both of Population I stars and of a dust streak across the middle of the galaxy tells us that active star formation has virtually ceased in these systems. The closest ellipticals have been resolved into older Population II stars. Globular clusters have also been photographed around the nearby giant elliptical galaxies. In Figure 17.2 globular clusters are faintly visible in the outer fringes of NGC 4486. The *dwarf ellipticals* have a sparser and much less compact distribution of stars than their larger counterparts, which tend to be more

FIGURE 17.1
Revised Hubble sequence of galaxies. The arm structures of the spirals are graded a to c, tight to open; nuclei, a to c, large to small. The ellipticals are graded from 0 to 7, spherical to most flattened.

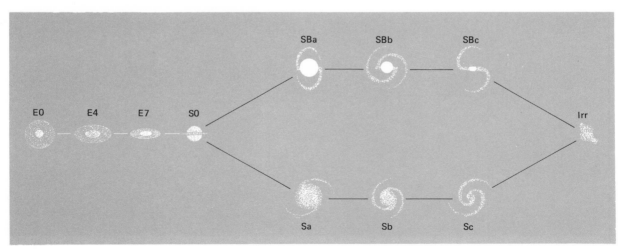

centrally concentrated. Dwarf elliptical galaxies appear to outnumber all other types of galaxies.

Among the conspicuous galaxies most are spirals, but put them among all other galaxies, including those of nondescript shape, and the spirals are not in the majority. There are many more elliptical and irregular galaxies with low luminosity that are difficult to observe and identify at great distances. Two distinct kinds of spirals are recognized: the *normal spirals* (S) shown in Figure 17.3 and the less numerous *barred spirals* (SB) shown in Figure 17.4. Normal spirals have two or more arms winding outward from opposite sides of the nucleus. In barred spirals, as you can see in Figure 17.4, the arms emerge from the ends of a straight arm or bar passing through the center.

Both kinds of spirals are graded by the tightness of the arms (a to c, from tight to open) and the size of their nuclei (again a to c, from large to small); the more loosely wound spiral has a smaller nucleus. The c and b subdivisions have clumps of bright gaseous knots, interspersed with highly luminous young Population I stars, strung along the arms like beads. In the central regions the older Population II stars predominate. Obscuring dust lanes look ragged out in the disk of the galaxy. In spirals that we see edge-on, such as NGC 4565 in Figure 17.3, this opaque interstellar material is striking—a long, dark streak threading across the galaxy's middle. The a category has a somewhat smooth and unresolved texture of older population stars and little evidence of bright nebulae or gaseous knots and bright stars. The halo is more conspicuous in a type spirals than in c type.

The SO type of galaxy, such as NGC 1201 in Figure 17.3, shows a smooth, abbreviated extension beyond the nucleus without spiral structure. From the edge the SBO galaxies look very flat, without the dark streak. In place of the elongated bar these galaxies have a stump, or faint suggestion of a disk without spiral arms, which is often surrounded by a faint halo or ring structure. Some astronomers have suggested that the SO and SBO galaxies may be what is left of spirals after their gas and dust are stripped away by encounters with other galaxies.

About 3 percent of the known galaxies are *irregulars,* exhibiting little symmetry, as can be seen in Figure 17.5. We distinguish two kinds of irregular galaxies. The group Irr I has some highly luminous blue stars, star clusters, and intermingling of gas with very little dust. A prime example is the nearby Magellanic Clouds, described in Section 17.3. The second

ELLIPTICAL GALAXIES

NGC 4486 Type E0

NGC 147 Type E6

Dwarf elliptical galaxy in Sextans

FIGURE 17.2
Elliptical galaxies. They are made up primarily of older stars. The number beside the E designation indicates the degree of flattening, from 7, the maximum, to 0, the minimum (spherical). Globular clusters are faintly visible in the outer fringes of NGC 4486. Seen edgewise, as NGC 147 illustrates, an elliptical galaxy shows no dark dust lane across the middle.

group, Irr II, shows more deformity, fairly conspicuous dust lanes, and a composite spectrum of white unresolved stars. The small companion to the spiral galaxy M51 is a good example (see page 371).

NORMAL SPIRAL GALAXIES

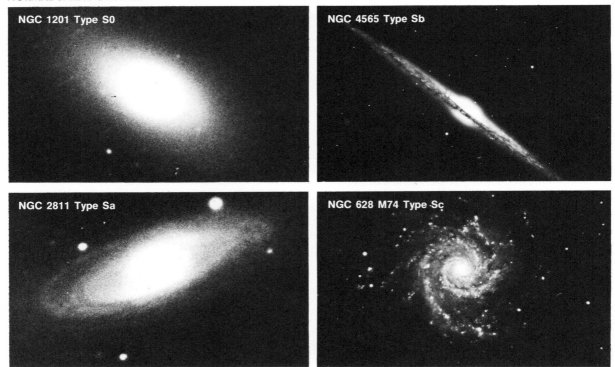

NGC 1201 Type S0

NGC 4565 Type Sb

NGC 2811 Type Sa

NGC 628 M74 Type Sc

FIGURE 17.3

Representative normal spiral galaxies. They range from SO, with only a hint of a disk, through the tight spiral Sa and Sb, to the loose, small-nucleus Sc. The Milky Way is an intermediate Sb type. Notice that the more loosely coiled the spiral arms, the smaller is the central portion. Viewed in profile, the Sa, Sb, and Sc spirals would all show the characteristic central dust streak.

WHAT DOES THE CLASSIFICATION MEAN?

Hubble's classification arranged the galaxies in a progressive morphological sequence, but the galaxies do not necessarily evolve from one form into another. We think it likely that the different forms in the sequence reflect differences in how galaxies evolve under a variety of conditions and environments, not different stages in evolution (see page 398).

What then is the Hubble sequence? It is, first, a *dynamic arrangement,* ordering the galaxies according to the degree of rotation and structure; and it is, second, a *population sequence* characterized by the progress of stellar evolution in each galaxy.

PECULIAR GALAXIES

Not all galaxies fit into Hubble's classification scheme. In recent years astronomers have discovered over ten thousand *peculiar galaxies* that do not fit the scheme because of either their optical appearance or their ab-

normal radio emission. Some of these galaxies are part of tattered strings of interacting galaxies; we can easily recognize them by their torn, fragmented appearance. Figures 17.6 and 17.7 show typical examples. Their malformations reveal great instability, possibly caused by the disintegration of a larger unstable mass into sections like the broken links of a chain. Other peculiar galaxies have tidally distorted forms, many of them connected by luminous bridges of stars and dust. According to computer-generated models for galactic interactions, encounters between grazing galaxies produce streams of tidal debris curving in opposite directions from the galaxies and many bridges between them. Many of these simulated appearances closely resemble actual forms, as shown in Figure 17.8.

Other galaxies with optical or radio peculiarities include the bright, extra large *elliptical galaxies;* the *N type of galaxies* with starlike nuclei; the *Seyfert galaxies* with small, active, bright nuclei exhibiting broad emission-line spectra; and other *active galaxies.* Nearly

BARRED SPIRAL GALAXIES

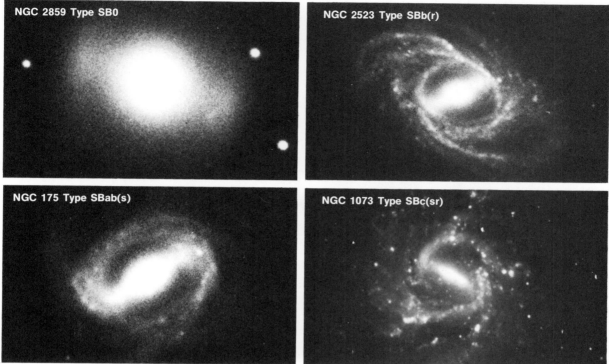

NGC 2859 Type SB0

NGC 2523 Type SBb(r)

NGC 175 Type SBab(s)

NGC 1073 Type SBc(sr)

FIGURE 17.4
Barred spiral galaxies. The arms of barred spirals trail from spindle-shaped spherical hubs. The arms may be thick or, sometimes, fine-drawn, encircling the hub. As is true in the normal spiral galaxies, the less tightly wound the arms of the galaxy are, the smaller is the central region.

all are strong radio emitters; some are strong X-ray and infrared emitters of the type connected with violent events. In Section 17.6 we will discuss the abnormalities associated with these unusual galaxies.

17.2 Properties of Galaxies

It is hardly surprising, in view of the enormous expanses involved (up to a hundred thousand times the diameter of our Galaxy), that the distances to the remote galaxies are at best crude estimates. Distances to neighboring galaxies, however, are fairly well established. The accuracy, not surprisingly, diminishes as the distance grows. Just as precise determination of the astronomical unit (the earth's mean distance from the sun) sets the scale for our solar system, so we must

IRREGULAR GALAXY

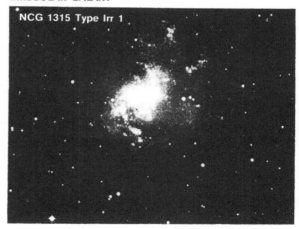

NCG 1315 Type Irr 1

FIGURE 17.5
Irregular galaxies are shapeless. Most contain turbulent gas clouds and brilliant blue stars, but some are poor in gas and have a few old red stars. (See also M82 in Figure 17.19.)

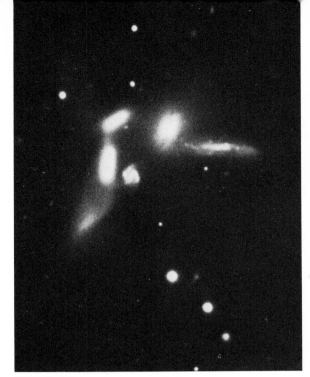

FIGURE 17.6
Interacting galaxies with connecting bridges in Serpens.

have an accurate yardstick for measuring distances of the galaxies in the universe. Essential cosmic data depend on how reliable this yardstick is. The linear dimensions, spatial distribution, intrinsic luminosities, and masses of the galaxies; the physical and evolutionary differences among them; the average density of matter in the universe; the rate of expansion of the universe; and the type of cosmological world model— all depend on the correct scale of distance.

STANDARD CANDLES: DISTANCE INDICATORS

To find the distance of galaxies, we assume that similar objects in our Galaxy and in others have the same physical characteristics. Suitably chosen objects can serve as *standard candles* or distance indicators. From investigations in our own Galaxy and the nearby galaxies, we know the absolute luminosity of these objects. To determine the distance to a galaxy, we take the absolute luminosity of a standard candle within the galaxy and its observed apparent brightness. Then, using the inverse-square law of brightness, we find the distance by way of the distance modulus, $(m - M)$, as described in Section 12.3.

Table 17.1 lists kinds of standard candles and the maximum distances to which they theoretically may be employed in estimating distances to the galaxies with the 5.1-meter Hale reflector. Except perhaps for the supernovae, in practice we cannot measure distances beyond roughly 50 million light years by using individual objects in the galaxy as distance indicators, because it is so hard to resolve these standard candles in the more remote galaxies. This is, of course, true for all the distance indicators. If they cannot be resolved in the galaxy, then they cannot be used. There are a variety of things that affect the resolution, such as orientation of the galaxy. The cepheid variables are still our most dependable criteria in finding distances. We can use them today for only the closer galaxies (less than 15 million light years away) where we can single them out.

OTHER DISTANCE METHODS

When we can not distinguish isolated objects in a galaxy, we often estimate the distance from the galaxy's total luminosity, which we obtain from its surface brightness, apparent size, or mass-to-light ratio. We must be careful to recognize exactly the class of galaxy involved, because galaxies differ greatly in brightnesses and sizes. Lumping all into one group would introduce large errors in the estimated distance. We can distinguish between the image of a galaxy at the threshold of visibility and a star's pointlike image by the galaxy's fuzziness or nonspherical shape. However, this still may not allow the astronomer to classify the galaxy, as the images in Figures 17.6 and 17.7 clearly show.

The distance to a large cluster of galaxies is much more reliable. We can select, say, the observed average apparent magnitude of the ten brightest galaxies as a criterion for luminosity instead of depending on one galaxy. If we adopt their mean absolute

The fires that arch this dusky dot—
  Yon myriad-worlded way—
The vast sun-clusters gather'd blaze,
  World-isles in lonely skies,
Whole heavens within themselves,
  amaze
  Our brief humanities.

Tennyson

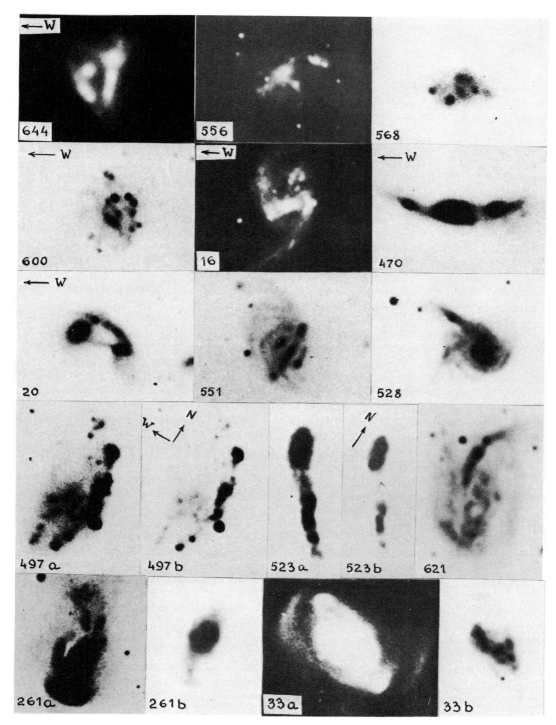

FIGURE 17.7
A composite of photographs of some interacting galaxies taken at the prime focus of the 6-meter telescope in the U.S.S.R. The interacting galaxies of the top two rows are referred to as nests, while the next two rows are chains of interacting galaxies. The last row is enigmatic as to structure and origin.

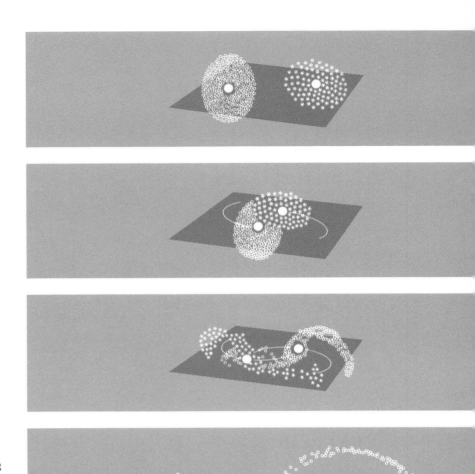

FIGURE 17.8
Computer simulation of two interacting galaxies. The near collision was made to simulate the appearance of NGC 4038/9 shown at the bottom. Circles are material from one galaxy, while the stars are material in the other. The interval between successive pictures is about 200 million years. Colliding galaxies can interact for as long as a billion years.

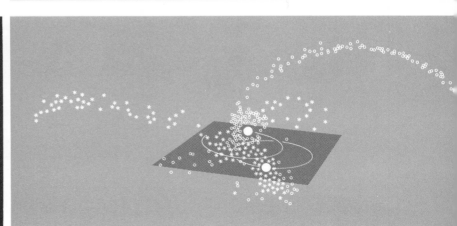

magnitude as $M = -21$, based on knowledge gained from similarly constructed nearer clusters, we can find the distance of the cluster. At the very least, we can make fairly good estimates of the distances of such clusters by comparing the apparent brightnesses of their brightest members or some other common luminosity characteristic.

COSMIC DISTANCE SCALE

Astronomers have set up a distance scale by building a chain of overlapping distances proceeding from the nearest objects to the farthest (see Figure 17.9). It begins with the distances of relatively close stars set by trigonometric parallax and by the distance of the nearby Hyades open star cluster. These distances serve as a basis for the next step, which comes from variable star data, chiefly from the cepheids, and from the spectroscopic and intrinsic brightnesses of stars in our Galaxy. This sequence in turn serves as a basis for distances of the neighboring galaxies, which are determined from characteristics of their brightest stars, cepheid variables, and other stellar data.

The next sequence takes distances of the more remote galaxies, using as criteria their brightest stars, surface brightness of the galaxy, and the apparent size of their bright gaseous nebulae. The next link in the chain is the cluster of galaxies, taking the brightest galaxy of the cluster or the cluster's luminosity type as a standard of comparison. Finally we connect this distance scale to distances of the most remote clusters of galaxies with the Hubble constant derived from expansion of the universe (discussed in Section 17.5).

SPACE DISTRIBUTION OF GALAXIES

After correcting for interstellar absorption in our Galaxy, astronomers originally thought that galaxies, at least to a first approximation, were scattered equally in all parts of the sky. But later, more detailed investigations of the numbers of galaxies still farther away revealed places where the galaxies seem up to a hundred times more densely packed than in other regions of galactic space.

There is a definite tendency for galaxies to group in clusters that contain anywhere from a handful to hundreds or even thousands of these enormous star systems. Astronomers have discovered countless clusters, ranging from dwarf systems to the enormous multiple systems with diameters up to 150 million light years, such as our local supercluster described on page 385. Inside the central region of a heavily populated large cluster, the average separation between galaxies may be close enough for them to collide.

Clustering is very noticeable when we examine the apparent distribution of galaxies brighter than thirteenth magnitude. Large clumps of galaxies are interspersed with a more or less random distribution of galaxies. It is not yet clear if we are seeing two kinds of superimposed distributions. Figure 17.10 shows the distribution of galaxies within about 300 million light years and that lie in the north Galactic hemisphere. It is obvious that they cluster together and are not uniformly distributed.

Over still larger regions, of the order of several billion light years in diameter, the average deviations from a uniform distribution are much reduced and the

TABLE 17.1
Distance Indicators

Brightest of Its Type (standard candles)	Absolute Magnitude (M)	Theoretical Maximum Distance (10^6 ly)
RR Lyrae variables	0	1.4
Population II red giants	−3	5
Cepheids	−6	20
Red supergiants	−8	50
Blue supergiants	−9	80
Novae	−9	80
Globular clusters	−10	130
H II emission nebulae	−12[a]	320
Brightest supernovae	−19	8,000

[a]Value depends on apparent size of emitting H II region, an excellent indicator of distance.

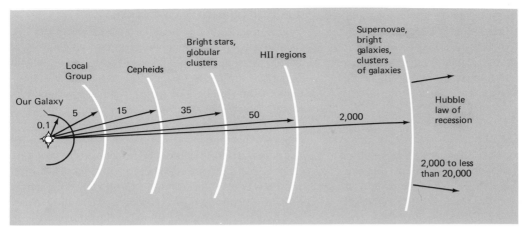

FIGURE 17.9
Practical yardsticks of distance measurements in units of 1 million light years (not drawn to scale).

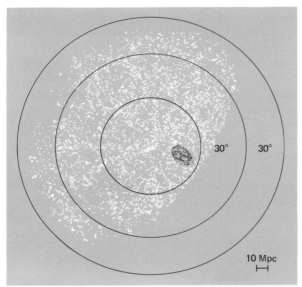

FIGURE 17.10
The distribution of galaxies with apparent magnitudes brighter than 15. These galaxies are typically nearer to us than 300 million light years. The plot is an equal-area polar projection, with the center being the north Galactic pole and the outer circle the Galactic equator. The line at the bottom is a distance of 30 million light years, seen from 300 million light years. The region outlined in black is the central region of the Virgo cluster, which is about 60 million light years from us. From the plot it is apparent that on scales of less than a billion light years the galaxies are not uniformly distributed but are definitely clumped together.

distribution of galaxies seems more homogeneous. Thus localized inhomogeneities apparently exist within the framework of a large-scale isotropic distribution of clusters in the universe. One of these inhomogeneities in a large aggregation known as the local supercluster is the *Local Group.* Our Galaxy belongs to this group, whose diameter is about 3 million light years.

The observed average separation between isolated galaxies is several million light years; if we allow for the suspected but as yet unseen dwarf galaxies, the average separation is even less. Within the Local Group the average separation is less than a million light years. An average cluster has a diameter of about 15 million light years and contains about 130 galaxies. The galaxies in a cluster are separated by only tens to thousands of their own radii, but the stars in a galaxy are separated by millions to tens of millions of their own radii from each other. Obviously, clustering of galaxies is a much more compact arrangement of matter than that of stars in a galaxy.

MASS AND LUMINOSITY
Galaxies have a very large range in size, brightness, and mass (see Table 17.2). Both the largest and the smallest galaxies are elliptical. The true dimensions of galaxies are derived from their apparent diameters and known distances. By combining their measured apparent magnitudes with their known distances, we obtain their absolute luminosities.

The easiest way to estimate the masses of galaxies is to find the Doppler shifts of spectral lines particularly

TABLE 17.2
Distinguishing Properties of Galaxies

Data	Irregular Galaxy	Spiral Galaxy	Elliptical Galaxy
Diameter range[a]	1/20–1/4	1/5–1	1/100–5
Luminosity range[a]	1/20,000–1/10	1/200–1	1/20,000–5
Mass range[a]	1/2,000–1/7	1/200–2	1/1,000,000–50
Mass-luminosity ratio (solar units)	1–10	10–30[b]	10–100[c]
Spectral class of central regions	A–F	F–K	K
Stellar population	I, some II	I (arms), II, and old I (center and halo)	II, some old I
Dust content	Some	Some	Very little or none
Percentage of neutral hydrogen to total mass	18	11–1[d]	Less than 0.1

[a]These figures are given in relation to our Galaxy, whose dimensions are as follows: diameter = 100,000 light years; luminosity = $20 \times 10^9 \, L_\odot$, mass = $170 \times 10^9 \, M_\odot$.
[b]Highest in the centers.
[c]Highest in the giant ellipticals.
[d]Decreases from the Sc to Sa spirals.

for spirals that we see more or less edgewise. We measure the difference in radial velocity at the opposite rotating edges (the edges as they approach us and recede from us along our line of sight). From this velocity difference and the known size of the galaxy, we can calculate the mass from Kepler's modified third law.

Binary galaxies whose members orbit each other provide another means of measuring masses of galaxies, just as we do for binary stars. The pair's difference in radial velocity, combined with their known separation, leads us to estimates of their individual masses, also with the use of Kepler's third law. This method is similar to the one we use in finding the masses of binary stars (see Section 12.6).

A third method makes use of a significant correlation between masses and luminosities for the various types of galaxies known as the *mass-to-luminosity ratio* (representative values are given in Table 17.2). The mass of the galaxy can then be estimated from its observed luminosity.

The mass-luminosity ratio is an important indicator of the kinds of stellar populations in the different galaxies. The variation in the value of the mass-luminosity ratio reflects the differences in the spectral types of the stars that make up different stellar populations in various classes of galaxies. It averages up to a hundred for the ellipticals, from about ten to thirty for the spirals, and from one to ten for the irregular galaxies. A high value means a larger proportion of the stars are faint dwarf stars whose mass contribution to the galaxy is much greater than their luminosity contribution. We must remember from the discussion of stars in our Galaxy that the luminosity of galaxies is provided by the intrinsically bright stars. Thus the distribution of light in a galaxy does not necessarily reflect the distribution of mass in that galaxy. Finally, the neutral hydrogen content decreases from left to right in Table 17.2, apparently an effect of the galaxies' rate of evolution.

17.3 The Local Group

COMPOSITION AND DISTRIBUTION

Over twenty known galaxies belong to the Local Group (see Table 17.3). Most are elliptical galaxies; of the rest, about half are spirals and half irregulars. The Local Group may have still more galaxies, hidden by obscuring material in our Galaxy or undetected because they are very faint dwarf or subdwarf systems. As an example, in 1978 three faint elliptical dwarf galaxies were discovered, which are believed to be part of the Local Group. However, their distances and luminosities have not yet been determined. Largest in the Local Group are the three spiral galaxies M31 (see Figure 16.3), our Galaxy, and M33 (see Figure 17.11), in that order. Smallest are the dwarf elliptical galaxies. The three-dimensional space distribution of the Local Group is such that our Galaxy is near one edge of a slightly flattened system with most others in the southern hemisphere (see Figure 17.12). One subgroup is centered around our Galaxy and another around the Andromeda galaxy.

MAGELLANIC CLOUDS

Our closest visible extragalactic neighbors are two naked-eye objects in the southern skies, the Clouds of Magellan (see Figure 17.13). They were named in honor of Ferdinand Magellan who died while attempting to circumnavigate the earth in 1522. The *Large Magellanic Cloud* (LMC) and the *Small Magellanic Cloud* (SMC) are a physically related double system immersed in a common envelope of neutral hydrogen that emits the characteristic 21-centimeter radiation. They are considered to be satellites of our Galaxy and

TABLE 17.3
The Local Group of Galaxies

Galaxy	Type	Distance (10^3 ly)	Diameter (10^3 ly)	Approximate Luminosity ($10^6 L_\odot$[a])	Approximate Mass ($10^6 M_\odot$[a])
Milky Way	Sb	—	100	20,000	170,000
Large Magellanic Cloud[b]	Irr I	170	30	2,900	25,000
Small Magellanic Cloud[b]	Irr I	190	25	630	6,000
Ursa Minor system[b]	E4 dwarf	220	3	0.4	0.1
Draco system[b]	E2 dwarf	220	3	0.8	0.1
Sculptor system[b]	E3 dwarf	270	8	2	3
Carina system[b]	E3 dwarf	500	4	?	?
Fornax system[b]	E3 dwarf	650	20	23	20
Leo II system[b]	E0 dwarf	700	4	0.7	1
Leo I system[b]	E4 dwarf	700	6	3	4
NGC 6822	Irr I dwarf	1,630	10	100	1,000
NGC 147[c]	E6	2,000	10	70	350
NGC 185[c]	E2	2,000	8	90	450
IC 1613	Irr I dwarf	2,150	15	65	250
NGC 205[c]	E5	2,250	16	330	3,000
NGC 221 (M32)[c]	E3	2,250	8	250	2,100
Andromeda I[c]	E0 dwarf	2,250	≈ 2	2	2
Andromeda II[c]	E0 dwarf	2,250	≈ 2	2	2
Andromeda III[c]	E0 dwarf	2,250	≈ 2	2	2
NGC 224 (M31, Andromeda galaxy)	Sb	2,250	130	25,000	300,000
NGC 598 (M33)[c]	Sc	2,350	60	4,000	39,000
GR 8	Irr dwarf	≈ 3,200	≈ 1	1.4	8

[a] $L_\odot = 4 \times 10^{33}$ erg/s; $M_\odot = 2 \times 10^{33}$ g.
[b] Satellites of our Galaxy.
[c] Satellites of the Andromeda galaxy.

FIGURE 17.11
The nearby spiral galaxy M33 in Triangulum. It is slightly more than 2 million
light years away.

probably move in orbits about the Milky Way. The central plane of the Large Magellanic Cloud is tilted nearly 90° to our line of sight; we also see the Small Magellanic Cloud at an oblique angle. Both galaxies have a ragged disklike structure somewhat flattened by rotation. In them there are stars of all descriptions and ages, including thousands of cepheid variables as well as gaseous nebulae, star clusters, and several supernova remnants. Very prominent in the Large Magellanic Cloud are numerous blue and red super-giants and much obscuring dust. Thus some star for-mation is still going on.

Two differences have been noted between similar objects in the Magellanic Clouds and in our Galaxy: (1) the distribution of the periods, colors, and varia-tion in magnitude of the cepheid variables, and (2) a concentration of blue giant stars in some globular clusters of the clouds, instead of the usual red giants of Population II. This means that these globular clus-ters formed within the last billion years or so. The Magellanic Clouds seem to be in a different, perhaps earlier, stage of evolution than is our Galaxy.

ANDROMEDA GALAXY, M31

The *Andromeda galaxy*, visible as a faint hazy patch of light with the unaided eye, is the largest member of

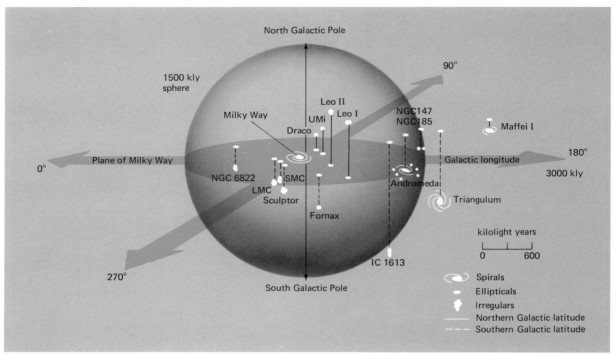

FIGURE 17.12
A three-dimensional plot of the locations of the members of the Local Group relative to the plane of our Galaxy.

the Local Group (see Figure 16.3). Its longest angular dimension covers nearly 5° on the sky, the same as the distance between the two end stars in the bowl of the Big Dipper. The central plane of the galaxy is inclined at an angle of about 12° to the line of sight.

On short-exposure photographs a small brilliant nucleus looks quite similar to the one detected in the infrared and microwave observations of our Galaxy's nucleus. Two well-delineated spiral arms wind out of the nuclear bulge for several turns. We observe many gaseous knots, open star clusters, and X-ray sources in the outlying disk. The galaxy's dust lanes exhibit chaotic patterns and obscure large regions of the stellar background. Neutral hydrogen gas extends well out beyond the optical image of the galaxy.

Population II red giants and a lesser number of planetary nebulae dominate the galaxy's central regions; the outlying portions are dominated by bright blue Population I supergiants (see Figure 13.9). In the intermediate areas between the arms, which are relatively free from obscuring material, is a mixture of both populations. These interarm regions are transpar-

ent enough that we can see remote galaxies through them. Completely surrounding the galaxy are more than two hundred globular clusters. About two dozen novae appear annually. Only one supernova has been detected; it was observed in 1885 in the nuclear region, rising to a peak ten thousand times brighter than the most luminous supergiants in the galaxy.

Doppler shifts in the spectrum of the central portion reveal gas moving outward from the center at speeds of 100 kilometers per second to 200 kilometers per second in an expanding ring, like the gas moving outward in the nuclear regions of our Galaxy. As in the Milky Way, the central bulge of the Andromeda galaxy

The world, the race, the soul—
 Space and time, the universes
All bound as is befitting each-all
 Surely going somewhere.

Walt Whitman

rotates like a solid body. The outer disk, however, has quasi-Keplerian motion, and the inner disk combines the two motions. We can estimate the mass of the system by applying Kepler's modified third law and assuming that most of the galaxy's mass is concentrated at the center. Its mass is about one and a half times the mass of our Galaxy. In almost everything our stellar system and the Andromeda galaxy are remarkably alike: major features, physical characteristics, and composition of celestial objects.

The two small satellites of Andromeda, NGC 205 and NGC 221, are typical elliptical galaxies with myriads of Population II stars. Three new satellites discovered in 1971 are extremely faint dwarf spheroidal systems, each possibly having several hundred thousand old Population II stars.

DWARF GALAXIES OF THE LOCAL GROUP

Each of the dwarf galactic systems—Ursa Minor, Sculptor, Draco, Leo, and Fornax—is a featureless aggregation of several million loosely concentrated stars symmetrically arranged about a sparsely populated center. These have mostly Population II red giants and a smaller number of RR Lyrae variable stars. The two irregular dwarf galaxies, NGC 6822 and IC 1613, present a different appearance. They are ragged systems of very young stars embedded in small gas pockets and a mixture of diverse population groups.

FIGURE 17.13
Magellanic Clouds. They are visible to the naked eye in the southern latitudes. In the photograph the Large Cloud, in the constellation Doradus, has an apparent diameter of 12°; the Small Cloud, in Tucana, has an apparent diameter of 8°.

17.4
Clusters of Galaxies

LOCAL SUPERCLUSTER AND RICH CLUSTERS

As we saw in Figure 17.10, clustering of galaxies seems to be the rule rather than the exception. Thus it is logical to ask whether there is any evidence for clusters of clusters. Within 20 million light years of us there are a number of small groupings much like the Local Group. And about 64 million light years away is the giant Virgo cluster (see Figure 17.14). It has approximately 2500 galaxies stretching about 13 million light years in diameter and containing many double and triple galaxies and subgroups of galaxies. The distribution of all these galaxies suggests that they may form an enormous flattened supersystem or cluster of clusters of galaxies, the *local supercluster*. The super-

cluster's center appears to lie in the region of the Virgo cluster. Its diameter is about 130 million light years and its collective mass is estimated to be about $10^{15} M_\odot$. The Local Group, which is near the edge of this supersystem, revolves around its center at about 400 kilometers per second.

One of the most impressive and rich clusters of galaxies is in Coma Berenices, about 370 million light years away (see Figure 17.15). The mass of this system is 16,000 times greater than that of the Milky Way. Its members are scattered over more than 25 million light years, with only a slight concentration toward the center of the cluster. It has an almost symmetrical distribution of many hundreds of galaxies, mostly elliptical. In general, the more compact the cluster, the higher is the percentage of elliptical galaxies it contains.

Another rich cluster of galaxies, about 570 million light years from us, is the Hercules cluster (see Figure 1.5). Both the Virgo and Hercules clusters have very little central concentration or symmetry. They contain a high percentage of spiral and irregular galaxies and many peculiar galaxies of the type associated with unstable configurations. Some clusters consist almost en-

FIGURE 17.14
Cluster of galaxies in the constellation Virgo. The two large elliptical galaxies are M84 and M86. Thousands of galaxies, beyond the edges of this photograph, form a rich loose irregular cluster which appears to have no central concentration. It possesses a number of small subgroups. All types of galaxies are represented in the cluster.

tirely of elliptical and SO galaxies. Table 17.4 lists some well-known rich clusters of galaxies. (The actual number of galaxies in a cluster may be much larger than noted because of the possible presence of many dwarf galaxies.) One catalog of rich clusters contains 2700 entries out to 4 billion light years.

MISSING MASS IN CLUSTERS?

Attempts to estimate the total mass of an entire cluster of galaxies have led to conflicting figures. One common method takes the Doppler line shifts arising from internal motions of the individual galaxies in the cluster and averages them. The spread of values about the average gives us a measure of the cluster's entire mass. Another method of estimating the mass uses the observed luminosities of the individual galaxies with their mass-luminosity ratios in Table 17.2. Knowing the types of galaxies in the cluster, we can use the total light emitted to find the cluster's mass.

The dynamical mass found by analyzing the observed radial velocity differences is many times greater than the luminous mass obtained from the light emitted. For the latter, the mass derived includes only the luminous matter within the cluster, but the former method includes matter that may be extraneous to the individual galaxies in the cluster. Perhaps a large amount of undetected matter exists as intergalactic material within the cluster, or as subluminous galaxies or as faint extended halos of the observed galaxies. This matter is believed to be needed to prevent the individual galaxies from dispersing, because the clusters appear to be gravitationally bound for indefinite periods of time.

In what form might this unseen matter be? If it were cool atomic hydrogen, then we should be able to detect it by its emission of 21-centimeter photons. If it were molecular hydrogen, there would be the possibility of detecting its ultraviolet spectrum with the orbiting ultraviolet observatories (see page 95). No evidence exists for extensive quantities of either atomic or molecular hydrogen in clusters of galaxies that would account for this puzzling discrepancy for the mass in clusters.

A third possibility is that the missing mass is in the form of a very hot intergalactic plasma that is almost completely ionized hydrogen. A hot gas would emit X-ray photons but no 21-centimeter radiation. Dozens of clusters of galaxies are powerful X-ray sources. The richer the cluster, the greater is the power output. The source of the diffuse X-ray emission apparently is intergalactic gas in the cluster whose temperature is from 10 to 100 million degrees Kelvin. How this tenuous, low-density gas is heated and radiates X rays is not well understood. There may be sufficient intercluster gas to bind the clusters gravitationally.

Early X-ray observations suggest that there is a diffuse background of X rays, of cosmological origin, in almost every direction of space. Does this radiation come from a hot gas that pervades the entire universe or from many unresolved discrete sources? The deep-sky survey by the *Einstein Orbiting Observatory,* or HEAO-2 (see page 96) seems to be finding large numbers of discrete X-ray sources. A preliminary estimate places the contribution to the diffuse background by discrete sources somewhere between one-third and two-thirds. Thus there may be more matter in both clusters and superclusters than makes itself evident in the visible and radio wavelengths. A great deal more data are needed before we can be sure.

17.5
Receding Galaxies

RED SHIFTS

The first clue to a remarkable discovery about galaxies was uncovered by Vesto Slipher early in this century. Going beyond his earlier work, he found unusually large red shifts up to 5700 kilometers per second in the absorption lines of all but 2 of the 43 galaxies he investigated. Even larger recessional velocities were found for the fainter galaxies observed by astronomer Milton Humason (see Figure 17.16).

Not long afterward, Edwin Hubble succeeded in estimating distances for a number of galaxies whose radial velocities had been measured. When the velocities were plotted against distances, he had a straight-line relationship such that the *farther* away the galaxy is, the *faster* it is moving away from us (see Figure 17.17). The only exceptions were several nearby galaxies, including members of the Local Group, which exhibited velocities of approach. A fundamental aspect of the universe had been found, one that overcomes the random motions of galaxies with increasing distance; it becomes evident only beyond the local neighborhood.

FIGURE 17.15
Central region of the cluster of galaxies in Coma Berenices, whose distance is estimated to be 370 million light years.

HUBBLE LAW OF RECESSION

If we interpret it literally, the proportional velocity-distance relationship—the *Hubble law of recession*—indicates that the universe is expanding. (We will discuss its cosmological implications in Chapter 18.) Hubble's original results have been extended by other investigators and now include hundreds of more distant galaxies and several dozen distant clusters.

TABLE 17.4
Representative Rich Clusters of Galaxies

Cluster	Approximate Distance (10^6 ly)	Approximate Diameter (10^6 ly)	Radial Velocity (km/s)	Estimated Number of Galaxies	Density of Galaxies (no./mly³)
Virgo	64	13	1,180	2,500	15
Pegasus I	212	4	3,700	100	30
Pisces	215	38	5,000	100	5
Cancer	260	14	4,800	150	15
Perseus	316	22	5,400	500	10
Coma	368	26	6,700	800	1
Hercules	570	1	10,300	300	570
Ursa Major I	880	10	15,400	300	3
Leo	1,000	10	19,500	300	5
Corona Borealis	1,100	10	21,600	400	7
Gemini	1,100	10	23,300	200	3
Boötes	2,100	10	39,400	150	3
Ursa Major II	2,200	8	41,000	200	10
Hydra	3,300	?	60,600	?	?

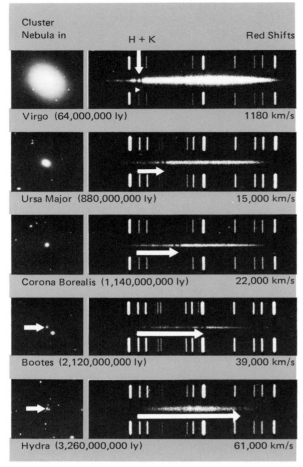

Cluster Nebula in	H + K	Red Shifts
Virgo (64,000,000 ly)		1180 km/s
Ursa Major (880,000,000 ly)		15,000 km/s
Corona Borealis (1,140,000,000 ly)		22,000 km/s
Bootes (2,120,000,000 ly)		39,000 km/s
Hydra (3,260,000,000 ly)		61,000 km/s

FIGURE 17.16
Relation between the red-shifted spectral lines of the galaxies and their distances. The length of the arrow indicates the amount of Doppler shift toward the red. The arrow tip marks the position of the shifted H and K lines of ionized calcium.

HUBBLE LAW OF RECESSION: *the farther away a galaxy is, the faster it is receding from us.*

In the mathematical expression of Hubble's law ($V = H \cdot r$) in which the radial velocity V is proportional to the distance r, the constant of proportionality H is called the *Hubble constant*. The value of Hubble's constant depends on the slope of the line in Figure 17.17. Taking a straight-line relationship between V and r, we see that a galaxy 2 billion light years away is

receding twice as fast as one that is 1 billion light years distant. When the distance indicators of Table 17.1 are unresolvable, we can use the velocity-distance relation to estimate the distance of a remote galaxy for a specified value of H from its red-shifted spectral lines.

As we will see in Chapter 18, the numerical value of H and any departure from the straight-line plot have a lot to do with the type of cosmological model that prevails in the universe. A recent estimate of the Hubble constant places its value at about 17 kilometers per second for each 1 million light years of distance, or about 55 kilometers per second for each million parsecs.

17.6
Active Galaxies

VIOLENT EVENTS IN THE COSMOS

Early explorations of the universe found a cosmos that seemed quiet, orderly, and predictable. It seemed to be filled with well-behaved galaxies, disturbed only by occasional outbursts of novae and supernovae. But when radio astronomers started probing the heavens in the late 1940s, our picture of the universe began to change abruptly. Evidence mounted that the universe is punctuated by the emission of extraordinary amounts of energy.

What are these violent events? In our Galactic niche matter behaves quite placidly, but there are some unusual events in other parts of the Galaxy, such as the occurrence of novae and supernovae. Other remarkable Galactic phenomena—gas flowing out of the nucleus, enhanced gamma ray, X-ray, ultraviolet, and infrared radiation, and nonthermal radio emission from various parts of the Galaxy—may be the after-effects of a more turbulent era in our Galaxy's early history.

Many supernovae have also been found in other galaxies. However, even more violent events have also been discovered: galaxies with extremely active nuclei, exploding galaxies, and quasars. Because the radiation from these active galaxies may be very intense throughout or in particular portions of the electromagnetic spectrum, they stand out from the normal galaxies. One of their telltale features is a continuous spectrum produced by nonthermal synchrotron radiation (page 181).

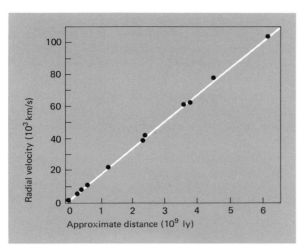

FIGURE 17.17
Hubble velocity-distance relation derived from clusters of galaxies. The constant of proportionality in this relation is known as the Hubble constant and its current value is about 17 km/s/mly or 55 km/s/mpc.

Astronomers distinguish between two kinds of radiation sources: thermal and nonthermal. The intensity of thermal radiation grows weaker with increasing wavelength in accordance with the blackbody distribution of energy (see Figures 4.12 and 11.3). This type of continuous distribution of energy, which is prevalent in our Galaxy and others, comes from stars and from interactions between free electrons and ions (mostly protons) of the interstellar medium. This is known as free-free radiation (discussed on page 181), and much of it lies in the radio wavelengths of the electromagnetic spectrum. We also obtain thermal radiation from planetary nebulae and bright gaseous nebulae like the Orion Nebula.

Energy from isolated Galactic sources like the Crab Nebula and other supernova remnants or from the Galactic nucleus is primarily nonthermal. Unlike thermal radiation, the intensity of nonthermal radiation grows stronger—or at least does not drop as rapidly—with increasing wavelength. Most often nonthermal radiation is synchrotron radiation emitted by rapidly moving free electrons spiraling around magnetic lines of force (see Figure 9.6).

We find the same kind of emission in many galaxies whose optical appearance and spectra are unusual, the so-called *active galaxies*. Even though they may have different forms, astronomers are convinced that the same energy source, whatever that is, is present in all of them in different degrees of intensity. The most luminous of these objects are the quasars, which emit up to a hundred times more light than does a normal galaxy like the Milky Way Galaxy. Indeed, there is strong evidence that the same energy source may operate at the center of our Galaxy but on a greatly reduced scale, by a factor of at least a million compared to the brightest quasars (see p. 359). So far the speculation on the nature of the source that best fits the observational data is one proposing that the central source is a massive black hole, with a mass of up to a billion solar masses, fueled by capturing gas and even stars from its crowded surroundings. Unfortunately, this model is not testable in any laboratory on the earth, nor is it capable of any clear-cut predictions. On the other hand, some astronomers think a new kind of physics may be required to unravel the mystery surrounding these energy sources. In the following sections we will discuss the various types of active galaxies whose general properties are summarized in Table 17.5. The Seyfert galaxies are one example.

SEYFERT GALAXIES

The typical *Seyfert galaxy* (named for Carl Seyfert, who discovered them in 1943) is a fairly large spiral with an unusually small and extremely bright nucleus. Several of these galaxies are powerful and variable radio, in-

> **TABLE 17.5**
> Some Properties an Active Galaxy May Possess
>
> 1. High luminosity
> 2. Nonthermal emission; possible excess emission in X-ray, ultraviolet, infrared, and/or radio spectral regions as compared with normal galaxies.
> 3. Variable emission; rapid variability suggesting small size for emitting region, possibly some slow variability.
> 4. Compact appearance; peculiar photographic appearance with high contrast between nucleus and rest of structure.
> 5. Explosive features; jetlike extensions out of galaxy's center or sometimes broad emission lines and general nonstellar spectrum.
> 6. Gravitational disturbance; high internal velocities and/or disturbed appearance at one or several points across galaxy.

frared, and X-ray emitters. Probably a very small percentage of all spirals are Seyfert galaxies. A photograph of one is shown in Figure 17.18.

Type I Seyfert galaxies have very broad emission and sharp forbidden lines, indicating that ionized matter is expelled at very high velocities (several thousand kilometers per second) from an extremely small nucleus. We do not know what causes this extraordinary activity in the nucleus. Except for their smaller luminosities, these Seyfert galaxies resemble the quasars in their emission of nonthermal radiation. (Quasars will be discussed in Section 17.7.) The less luminous type II Seyfert galaxies exhibit narrower emission lines in their spectra, along with relatively stronger forbidden lines. Their nucleus is somewhat larger, more nearly resembling a normal galaxy, and they contain some dust. The luminosity of Seyfert galaxies is compared to that of other active galaxies in Table 17.6. There is some thought that Seyfert galaxies are in an early disruptive stage through which all spirals, including our own, must pass. Thus there may be an evolutionary sequence for active galaxies leading to normal galaxies like our Galaxy.

N GALAXIES

N galaxies have some properties in common with Seyfert galaxies and quasars, although they are somewhat less luminous. They have a bright sharp nucleus surrounded by a small nebulous envelope. About a dozen have been identified, all are at great cosmological distances according to the large red shifts exhibited by their spectral lines. A very distant Seyfert galaxy might resemble an N galaxy. N galaxies fluctuate rapidly in brightness and color, at times within a few days, suggesting that the variable source must be less than a light week (the distance light travels in a week) in diameter. Other similarities to the quasars and Seyfert galaxies are their strong, broad emission lines and their large output of radio energy.

BL LACERTAE OBJECTS

A small group of compact extragalactic objects that are closely related to the Seyfert and N galaxies are the *BL Lacertae objects*. They are named after their prototype BL Lacertae, which was first identified as a variable star in 1929. Its true nature was not revealed

FIGURE 17.18

Seyfert galaxy NGC 1275, or Perseus A. It is a member of the Perseus cluster of galaxies and radiates strongly throughout the electromagnetic spectrum. It appears to resemble an EO or SO galaxy. In the light of the Hα emission line in the Balmer series of hydrogen, the galaxy has a filamentary appearance, like the Crab Nebula in Figure 14.16. Two sets of emission lines with different Doppler shifts suggest that NGC 1275 may be in collision with another galaxy or that there is an intervening galaxy along the line of sight.

TABLE 17.6
Luminosities of Active Galaxies

Object	Spectral Region: Rate of Radiation (erg/s)					Total Expended (erg)
	Radio	Infrared	Optical	X-Ray	Gamma Ray	
Milky Way Galaxy	10^{38}	3×10^{42}	4×10^{43}	?	?	10^{61a}
Supernovas (Crab Nebula)	10^{34}	?	10^{36}	10^{37}	6×10^{33}	10^{51}
Radio galaxy (M87)	10^{42}–10^{45}	$< 10^{43}$	10^{44}	10^{41}–10^{43}	?	10^{58}
Seyfert galaxy (NGC 1275)	10^{40}–10^{46}	10^{46}–10^{47}	10^{43}–10^{45}	10^{42}–10^{45}	?	10^{59b}
Disturbed galaxy (M82)	10^{39}–10^{41}	10^{44}	10^{55}	?	?	10^{59b}
Quasar (3C 273)	10^{44}–10^{46} (variable)	10^{47}–10^{48}	10^{45}–10^{47}	10^{45}–10^{46}	10^{46}	10^{60b}

[a]Nuclear energy production in stars during the last 15 billion years.
[b]These violent events occur over a much shorter interval (10^5 years to 10^8 years).

until 1969 when radio astronomers singled it out as a very active radio source. About forty such objects are presently known. They are characterized by a sharp, brilliant nucleus emitting strong synchrotron radiation whose continuous spectrum is crossed by weak emission lines. Surrounding the bright nucleus is a faint halo whose stellarlike spectrum resembles that of a typical elliptical galaxy—at least this is true in the case of BL Lacertae.

Although their emission is strong in all wavelengths, the BL Lacertae objects radiate the most energy in the optical and infrared wavelengths. They undergo rapid changes in brightness, which are accompanied by variable polarization. At peak brightness their luminosity rivals that of the brightest quasars. Very long baseline interferometry radio measurements point to a central source, which is at most only a few light years in diameter. This is also inferred from the rapid fluctuations in brightness, since an object that changes in brightness by as much as 30 percent in a single night, as BL Lacertae does, must be very small indeed (see the explanation on p. 330).

RADIO GALAXIES

Catalogs now list more than ten thousand discrete radio sources that are sources of nonthermal radiation. They are divided into two groups according to their physical appearance: compact sources and extended sources. Compact sources are very small and most are identified as *quasars* (see the next section); most of the extended sources are *radio galaxies.*

Radio galaxies may emit 10 million times more nonthermal radio energy than the thermal radio radiation emitted by a normal galaxy. At least four kinds of radio galaxies are recognized (see Figure 17.19):

1. Giant ellipticals with stringlike protrusions in the form of jets or other abnormal extensions, such as in M87.

2. Curious-looking galaxies whose radio-emitting regions are displaced from their optical source, such as Cygnus A and Centaurus A.

3. Highly distorted galaxies that suggest past explosions, such as M82.

4. Tail galaxies in which the emitting regions appear to have been swept back from the optical galaxy, such as in NGC 1265.

The positions of radio-emitting regions vary with the different galaxies. A small, intense radio-emitting center may coincide with the optical center; a tiny, strongly radiating core may coincide with a weaker large envelope, as in M87 (see Figure 17.19). More often, the emitting source, in the form of a lobe, is on both sides of the optical region, as in Cygnus A (Figure 17.19). Optical emission has also been detected within

FIGURE 17.19

Radio galaxies. These illustrations show the four kinds of radio galaxies discussed in the text.

Radio galaxy M82 (*top left*) seems to resemble an exploding galaxy. It shows filaments extending 25,000 light years outward from the center. Its abnormal appearance may have been caused by gravitational distortion from its neighbor M81.

The optical image of radio galaxy Cygnus A (*top right*) shows a fuzzy double object; the radio map reveals two large external sources within which there is another structure that suggests multiple explosions. Cygnus A, about 500 million light years away, was the first discrete radio source outside our Galaxy to be discovered (1945).

Virgo A is a giant elliptical galaxy about 64 million light years away. In the optical picture (*middle left*), taken with

M82

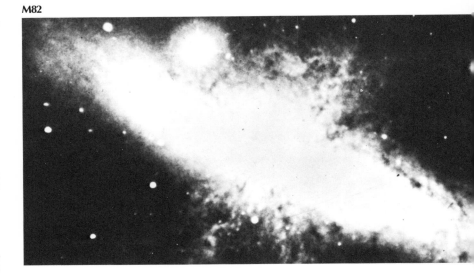

Giant elliptical galaxy in Virgo, M87; also known as Virgo A

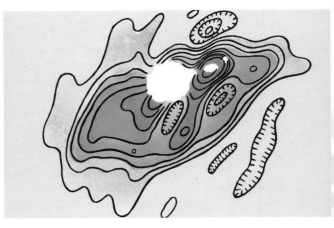

Radio galaxy Centaurus A

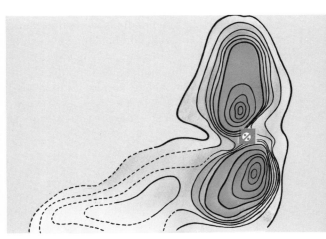

Radio galaxy Cygnus A

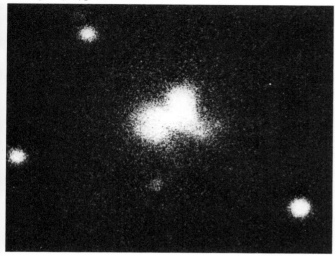

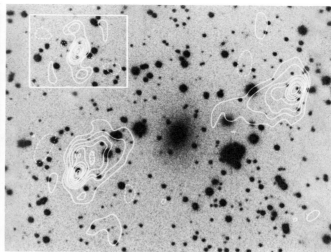

NGC 1265

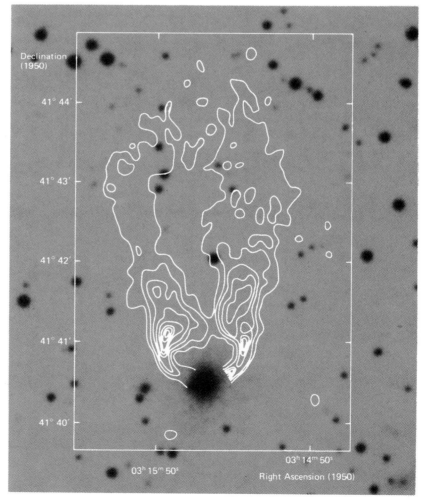

a short exposure, we see a small intense core and a luminous jet that emits synchrotron radiation; the new, computer-enhanced photograph in the inset shows that not a single jet but a series of objects is being ejected. Long-exposure photographs show that the galaxy is surrounded by several thousand globular clusters. The radio contour map (*middle right*) reveals a small bright core at the center of the superimposed optical image surrounded by a weaker halo that extends well beyond the optical source. Notice that the jet streamer is also a region of strong radio emission.

Radio galaxy Centaurus A (*bottom left*), one of the strongest radio sources, is about 16 million light years away. The optical picture shows a strong obscuring lane cutting across the middle of the galaxy. A small double radio source appears on either side of the obscuring dust lane of the galaxy, drawn in the rectangle at the center of the radio map. The central compact source fluctuates in brightness over a wide range of wavelengths. Farther out, on a line perpendicular to the dust band, appear two intense, oval-shaped radio cores surrounded by a fainter, large elliptical region.

Radio tail of the radio galaxy NGC 1265 in the Perseus cluster (*bottom right*). The region responsible for the radio emission of the visible galaxy trails behind the galaxy somewhat like the earth's magnetosphere.

the radio-emitting lobes of several radio galaxies. Bright threadlike features connect the optical galaxy to the radio lobes. Possibly, high-energy subatomic particles are violently ejected along these features from the central source. The sequence of jets or blobs along the lobes suggests that a number of explosions have occurred at different times. In some radio galaxies a long, fan-shaped tail follows the central source. Its appearance, somewhat like that of the earth's magnetosphere, suggests that a bow shock wave has developed and is plowing through the intergalactic medium leaving matter in its wake (see NGC 1265 in Figure 17.19).

The galaxy M87 is a giant elliptical galaxy that coincides with the radio source known as Virgo A. It is about 64 million light years away and is the brightest member of the Virgo cluster (see page 386). Photometric and spectroscopic measures show that M87 is a normal-looking galaxy except in and around its center. In the center, as can be seen in Figure 17.19, a brilliant spike of light is surrounded by a much fainter concentration of stellar matter, estimated to be about 5 billion solar masses packed into a space no larger than a few hundred light years across. From its mass, based on the observed internal Doppler shift velocities and the amount of light emitted, astronomers derive a very high mass-luminosity ratio. Within this region the ratio appears to increase very rapidly as we move toward the center of the galaxy. The straightforward conclusion is that there is a supermassive black hole at the center with an accretion disk around it. As matter swirls into the black hole, the energy released appears as strong X-ray, optical, and radio emission. However, it is not clear how a black hole of this enormity could accrete so much matter in a time equivalent to the age of the universe.

A rather complex structure can be seen in the elliptical galaxy Centaurus A (see Figure 17.19). Here a wide obscuring lane passes through the middle of the galaxy, but the radio picture is that of two small, intense radio sources on either side of the dust lane and two weaker and much larger radio sources well separated on opposite sides of the optical image. Luminous jets appear to extend nearly half a degree out from the optical image toward the upper radio lobe. This was the first optical evidence of a direct physical connection between an optical image and an extended radio source. It gives support for the idea that explosive material ejected from the optical galaxy is flowing into the radio source.

One of the most unusual radio sources is the peculiar galaxy M82, which lies not far from the pointers in Ursa Major and is about 10 million light years away (see Figure 17.19). M82 and M81, a spiral galaxy, are apparently the central members of a small group of galaxies similar to our Local Group. Doppler-shifted emission lines in the spectrum of M82 come from the drawn-out filaments perpendicular to the principal axis of the galaxy. This luminous material emits strongly polarized synchrotron radiation, which ionizes the hydrogen and other atoms in the filaments producing the bright-line spectra. M82 appears to be connected to two neighboring galaxies, M81 and NGC 3077, by long narrow bridges of neutral hydrogen. These galaxies probably experienced close encounters in the past. M82 seems to have undergone a violent tidal interaction about 100 million years ago; subsequently, it appears that an explosion may have taken place that has expelled matter along its minor axis with velocities up to 1000 kilometers per second. It has also been proposed that the polarization and Doppler shifts may arise from scattering of visible photons from the galaxy off of a slowly drifting extragalactic dust cloud in its vicinity. In this case M82 would not necessarily be an exploding galaxy.

How do astronomers account for the intense radio emission in the strongest radio sources? Something like a trillion solar masses of hydrogen, equivalent to a hundred average-sized galaxies, would have to be converted into energy by thermonuclear fusion to equal the radio energy being emitted by these sources. Two possibilities are either that matter is being accreted by a massive collapsed object such as a black hole at the center of the galaxy, or that gravitational collapse of many stars is releasing gravitational potential energy. However it is done, the energy is being produced more or less continuously over many millions of years.

17.7
Quasi-Stellar Objects

DISCOVERY

By 1960 positions of radio sources could be determined very accurately by radio interferometry, so that astronomers could identify some compact radio sources with visible objects, which they named *quasi-*

stellar objects (QSOs), popularly called *quasars*. The QSOs, which resemble ordinary stars on a photograph, had been photographed many times before they were identified, but no one suspected how peculiar they were. The first optical recognition of their unusual nature came in 1960, with an object labeled 3C 48, the forty-eighth entry in the third Cambridge Radio Catalog (see Figure 17.20). It looked like an ordinary star of the sixteenth magnitude, but its bright-line spectrum superimposed on a continuous spectrum was undecipherable until about 1962.

In 1962 Maarten Schmidt of the Hale Observatories made a spectrogram of the small thirteenth-magnitude radio source 3C 273 (it's shown in Figure 17.20), which had been accurately pinpointed by Australian radio astronomers with the 64-meter radio telescope at Parkes, Australia. Within weeks he untangled the puzzle, recognizing in the emission spectrum the characteristic spacing of the second, third, and fourth lines of the Balmer series of hydrogen (see Figure 17.21). They were displaced to the red by a large Doppler shift corresponding to a velocity of 15 percent of the speed of light. The puzzle solved, astronomers soon decoded spectrograms of similar objects, also revealing very large red shifts.

APPEARANCE AND DISTRIBUTION

More than eight hundred quasi-stellar objects have now been examined spectroscopically and the number continues to grow. However, they are still very much an enigma. The majority have a starlike optical appearance. Figure 17.20 shows four quasars.

Twinkle, twinkle, quasi-star
Biggest puzzle from afar
How unlike the other ones
Brighter than a billion suns
Twinkle, twinkle, quasi-star
How I wonder what you are.

George Gamow

One group, the *quasi-stellar* radio sources (QSSs), has strong radio emission that distinguishes them from a second group, *blue stellar objects* (BSOs), which are not radio emitters. The latter are identified by their unusually bright ultraviolet color and large red shifts.

These two classes of inconspicuous, apparently starlike bodies have extensively red-shifted spectral lines and emit extraordinarily large amounts of radiant energy from a very small region. Many of their angular diameters are less than 0.002 arc seconds. In a few of the nearer quasars some nebulous extensions have been photographed, whose faint stellarlike spectra have the same red shifts as the quasar. This suggests that the visible quasar may be the active center of a galaxy or of an object similar to a galaxy.

Despite their apparent tiny sizes, the radio images are frequently structured and noncircular. Quite a few have a hierarchy of minute discrete components; others have extended regions on either side of the optical center emitting radio frequency radiation, somewhat like radio galaxies. Measurements by very long baseline radio interferometry disclose that in some of

FIGURE 17.20
Four quasi-stellar radio sources photographed with the 5.1-meter Hale reflector. From left to right, 3C 48, 3C 147, 3C 273, 3C 196.

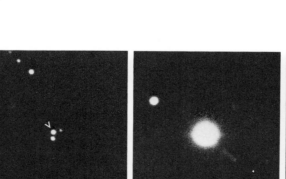

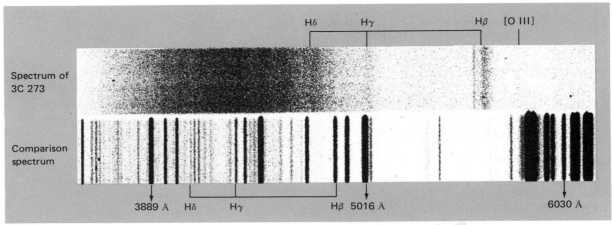

FIGURE 17.21
Spectrogram (negative) of the quasar 3C 273. Notice the pronounced red shift of the Balmer hydrogen lines (Hβ, Hγ, and Hδ) in relation to the comparison spectrum.

the quasars there is apparently a rapid separation, exceeding the speed of light, of close, minute sources. There is as yet no satisfactory explanation of these measurements.

The only way of estimating the distances of remote quasars is to use the Hubble law of recession. Sample counts of QSOs show that their numbers increase much faster with distance than would be the case if their distribution in space were uniform. A plausible interpretation is that the quasi-stellar objects were formed in large numbers within a couple of billion years after the universe began to expand. The universe may then have had a thousand times more quasars than now. Of these 15 million objects (an estimate, of course), which some believe are the short-lived brilliant cores of newly formed galaxies, most would by now have evolved into normal galaxies. Therefore only those very remote quasars, which represent an earlier epic in the universe, would still be observable.

ENERGY FLUX FROM QSOs

Quasi-stellar objects are striking in that an incredible flood of energy gushes out of a source no bigger than a fraction of a light year in diameter. Some give off more energy—by a hundred times—than our whole Galaxy. That the emitting region must be small is shown by the very rapid variation in brightness by many of them and by the twinkling of their radio

emission (like the scintillation of pulsars). The scintillation effect is superimposed on a slower variation in brightness observed in many QSOs which is apparently coming from a region that may be only light days in diameter.

The variability in the QSOs' optical and radio emission is unpredictable. Old photographs show changes in some of them amounting to only a few tenths of a magnitude in days to weeks and even in years; others show larger changes in days or months. Optical and radio fluctuations do not seem to be correlated. The emitted radiant energy appears to peak in the infrared; it is mainly synchrotron radiation.

Spectra of the QSOs exhibit broad emission lines of familiar ionized elements (such as carbon, magnesium, oxygen, neon, silicon, helium) and of hydrogen, overlying a continuous spectrum. The bright-line spectrum apparently originates in hot gases surrounding the source of energy. The lines are highly displaced toward the red. We can identify the lines from their relative spacings that are the same as those in the spectra of planetary nebulae, novae, and other hot sources, as well as laboratory sources. Measured radial velocities for quasars range from a few percent up to 91 percent of the velocity of light. Why no quasars have so far been discovered beyond this limit is puzzling, since the technique for measuring their red shift, at least up to 95 percent of the velocity of light, is available.

About 10 percent of the quasi-stellar objects also have narrow absorption lines in their spectra, usually with red shifts less than or equal to those of the emission lines. Another puzzle is the discovery of one or more absorption-line systems having slightly different velocities. An obvious interpretation is that the absorbing is done by cool shells or clouds moving outward at different speeds up to appreciable fractions of the speed of light. On the other hand, if the absorption is not physically related to QSOs, it presumably must be intergalactic. Since the absorption lines are close to the wavelength of the emission lines, the first of these hypotheses seems more acceptable, but the question is far from settled.

ARE THE RED SHIFTS COSMOLOGICAL?

Most astronomers believe that the Doppler shifts of the spectral lines in the spectra of quasi-stellar objects correspond to true velocities resulting from the expansion of the universe, a view known as the *cosmological interpretation*. However, some astronomers think that these objects may not be so distant after all. If they are closer, their energy output can be smaller, bringing them more in line with other celestial objects. One noncosmological interpretation, the *local Doppler hypothesis*, is that these bodies were violently expelled at high speeds from nearby galaxies or from the center of our Galaxy. Looking at strings of galaxies that include some quasars, it looks like they could have been ejected in opposite directions from the nuclei of large active galaxies. If the quasars were ejected from nearby galaxies, though, some should have velocities of approach. However, no blue-shifted quasars have been observed.

Very different red shifts have been measured for a few quasars and their apparently allied galaxies or groups of galaxies; they vary by as much as many thousands of kilometers per second (see Figure 17.22). In each of these cases the quasar is always red-shifted more than the allied galaxy. If enough discrepancies are substantiated (for reasons other than chance alignments of bodies at different distances), the cosmological interpretation that red shifts are caused *entirely* by the expansion of the universe could be questionable. Some unknown secondary effect, possibly depending on the type of galaxy, may also be involved. Yet several quasars that appear to belong to clusters exhibit the same red shifts as the galaxies in the cluster.

If the cosmological explanation for the quasar red shifts is correct, how can we account for the vast

FIGURE 17.22

Different redshifts for galaxies and quasi-stellar objects which are apparently related. The dark pair at the center of this negative image are quasars. They lack any detailed structure, while the cluster of galaxies to the right of them does show a complex structure. The cluster of galaxies and the right quasar have the same redshift, but the quasar to the left, which is apparently associated with the other two images, has a redshift over four times larger.

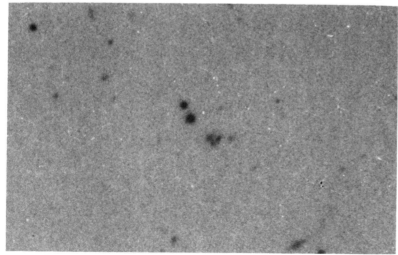

amounts of energy pouring out of such distant but incredibly small regions? Gravitational collapse of a cluster containing massive objects whose combined mass is a trillion solar masses could be converted into radiation agreeing with the observed amounts (10^{45} ergs per second to 10^{48} ergs per second). The high mass required by millions of QSOs could raise the average density of matter in the universe enough to change our thinking about its expansion. We can get around the difficulty by imagining that the energy could be supplied by multiple supernova outbursts within dense galactic nuclei. Other possibilities are radiation produced by matter being dragged into a massive black hole; unstable plasma clouds radiating in a magnetic field; or a giant pulsar ejecting high-energy electrons.

Based upon what we know, the best quasar model for the emission of radiation is the following. There is an energy source, which is only a few light weeks in diameter, producing synchrotron radiation at the center of the quasar. Ultraviolet light from the synchrotron source is absorbed in a hot plasma cloud extending a few light years from the center and expanding outward; it heats and ionizes the expanding gas. The broad emission lines are due to the light of these gases being shifted to the blue or red, depending on whether or not the material is moving toward or away from us. Absorbing cool patches of matter located several thousand light years outside the light-emitting region account for the observed absorption features in the spectrum.

17.8 Evolving Clusters and Galaxies

FORMATION OF CLUSTERS

How did the clusters of galaxies form? One current idea is that large unstable aggregates of material condensed out of the turbulent primordial gas clouds that were formed in the early expanding universe. These huge unstable condensations presumably collapsed gravitationally into one of two forms: either a symmetrical, dense cluster of galaxies resembling the Coma cluster, which is stabilized by its slowly rotating elliptical members and their small internal random motions; or an elongated, semistable cluster resembling

My suspicion is that the universe is not only queerer than we suppose, but queerer than we can suppose.

John Haldane

the Virgo cluster, with its large complement of fast-rotating spirals and their large internal motions. Clusters of galaxies may still be forming from intergalactic material, most of which remains undetected. Galactic evolution in clusters may be affected by forces between galaxies, by merging of colliding galaxies, by cannibalization of one galaxy by another, and by movement through an intergalactic medium. If hydrogen gas is pulled out of a galaxy, then star formation will cease and the galaxy will have prematurely ended that phase of its existence. This will certainly influence subsequent evolution of the galaxy.

EVOLUTION OF GALAXIES

Evolution of individual galaxies is influenced by the density and temperature of the primordial gas, its turbulence, magnetic fields, the rate at which intergalactic matter accretes, the local radiation environment, and the rate of star formation. The gaseous material from which an individual galaxy forms, the protogalaxy, isolates itself from its surroundings before stars begin to take shape.

High gas density in a protogalaxy favors rapid star formation before gravitational contraction can occur, and the end result is possibly an elliptical galaxy. In the gas-impoverished ellipticals, star formation has virtually ceased. Star formation in a protogalaxy with lower gas density is slower, leaving time for gas to settle toward an equatorial plane about the axis of rotation of the protogalaxy. This gas forms the disk portion of a spiral galaxy, which is surrounded by a spherical halo of stars created before the collapse. Star formation then migrates outward from the center of the disk and eventually spreads into the outlying spiral arms. The degree of flattening in the elliptical galaxy or in the disk of the spiral galaxy depends on the amount of spin the protogalaxy had before the contraction started. Also, the rate of rotation by a protogalaxy is probably important in determining whether it evolves into an elliptical or spiral galaxy, since the rate of rotation for the spirals appears to be larger than that for elliptical galaxies. One as yet unanswered question about the way galaxies evolve concerns the important physical processes that cause the great variation in mass and structure of galaxies.

In many respects, the study of galaxies is only beginning. Each new observation seems to raise more questions than answers. A great deal of effort awaits astronomers before many of the riddles can be solved in our quest for knowledge about nature's cosmic building blocks—the galaxies.

SUMMARY

From extensive photographic data Hubble showed that almost all galaxies can be classified as spiral, elliptical, or irregular. The dwarf elliptical galaxies are the most numerous. Our Galaxy is classed as a large spiral. However, there are galaxies which do not fit into the Hubble classification sequence.

In the last fifty years information on the galaxies has mounted rapidly: numbers and types, distribution in space, sizes, luminosities, motions, masses, rotation, structure, and composition of stars, gas, and dust. Surveys show many regions with scores or thousands of galaxies bunched into clusters. Our own Galaxy belongs to a small cluster with nearly two dozen galaxies, the Local Group. It is about 3 million light years across. Some clusters of galaxies may in turn belong to supercluster systems spread over enormous regions of space—regions up to 150 million light years in diameter.

A great scientific discovery in this century came from Hubble, who found that the galaxies are receding from us at speeds proportional to their distance from us; this is interpreted as a general expansion of the universe. For each 1 million light years of distance, the recessional speed increases 15 or 16 kilometers per second.

Early explorations of the universe in the visible wavelengths indicated a quiet, orderly cosmos, except for an occasional outburst in the form of a nova or supernova. When radio astronomers started probing the heavens in the late 1940s, the picture of the universe changed. Powerful new sources of radio radiation were discovered farther out in space than ever before. While radio astronomers and optical astronomers continued their explorations, the evidence mounted: the abnormal emission from radio galaxies, certain galaxies with highly active cores (Seyfert galaxies, N galaxies, BL Lacertae objects), and other active galaxies indicated a universe punctuated with violent cosmic events.

This impression became more firmly rooted following the radio discovery of the very energetic quasi-stellar objects (quasars) and the disclosure that, according to their large red shifts, these mysterious sources appear to be at great cosmological distances. If this interpretation is correct (a view not shared by all astronomers), we are looking back in time toward an early period in the history of the universe when it was filled with great unrest.

REVIEW QUESTIONS

1. What, if anything, does the Hubble sequence of galaxies represent?

2. Describe briefly the general appearance of the three principal classes of galaxies (spiral, elliptical, and irregular).

3. Discuss the distribution of the population types and interstellar material within the three types of galaxies in the Hubble sequence.

4. What is the range in the diameters, luminosities, and masses of the galaxies? Where does our Galaxy fit into the picture?

5. Explain how certain distance indicators can be used as standard candles in estimating the distances to the galaxies.

6. Discuss the findings that have been made regarding the clustering of the galaxies.

7. How do we know that there is a relatively small clustering of galaxies known as the Local Group? Where does the Local Group fit within the present picture of the local supercluster?

8. What is the Hubble law of recession of the galaxies? What is its significance?

9. What is the Hubble constant? What does its numerical value tell us?

10. How is the stepping sequence of distances, which links the distances of remote objects to the local distance scale, derived by astronomers?

11. Discuss the possibility that a considerable portion of the total mass of a cluster of galaxies may be locked up in invisible intergalactic matter.

12. How does a spiral galaxy manage to preserve its spiral structure?

13. How does stellar evolution proceed within a spiral galaxy? Use our Galaxy as a basis for discussion.

14. How can an astronomer identify a certain object as a quasar rather than as a pulsar or a faint star?

15. How have astronomers determined that quasars are very small bodies?

16. What evidence is there that the nuclei of Seyfert galaxies are unstable structures of high-energy content?

17. What are the arguments in favor of the thesis that the quasi-stellar objects are cosmologically very distant?

18. What are the arguments that the quasi-stellar objects are not as far away as their red-shifted lines indicate?

19. Why do we have difficulty accounting for the high-energy content of the QSOs?

20. Describe the optical and radio appearances of the radio galaxies.

18
Cosmology

18.1
Study of the Universe

In our study of the planets and the stars, their evolution and their motions, we have repeatedly talked of the force of gravity. It is, however, only one of four basic forces in nature (the others being the strong nuclear, the weak interaction, and the electromagnetic force; see page 73), and it is the weakest of the four. It becomes the most powerful in the universe at large because its strength increases with increasing mass. Did gravity, then, shape the universe? If so, was there a beginning and will there be an end? How has the universe evolved under the influence of gravity?

When we study *cosmology*, we question the origin, structure, and evolution of the universe. If light had an infinite velocity, then it would bring us information about all parts of the universe instantaneously. Since light has a finite velocity, we observe the universe at different stages in its evolution depending upon the distance various parts are from us. That is, the more distant regions are observed earlier in time than the nearer regions. It is like trying to understand the properties and characteristics of human evolution, but never being allowed to see all the fossil remains of a human being that are the same age. The challenge cosmologists face is developing—from a theory for gravity (such as general relativity) and some reasonable simplifications (principles)—a theoretical model of the universe that can fit the observational data from different ages of the universe.

One assumption we start with in developing cosmological models of the universe is the following postulate:

COSMOLOGICAL PRINCIPLE: *All observers in space see the universe in its essential features in the same way, i.e., the universe is homogeneous (uniform) and isotropic (the same in all directions).*

◄ Edwin Hubble (circa 1925) sitting at the observer's station of the Mount Wilson telescope, with which he discovered the cepheids in M31 and the expansion of the universe.

> Man is not born to solve the problems of the universe, but to find out where the problems begin, and then to take his stand within the limits of the intelligible.
>
> J. W. von Goethe

This assumption says that our sample of the universe, except for local variations, is no different from another sample selected at random at a different place in the universe. Is there any evidence to support this assumption? The apparent large-scale isotropic distribution of the galaxies and of the cosmic blackbody radiation (the dying embers of the fireball explosion that supposedly started the big bang) appear, at least, not to contradict this postulate.

18.2
Static Cosmological Models

Using the basic postulates, astronomers can develop several models of the universe from the *field equations* of the theory of relativity. Field equations are mathematical expressions telling us about the properties of space-time and its relationship to matter.

In 1917 Einstein solved his field equations for a static, or nonexpanding, universe. Einstein's solution was a reasonable approach before the red shift was discovered. He naturally assumed that the random motions of the galaxies cancel out to zero, leaving the universe in a static condition, and that the average density of matter spread out over the universe remains constant.

If the universe is static, if the radius does not change with time, what does it look like? The model Einstein preferred was a spherically closed universe with matter thinly spread out, rather than a flat and infinite universe. He pictured the cosmos as finite in extent but boundless, just as a sphere's surface is limited in size but does not have an edge or boundary.

Einstein found he had to add a slight repulsive force between material particles, which acted over the distances separating galaxies, to keep the universe from collapsing by its self-gravitation, so he introduced a *cosmological constant* in his equations. He found that the radius of the universe was inversely proportional to the square root of the mean (spread-out) density of matter. For reasonable estimates of the mean density, which are between 10^{-29} and 10^{-31} grams per cubic centimeter, the radius of Einstein's model universe is between 10 billion and 100 billion light years.

Einstein's model universe was truly static and predicted no red shifts (see below) for galaxies. Soon after Einstein's work, another apparently static solution was found by the Dutch astronomer Willem de Sitter. His model universe had the remarkable property of predicting a red shift proportional to distance, and it drew a great deal of attention as red shift data were obtained for galaxies. However, in retrospect, de Sitter's model was not really static (for reasons that are too complicated to discuss here); it was in reality a forerunner of the nonstatic models.

18.3 Nonstatic Cosmological Models

Nonstatic, or time-dependent, models can be such that either the universe is infinite, curved, and open; or it is finite, curved, and closed; or it is flat. Which model is correct? The answer depends on several crucial physical factors: the average density of matter in the universe, the exact Hubble law of recession, and the rate for expansion of the universe.

EXPANDING UNIVERSE

Even before Hubble discovered the law of receding galaxies in 1929, he had measured red shifts for a number of galaxies. Cosmologists consequently became more interested in nonstatic or expanding models of the universe. The nonstatic solutions do not need a cosmological constant, and this was a great relief to cosmologists—and to Einstein in particular, who was aesthetically displeased when he adopted it. He is said to have felt that introducing the constant was the greatest blunder of his life.

Let us reexamine the Hubble law of recession in an expanding universe. We showed in Figure 17.17 that a galaxy's velocity of recession is proportional to its distance; the farther away the galaxy is, the faster it is receding from us. This is interpreted as a true Doppler effect caused by the expansion of the universe.

To help us visualize this law on a cosmological scale, consider a two-dimensional analogy: tiny, flat microbes on an ink spot (their galaxy). The microbes are distributed more or less uniformly with other ink spots (other galaxies) on the surface of a balloon that is being inflated (their expanding universe). They observe themselves to be living in a flatland. Suppose that at one time four galaxies, denoted by G_1, G_2, G_3, and G_4, are observed by the microbe astronomers from

RELATIVISTIC DOPPLER EQUATION

The Doppler effect must be modified when the relative speed of recession or approach is an appreciable fraction of the speed of light. The correspondence between the nonrelativistic and relativistic forms is

$$z = \frac{\Delta\lambda}{\lambda} = \frac{v}{c} \qquad \text{(nonrelativistic)}$$

$$z = \frac{\Delta\lambda}{\lambda} = \sqrt{\frac{1 + (v/c)}{1 - (v/c)}} - 1 \qquad \text{(relativistic)}$$

where $\Delta\lambda$ is the measured wavelength shift at λ, v is the radial velocity of the object, and c is the velocity of light. If a shift toward the red from 3000 angstroms to 9000 angstroms is observed in a quasar, $\Delta\lambda/\lambda = (9000 - 3000)/3000 = 2$, and its recessional velocity is 4/5 c, the correct value calculated from the relativistic equation. If we insert these values in the nonrelativistic equation, the velocity is twice the speed of light ($v/c = 2$), a physical impossibility. It is often convenient to express the red shift as the proportional shift in wavelength, z or $\Delta\lambda/\lambda$.

their home galaxy G. They occupy the positions shown in Figure 18.1a, and their initial distances are given in the second column of Table 18.1. We assume that the four galaxies are originally confined to a region of space small enough to be considered essentially flat. One second later, the balloon has doubled in size (Figure 18.1b) and the distances have doubled, as can be seen in the table. In the last column of the table we see that the speed of recession of the four galaxies as observed from the home galaxy is proportional to their distance. This is the Hubble law of recession, and it is the same from whichever galaxy we consider it, as implied in the cosmological principle.

If, as the balloon continues to expand, our discriminating microbe astronomers measure the velocities of the very remote galaxies, they might discover that the straight-line relation between velocity and distance no longer holds. Suppose the little astronomers find that for the most distant galaxies, velocity increases faster than distance, causing a distance-velocity graph to curve upward slightly. This could be one indication that their supposedly flat, two-dimensional universe is positively curved (i.e., it turns back on itself), leading to a *closed universe*.

CURVED SPACE

We can think of space as a three-dimensional surface in a four-dimensional hyperstructure, just as our imaginary flat microbe astronomers might see their flatland as the two-dimensional surface of a three-dimensional sphere. If the two-dimensional creatures knew how to evaluate the properties of their space geometry, they would know on what kind of surface they lived. They might find out by traveling great distances along the surface of their world and noticing how the curvature changes in value from point to point. They could then decide whether they lived in a Euclidean world of zero curvature or whether their world is non-Euclidean and on the average exhibits, in a global sense, positive or negative curvature. These

FIGURE 18.1
Recession of the galaxies on a two-dimensional surface. a. A partially inflated balloon has a random distribution of ink spots (galaxies) at a certain time. b. The same balloon after inflation shows the apparent distribution of ink spots (galaxies) one second later. The analogy between the balloon and receding galaxies is inexact in one way; although the space between the galaxies stretches, the galaxies themselves do not expand.

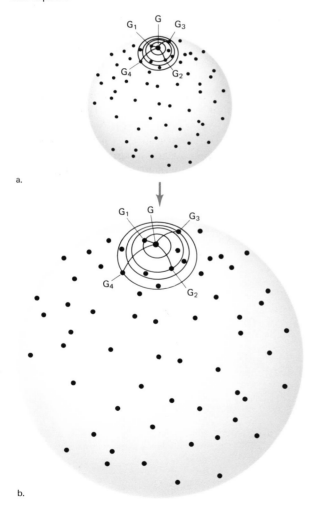

TABLE 18.1
Recessional Motion of Two-dimensional Galaxies

Galaxy	Initial Distance from Home Galaxy G (cm)	Distance from Home Galaxy One Second Later (cm)	Speed of Recession (cm/s)
G_1	1	2	1
G_2	2	4	2
G_3	3	6	3
G_4	4	8	4

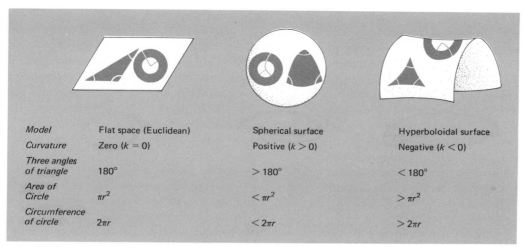

FIGURE 18.2
Two-dimensional world models.

Model	Flat space (Euclidean)	Spherical surface	Hyperboloidal surface
Curvature	Zero ($k = 0$)	Positive ($k > 0$)	Negative ($k < 0$)
Three angles of triangle	180°	$> 180°$	$< 180°$
Area of Circle	πr^2	$< \pi r^2$	$> \pi r^2$
Circumference of circle	$2\pi r$	$< 2\pi r$	$> 2\pi r$

are the only global options open to them. The three choices are shown in Figure 18.2.

Do we three-dimensional beings, living on the surface of a four-dimensional hyperstructure, have a way of determining the curvature of this hyperstructure? We too have three choices for the large scale structure of the universe.

1. The universe has zero curvature ($k = 0$), and it is flat, infinite in extent, and Euclidean, the space geometry most familiar to us.

2. The universe has positive curvature ($k > 0$), and it is finite in extent but unbounded, like the two-dimensional surface of a sphere.

3. The universe has negative curvature ($k < 0$), and it is hyperbolic and open-ended.

Some relativists have raised the specter of a more complicated universe, one in which the four-dimensional system of space-time is immersed in a kind of superspace that could have ten or more dimensions.

GEOMETRY OF THE UNIVERSE

We live in a world of Euclidean geometry, dictated by our common everyday experience—or so we think. For centuries we took for granted the famous axioms and postulates of Euclid (from which the theorems of our high school geometry were developed). During the nineteenth century, mathematicians demonstrated that other kinds of geometry are conceivable.

For example, Euclidean geometry postulates that through any given point, one and only one line can be drawn that will never intersect a given line (that is, a parallel line). Other assumptions are possible. For example, no line can be drawn through the external point that will *not* intersect the given line. Or, through any point not on a given line, any number of lines can be drawn that will *never* intersect the given line. The former geometry possesses what we call *positive curvature*, as on the surface of a sphere, where "straight" lines curve outward; the latter geometry possesses *negative curvature*, as on a saddle-shaped surface, where "straight" lines curve inward. Both contrast with Euclidean geometry, in which straight lines on a flat surface exhibit no or *zero curvature*. No matter which kind of space curvature accurately describes our world model, within our small domain of the universe the appearance is essentially Euclidean. But at very great distances, our viewpoint might change as we develop principles for properly surveying the large expanses of the universe. That is our current dilemma: we do not yet know whether the space curvature of the universe is positive, negative, or zero.

FRIEDMANN MODELS

Beginning in 1922 with the Russian mathematical physicist Alexander Friedmann (1888–1925), cosmologists have derived world models based on nonstatic solutions of Einstein's field equations. When the cosmological constant is set equal to zero, only three solutions are possible. The different types of models depend upon the average density of matter in the universe. If the average density is *less* than a certain critical value, then the universe will be infinite spatially, or an open universe of negative curvature (curve *A* in Figure 18.3). If the average density is *equal* to the critical value, then the universe is a flat Euclidean universe of zero curvature (curve *B* in Figure 18.3). Finally, if the average density is *greater* then the critical value, then the gravitational field produced by the matter in the universe curves the universe back on itself. It is a closed universe of positive curvature (curve *C* in Figure 18.3).

In these models the value of the Hubble constant (*H*) decreases with cosmic time at different rates, depending on the model, as the expansion is slowed by self-gravitation of the universe. The critical density turns out to be proportional to the square of the Hubble constant, and for present values of *H*, the critical density is about 5×10^{-30} gram per cubic centimeter. More complex Friedmann models result when the cosmological constant is different from zero. In Figure 18.3, all the models must agree with the radius of expansion and the rate of expansion at the present epoch, as shown by the point at which the curves coincide labeled "now".

OPEN UNIVERSE

In an *open universe* of negatively curved space, expansion settles down to increase proportionately to time ($R \propto t$). There is but one expansion starting from a singularity (see Figure 18.4), beginning supposedly with a big bang in a superdense core of superhot matter. The amount of matter in the universe is not sufficient to halt the expansion by self-gravitation; the expansion proceeds inexorably toward infinity. While space stretches out with the expansion, the average density of matter decreases as the distances between the galaxies increase.

Light arriving from the most distant objects will just manage to reach us and will be red-shifted almost

AGE OF THE UNIVERSE

The reciprocal of the Hubble constant can be used to estimate an age for the universe, assuming a uniform expansion over its lifetime. Since the relation between distance, velocity, and time is

$$d = vt, \quad \text{or} \quad t = \frac{d}{v}$$

for non-accelerated motion, then by analogy from the Hubble law of recession (p. 388),

$$V = H \times r, \quad \text{or} \quad \frac{1}{H} = \frac{r}{V}$$

Thus, the reciprocal of the Hubble constant has units of time and it is known as the *Hubble time*. For a value of *H* equal to 15 kilometers per second per megalight year or 50 kilometers per second per megaparsec, the Hubble time is

$$\frac{1}{H} = 20 \times 10^9 \text{ years}$$

and for *H* equal to 18 kilometers per second per megalight year or 60 kilometers per second per megaparsec, the Hubble time is

$$\frac{1}{H} = 16 \times 10^9 \text{ years}$$

These values are compatible with the ages of the oldest globular clusters in our Galaxy (see Table 13.3) of 10 to 20 billion years. If the expansion has slowed with time, then the universe must be younger than the above values.

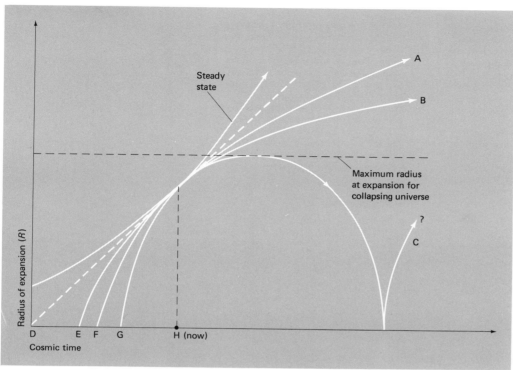

FIGURE 18.3

Friedmann models (*A, B,* and *C*) in which the cosmological constant equals 0: *A, k = −1,* open universe; *B, k = 0,* Euclidean universe; *C, k = +1,* closed universe, where *k* is a measure of the space-time curvature. The steady state model, which has no beginning and no ending, is shown for comparison. The straight dashed line corresponds to a universe whose rate of expansion has always remained constant. Point *H* indicates the present epoch; *EH, FH, GH* correspond to the time that has elapsed for each model since the expansion began. *DH* corresponds to the longer Hubble time if the universe has constantly expanded at its present rate.

beyond perception. This outermost limit of observability is our *event horizon.* It is *not* an edge to space; the universe no more has an edge than it has a center.

FLAT EUCLIDEAN UNIVERSE

The scenario for a *Euclidean universe* of zero curvature (also known as the Einstein–de Sitter universe) begins at a singularity with the supposed hatching of the cosmic egg. The expansion is proportional to the two-thirds power of time ($R \propto t^{2/3}$). This universe is therefore expanding more slowly than the open universe model. Self-gravitation of the universe forces the expansion to come to rest at infinity as the average density falls from an infinite density when time began to zero at the end.

CLOSED UNIVERSE

In a *closed universe* of positively curved space, there exists a finite amount of matter, a finite limit to space, and a finite time for closure (see Figure 18.4). As its expansion is slowed by self-gravitation, the universe eventually reaches a maximum size depending on the average density of matter. Contraction takes over, slowly at first, then accelerating toward a spectacular climax called the *big crunch,* as all matter collapses toward a superheated, superdense state. The greater the average density of matter, the less time it would take to reach this state. Some cosmologists suggest that the universe would then rebound into a new cycle and that the recycled universe need not have the same physical details as the previous one. It could have different physical and chemical properties, in

FIGURE 18.4
Cosmological models of the universe. *Top:* In the big bang, open-ended version, the universe was hatched from an exploding, superdense, superhot primeval atom some 15 to 20 billion years ago. As the universe expands toward infinity, the galaxies spread out and the average density of matter diminishes. *Middle:* In this version, the universe expands and contracts rhythmically in a period of many billions of years. There is enough matter so that self-gravitation of the universe brings the expansion to a halt, after which contraction sets in. At some unspecified distance from the point of origin, a new cycle of expansion and contraction may begin. *Bottom:* In the steady state version, as the universe expands, the distribution of the galaxies remains constant. Old galaxies disappear and new galaxies are formed, so that the universe presents an unchanging appearance. This requires that new hydrogen be created spontaneously, in order that the average density of matter remain the same as the universe expands.

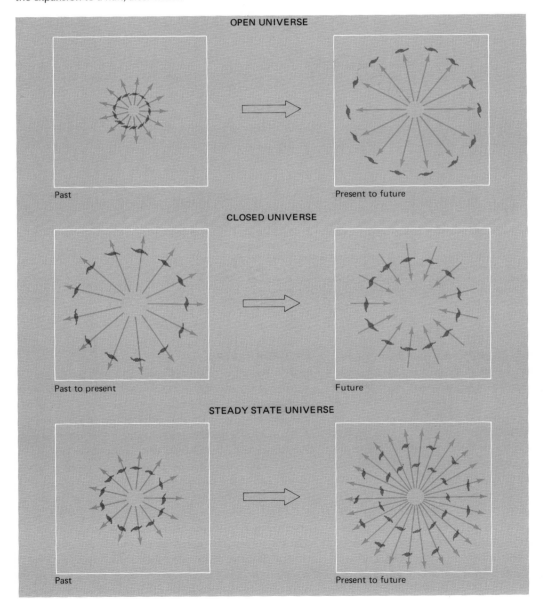

OPEN UNIVERSE

Past

Present to future

CLOSED UNIVERSE

Past to present

Future

STEADY STATE UNIVERSE

Past

Present to future

Hoyle attended Cambridge University and received his master's degree there in 1939. He was one of the prime developers, along with Herman Bondi and Thomas Gold, of the steady state model of the universe. In the late 1940s Hoyle modified Einstein's field equations to take into account what he called the perfect cosmological princi-

ple: If the universe is homogeneous in space, it must be so in time. This required that the universe must always look the same at all times. In an expanding universe, then, new matter (hydrogen) must be spontaneously created to maintain a constant space density for the galaxies, since expansion tends to thin them out. With the discovery in 1965 of the isotropic 3-K microwave radiation—the cooled relic of the fireball radiation that initiated the expansion—the steady state theory, which postulates no beginning and no ending for the universe, fell into disfavor.

Hoyle has made notable contributions in astrophysics, particularly in nucleosynthesis. In collaboration with his colleagues, he described a scheme of thermonuclear reactions within stellar interiors that accounts for the production of the heavier elements beyond hydrogen and helium. He showed that by the time iron was synthesized in the massive stars, since the nuclear burning of iron requires an input of energy, the stars must collapse gravitationally and possibly blow up as supernovas. Low-

mass stars like the sun, he showed, would degenerate into white dwarfs.

In 1959 Hoyle put forth his hypothesis that the tenuous hydrogen intergalactic gas could have a temperature of 100 million degrees. This has since been borne out by the discovery of X-ray emission from the hot hydrogen gas distributed between the galaxies. Hoyle has recently advanced a new theory for the origin of life on the earth. He proposes that the primitive earth was bombarded by numerous comets and meteorites containing living cells created from previous biochemical evolution. Thereafter Darwinian evolution took over.

Hoyle is one of the great popularizers of science and has written a number of books, including science fiction, for the lay reader. Throughout his career Hoyle has remained an individualist in his thinking and has thrived on controversy, often in the face of adverse criticism of his views by fellow astronomers. The bulldog tenacity of his convictions and his great powers of intuition have made him one of the outstanding figures of this century.

which only the fundamental constants of nature—the velocity of light, the gravitational constant, and the Planck constant of radiation—might remain unchanged; or it might collapse into a black hole and be transformed into a strange new universe.

18.4 Steady State Model

Astronomer Fred Hoyle tried a novel way of deploying the field equations of general relativity. Starting with original research by Herman Bondi and Thomas Gold in 1948, he devised a nonstatic model of the universe whose general appearance remains unaltered *forever*. This is the *steady state model* of contin-

uous creation (see Figure 18.4). Hoyle extended the cosmological principle to arrive at the following:

PERFECT COSMOLOGICAL PRINCIPLE: *Not only does the universe appear the same to all observers but it looks the same in perpetuity.*

The steady state model has no singularity, no beginning, and no end. Space expands exponentially with time toward infinity. The Hubble constant (H) does not vary with time as in the evolving models, where it decreases with time. Galaxies form, evolve, and disappear, while the average density of matter in space remains constant. To keep the population of galaxies, or the average density of matter, constant, we have to assume that new matter—hydrogen—is continuously being created. This creation compensates for the expansion that thins out matter. The average rate of creation in a large classroom would be about one hydrogen atom in 50 million years (2.8×10^{-46} gram per cubic centimeter per second), a rate hopelessly beyond detection. Because galaxies are being formed at a steady rate to keep pace with the expansion, the average separation between them remains unchanged with the passage of time.

18.5
Which World Model?

EVOLUTIONARY OR STEADY STATE?

Most astronomers favor an evolutionary universe in one form or another. If the amount of undiscovered nonluminous material in the form of intergalactic matter, dead galaxies, and black holes turns out to be appreciable, it would raise the average density of matter in the universe and increase the universe's self-gravitation. Then the closed evolutionary model might become the appropriate cosmology.

The steady state model in its original form now seems inconsistent with the observed properties of the universe. The counts of several thousand radio sources at great distances reveal that these remote sources seem more numerous in a given volume of space than are the closer sources. This result contradicts the steady state prediction, which requires the counts remain the same in all regions of space. The steady state theory also implies that, on the average, the types of galaxies should be the same for remote galaxies as well as for nearer ones. Among the several hundred quasi-stellar objects spectroscopically examined so far, none fits Hoyle's original description of widely assorted types of galaxies of different ages. Like the distribution of radio sources, the distribution of quasi-stellar objects appears to strongly support the evolutionary models. However, the most telling argument in their favor is the presence of low-temperature, isotropic, microwave background radiation filling the universe, which the steady state theory, in its original form, cannot account for.

THE 3-K COSMIC BACKGROUND RADIATION

Tracing our way back through the expanding universe's history, we cannot help but think that mat-

ter was at one time more densely packed. In the beginning it must have congregated into a hot, super-dense state that exploded violently (the big bang), accompanied by a high-powered blast of high-energy photons (the primeval fireball). Expansion of the universe has cooled this radiation so that most of its energy lies in the microwave spectral region. According to the postulates we laid down in Section 18.1, observers throughout the universe should detect it as low-energy *cosmic background radiation* coming from all directions in space.

In 1934 George Gamow was the first to theoretically predict the existence of this radiation. In 1965 this low-temperature radiation was discovered accidentally by Bell Telephone physicists Arno Penzias and Robert Wilson.[1] They were using a horn-shaped antenna designed to pick up radio signals from earth-orbiting communication satellites. (The discovery reminds us of another Bell Telephone scientist, Karl Jansky, who thirty years earlier accidentally detected radio noise coming from our Galaxy while he tried to find out what caused radio interference on the company's transmission lines.) Cosmic background radiation was also, as it happened, being theoretically reinvestigated by Robert Dicke and his Princeton co-workers. They soon learned that Penzias and Wilson had detected the background radiation—a coincidence not too uncommon in science. The discovery of the cosmic background radiation is the most significant cosmological discovery since Hubble showed that the universe is expanding.

The theory predicts that the radiation of the fireball should correspond to blackbody radiation at a temperature several degrees above absolute zero. The radiation should be isotropically distributed—the same in all directions. It appears that these predictions have been confirmed by observations. The microwave spectral distribution showing the observational points on the blackbody curve is plotted in Figure 18.5. The dots correspond to direct microwave measurements made from radio telescopes on the ground. The open circles represent indirect optical data obtained from interstellar cyanogen (CN) absorption in lines observed in the visible spectrum of several early type stars. Cyanogen's absorption of fireball photons at wavelengths of 0.13 centimeter and 0.26 centimeter raises the molecule to one or the other of the two levels slightly above the

---

[1] For their discovery Penzias and Wilson shared the Nobel Prize in physics in 1978.

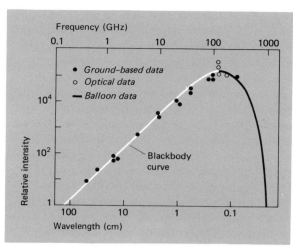

FIGURE 18.5
Microwave cosmic background radiation. The observed distribution of the background radiation closely matches the blackbody energy curve at a temperature of 3 K. In another 15 billion years the temperature will drop to about half its present value. The most recent data suggest that the cosmic background radiation is about ten percent brighter near the peak and twenty percent fainter on the short wavelength side than a 3 K blackbody curve.

ground level. From these excited states the molecules may absorb stray starlight photons to produce the observed lines. The critical part of the curve was first derived from balloon measurements in 1975 in the range of 0.13 centimeter to 0.03 centimeter, made with a spectrophotometer from 30 kilometers above the earth. (The short side of the peak must be observed from outside the earth's atmosphere because of water vapor and oxygen molecules, which absorb microwave radiation in this region.) Later balloon measurements suggest that the blackbody curve is slightly distorted for reasons yet unknown.

NEW ETHER DRIFT EXPERIMENT

The omnipresent, isotropic background radiation can be used as a backdrop for determining the absolute motion of the earth, or the solar system, or the Galaxy. Unlike Michelson and Morley in their celebrated experiment to detect the ether drift of the earth (see page 43), astronomers aboard a *U-2* aircraft over California at 15,000 meters (50,000 feet) have apparently succeeded in measuring the drift of the earth in the sea of cosmic background radiation photons.

In the direction of the earth's motion the background radiation should be slightly hotter according to theory; in the opposite direction, slightly cooler. Consequently, there should be a slight departure from isotropy (anisotropy) in the background radiation due to the earth's motion. The sensitive detectors used to measure the anisotropy were carefully designed to differentiate it from various extraneous instrumental, terrestrial, and extraterrestrial effects. Observed was a minute anisotropy recorded as a temperature difference of 0.0035 K from the average value. The maximum (hottest) and minimum (coolest) differences are in the directions of the constellations of Leo and Aquarius, respectively. Also the velocity of the sun (or solar system) in the direction of Leo differs somewhat from the data given in Chapter 16. The data in this experiment reveal that our Milky Way Galaxy is traveling through the sea of background photons at about 600 kilometers per second and the earth at about 400 kilometers per second. These are astounding results, which need confirmation. Since the early days of relativity theory it was never thought possible to observe absolute motion—that is, motion with respect to any standard of rest. Or is what is observed an illusion? Further experiments are planned to check on the reality of the results and their interpretation.

REVISED STEADY STATE MODEL

The low-temperature microwave background radiation is a logical consequence of a fireball explosion. If it is a relic of the primeval fireball explosion, then the background radiation cannot be reconciled with the steady state theory. That theory assumes that the universe was never in the superdense condition but has always been pretty much the same as it is today.

Hoyle has come up with a theory that abandons continuous creation and includes the microwave background radiation. The red shift of the galaxies is attributed to a change in the mass of subatomic particles with time, not to an expansion of the universe. By this theory, masses of atoms in distant galaxies whose light we now observe used to be less than they are today. The old radiation is red-shifted in comparison with radiation from present-day atoms. Extrapolating time backward, we find a moment in the history of the universe when the mass of the electron was zero. The big bang does not signal an abrupt beginning of creation; but rather it is the moment when electrons of negative mass became electrons of positive mass.

Half the universe existed prior to that time and remains invisible to us. Light from galaxies in that half is blurred near the boundary. The blurred radiation that passes into our half of the universe is identified as the cosmic background radiation. Hoyle concludes: "We may owe much of our present world to the situation on the far side of the 'surface of zero mass' which has hitherto been thought to represent the origin of the universe." Although Hoyle's ideas represent a splendid example of the creative process in a scientist's mind, his proposals have very few adherents.

OBSERVATIONAL TESTS OF COSMOLOGICAL MODELS

Cosmologists face a dilemma: they cannot get sufficiently reliable data on the furthermost galaxies and quasars for a definitive observational test of the various cosmological models. But some testing can be done. An equation that permits us to test cosmological world models by observation, can be derived from Einstein's field equations. This equation connects the bolometric apparent magnitudes (see Section 12.3) of the galaxies with the red shift factor z, or $\Delta\lambda/\lambda$, the Hubble constant H_0, and a related deceleration parameter called q_0. (The zero subscript refers to the present epoch.) The factor q_0 characterizes the rate at which the expansion is slowing because of the self-gravitation of the universe, which brakes the expansion. Graphically, the equation is a red shift $(\Delta\lambda/\lambda)$–against–magnitude plot (see Figure 18.6) instead of the red shift (velocity)–against–distance plot, as in the original Hubble diagram for the law of recession (Figure 17.17). For both of these we assume that galaxies with similar characteristics have about the same luminosity. Distance estimates for the remote galaxies are replaced by their apparent magnitudes, which are easier to derive. Distance and magnitude, as

FIGURE 18.6
Hubble diagram for galaxies and quasars (QSOs). The curves show the values predicted by the four cosmological models. The radio quasars and radio-quiet quasars are too bright compared with the radio galaxies of the same red shift. A = closed spherical universe; B = closed spherical universe; C = Euclidean universe; D = open hyperbolic universe; E = steady state universe.

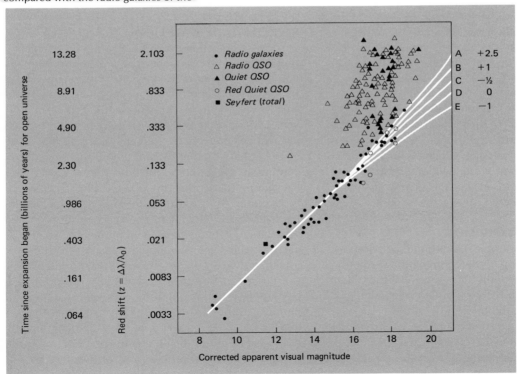

we saw, are related to each other by the inverse-square law of light.

The equation for several values of q_0 is plotted in Figure 18.6. The type of world model follows from the numerical values of two parameters: the Hubble constant H_0 and the factor q_0 or the mean density of matter, which is related to the factor q_0. The world model's dependence on the factor q_0 is shown in Table 18.2. The constant k defines the type of world curvature, as discussed on page 405.

For models in which q_0 is greater than 1/2 ($k = +1$), the expansion will eventually cease and contraction will take over. For $q = 1/2$ ($k = 0$), the expansion decelerates to infinity, where it comes to rest as the mean density reaches zero. For q_0 between 0 and less than $1/2(k = -1)$, the universe keeps expanding forever with finite velocity as the density approaches zero. In the steady state model, $q_0 = -1$ and the mean density remains constant as the universe expands to infinity.

Once astronomers can fit sufficiently reliable observational data to one of the theoretical curves in the plot of Figure 18.6, the type of world model follows. Currently, the observational data agree with the theoretically calculated slope of the curve along the beginning portions. Farther along, however, it is not possible to differentiate unambiguously between the models by departures from linearity. Although the quasars have about the same slope as the galaxies, they do not conform to the plot of the galaxies. They are much too bright for their measured red shifts. A critical review in 1976 by a group of astronomers—based on estimates of cosmological expansion, average density of matter in the universe, age of chemical elements, and ratio of deuterium to hydrogen—leaned toward an open universe. Thus there does not seem to be enough matter to close the universe. According to Table 18.2 the critical density for closing the universe is around 5×10^{-30} gram per cubic centimeter, which is equivalent to about three hydrogen atoms per cubic meter. According to our best estimates, the average density is at most only one-tenth that needed to close the universe.

Recently evidence has arisen supporting a closed universe. Astronomers find that the strength of the triply ionized carbon spectral line at 1549 angstroms correlates with the energy in the nearby continuous portion of the spectrum for quasars exhibiting a flat (constant intensity) radio spectrum. This makes it possible to calibrate quasar luminosities and hence determine distances. A plot of 38 quasars with red shifts from 1.1 to 1.4 on the Hubble diagram suggests a closed universe. Perhaps there is more invisible matter in the universe than suspected. It is estimated that the mass required to account for the diffuse X-ray brightness of the sky exceeds by a large factor the amount of visible matter present in the universe, but even this does not preclude an open universe.

Evaluating the fundamental parameters is slow work. The Hubble constant H_0 and the deceleration factor q_0 must be derived bit by bit, because optical research can be done only with a handful of very large telescopes. Twenty years might be enough time to get reliable values for H_0 and q_0 from better

TABLE 18.2
Cosmological Models

Model	Curvature k	Deceleration Factor q_0	Mean Density[a] ρ_0 (g/cm³)	Age of Universe (yr)	Type of Universe
Evolutionary[b]	+1	Greater than 1/2	Greater than 5×10^{-30}	Less than 13×10^9	Closed spherical (finite)
Evolutionary[b]	0	1/2	5×10^{-30}	13×10^9	Euclidean (flat, infinite)
Evolutionary[b]	-1	0 to less than 1/2	Less than 5×10^{-30}	Greater than 13×10^9	Open hyperbolic (infinite)
Steady State	0	-1	5×10^{30}	Indeterminate	Timeless Euclidean

[a] For $H_0 = 17$ km/s/Mly or 50 km/s/Mpc.
[b] Friedmann model with cosmological constant equal to zero.

observational data. Then we may know the truth about our physical world—if there are no upsetting discoveries or unexpected developments in the meantime.

18.6
Big Bang and
the Evolving Universe

BIRTH OF THE UNIVERSE

The early stages of the big bang model require high-energy physicists to explore theoretically how matter and radiation act under extreme physical conditions unattainable on the earth. We will follow the evolution of the universe as those who have studied the problem lay it out. See the timetable of events up to the present given in Table 18.3.

The initial state of the universe defies analysis. Supposedly it began as unimaginably chaotic and superdense, with a high temperature to match (greater than 1500 billion degrees Kelvin), about 15 to 20 billion years ago when the fireball erupted. In the first one-hundredth of a second the constituents may have been the quarks, which are the possible building blocks of the subatomic particles like protons and neutrons. Thereafter, the superdense, hot cosmic fluid was a mixture of the family of strongly interacting elementary particles called *hadrons*—protons, neutrons, and others—and a smaller proportion of photons and a family of lighter particles called *leptons*—muons, electrons, neutrinos, and others. Matter and antimatter were almost evenly divided, with matter somewhat more plentiful than antimatter by one part in a billion. If matter and antimatter were equally proportioned at the start, some theorists say, matter would be almost completely annihilated and the universe would have consisted of radiation and very little matter, far less than is still around. As far as we know now, nothing says the universe is not fragmented into islands of matter and antimatter.

At nearly the instant when time began, the fireball erupted from the annihilation and conversion of the hadrons into powerful gamma ray photons. Events occurring then (the *hadron era*) are unclear mainly because they are dominated by incompletely understood interactions among strongly interacting nuclear particles.

The *lepton era* began after the hadrons of lowest mass annihilated each other. It continued as the lighterweight particles were destroyed, and it ended with annihilation of the electrons and positrons as the neutrinos produced in these reactions broke away to form a ghost world of their own, moving about eternally and independently of the other constituents of the universe. At their present temperature of about 2 K, these relic neutrinos number about 1200 per cubic centimeter, but they cannot be detected with present technology.

About 1000 seconds after the start of the fireball explosion, the *radiation era* began, and expanding space was filling mostly with photons and neutrinos. The powerful gamma ray radiation was decoupled from matter. These photons were set free to move about in the expanding universe forever. In the words of physicist Edward Harrison: "Matter was like a faint precipitate suspended in a world of dense light."

During the early radiation era, matter consisted mostly of protons and neutrons; the neutrons combined with protons to produce deuterium, and then helium was synthesized. By this time, with the lowered temperature, the electrons had been captured by the protons to form atoms; the protons and the electrons could no longer scatter photons. The universe then became transparent. The fireball radiation that flooded the expanding universe for the first million years now appears as a relic—the cosmic background radiation.

DEUTERIUM AND HELIUM

Most of the deuterium and helium and some lithium and boron were synthesized during the first several hundred seconds or so of the radiation era, at a fireball temperature of about a billion degrees, when the rapidly expanding universe had cooled enough to allow protons and neutrons to combine. The calculated percentage of deuterium and helium created depends critically on the values we adopt for the early matter density and present temperature of the microwave radiation. The results should be consistent with the

The astronomers said: "Give us matter, and a little motion, and we will construct the universe."

Ralph Waldo Emerson

TABLE 18.3
Evolution of the Universe Since Birth

Epoch	Time	Density (g/cm³)	Temperature (K)	Approximate Radius (ly)	Event
Big bang	Zero	Infinitely high	Extremely high	0	Fireball erupts
Hadron era	Less than 10^{-4} s	Greater than 10^{14}	Greater than 10^{12}	< 0.02	Strong interactions by elementary particles
Lepton era	10^{-4} s to 10^2 s	10^{14} to 10^3	10^{12} to 10^9	0.02 to 200	Rapid expansion and cooling; thermal equilibrium of electrons, positrons, neutrinos, and photons; helium nuclei forming
Radiation era	10^3 s to 10^6 yr	10^3 to 10^{-21}	10^9 to 3000	200 to 2×10^7	Radiation uncouples from matter; deuterium and helium formed
Matter era	Greater than 10^6 yr	Less than 10^{-21}	Less than 3000	2×10^7 to $< 2 \times 10^{10}$	Quasars and clusters of galaxies condense
Present era	15–20×10^9 yr	5×10^{-30} to 5×10^{-31}	3	2×10^{10}	Galaxies and stars have formed; stars still forming

present estimated ratios of deuterium and helium to hydrogen.

The ratio of deuterium to hydrogen in the earth's oceans, the meteorites, the Jovian atmosphere, the interstellar clouds, the Orion Nebula, and different parts of our Galaxy is about a few times 10^{-5}. This ratio is much higher than could have been created in stellar interiors during normal stellar evolution. The most likely explanation is that it was produced within the first few minutes after the big bang. This density of deuterium leads to estimates of the present average matter density in the universe that are less than the critical density needed to close the universe. Thus we again find evidence suggesting an open universe that will expand forever.

The theoretically calculated abundance ratio of helium to hydrogen is about 25 percent by weight, or one helium atom to ten hydrogen atoms—about the same as the observed ratio in most stars, including the sun, and in planetary nebulae and H II regions. The predicted abundances for the heavier elements, however, are low in comparison with the observed values. It seems, then, that they were primarily created after the big bang by thermonuclear reactions inside the stars and in supernova explosions. Had the early stage of the universe been more dense, or had the temperature not dropped so rapidly, much of the deuterium would have been cooked into heavier nuclei, which does not appear to be what we actually observe.

THE GALAXIES FORM

The era of galaxies began when matter was more plentiful than radiation, when the universe was about a million years old and about a thousandth of its present size. How was this primordial matter (with a density of less than 10^{-21} gram per cubic centimeter and a relatively low temperature) distributed in space? How was it formed into clusters of galaxies and individual galaxies? As British astronomer James Jeans once expressed it, did the finger of God as Director of the Universe stir this matrix to form eddies or pockets of condensed gas? Or did local inhomogeneities in the chaotic gas develop because of random turbulence or shock waves within the medium? The galaxies supposedly condensed out of these churning primordial eddies. The most favorable period when very large turbulent eddies could condense occurred at the beginning of the *matter era* when the temperature had dropped to about 3000 K. The estimated accumulated mass was about 10^{15} to 10^{18} solar masses, a mass in excess of the mass of a typical rich cluster of galaxies. The galaxies in turn were created out of the fragments within the protocluster.

An alternative explanation for the formation of galaxies starts with an initially homogeneous isotropic universe. Gravitational instabilities produced density enhancements from which clumps of matter were formed. However, the clumps could not hold together unless the gravitational force exceeded the kinetic pressure exerted by matter and radiation. This occurred at a temperature of about 3000 K. The critical mass for this to happen turns out to be about 10^5 solar masses, the mass of a typical globular cluster. The galaxies then formed from the accretion of protoglobular clusters. The self-gravitation of these overdense regions where galaxies were created locally retarded the expansion of the universe sufficiently to permit numbers of galaxies to accrete into a cluster of galaxies.

Summarizing the two accounts above of galaxy formation, we theorize that the protoclusters of galaxies formed first and then fragmented into individual galaxies and globular clusters; or that the protoglobular clusters formed first and accreted into galaxies, which in turn accreted into the clusters in the overdense regions. These somewhat represent the two ends of the spectrum of ideas for the formation of galaxies in the big bang cosmology.

> Of the real universe we know nothing, except that there exist as many versions of it as there are perceptive minds.
>
> Gerald Bullitt

ARE THERE ANTIGALAXIES?

There is no reason why antimatter cannot exist in the universe so long as it does not collide with ordinary matter. Laboratory experiments have demonstrated the symmetry between particles and antiparticles and their destructive annihilation. For every particle (proton, neutron, electron) there is an antiparticle [antiproton, antineutron, antielectron (positron)]. An antihydrogen atom would therefore contain a negatively charged proton and a positively charged orbiting electron. Its spectral signature would be the same as that of the normal hydrogen atom because the photon of radiation is its own antiparticle. If the Andromeda galaxy is an antigalaxy, we have no way of detecting that fact because the spectral lines of matter and antimatter would appear at identical wavelengths with the same intensities.

MINI BLACK HOLES

Combining gravity with quantum mechanics has enabled us to probe much deeper into the bizarre complexities of the rotating black hole (see page 332). Since the laws of classical physics, including relativity theory, break down at a singularity, Stephen Hawking has developed a quantum theory of gravitation (a difficult mathematical feat) to analyze the properties of a black hole. His calculations led him to the concept of mini black holes capable of emitting particles and radiation. It turns out that particles from inside the black hole can tunnel their way out into open space as if there were no event horizon, thus unclothing the singularity (see p. 335). The naked singularity stands exposed as a region of infinite curvature—a highly warped kink in the space-time geometry of the mini black hole.

Possibly the enormous pressures present during the early stages of the expanding universe could have compressed pockets of matter into mini black holes. A typical mini black hole could be the size of a proton ($\approx 10^{-13}$ centimeter) having the mass of an asteroid (≈ 1 billion tons). If mini black holes were formed in large numbers after the big bang, their gamma ray emission might be detectable with today's instruments. On the basis of present-day estimates of the observable gamma ray background radiation in space, the number of proton-sized black holes is not greater than 300 per cubic light year. The number throughout our Galaxy would be staggering. Compare this number with the number of ordinary black holes arising from the collapse of massive stars, estimated to be about 20,000.

Calculations show that when two black holes collide and merge into a single black hole, the surface area of the resulting event horizon is *greater* than the sum of the individual areas of the original event horizons. The collision generates a very violent outburst of gravitational waves (see page 417), with a subsequent loss of mass. Hawking's theory predicts that pairs of particles and their antiparticles—such as electron-positron pairs, neutrino-antineutrino pairs, photon-antiphoton pairs—are created in the powerful gravitational field of a rotating black hole. If a particle or antiparticle falls into a black hole, its opposite may be created and escape from the black hole. The black hole would thus appear to be emitting particles and antiparticles. The resulting emission decreases the mass of the black hole, while its temperature rises. Eventually a runaway effect de-

GRAVITATIONAL RADIATION: TESTING RELATIVITY THEORY

One of Einstein's predictions lay idle for nearly half a century as too difficult to verify: extremely weak *gravitational waves* are radiated in space with the velocity of light by rapidly accelerated or spinning bodies. These waves might be detectable with sensitive apparatus, and large astronomical objects undergoing violent activity, such as supernova explosions or the nucleus of an active galaxy, may be the best places to uncover them. Any gravitational wave passing through an object momentarily deforms the space around it and causes the object to vibrate slightly.

More than a decade ago Joseph Weber tried to pick up, on the surfaces of large suspended aluminum cylinders, infinitesimal oscillations that would be produced by gravitational waves striking the cylinders. The apparatus had to be carefully isolated from nongravitational disturbances. According to Weber, he succeeded in detecting gravity waves simultaneously in his Maryland laboratory and at the Argonne National Laboratory near Chicago. Most of these waves were reported to be coming from the Galaxy's center in Sagittarius. It now appears that the observed oscillations are much too large to be consistent with current physical theory. So far, other and far more sensitive gravity wave detectors, which can detect deformations as small as 10^{-17} centimeters, have failed to find any evidence of gravitational radiation coming from the center of our Galaxy or anywhere else. However, indirect evidence of gravitational radiation has been found in the radio observations of the binary pulsar (see page 330) whose or-bital period is 7.75 hours. The gravitational interaction between the pulsar and its close companion, believed to be a neutron star or white dwarf, results in part of the orbital kinetic energy being radiated away in the form of gravity waves, according to relativity theory. The loss in energy decreases the orbital separation between the components. Radio monitoring during the period 1974–1979, covering some 1,000 orbital revolutions, shows a decrease in the orbital period of about 101 microseconds per year. Allowing for the uncertainties in the mass of each component and the inclination of their orbital plane, the result is in pretty good agreement with general relativity's prediction of 76 microseconds per year.

---

velops that results in a rapid and an enormous release of energy, followed by the complete evaporation of the black hole down to zero mass within a finite time. The constant drain of energy of the mini black hole slows down its spin until it eventually comes to a stop. In its final death throes the black hole might look like a white hole from which energy pours.

For a black hole of one solar mass, the process described above is too slow and unobservable; the temperature is a tiny fraction above absolute zero and its calculated lifetime is 10^{66} years—trillions of times the accepted age of the universe (15–20 billion years). A mini black hole of about 1 billion tons could develop a temperature of 120 billion degrees and have a lifetime expectancy roughly equal to the age of the universe. The flash of gamma radiation arising at the time of its rapid implosion down to zero mass might now be observable. Perhaps it is conceivable that we could look for the powerful gamma radiation now pouring out of exploding primordial mini black holes that were created during the early history of the universe as another means of verifying the big bang cosmology.

A FINAL NOTE

We end with a word of caution. We have followed Occam's razor by introducing certain simplifying assumptions: the local physical laws are valid everywhere; the universe is isotropic and homogeneous; and the observed red shift of the galaxies is due to the expansion of the universe. But we do not know the real distribution of matter and its various forms with sufficient accuracy, nor do we fully comprehend its astrophysical nature and evolution, which would enable us to determine its intrinsic properties at the time its radiation was first emitted. And finally, we cannot directly observe space-time curvature in the laboratory and can only infer from theory and indirect observation its behavior in the universe at large.

SUMMARY

Starting with the theory of gravitation and a few assumptions, we can construct several cosmological models that describe the beginning, the present, and the future of the universe—of everything everywhere. One basic precept common to all developed models is the cosmological principle: the large-scale structure of the universe looks the same to all observers in space. It is further assumed that the universe is more or less uniform (homogeneous) and the same in all directions (isotropic).

The static model, which does not depend on time, was proposed by Einstein. This model was abandoned before Hubble discovered that the universe is expanding. It was replaced by several nonstatic or time-dependent models: an infinite, open universe of negative curvature; a finite, closed universe of positive curvature; a flat Euclidean universe of zero curvature; a steady state universe that presupposes no beginning and no ending. All models except the steady state model begin with a big bang when all matter was crunched into a superhot, superdense condition. In the open universe, space expands forever at a decreasing rate due to the self-gravitation of the universe, and matter thins out gradually with the expansion. In the flat Euclidean universe, space expands less rapidly, coming to rest at infinity in a virtually empty universe. In the closed universe, space expands even less rapidly because more matter is present, which brings the expansion toward an accelerating, superhot, superdense crunch whence it all began.

The cosmos is filled with low-temperature (3 K) microwave radiation. It seems to confirm that the universe as we now see it began with an explosion of a primeval fireball (the big bang) 15 to 20 billion years ago; its dying embers constitute the low-temperature radiation. Cosmologists cannot at present choose the best model because they can distinguish the critical differences among the time-dependent models only at extreme distances. That far away, the galaxy or quasar data (the relationship between recessional velocities and distances of the galaxies from us) are too uncertain to support a confident decision. Cosmologists will therefore have to wait for more precise observations before they can decide how the universe began and how it may end.

REVIEW QUESTIONS

1. What four basic natural forces rule the cosmos? Describe their roles.

2. If gravity is the weakest of the four forces, why is it so dominant within the universe at large?

3. What is modern cosmology all about? What was ancient cosmology (Chapter 1) concerned with?

4. What is the difference between a static and a nonstatic universe? Why is a static universe physically impossible?

5. Why is the red shift of the galaxies interpreted to be a consequence of the expansion of the universe?

6. Describe the possible variations of the time-dependent world models: open, closed, or Euclidean.

7. Why has it been so difficult to decide from observational data what the world model of the universe is?

8. What is the cosmological principle? The perfect cosmological principle?

9. Describe the steady state model. What are its drawbacks?

10. What two important parameters must be evaluated in order to determine the correct world model?

11. What evidence favors the big bang version?

12. Explain the origin of the low-temperature (3 K) isotropic cosmic background radiation.

13. Describe the probable course of events *within* the first million years after the big bang according to the present interpretation.

14. Describe the probable course of events *after* the first million years following the big bang according to the present interpretation.

15. Why do we believe that most of the deuterium and helium present in the universe was formed within the first few minutes after the big bang?

16. How would you disprove the claim that the universe was created one second ago with everything in place, including our conception that the universe existed long before this?

17. Describe the properties of the three forms of space curvature.

18. How is it possible for two travelers who start out on parallel tracks to meet at a distant point in curved space?

19
Exobiology: Life on Other Worlds

People have thought since ancient times that other worlds may have life. Many Greek philosophers believed they did. Following the remarkable revolution in scientific thought and experimentation that began in the seventeenth century, the notion of extraterrestrial life was advocated by many eminent scientists and philosophers. In succeeding centuries descriptions of extraterrestrial beings and accounts of their activities appeared more and more often.

For a decade or so the question, "Are we alone?" has brought the reply, "We are not alone"—in space. Yet we have no more direct proof for that belief than our predecessors had, though we have found a few promising leads. Even people without training, and certainly the student who has studied some astronomy, doubt that we are the sole intelligent inhabitants of this vast universe. Scientific enlightenment has helped us cast aside the ancient anthropocentric convictions of our uniqueness in nature.

In Chapter 10 we explored the possibilities of other life forms in the solar system after discussing the development of life on the earth. Let us continue this theme by considering the possibilities of life beyond the solar system.

19.1
Beyond the Solar System

ORGANIC COMPOUNDS AND ORGANISMS IN INTERSTELLAR SPACE

That the seeds of life are eternally present in the universe is an idea that became popular in the late nineteenth and early twentieth centuries. According to *panspermia* (omnipresent life), living organisms are not spontaneously generated from nonliving matter but are transmitted from planet to planet. Swedish chemist Svante Arrhenius elaborated on this idea in 1907, suggesting that microorganisms (spores or bacteria) attached to cosmic dust particles were dispersed throughout interstellar space. They were readily available to fertilize any hospitable world they found

◀ "To consider the earth as the only populated world in infinite space is as absurd as to assert that on a vast plain only one stalk of grain will grow." (Greek philosopher Metrodoros of Chios, fourth century B.C.)

in their wanderings, propelled about by pressure of light from the stars. We find it hard to imagine how such isolated organisms floating around in space could survive bombardment by cosmic ray particles, or gamma ray, X-ray, and ultraviolet photons. Perhaps they could find sanctuary deep inside dense interstellar clouds (see Figure 19.1), which might protect them from most damaging electromagnetic radiation. In this respect, we do not even know how the recently discovered organic compounds in interstellar clouds persist against the relatively high odds of dissociation or radiation damage.

We hope to learn whether the earth's biology evolved directly from interstellar organic molecules that may have been present initially in the solar nebula, or whether it developed from the earth's primitive ingredients as discussed in Section 10.1. Curiously, formaldehyde, one of the organic molecules found in the interstellar clouds, also turned up in the large stone meteorite that fell in northern Mexico in 1969. There is now little doubt that a link exists between the clouds of interstellar dust and meteorites. One suggestion is that the intense asteroidal and meteoroidal bombardment that battered the terrestrial planets 4 billion years ago brought to the earth the earliest organic compounds of hydrogen, carbon, and nitrogen. From these were synthesized, when conditions were right, the more complex organic molecules that chemically evolved into the first living organisms.

LIFE AMONG NEARBY STARS

If we decided to select the stars in the solar neighborhood where life might be found, how would we proceed? Some criteria have been proposed that could be used when looking for stars with planetary systems. The vast majority of binary and multiple stars—about

I have felt that man is a stranger on this planet. A total stranger. I always played with the fancy maybe a contagion from outer space is the seed of man. Hence our prior occupation with heaven, with the sky, with the stars, with the gods, somewhere out there, in outer space.

Eric Hoffer

FIGURE 19.1

Cone nebula in Monoceros, photographed in red light with the Hale reflector. It is located on the southern fringes of the young star cluster NGC 2264.

half the stellar population—would be ruled out because orbits of planets around them presumably would not be stable enough to keep them in a thermally habitable zone. The exception might be stable binary systems in which a planet is very close to one of the components, or a planet orbiting at a great distance from a close pair. However, this would not necessarily place such planets in the habitable zone nor ensure that the zone remained thermally constant due to the presence of two orbiting suns. A spectroscopic analysis of 123 solarlike stars out to 85 light years, discussed in Section 12.6, led to the speculation that 20 percent of these stars may have planetary systems rather than stellar companions.

Another large stellar group, the hotter stars of spectral classes O, B, and A, would be rejected because their time on the main sequence is much too short (less than a billion years) to permit chemical and biological evolution (which would probably take several billion years). Also, all stars in the final stages of evolution are probably not good candidates.

We are left with main sequence stars between spectral classes F2 and K5 as the best possibilities for supporting advanced forms of life. These stars do not rotate as fast as the hotter main sequence stars, possibly because of the angular momentum (rotational energy) transferred from star to planets as a planetary system was formed.

The habitable zone (or ecoshell) around the cool red dwarf stars of spectral class M would be too small and too close to the star to have a significant number of planets. Also, a planetary body that close to its parent star might be forced into a synchronous rotation with it, exposing the same hemisphere continuously to the star—it would be too hot on one side and too cold on the other side of the planet to support life. Furthermore, red dwarfs lack sufficient ultraviolet radiation to generate organisms through a sequence of chemical evolution. Possible ecoshells are illustrated in Figure 19.2 for late main sequence stars.

Only one technique is presently available to astronomers for searching for planetary bodies, and that is picking up minute periodic deviations or wobbles in the star's proper motion. These are the deviations caused by orbital motion of the planets. This periodic motion is, in reality, the movement of the star around the barycenter of the star-planet system, and it is detectable only where the ratio of the planet's mass to that of the star is not insignificant.

A few nearby red dwarfs of spectral class M have a wavy proper motion, suggesting that an unseen companion may be there. These are astrometric binaries (see Section 12.6). It is possible to derive an estimate for the mass of the invisible companion from the amplitude of the wiggle of the visible star. Best known of this class is Barnard's star, a red dwarf with the largest observed proper motion (see Figure 19.3). After half a century of studying its proper motion on 3800 photographs, American astronomer Peter van de Kamp has found a wavy pattern of minute amplitude in its proper motion (see Figure 19.4). He interprets this wiggle as caused by two invisible, planet-size bodies revolving in almost circular orbits with periods of about 12 years and 22 years. Both, however, are well

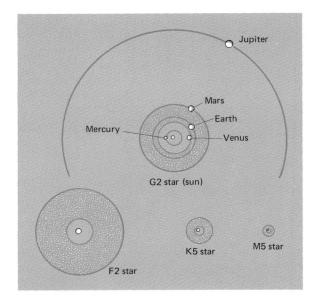

FIGURE 19.2
Possible ecoshells around stars of different spectral types are shown in solid color circles.

FIGURE 19.3
Proper motion of Barnard's star. The left image of each pair of stars was secured one year earlier than the right image. Barnard's star is in the center. Notice that it shows a change in position (mostly north), whereas the other stars show no evidence of proper motion. The proper motion of Barnard's star has a minute periodic wiggle caused by the presence of its planetary companions, as shown in Figure 19.4.

outside the ecoshell of the parent star. Another analysis using less extensive but independent photographic data shows no evidence for planetary motion. The disagreement tells us of the great difficulty in making such delicate measurements.

Of the 54 stars within about 16 light years of the sun, only 3 meet the criteria needed for an ecoshell. These are the main sequence stars Epsilon Eridani (K2), Epsilon Indi (K5), and Tau Ceti (G8) (see Table 12.1). If we take the solar neighborhood as a representative sample (in which 3 stars out of 54 within 16 ly of the sun have potentially habitable planets), then the average distance between biologically suitable stars is about 18 light years. Within a radius of 1000 light years, then, we would expect to find 7 million stars having suitable planets harboring some kind of life. Even if only 1 in 1000 of these planetary systems has an intelligent species, that still leaves 7000 sites of intelligent life within 1000 light years. If we conservatively estimate a million civilizations distributed randomly throughout the Galaxy with a technology at least equal to ours, then the average separation between them would require 600 years even for sending and receiving messages—hardly a hurried conversation.

Let us examine the possibilities for interstellar communication. We can start by making an intelligent inquiry into the possible number of *communicative civilizations* in our Galaxy. That is the name we give to societies technologically competent and motivated enough to engage in an interstellar dialogue. We have a general formula expressing the number of such galactic communities in terms of several uncertain factors and probabilities. It compresses a great deal of ignorance into a small amount of space.

$$\text{number of communicative societies in our Galaxy} = \left(\text{astronomical factors}\right) \cdot \left(\text{biological factors}\right) \cdot \left(\text{sociological factors}\right)$$

$$N = (R_s \cdot f_p \cdot n_p) \cdot (f_l \cdot f_i) \cdot (f_c \cdot L)$$

R_s is the average annual rate of star formation during the Milky Way's existence, or about 20 stars per year. This figure is derived from the estimated number of stars in the Galaxy—200 billion—that formed during its

FIGURE 19.4

Wavy proper motion of Barnard's star. The barycentric motion, motion of the observable star about the barycenter, in both right ascension (east-west on the sky) and declination (north-south on the sky) is caused by the presence of one or more invisible companions. Each point is a yearly mean of an average of ninety-three photographs. The length of the vertical bar is twice the probable error.

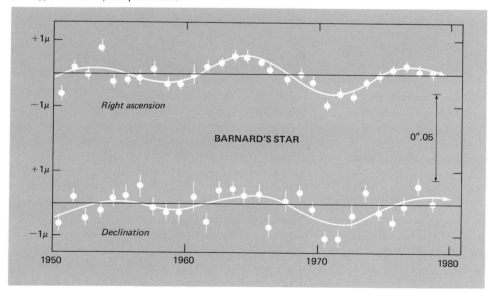

approximately 10 billion years. The fraction of stars with planets, f_p, is roughly 1/2, which comes from ruling out the multiple star systems, about half the stellar population. The factor n_p is the number of planets in each planetary system suitable for life; for our only example, the solar system, that equals 1. Such planets must be in the thermally habitable zone surrounding the parent star where physical conditions, temperature, atmosphere, water, and the rest, are conducive to supporting and maintaining life. In the solar system two such planets, earth and Mars, are inside the life zone. Conservatively, we estimate one planet for each system as the average number with a suitable environment for spawning life.

Among the biological factors, f_l is the fraction of planets on which life actually does appear. Life is possible on only those planets for which the parent star is an F, G, or K main sequence star, or about 20 percent of the stellar population. The factor $f_l = 1/5$ is arrived at on the assumption that under the proper conditions, life will take hold, flourish, and evolve into a myriad of thriving forms sooner or later—every time. On the other hand, the factor f_i is the fraction of biological species that evolve into a technically competent culture ultimately able to engage in interstellar communication. We are going to guess that the probability that nature, with 2 or 3 billion years of effort, will create at least one intelligent and communicative species on a planet is 100 percent. Therefore we set the factor f_i equal to 1.

Now for the sociological factors. The factor f_c is the fraction of galactic societies technologically able to take part in interstellar communication; this factor we guess is equal to 1/2. These are the land-based creatures who can develop the technology for interstellar communication and are motivated to undertake it. The factor L is the length of time the civilization continues in its communicative phase. Our own interest in interstellar communication dates back only a few decades in a period of more than six thousand years of civilization.

The number of intelligent communicative societies, then, is

$$N \approx 20 \times 1/2 \times 1 \times 1/5 \times 1 \times 1/2 \times L = L$$

In other words, the number of communicative civilizations in our Galaxy approximates the average number of years spent in the communicative phase. And the factor L is probably the most uncertain of all to evaluate.

When we think about the possibilities of our own destruction by nuclear holocaust, by biological disasters from ill-conceived genetic engineering, by changes in the planet's ecology and climatology due to human stupidity and blunders, by terrestrial and extraterrestrial catastrophes, and by other calamities that could befall a civilized society, it is tempting to predict that the moment of civilized glory may indeed be brief in the span of an intelligent species. Who can assure us that the prevailing good sense of even a superior intelligence will be able to control its own destiny indefinitely? Indeed, a succession of technological ages might struggle from the ashes of the preceding civilization during a planet's history. Nevertheless, there should be a sufficient number of communicative societies present to make interstellar communication worth a try.

Assuming we have a fair grasp of the values for the product $R_s f_p n_p$, then we can calculate the average separation between communicative societies for various values of the product $f_l f_i f_c$ and L. Table 19.1 lists values that cover a range of reasonable values for the two variables of the product $f_l f_i f_c$ and L. Note that in the first row, if a communicative species survives in a communicative phase for only a thousand years, then the length of time for messages to travel the distance between civilizations exceeds the lifetime of the communicative phase. Thus only the combinations below and to the left of the line in the table present any reasonable chance for an exchange of messages.

19.3
Possible Forms of Life

WHAT KIND OF INTELLIGENT SPECIES?

During the course of the expanding universe, life became possible only after the galaxies had formed and their stars had existed long enough to synthesize the heavy atoms needed to produce life. Had the primeval condensate that immediately followed the big bang not cooled rapidly in the first few minutes, life would not have materialized in the universe, because the hydrogen would have been converted into helium, leaving little available for stellar nuclear synthesis, and the heavy elements would not have been formed subsequently in stellar interiors.

TABLE 19.1
Approximate Distance in Light Years to Nearest Communicative Civilizations

L (yr)		$f_l f_i f_c$ [a] High density of civilizations in the galaxy				
		10^{-1}	10^{-2}	10^{-3}	10^{-4}	10^{-5}
10^3		3,000	10,000	30,000	100,000	300,000
10^4	Long-lived civilizations	1,000	3,000	10,000	30,000	100,000
10^5		300	1,000	3,000	10,000	30,000
10^6		100	300	1,000	3,000	10,000

[a]Assuming $R_s f_p n_p = 20 \times 1/2 \times 1 = 10$, so that $N = 10(f_l f_i f_c) L$.

We do have logical reasons for believing that any extraterrestrial form of life would arise from some kind of autocatalytic chemical reactions that, in a favorable ambient medium, would push chemical evolution toward producing proteinlike molecules. One product in the autocatalytic reaction accelerates the reaction, speeding up the whole proceeding. In DNA this is the activity associated with the enzyme protein (see Section 10.1). The amassed complex molecules, in turn, would synthesize macromolecular systems that, in time, would develop into primitive multicellular biological organisms. Sophistication in living organisms increases in proportion to their ability to react to changes in their environment, which is partly due to their ability to store and recall experiences.

The biological conditions that provide the widest latitude for developing life would be found on those planets containing water as a solvent and an oxidizing atmosphere with the proper optical windows. These living organisms would most likely originate in a liquid system enriched by chemical contributions from the solid and gaseous interfaces. That they might come from a purely solid or gaseous medium is highly improbable.

On the physical side, too large a planet would hold an oversupply of the reducing gaseous molecules and hydrogen. Too high a surface gravity would require a strong skeletal supporting structure and might make sluggish creatures. This kind of reasoning leads to a planet with an upper limit of about twice the earth's size as having the best physical conditions for life.

It is quite possible that highly evolved biological species on other planets would differ in many ways from terrestrial life forms, even if the key molecule DNA were there, because of the enormous number of chance combinations of nucleotides possible in the structure of DNA and because of differences in environment. If we use the earth as a representative sample, in which living forms have developed many similar structural features, we might suspect that a highly evolved, land-based, extraterrestrial species, subject like ourselves to gravity and electromagnetic radiation, might have some of our biological characteristics. It need not, however, resemble the human being too closely. We have reason to believe that a marine environment would be less suitable for evolving an organism with intelligence. And airborne species of any size would expend too much energy remaining aloft—there would not be enough energy left for a highly developed intelligence.

If there are places in the universe where people make genuine sense of their existence as a species and where they comprehend the delicate connections between the individual and collective existence, all the treasures on earth would be a small price to pay for the clues.

Norman Cousins

Evolved intelligent creatures would have to have a somewhat symmetrical body structure with a central nervous system, a brain, and sensory organs up front. The type of sensory organs would be dictated by the environment and by the electromagnetic radiation to which the species was exposed. They would enjoy the advantages of mobility and manipulative ability, but they need not have the same symmetrical appendages as earth creatures. There would be a biological upper limit to size, for too bulky a creature demands a large source of energy for locomotion. Also, in a large creature a nerve pulse might take too long to travel from a remote area of its body to the brain, and the creature might not react quickly enough to bodily danger. On the other hand, very small beings might evolve to high intelligence if their ancestors were not forced to use all their energies in competing against larger and more powerful adversaries for survival.

EVOLUTIONARY TRACK IN HUMANS

Evolution is nonrepeatable, some biologists insist, implying that if life were wiped out on the earth, nature would not repeat itself by fashioning another human being. The chain of cause and effect in evolutionary sequences would be quite different the second time around. They say, too, that evolution could have taken a different track, and something different from humans might have been at the end of the line. Extraterrestrial beings need not have humanoid characteristics, then. But this notion is challenged by other biologists who feel that different evolutionary paths are possible but they would eventually converge toward approximate humanoid form.

Our great leap forward came in the last few million years when a humanoid creature learned to walk upright, freeing its hands for the delicate manipulation of tools. As the brain evolved and mental capacity expanded, a collective culture and civilization set people apart from all other living creatures—a benefit not without its price.

AN INTELLIGENT SPECIES EVOLVES

If our cosmic environment is duplicated in many regions of our Galaxy, other civilizations could have developed much as ours has. The idea is called the assumption of mediocrity; our only guide is our own provincial example. And why not other intelligent societies with a long biological evolution toward increasing complexity, followed by a sudden phenomenal rise in social, cultural, and technical development?

Contract human existence to one day, and civilization would come along only in the last minute of that day. We wonder, of course, if sophisticated societies more advanced than ours might exist in parts of the Galaxy where stellar creation began long before our sun was formed. It is also possible, however, that sophisticated societies may have been able to extend their evolutionary possibilities by freeing themselves from their fragile environments and taking up residence in space stations, wandering through the galaxy. If so, then intelligent species may exist in locations other than the immediate vicinity of stars—perhaps much closer to us. But how possible is it for intelligent beings to exist permanently in an artificial environment?

A design study participated in by a group of scientists in various disciplines in 1977 concluded that there are no insurmountable obstacles (except funding, amounting perhaps to $100 billion) to prevent humans from living in space; that space provides great reserves of matter and energy that can be used for the benefit of inhabitants on the earth and in space; and that we have both the knowledge and ability to colonize space. Thus, we should consider the possibilities of other intelligent beings living in such an environment.

SPACE COLONIES

Princeton physicist Gerard O'Neill believes it is technologically possible to construct a number of cylindrical terrariums in coupled pairs, stationed at one of the Lagrangian earth-moon points[1] in orbit around the earth. Each space structure would house a complete ecological system imitating the earth's environment with land and water areas, animals, birds, and even an artificial blue sky. It would rotate at 1 g around its cylindrical spin axis, which would constantly point to the sun. With the proper positioning of arrays of windows and mirrors, the sun's light could be angled to provide both daily and seasonal cycles. A large solar collector mirror at each end of the cylinder would furnish sufficient heat to run a steam turbogenerator electric power plant. A transport system would link the various cylinders with each other and the earth.

All but about 2 percent of the construction material would be taken from the moon and possibly the asteroids. A transport mass-driver employing electromagnetic energy would be used to mine the material by remote control and deliver it to the site of construction. The first proposed model would house 10,000 res-

[1]There are two such stable points located at one of the vertices of the two equilateral triangles whose other vertices are occupied by the moon and the earth.

idents in a cylinder 1 kilometer long and 200 meters wide with a rotation period of 21 seconds (see Figure 19.5). The total mass is 500,000 tons, equivalent to that of a present-day tanker. Later, larger units could be constructed with cylinders up to 32 kilometers long and 6.4 kilometers in diameter, capable of accommodating from 200,000 to 20 million inhabitants.

19.4 Extraterrestrial Communication

INTERPLANETARY ATTEMPTS

Before we started radio communication in the last century, people had suggested ingenious ways of signaling our presence to other worlds in the solar system: huge bonfires in simple geometric patterns such as squares or triangles; planting a 16-kilometer-wide strip of pine forest in Siberia in the form of a right triangle; huge mirrors to reflect sunlight; a 30-kilometer circular ditch filled with water over which kerosene would be poured and set burning; a powerful concave mirror to focus sunlight on Mars and burn simple numbers on the desert sands of the planet; a network of large sunlight-reflecting mirrors strategically positioned in several European cities forming the shape of the Big Dipper in Ursa Major.

In 1899 eccentric electrical pioneer Nikola Tesla undertook to transmit a powerful electrical signal into space from his Colorado laboratory and to detect any replies. His apparatus was a large primary coil 23 meters in diameter with a 60-meter mast topped by a 1-meter copper ball. He alternately introduced powerful surges of electricity into the copper ball and into the ground, believing that the earth's magnetic field would increase the power of his signal. Incandescent

FIGURE 19.5
Exterior of a proposed space habitat for about ten thousand people. The large sphere in the center is nearly a mile in circumference and rotates to provide a gravity comparable to that of the earth.

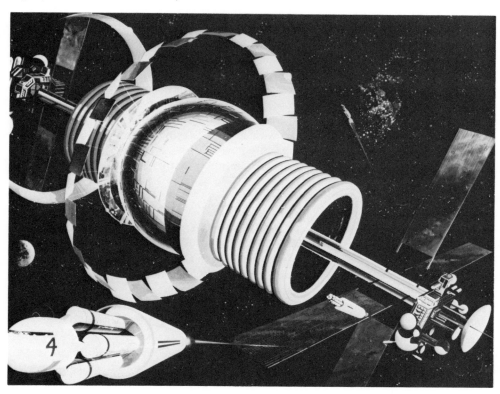

lights were set glowing 40 kilometers away, but no detectable extraterrestrial responses came in at the time, although a year later he claimed to have picked up interplanetary signals.

In 1921 Guglielmo Marconi, while conducting atmospheric tests aboard his experimental communications yacht, believed he had detected regular pulsed signals from space.

When Mars was in close opposition to the earth in 1924 (56×10^6 kilometers), Professor of Astronomy David Todd at Amherst College persuaded the federal government to turn off its high-powered transmitters for five minutes before every hour between August 21 and August 23. During these silent periods he tuned a special receiver to a wavelength between 5 kilometers and 6 kilometers to record on tape any signals coming through. He was aided in his efforts by other listeners throughout the country. Out of a melange of dots, dashes, and jumbled code groups, nothing definite could be ascribed to an extraterrestrial source. We have since learned that such long waves are reflected back into space by the ionosphere, so that his efforts could not have produced a positive result.

Todd had earlier proposed converting an abandoned mine shaft in Chile into a telescope by filling a 15-meter bowl at the bottom with mercury. The heavy liquid would be set in rotation to form a natural parabolic reflector; a powerful light source at its focus would transmit an intense beam of light to Mars when the planet passed overhead. Astronomers duly criticized this arrangement as unworkable. Today's optical systems would use lasers for signaling; they may be powerful enough for communication beyond the solar system.

EXTRATERRESTRIAL ATTEMPTS

Radio telescope design and observing techniques were so spectacularly improved after World War II that a few astronomers and physicists privately considered the feasibility of detecting extraterrestrial signals from intelligent sources. The subject finally surfaced in the British scientific journal *Nature* in September 1959, when physicists Guizeppe Cocconi and Phillip Morrison presented logical reasons why some effort should be put into searching for interstellar signals generated by intelligent life.

For now it seems more practical for us to listen for signals than to transmit them. Presumably there are advanced galactic societies that have the ability and the desire to transmit as well as to receive; if not, interstellar communication may never happen. Perhaps messages that older, more accomplished cultures have been transmitting for centuries have by now reached the solar system. The most advanced celestial communities could avail themselves of energy sources far more sophisticated and powerful than any we can dream of today, perhaps even using the energy output of their parent star by modulating its light as a signal. We may be no more aware of them, it is said, than the New Guinea aborigines, who use drums for communication, are aware of the international radio traffic constantly passing overhead.

The first attempt in this country to detect artificial signals from space was conducted by Frank Drake at the National Radio Astronomy Observatory at Green Bank, West Virginia. This undertaking was called *Project Ozma*, after the legendary princess in the imaginary land of Oz. The 26-meter radio dish was aimed at Tau Ceti and Epsilon Eridani for 150 hours of observation from May through July 1960. The observation proved to be unrewarding. Since then intermittent attempts to locate intelligent signals have been carried out from several radio observatories. None have succeeded, but the several hundred stars examined are a very tiny sample of the possible sources. Table 19.2 lists current and past efforts. The most unusual effort is the third from bottom: the attempt to spot intelligent signals from a supercivilization in another galaxy. Even if a communication were received from another world, it would take inordinate amounts of time to exchange messages. Imagine asking the U.S. Congress for money to finance a thousand-year experiment! Perhaps the first step in acknowledging contact with an extraterrestrial society would be to transmit a duplicate of the received message back to the source to inform it that its inquiry had been received and recognized as originating from an intelligent source.

OUR NOISY EARTH

In terms of radio signals the earth is a very chatty planet. It pours out a constant flood of signals into space from FM, TV, radar, and commercial stations above 30 MHz. This radio frequency radiation might be intercepted by a nearby civilization up to about 35 light years away. That is as far as the radio waves have traveled in space since the early days of very high frequency (VHF) propagation. We, in turn, are looking for radio signals from nearby stars that might have advanced civilizations on habitable planets.

TABLE 19.2
The Search for Intelligent Signals

Investigator	Observatory	Date	Frequency	Targets
Drake	National Radio Astronomy Observatory (U.S.)	1960	1420 MHz	Epsilon Eridani, Tau Ceti
Troitsky	Gorky University (U.S.S.R.)	1968	927 MHz	12 nearby solar type stars
Verschuur	National Radio Astronomy Observatory (U.S.)	1972	1420 MHz	10 nearby stars
Troitsky	Eurasian Network (U.S.S.R.)	1970 to present	600 MHz, 1000 MHz, 1875 MHz	Pulsed signals from entire sky
Zuckerman, Palmer	National Radio Astronomy Observatory (U.S.)	1972 to present	1420 MHz	About 600 nearby solar type stars
Kardashev	Eurasian Network (U.S.S.R.)	1973 to present	Several	Pulsed signals from entire sky
Bridle, Feldman	Algonquin Radio Observatory (Canada)	1974 to present	22.2 GHz	About 500 nearby stars
Drake, Sagan	Arecibo Observatory (Puerto Rico)	1975 to present	1420 MHz, 1653 MHz and 2380 MHz	Several nearby galaxies
Dixon, Cole	Ohio State University (U.S.)	1973 to present	1420–1700 MHz	Sky survey of stars
Bowyer, Lampton et al.	Hat Creek Observatory, University of California (U.S.)	1972 to present	Variable; supplemental to other research	Random directions

The total amount of power radiated on the earth is a staggering 300 million watts, most of it contributed by television and radar transmitters, with the greatest concentration in the United States and Europe. Extrasolar eavesdroppers with sophisticated antennas and receivers could detect a brief daily crescendo of radio noise when eastern North America or West Europe are presented edgewise viewed from a point in space. This is the most favorable condition for detecting the signals since they are radiated parallel to the earth's limb at this time. A distant observer would also observe an annual effect superimposed on the daily signals monitored over a long period of time. A distant observer, interpreting their frequency, modulation, and intensity, could gain an insight into the cultural and scientific values of our civilization.

CHOOSING THE FREQUENCY

Deciding which microwave region to use for interstellar discourse is not difficult. Radio astronomers suggest that the wavelengths between 3 centimeters and 30 centimeters (1 gigahertz to 10 gigahertz) would be suitable (see Figure 19.6). There the atmospheric absorption and Galactic noise from synchrotron radiation, the 3-K cosmic background radiation, and other unwanted sources are at a minimum (the curve in color in Figure 19.6). A logical frequency with universal significance would be about 1.420 gigahertz (21 centimeters), the emitting frequency of the dark neutral hydrogen clouds outlining the spiral arms of our Galaxy. We assume that other galactic inhabitants are running similar surveys of the hydrogen distribution in our Galaxy. With both our receivers and theirs tuned to the same spectral region, coded transmissions by either side would have a good chance for successful detection.

The quietest part of the microwave spectrum is very nearly between the lines of hydrogen (H) at 1.420 gigahertz and of hydroxyl (OH) at 1.668 gigahertz. The region between the two frequencies is called the water hole, because H and OH are the products into which water (H_2O) dissociates. If water-based life is present elsewhere, it might be reasonable to expect advanced societies to use a frequency at the water

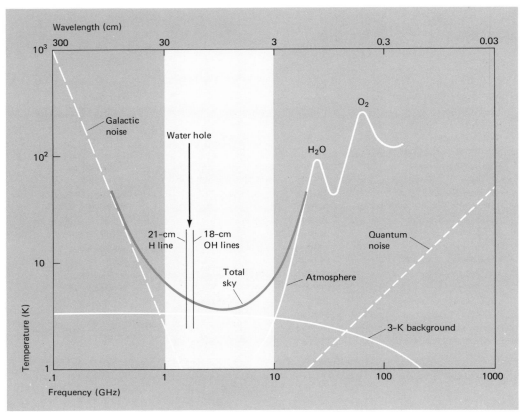

FIGURE 19.6
The microwave window into outer space from the earth is clearest where the total sky radiation is a minimum, between 1 gigahertz and 10 gigahertz.

The contributions from the different noise sources have been added to give the curve marked *Total sky*.

molecule's center of mass. This communication frequency is 1.65242 gigahertz.

A MESSAGE FROM THE EARTH

The 305-meter radio telescope at Arecibo, Puerto Rico, can transmit a message that could be received by an identical radio telescope many thousands of light years away. After the great dish was resurfaced and its efficiency was improved in late 1974, a powerful signal was beamed toward the globular cluster M13 in Hercules (see Figure 13.7). A message was transmitted in binary code form, a long string of zeros and ones. Though the cluster is 23,000 light years away, it was chosen because the beam width of the arriving signal would just span the cluster's diameter, thus wasting no power. Also, of the several hundred thousand stars in the cluster, there might be several civilizations able

to intercept the astrogram. The binary-coded message contained information about the earth's biology—the composition of DNA and the human population—an inventory of the solar system, and a diagram of the radio telescope that transmitted the coded signal (see Figure 19.7). There is no expectation of our ever receiving an answer. The message was sent as a token of our recognition of the existence of intelligent life somewhere in the Galaxy. Even if a response were forthcoming, we cannot expect it until approximately 50,000 A.D. Partners in an interstellar exchange will need eternal patience.

PROJECT SETI

The groundwork has been laid for developing the search strategy that could ultimately bring us into communication with advanced societies. One pro-

FIGURE 19.7

Arecibo message in pictures and a translation showing the binary version of the message decoded. Each number is marked with a label that indicates its start. When all the digits of a number cannot be fitted on one line, the digits for which there is no room are written under the least significant digit. (The message must be oriented in three different ways for all the numbers shown to be read.) The chemical formulas are those for the components of the DNA molecule: the phosphate group, the deoxyribose sugar, and the organic bases thymine, adenine, guanine and cytosine. Both the height of the human being and the diameter of the telescope are given in units of the wavelength that is used to transmit the message: 12.6 centimeters.

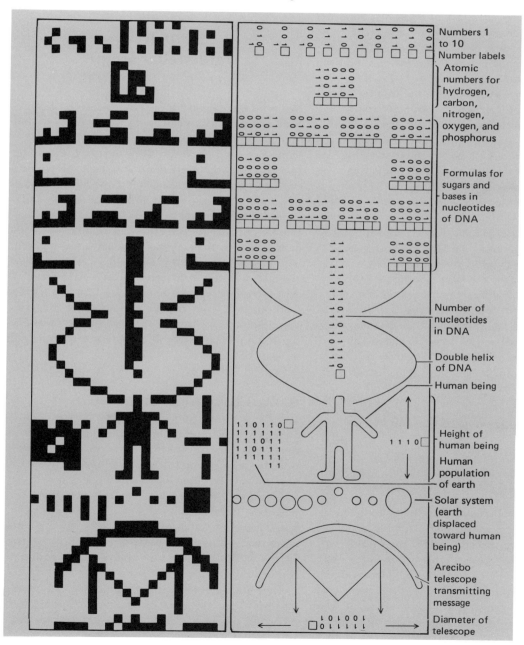

gram that has been proposed to do this is *Project SETI* (Search for Extraterrestrial Intelligence). This project involves a joint five-year search by NASA's Ames Research Center and the Jet Propulsion Laboratory for radio leakage emanating from planets harboring communicative civilizations. Present radio installations would be equipped with the most advanced receiver technology, and the total amount of funding is expected to amount to less than $20 million, no more than the cost of a couple of miles of a superhighway. Both groups would be able to monitor from one million to one billion narrow-band microwave channels *simultaneously* by means of a multichannel spectral analyzer.

The Ames effort would be confined to a selected list of solar types of stars within 1000 light years. Three existing radio telescopes, two in the Northern and one in the Southern Hemisphere, would be used for small periods of time for the monitoring. The Jet Propulsion Laboratory would conduct an all-sky survey, employing their Goldstone 26-meter dish located in the Mojave Desert and several smaller horn antennas.

There are, of course, many ways we could attempt to communicate with other worlds. One possibility is to orbit a large antenna dish, shielded from the earth, in a 28-day orbit around the earth (see Figure 19.8). A relay satellite in orbit around the dish would transmit data back to the earth. Still another idea is to construct one or more large dishes inside the lunar craters on the back side of the moon, as shown in Figure 19.9. This would provide complete shielding from radio interference from the earth.

> There is no easy way to the stars from the earth.
>
> Seneca

While we wait for better tools, serendipitous contact is a slim possibility. We might stumble onto artificial signals randomly directed to other worlds; we might intercept signals being exchanged between two communicating groups; perhaps we could locate signals between a cruising spaceship and its parent planet; or it is even possible we could intercept signals being sent from an automated probe monitoring our solar system back to its home station; or

eavesdropping on a galactic network exchanging information among societies is another possibility. Once interstellar communication is begun, however, it will be a dramatic end to our cultural isolation and it will bring us into a galactic club of intelligent societies. Social scientists have only begun to think about the sociological ramifications of membership in such a prestigious club.

19.5 Interstellar Flight

No more than a generation ago, only those addicted to science fiction would speak publicly of communicating with civilizations in outer space. Now, with unmanned spacecraft flying beyond the solar system and carrying information about ourselves (see page 435) and with a radio telescope having already broadcast a message beyond our solar system and currently being tuned to receive intelligent signals from anywhere, the subject is respectable. More and more scientists believe that someday we will chat with another advanced civilization. A galaxy with 200 billion stars is very likely to have many intelligent beings whose technologies are further advanced than our own. They may be signaling that they are there; perhaps we simply have not picked up the signal. Or perhaps it is still on its way after hundreds of years of travel.

IS TRAVEL TO OUTER SPACE POSSIBLE?

The stars are so far from us and the technological and biological difficulties in traveling to them are so great that a trip even to the closest star seems hopeless—now, for a long time to come, and maybe forever. Just consider the time it would take. Begin with a modest round trip to Alpha Centauri, our nearest neighbor. The distance is 4.3 light years; our ship's speed is constant at 50 kilometers per second, slightly more than we need to escape from the solar system (at the earth's distance from the sun). Our round trip will take about 52,000 years.

How, then, do we get around time? The only alternative is to use a ship that can move at something approaching the speed of light, say, 95% of the speed of light. That will certainly shorten the trip and also let us

FIGURE 19.8
A large antenna constructed in space
and shielded from radio radiation
emitted on the surface of the earth
could be an economical and effective
cosmic radio signal detector.

take advantage of relativistic time dilation (see Section 3.7). The implications of time dilation for space travel can be illustrated by the following example showing what is known as the *clock paradox.*

Astronomer Tom is planning to leave the earth on a round-trip space flight to star *S*, 12 light years away, at a rocket speed of 3/5 the velocity of light (*v* = 3/5 c) (see Figure 19.10). At the same time, astronomer Dick is to leave in the opposite direction on a round-trip space voyage to star *T*, also 12 light years away, at a rocket speed of 4/5 the velocity of light (*v* = 4/5 c). The third man, Harry, will remain on the earth to monitor their flights. Before takeoff the three men synchronize their clocks. To avoid complications in recording the travelers' clock times, we assume that the periods of acceleration at the beginning of the outward and return paths and the periods of deceleration in approaching each star or the earth are extremely brief compared with the time spent in

moving at the constant velocity. They may therefore be neglected.

Harry predicts that Tom will be gone 40 years and Dick 30. Once back on the earth, however, Tom will claim that he has been away 32 years by his clock time, and Dick will say he has been gone only 18 years according to his clock. To clarify these conclusions, consider Tom's situation. It takes a light ray 24 years to make the round trip between the earth and star *S*. Tom, traveling at 3/5 c, will accomplish the feat in 24 ÷ (3/5) = 40 years, judged by Harry's clock. But Tom's clock runs slow by the contraction factor:

$$\sqrt{1 - (3/5)^2} = 4/5$$

so that Tom's round-trip time will be (4/5) × 40 = 32 years, according to Tom's clock.

Suppose all three astronomers are 20 years old at the start. When Dick returns, he finds that Harry is 50 and he is 38. When Tom returns, Harry is 60 and Tom is

FIGURE 19.9
A large dish which could be built in a lunar crater on the far side of the moon could provide a large collection area with complete shielding from radio emission from the earth.

52. When Tom and Dick meet again on the earth, Tom will be only 4 years older than Dick because Dick came back sooner than Tom by 10 earth years. Because biological aging includes the time element in molecular cell growth, we presume the ages here are biologically correct.

Some have argued that we can think of the earth as moving relative to the fixed inertial frame of the spaceship, so that the earth rider could end up younger than the space traveler. But when we apply the Doppler principle and analyze signal flashes recording the passage of time from one frame of reference to another, we can show that the space traveler does indeed age more slowly than the stay-at-home.

The recipe for living forever is not, however, simply to move at a speed close to the velocity of light in one direction. An observer who leaves an inertial frame of reference with a uniform or accelerated velocity relative to it must return to it so that his time lags behind that of the observer who remains in the inertial frame.

From the results of the above example let us refigure the trip to Alpha Centauri discussed earlier. With this trip, we will need to consider acceleration as a factor, since we cannot expect to accelerate rapidly to such high speeds and survive. If we could accelerate the spaceship continuously at a constant rate of 1 g (equivalent to the acceleration due to gravity that we experience on the surface of the earth), the passengers would feel no great discomfort. The best technique then would be to reverse the acceleration halfway out and come to rest in the vicinity of Alpha Centauri; then on the return trip, accelerate at 1 g halfway and decelerate at 1 g toward the earth the rest of the way. Taking acceleration and deceleration into account, the round-trip earth time

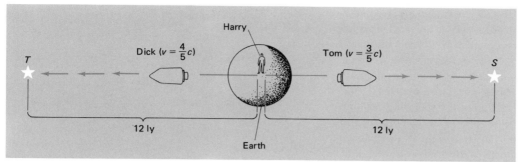

FIGURE 19.10
Astronomers' space trips.

would be 12 years; but the contracted time aboard the spacecraft, with a maximum speed of 95 percent of the velocity of light, would be 7.2 years.

Some representative travel times, with constant acceleration and deceleration of 1 g up to maximum velocity of 98 percent of the velocity of light, are shown in Table 19.3. From the table it appears that a round-trip journey to the Andromeda galaxy (2.2 million light years) could theoretically be made during an astronaut's life span. Unfortunately, the power a spacecraft would need for even less ambitious interstellar flights than this is overwhelmingly prohibitive.

FLIGHT TO TAU CETI

Let us see how much power a spaceship would need for a round-trip flight to a star 12 light years away at a maximum rocket speed of 99 percent of the velocity of light. This is approximately the distance to Tau Ceti, one of the biologically suitable stars. The most efficient energy device conceivable (it may never be achieved in practice) is a *photon rocket engine* powered by controlled annihilation of equal amounts of matter and antimatter. How much propellant would a rocket ship need for a payload weight of 100 tons (about twice as heavy as the *Apollo* ships) to make a round-trip flight to Tau Ceti, 12 light years away, at a maximum speed of 0.99c?

The spaceship will leave the earth at rest velocity and (1) accelerate up to 0.99c at the halfway point; (2) decelerate to 0 velocity near Tau Ceti and turn around; (3) accelerate to 0.99c up to the midway point on the return trip; and finally (4) decelerate to 0 velocity on arrival back at the earth. Solving the problem according to relativistic mechanics, these four stages call for a total initial spaceship weight of nearly 4 million tons,

VOYAGER'S MESSAGE

Once past the Jovian planets, which are their primary goals, the *Voyager* spacecraft will head out of the solar system into interstellar space. It will not be possible, given our present technology, to monitor their interstellar voyage. But on the chance that the spacecraft may be intercepted at some point in the distant future by intelligent life beyond our solar system, the two craft carry messages from the earth. The chance of this happening in the next several hundred thousand years is quite remote, but scientists feel that including the messages is well worth the effort in terms of learning how to communicate with other beings.

Each *Voyager* carries a 12-inch gold-plated phonograph record with cartridge and stylus plus pictorial instructions on their use. The records contain information about the solar system; the biological makeup of humans, animals, and plants; geographical points of interest; natural and human artifacts; and various human activities. Also included is a message from President Carter, followed by greetings in 55 languages and a "Hello" in a child's voice. Next there are sounds of the earth, such as rain falling, laughter, a kiss, a train whistle, bird calls, a horse and cart, a baby's voice, and others. The record ends with music that ranges from Bach to the blues, including various instrumental sounds and chants by peoples of the world.

all of which (except the 100-ton payload) is expended by the time the crew returns home 28 earth years later. The astronauts, however, are only 10 years older.

The large initial quantity of fuel and the weight of the life-support system for the crew are not the only physical difficulties. At its maximum speed of $0.99c$, the exposed frontal section of the spaceship will be blasted by several million billion atoms each second on each square meter, even though the density of interstellar space (mostly hydrogen) is about one atom per cubic centimeter. This count is high mainly because of the relative velocities of encounter between the moving ship and the atoms and partly from the Lorentz contraction factor, which decreases the spacing between the atoms. The surface is bombarded by several hundred times the flux of a high-intensity proton accelerator. Only massive concrete shielding could protect the astronauts inside. Also, a two-stage chemical-nuclear propulsion system would be necessary to move the spaceship far enough from the earth so that it could be launched without causing radiation damage to the earth. The gamma rays produced by the annihilation of matter in launching the ship would be equivalent to detonating a few hundred 1-megaton hydrogen bombs.

If we equipped ourselves with today's most efficient chemical propulsion system for the interstellar flight, the fuel requirements would be beyond belief. Maximum rocket exhaust velocities now are approximately 4 kilometers per second. To achieve the final velocity of $0.99c$, a 1-ton payload would require a propellant weight of about $10^{100,000}$ tons. With the estimated total mass of the universe at about 10^{50} tons, the entire idea crumbles as a mathematical exercise in futility. We can only conclude from these findings that physical contact with extraterrestrial beings may not be feasible ever.

TABLE 19.3
Round-Trip Interstellar Spaceflight Times and Corresponding Distances

Astronauts' Dilation Time (yr)	Earth Time (yr)	Distance Reached (ly)
5	6.5	1.7
10	24	9.8
20	270	137
30	3,100	1,565
40	36,000	17,600
50	420,000	208,640
60	5,000,000	2,479,000

HAVE UFOs FROM SPACE BEEN HERE?

Controversy over unidentified flying objects (UFOs) bestirs many—from the serious-minded scientist seeking a rational explanation for seemingly inexplicable phenomena to the charlatan preying on a gullible public. Reports of UFO sightings are as old as the Bible. *Ufology* attracts astronomers, physicists, meteorologists, psychologists, social scientists, saucer cultists, organizations like the Center for UFO Studies, antiestablishment groups, and societal derelicts. But those who extrapolate from the very few unexplained sightings to visitations from outer space strain credulity. The proofs will have to be much more convincing before scientists will accept the notion that our planet, in the Galactic boondocks, has been singled out for attention by superintelligent creatures.

In the previous paragraphs we saw that our only reasonable stratagem is to try electromagnetic exchanges with other communicative societies. So assuming that the economical use of power is universal, we may conclude that the least expensive and most feasible way for any intelligent civilization to obtain interstellar discourse is by long-distance radio signals, not by interstellar flights.

FINAL COMMENTS

There is little doubt in the minds of most astronomers that intelligent beings outside the earth do exist and that they are probably spread throughout our Galaxy and the universe. This very concrete realization is one of the best examples of the creative mind in science. Creative in the context of science means the creation

> For a moment of night we have a glimpse of ourselves and of our world islanded in its stream of stars—pilgrims of mortality, voyaging between the horizons across the eternal seas of space and time.
>
> Henry Beston

of an order; the finding of likenesses, links and hidden patterns which the human mind by an act of imagination picks out of nature and arranges. It is also an example of the faith that science has in the ability of the mind of man. Faith is so very important to the scientist for ". . . faith is the substance of things hoped for, the evidence of things not seen."

> The highest wisdom has but one science—the science of the whole —the science of explaining the whole creation and man's place in it.
>
> Leo Tolstoy

FIGURE 19.11
"We have found a strange foot-print on the shores of the unknown. We have devised profound theories, one after another, to account for its origin. At last, we have succeeded in reconstructing the creature that made the foot-print. Lo! it is our own." (Sir Arthur Eddington)

SUMMARY

The idea that life exists on other worlds is a notion that is centuries old. An ancient theory, panspermia, claimed that the seeds of life are in interstellar space in the form of microorganisms, ready to invade any hospitable planet they encounter. But the discoveries we have made of amino acids in meteorites and of organic molecules in the interstellar clouds do not prove that life exists elsewhere. Nevertheless, though we have no solid evidence of intelligent life in extraterrestrial space, it would be scientifically absurd to deny it is somewhere else. In the last quarter century, speculation on the possible existence of intelligent life elsewhere in the universe has been heightened by dramatic discoveries.

The search for stars in our Galaxy that show promise of supporting life has led us to set up criteria for such stars. On the basis of these criteria, we have rejected all but the main sequence stars of a solar type with spectral classes between F2 and K5. Possibly millions of these stars in the Milky Way harbor planets that are congenial to higher orders of life. All we can do is vaguely guess at the forms into which extraterrestrial beings may have evolved. For now technological and biological difficulties make interstellar spaceflight look impossible. A round trip to Tau Ceti, a biologically suitable star about 12 light years away, with the ideal matter-antimatter propellant pushing the spaceship and its 100-ton payload to 99 percent of the speed of light, would mean an initial weight at lift-off of 4 million tons. And the ship's high speed through the interstellar medium would require many tons of shielding to protect the crew against deadly radiation. But the fuel requirements for achieving the necessary $0.99c$ speed pose the most difficult problem. Using today's most efficient propulsion systems, we would need far more fuel to accelerate even a 1-ton payload to $0.99c$ than there is estimated mass in the universe. Thus, for now, only one way of communicating with extraterrestrial societies seems possible: electromagnetic radiation.

Astronomers and others have tried since this century began to detect extraterrestrial signals. Project Ozma, the first modern American effort, was begun in 1960, using the 26-meter radio dish at the National Radio Astronomy Observatory in West Virginia. No intelligent signals were detected from the two stellar possibilities the dish was aimed at, Tau Ceti and Epsilon Eridani. All attempts to find signals from other stars have been unsuccessful, though reasonable speculation tells us our Galaxy may have a million civilizations that could carry on interstellar communication.

The communicative phase of our own civilization is only a few decades old; and because of our limited technology and energy sources, for now it seems more practical to listen for extraterrestrial signals than to try to transmit them. We could search for normal transmissions in and among other planets. We might try to intercept signals transmitted on frequencies close to that of neutral hydrogen, at 21 centimeters: advanced societies in other galaxies could be using this part of the spectrum to broadcast. Perhaps, if we listen, we will hear them.

REVIEW QUESTIONS

1. Why does interstellar flight appear to be an impossible dream?

2. What advantage does a space voyager gain if he can make a round trip flight from the earth to another world at relativistic speeds?

3. Discuss the early attempts to communicate with extraterrestrial beings.

4. What was Project Ozma? What has followed along similar lines since then?

5. What is a reasonable guess as to the possible number of communicative societies inhabiting the Galaxy? How is this figure arrived at?

6. Should we try to transmit coded signals to other worlds or should we instead listen for them? Give reasons for your answer.

7. Explain why the wavelength spectral region between 3 centimeters and 30 centimeters might be suitable for communication beyond the solar system.

8. When the pulsars were first discovered, the

thought naturally occurred that other worlds were trying to talk to us. Why was this notion soon abandoned?

9. How might artificial signals be picked up accidentally?

10. Do you believe that interstellar communication may be going on somewhere in our Galaxy, or do you dismiss this idea as too preposterous for serious discussion?

11. Is there any reason why an advanced communicative society might prefer to transmit a beamed message in a nonmicrowave spectral region?

12. Describe the SETI project and its objectives.

13. How has it been shown that Barnard's star probably has planets circling it?

14. What are reasonable criteria for selecting a stellar candidate with a suitable and habitable zone from among the sun's neighbors?

Appendix 1
Astronomical Symbols and Abbreviations

⊙ sun

● new moon

☽ first quarter moon

○ full moon

☾ last quarter moon

☿ Mercury

♀ Venus

⊕ Earth

♂ Mars

♃ Jupiter

♄ Saturn

♅ Uranus

♆ Neptune

♇ Pluto

④ Vesta asteroid (number is order of discovery)

♈ Aries or vernal equinox

♉ Taurus

♊ Gemini

♋ Cancer

♌ Leo

♍ Virgo

♎ Libra or autumnal equinox

♏ Scorpius

♐ Sagittarius

♑ Capricornus

♒ Aquarius

♓ Pisces

α right ascension

δ declination

a semimajor axis of ellipse; aperture of telescope; acceleration

Å angstrom

AU astronomical unit

c velocity of light

cm centimeter

d or r distance

D diameter

e orbital eccentricity

eV electron volt

f frequency

g gram; acceleration at Earth's surface

G gravitational constant

h Planck constant

H Hubble constant

Hz hertz

λ wavelength

L luminosity

ly light year

m apparent magnitude; mass; meter

M absolute magnitude; mass

μ proper motion

μm micron or micrometer

p parallax

pc parsec

P orbital period

R radius

ϱ density

s second of time

t time

v velocity

w watt

y year

°C degrees Celsius

K degrees Kelvin or absolute

° degree of arc

′ minute of arc

″ second of arc

440

Appendix 2
Measurement and Computations

METRIC AND ENGLISH UNITS

English and metric systems of measurement are the two principal systems used today. The metric, based on the decimal system, is more widely used because of its logical and computational simplicity. It has long been employed by scientists and by most non-English-speaking peoples. The English system is gradually being phased out even by English-speaking countries.

Both systems begin with defined standards of length, mass, and time: the foot, the pound, and the second in the English system; the meter, the gram, and the second in the metric system. All other quantities of measurement can be expressed in these units. Where did the English units come from? You might want to answer the question as a research project.

The meter was originally chosen as one ten-thousandth of the distance along the Paris meridian between the equator and the North Pole, as determined late in the eighteenth century. More recent measurements of earth's dimensions have slightly altered the original value. The meter actually is an arbitrary unit of length spaced off by two marks on a bar of platinum-iridium alloy housed in the International Bureau of Weights and Measures at Sèvres near Paris, France. It is presently specified as a wavelength of light: 1,650,763.73 wavelengths of the orange red spectral line emitted by radiating atoms of krypton 86 in a vacuum.

The unit of mass is the gram, equal to the mass of one cubic centimeter of water at 4° Celsius, the temperature at which water is most dense. The mass standard is a platinum-iridium cylinder equal to 1000 grams or 1 kilogram, an arbitrarily chosen base quantity. The standard unit of volume is one cubic centimeter. A volume of 1000 cubic centimeters is 1 liter. The standards of mass and volume are also kept in the International Bureau of Weights and Measures. Exact copies are maintained in national depositories throughout the world.

The unit of time, the second, was originally defined as 1/86,400 of the period for one complete rotation of earth relative to the sun. There are 86,400 seconds in one mean solar day. Because earth's rotation varies slightly, a new definition of the second was introduced in 1964: the time it takes the cesium 133 atom to make 9,192,631,770 vibrations. This is a precise natural frequency emitted by the atom, corresponding to a wavelength of 3.263483271 centimeters.

Conversion Factors—Metric to English

To Find	Multiply	By
inches	centimeters	0.39370
feet	meters	3.28084
miles	kilometers	0.62137
ounces	grams	0.03527
pounds	kilograms	2.20462
short tons	metric tons	1.10231

Conversion Factors—English to Metric

To Find	Multiply	By
centimeters	inches	2.54
meters	feet	0.3048
kilometers	miles	1.60934
kilograms	pounds	0.4536
metric tons	short tons	0.9072

Conversion Factors Between Units

To Find	Multiply	By
angstroms	centimeters	10^{-8}
centimeters	angstroms	10^8
angstroms	microns	10^4
kilometers	light years	9.4605×10^{12}
kilometers	parsecs	3.086×10^{13}
parsecs	light years	0.3066
light years	parsecs	3.2617
light years	kilometers	1.0570×10^{-11}
metric tons	kilograms	10^{-3}
seconds (time)	years	3.156×10^7
ergs	electron volts	6.242×10^{13}
electron volts	ergs	1.602×10^{-12}
watts	ergs/second	10^{-7}
ergs/second	watts	10^7

EXAMPLES OF UNIT RATES

0.1 gram per cubic centimeter:
0.1 g/cm³ or 0.1 g·cm⁻³

0.1 gram per cubic centimeter:
0.1 g/cm^3 or $0.1 \text{ g} \cdot \text{cm}^{-3}$

200 kilometers per second:
200 km/s or $200 \text{ km} \cdot \text{s}^{-1}$

1/1000 gram per square centimeter:
0.001 g/cm^2 or $0.001 \text{ g} \cdot \text{cm}^{-2}$ or $10^{-3} \text{ g} \cdot \text{cm}^{-2}$

9.8 meters per second per second:
9.8 m/s^2 or $9.8 \text{ m} \cdot \text{s}^{-2}$

100 ergs per square centimeter per second:
$100 \text{ erg/cm}^2 \cdot \text{s}$ or $100 \text{ erg} \cdot \text{cm}^{-2} \cdot \text{s}^{-1}$

ARC MEASUREMENT AND GEOMETRICAL RELATIONS OF SPHERES

A circle contains 360 degrees (360°)

1 degree (1°) = 60 arc minutes (60')

1 arc minute (1') = 60 arc seconds (60")

1 radian = 57.29578° = 206,264.8"

1 arc second = 4.848×10^{-6} radians

Arc length = radius of arc × subtended angle in radians

$\pi = 3.14159$

Area A of a circle of radius r or diameter d:
$A = \pi r^2 = 3.14 r^2 = \pi d^2/4 = 0.785 d^2$

Circumference C of a circle of radius r or diameter d:
$C = 2\pi r = 0.628 r = \pi d = 3.14 d$

Surface area S of a sphere of radius r:
$S = 4\pi r^2 = 12.56 r^2$

Volume V of a sphere of radius r:
$V = (4/3)\pi r^3 = 4.19 r^3$

SIMPLE COMPUTATION

We simplify arithmetical operations with very large and very small numbers by a few easy rules as shown in these examples:

$584,000 = 584 \times 10^3 = 58.4 \times 10^4 = 5.84 \times 10^5$
$= 0.584 \times 10^6$
$0.00000485 = 485 \times 10^{-8} = 48.5 \times 10^{-7}$
$= 4.85 \times 10^{-6} = 0.485 \times 10^{-5}$
$3 \times 10^{10} = 30,000,000,000 = 30 \times 10^9 = 300 \times 10^8$
$= 3000 \times 10^7$

Rules:
Multiplication (add exponents):
$10^{-3} \times 10^2 \times 10^4 = 10^{-3+2+4} = 10^3$

Division (subtract exponents):
$10^4/10^2 = 10^{4-2} = 10^2$

Raising power or extracting root (multiply exponents):
$(10^2)^5 = 10^{10}; \ (10^{-2})^4 = 10^{-8}; \ (10^{-10})^{1/2} = 10^{-5}$

Physical Constants		
Quantity	Symbol	Value
Velocity of light in vacuum	c	2.998×10^{10} cm/s
Gravitational constant	G	6.667×10^{-8} cm³/g·s²
Planck's constant	h	6.626×10^{-27} erg·s
Mass of hydrogen atom	m_H	1.673×10^{-24} g
Mass of proton	m_p	1.6725×10^{-24} g
Mass of electron	m_e	9.109×10^{-28} g

Complete Examples:

$$\frac{(4 \times 10^2)^2}{2 \times 10^4} = \frac{4^2 \times 10^4}{2 \times 10^4} = \frac{16}{2} \times 10^0 = 8$$

$$\frac{(64 \times 10^{-2})^{1/2}}{\sqrt{16} \times 0.02} = \frac{64^{1/2} \times 10^{-1}}{4 \times 2 \times 10^{-2}} = \frac{\sqrt{64} \times 10^1}{8}$$

$$= \frac{8}{8} \times 10^1 = 10$$

$$\frac{0.662 \times 2,400,000}{0.0007 \times 120} = \frac{6.62 \times 10^{-1} \times 2.4 \times 10^6}{7 \times 10^{-4} \times 1.2 \times 10^2}$$

$$= \frac{6.62 \times 2.4 \times 10^5}{7 \times 1.2 \times 10^{-2}} = 1.89143 \times 10^7$$

$$\frac{2187 \times 0.00275}{1.43 \times (0.91)^2} = \frac{2.187 \times 10^3 \times 2.75 \times 10^{-3}}{1.43 \times 0.8281}$$

$$= \frac{6.01425}{1.184183} = 5.07882$$

$$\frac{3 \times 10^5 \times 3.5 \times 0.052}{55 \times 10^{-2} \times (8.6 \times 10^{-1})^2} = \frac{3 \times 10^5 \times 3.5 \times 5.2 \times 10^{-2}}{0.55 \times 73.96 \times 10^{-2}}$$

$$= \frac{54.6 \times 10^3}{40.678 \times 10^{-2}}$$

$$= 1.34225 \times 10^5 = 134,225$$

Mathematical Symbols:

$\propto$ proportional to

$\approx$ very nearly equal to

$\sim$ roughly or approximately equal to

$>$ greater than

$<$ less than

$=$ equals

Ratio and Proportion:

$y \propto x$: y is proportional to x, or y varies directly as x

$b \propto 1/d^2$: b is inversely proportional to the square of d, or b varies inversely as the square of d

$F \propto M \cdot m/r^2$: F varies directly as the product of M and m and inversely as the square of r

The proportionality sign $\propto$ is removed by introducing a constant of proportionality. In the above examples:

$$y = cx \qquad b = \frac{k}{d^2} \qquad F = \frac{GMm}{r^2}$$

These are different forms of the same quantities:

$$pv = RT, \; p = \frac{RT}{v}, \; v = \frac{RT}{p}, \; T = \frac{pv}{R}, \; \frac{p}{R} = \frac{T}{v}, \; \frac{v}{R} = \frac{T}{p}$$

Astronomical Constants

Object or Quantity	Symbol	Value
Mass of earth	$M_\oplus$	5.977×10^{27} g
Mean radius of earth	$R_\oplus$	6.371×10^{8} cm
Mass of sun	$M_\odot$	1.989×10^{33} g
Radius of sun	$R_\odot$	6.960×10^{10} cm
Solar luminosity	$L_\odot$	3.82×10^{33} erg/s
Astronomical unit	AU	1.49598×10^{13} cm
One sidereal year		365.256 days = 3.156×10^{7} s

Powers of Ten

Word	Number	Power	Prefix	Symbol
trillion	1,000,000,000,000	10^{12}	tera	T
billion	1,000,000,000	10^{9}	giga	G
million	1,000,000	10^{6}	mega	M
thousand	1,000	10^{3}	kilo	K or k
hundred	100	10^{2}	hecto	h
ten	10	10^{1}	deca	da
unit	1	10^{0}		
tenth	0.1	10^{-1}	deci	d
hundredth	0.01	10^{-2}	centi	c
thousandth	0.001	10^{-3}	milli	m
millionth	0.000,001	10^{-6}	micro	μ
billionth	0.000,000,001	10^{-9}	nano	n
trillionth	0.000,000,000,001	10^{-12}	pico	p

Appendix 3
Astronomical Coordinate Systems

INTRODUCTION

In order to locate an astronomical object on the sky, astronomers employ coordinate systems similar in concept to that used to plot a point in the *XY* plane. The intersection of the horizontal *X* and vertical *Y* distances reckoned from a common origin mark the position of the point in the *XY* diagram. However, the *celestial sphere,* which is the apparent sphere of the sky, is not a plane but is the inside of a sphere and one that rotates relative to our location on the surface of the earth. On the celestial sphere distances are measured along the arc of a great circle in degrees. A *great circle* on a sphere is the arc formed by passing a plane through the center of the sphere. Thus the great circle divides the sphere into two equal halves. All other circles on a sphere that are not great circles are called *small circles.*

There are several features in common among all astronomical coordinate systems. Each has a principal axis or polar axis about which the system rotates. The points of intersection of this axis and the celestial sphere are the poles of the system. Perpendicular to the principal axis is a great circle on the celestial sphere, which is the principal reference circle along which one coordinate is measured. Finally, there are secondary reference circles on the celestial sphere, which are great circles perpendicular to the principal reference circle. There are an infinite number of these secondary reference circles that meet at the poles of the principal axis, as shown in Figure A3.1.

HORIZON COORDINATE SYSTEM

The horizon coordinate system is probably the most natural one and the one most easily visualized. The principal axis of the system is defined as being parallel to the direction of gravity. Extended upward it intersects the celestial sphere at the point known as the *zenith,* which is directly overhead for an observer. The opposite end of the axis underneath the observer is called the *nadir* (see Figure A3.2). The principal reference circle, the *astronomical horizon,* is the great circle marked on the celestial sphere by a plane perpendicular to the zenith-nadir axis and tangent to the earth at the point of the observer. There are four cardinal points of reference on the astronomical horizon — north, east, south, and west. They are each 90° from each other.

If we project the terrestrial meridian of the longitude passing through the observer's position onto the celestial sphere, it defines the great circle called the *celestial meridian.* The meridian passes through the north and south points of the horizon, as well as the zenith and nadir. It also contains the two points that are the intersection of the

earth's axis of rotation extended to the celestial sphere—the *north and south celestial poles.* The position of the celestial poles relative to the north and south points of the horizon depends upon the observer's latitude. For an observer in the Northern Hemisphere the angular distance of the north celestial pole above the north point of the horizon is equal to the observer's latitude.

The coordinates in the horizon system are known as azimuth and altitude. *Azimuth* is the angular distance measured from the north point of the horizon along the astronomical horizon. The secondary reference circles in this system are known as *vertical circles.* Thus the azimuth is measured to the foot of the vertical circle passing through the object of interest. As an example, the azimuth of the east point of the horizon is 90°; the south point, 180°; and

FIGURE A3.1
Basic plan of the various astronomical coordinate systems. Beginning with the principal axis, the principal reference circle is everywhere 90° from the poles, while the infinity of secondary reference circles are perpendicular to the principal reference circle.

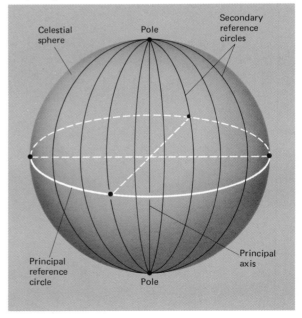

the west point, 270°. *Altitude* is the angular distance of the object of interest above or below the horizon measured along a vertical circle. The altitude of the zenith, for example, is +90°, while that of the nadir is −90°. The angular distance from the zenith along a vertical circle to the object is called the *zenith distance*. Hence the sum of the altitude and the zenith distance for an object is equal to 90°. In Figure A3.2 the altitude of the star is about 60° (its zenith distance is thus 30°) and its azimuth is about 225°.

The major disadvantage of the horizon coordinate system is that it is peculiar to the observer and not at all a general system. Also, since the altitude and the azimuth of an object, such as Jupiter, continually change as the earth rotates, one must know the exact location and time at which an altitude and azimuth are measured in order for them to have any meaning to anyone beside the observer. A more general system is the equatorial coordinate system.

EQUATORIAL COORDINATE SYSTEM

The principal axis of the equatorial coordinate system coincides with the axis of rotation of the earth. Its pole's are the north and south celestial poles, as defined in the preceding section. The principal reference circle is the *celestial equator,* while the secondary reference circles are called *hour circles,* as shown in Figure A3.3. This system differs from the horizon system in that the reference circles are on the celestial sphere and thus rotate with it relative to the observer. In the horizon system the astronomical horizon and the vertical circles remain fixed relative to the observer, and the celestial sphere moves relative to them.

Figure III.3 shows the relationship between the equatorial and horizon systems for one particular time.

As the earth rotates from west to east, a star traces a path across the sky called a *diurnal (daily) circle* from east to west. For an observer located at the north geographic pole, his astronomical horizon coincides with the celestial equator and the north celestial pole coincides with his zenith, as shown in part a of Figure A3.3. There, diurnal circles are parallel to the horizon and the stars neither rise nor set. One-half of the sky is always above the horizon and the other half is never visible. On the geographic equator (part b in Figure A3.3) the celestial equator passes through the zenith and it intersects the astronomical horizon at right angles. The star's diurnal path comes straight up from the horizon and sets perpendicular to the horizon. Since the north celestial pole coincides with the north point of the horizon, all stars will rise and set. At intermediate geographic latitudes, the sky rotates at an oblique angle, with the value depending upon the value of the latitude. The celestial equator passes through the east and west points of the horizon regardless of the geographic latitude of the observer.

The coordinates in the equatorial system are called right ascension and declination. *Declination* is the angular distance of an astronomical object north or south of the celestial equator measured along the hour circle through the object. Thus the declination of the north celestial pole is +90°, while that of the south celestial pole is −90°. In Figure A3.4 the declination of the star is about +60°. The other coordinate is called *right ascension* and is mea-

FIGURE A3.2

The horizon system of coordinates. For a star in the southwestern part of the sky, the measurement of the azimuth is along the astronomical horizon and the measurement of the altitude is along a verticle circle.

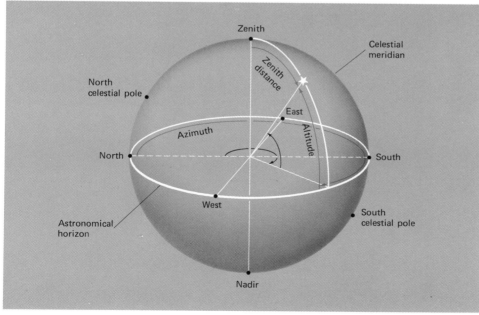

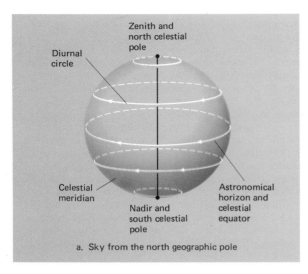

a. Sky from the north geographic pole

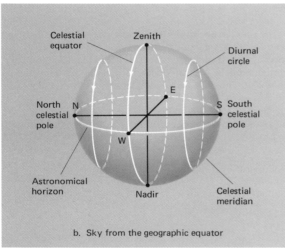

b. Sky from the geographic equator

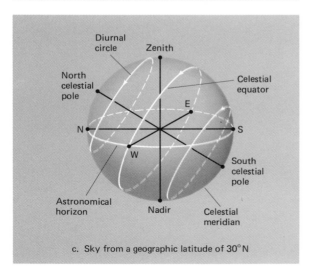

c. Sky from a geographic latitude of 30°N

sured along the celestial equator. However, to actually make a measurement, the astronomer needs a reference point on the celestial equator. Such a point is the one formed by the intersection of the annual apparent path of the sun on the celestial sphere, called the *ecliptic,* and the celestial equator. Since both the ecliptic and the celestial equator are great circles, they intersect at two points. One, at which the sun passes from south to north of the celestial equator, is called the *vernal equinox* (♈), since the sun moves through this point on or about 21 March each year. The other point, at which the sun passes north to south relative to the celestial equator, is called the *autumnal equinox.* The reference point for measuring right ascension is the vernal equinox, and it is measured eastward in units of time rather than degrees of arc to the foot of the hour circle passing through the object of interest. Since a 360° rotation by the earth corresponds to 24 hours, then 1 hour equals 15° of arc, or 1° equals 4 minutes, and so on. The star in Figure A3.4 has a right ascension of about 6 hours, or 90°. The angular distance of the hour circle from the observer's celestial meridian is called the *hour angle* of the star. For the example in Figure A3.4 the hour angle shown is about 1.5 hours, or 22.5° of arc.

It is obvious that the right ascension and declination of a celestial object remain fixed as the sky rotates. This makes it possible to catalog astronomical objects by their right ascensions and declinations in the same way in which places on earth are cataloged by their longitudes and latitudes. Because of precession (see p. 18), the vernal equinox slides westward along the ecliptic by about 50 seconds of arc per year. Thus over a number of years a star's right ascension and declination change, so that cataloged positions are not accurate indefinitely and must be updated periodically to correct for the effect of precession.

In addition to the horizon and equatorial systems, there are two less frequently used systems of astronomical coordinates, the *ecliptic* and the *galactic* systems. The principal reference circle for the ecliptic system is the ecliptic; for the galactic system it is the central plane of the Milky Way. These four systems are summarized in Table A3.1.

STAR MAPS

Although astronomical coordinates systems are necessary in order to make precision observations, a general knowledge of the sky can be obtained with the aid of star charts (see Figures A3.5, A3.6, A3.7, and A3.8). The starting point for learning the constellations is Polaris, the North Star, which is located about one degree from the north celestial pole.

Five prominent constellations revolve around Polaris—Ursa Major the Big Bear, Ursa Minor the Little Bear, Cepheus the King, Cassiopeia the Queen, and Draco the Dragon. These are called circumpolar constellations because they are visible all night and every night to observers located north of about 40° north latitude.

To locate the five circumpolar constellations, refer to Figure A3.5 in which Polaris is in the upper part of the

FIGURE A3.3
Relation between equatorial and horizon coordinate systems.

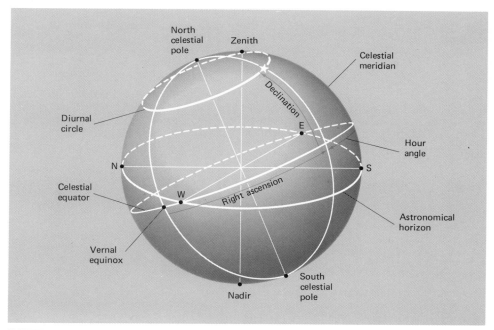

North celestial pole

Zenith

Celestial meridian

Declination

Diurnal circle

Hour angle

E

N

S

Celestial equator

Right ascension

Astronomical horizon

W

Vernal equinox

Nadir

South celestial pole

FIGURE A3.4

The equatorial system of coordinates, including its relation to the horizon system. As the earth rotates, the hour-angle of the vernal equinox continually increases, repeating after a 360° rotation of the earth.

figure and the five constellations are around it. Polaris, which marks the end of the handle of the Little Dipper and the tip of the tail of the Little Bear, locates the constellation of Ursa Minor. This constellation is easy to locate, because its clear, bright asterism is oriented so that either dipper can always pour into the other. The two end stars in the bowl of the Big Dipper, Merak and Debhe, are called the "pointers" because a line drawn through them always points to Polaris. When this line is extended to the first star beyond Polaris, the asterism of Cepheus is located. It is a house with a steep roof, and this star marks the apex of the roof. The asterism of Cassiopeia, which is the letter M or W, is located by extending the line which joins the ends of the handles of the two dippers to the first star beyond Polaris. This star marks one end of the asterism. When another star is included in the asterism, the outline of the chair on which Cassiopeia is seated can be easily visualized. The constellation of Draco is located by dividing the line which joins the ends of the handles of the two dippers into three equal parts. Two-thirds of the way on the line from Polaris is the star Thuban, which marks the tip of the dragon's tail. Draco has no asterism, because its arrangement of stars very clearly outlines the head, body, and tail of the dragon.

Autumn Constellations
Three beautiful constellations are visible almost directly overhead in the autumn months (see Figure A3.5). They are Cygnus the Swan, Lyra the Harp, and Aquila the Eagle, and they all lie in the Milky Way. The brightest star in each of these constellations—Deneb in Cygnus (the tail of the swan), Vega in Lyra (one corner of the small triangle), and Altair in Aquila (the middle star in the straight line)—form the summer triangle, which is almost a right triangle. The asterisms are the Northern Cross for Cygnus, a rectangle and triangle for Lyra, and a straight line of three stars for Aquila.

The three zodiacal constellations—Capricornus the Sea Goat, Aquarius the Water Carrier, and Pisces the Fishes—are visible low in the autumn sky between the southern and eastern points on the horizon.

Above the eastern horizon, the constellation of Pegasus the Winged Horse can easily be located by its asterism, the Great Square. The square represents the body of the horse, which is upside down, and only the front half is visible. The dim stars to the west of the square form the head and the two front legs of the animal. The star Alpheratz, which is located at the northeast corner of the square, is common to two constellations—Pegasus and Andromeda the Chained Lady. The asterism of Andromeda is two rows of stars diverging from Alpheratz and curving toward the constellation of Cassiopeia.

Winter Constellations
The winter hunting scene, the most outstanding feature of the winter sky (see Figure A3.6), comprises several constellations—Orion the Mighty Hunter; his two dogs, Canis Major the Big Dog and Canis Minor the Little Dog; his adversary, Taurus the Bull; and his prey, Lepus the Hare. The asterism of Orion is the Hour Glass. The star Betelgeuse,

TABLE A3.1
Astronomical Coordinate Systems

Coordinate System	Principal Axis	Principal Reference Circle	Coordinate (units)	Secondary Reference Circles	Coordinate (units)
Horizon	Zenith-nadir	Astronomical horizon	Azimuth (0°–360°)	Vertical circle	Altitude (±0°–90°)
Equator	North-south celestial pole	Celestial equator	Right ascension (0–24^h)	Hour circle	Declination (±0°–90°)
Ecliptic	North-south ecliptic pole	Ecliptic	Celestial longitude (0°–360°)	No name	Celestial latitude (±0°–90°)
Galactic	North-south galactic pole	Plane of Milky Way	Galactic longitude (0°–360°)	No name	Galactic latitude (±0°–90°)

(bright red), marks the right shoulder. The star Rigel marks the left foot. Orion wears a belt of three stars, with the top star, Mintaka, lying almost on the celestial equator. The curved row of stars in front and to the west of Orion represents the lion, which Orion is holding with his left hand. Below Orion is Lepus, and following the mighty hunter, to the east, are his two dogs. Canis Major has no asterism, because the outline of a dog can be easily visualized from the arrangement of the stars. The brightest star in the sky is Sirius, which marks the nose of the Big Dog. The constellation of Canis Minor is difficult to visualize, because most of its stars are very dim. Its brightest star, Procyon, together with Sirius and Betelgeuse form the winter triangle. Completing the scene is Taurus, a zodiacal constellation, which is in front and to the west of Orion. Its asterism is the open cluster of stars, the Hyades, which marks the head of the bull. Only the front part of the bull's body is visible, because the Greeks imagined that the bull was swimming in the Mediterranean Sea. The right eye of the bull is the bright star Aldebaran. The open cluster of the Pleiades is located in the left shoulder of the bull.

The three zodiacal constellations are Aries the Ram, Taurus the Bull, and Gemini the Twins. The asterism of Aries is a small triangle, and Hamal is its brightest star. The asterism of Gemini is two nearly parallel rows of stars. The twin sons of Jupiter are represented by the stars Castor and Pollux. Castor is located at the top of the western row of stars; Pollux, at the top of the eastern row.

Above the winter hunting scene are the two beautiful constellations of Auriga the Charioteer and Perseus the Hero, which contains the open clusters h and χ Persei. The asterism of Auriga is a five-sided figure, with Capella its brightest star. The asterism of Perseus resembles a wishbone, and its most interesting star is Algol, the "Blinking Demon". In the southern part of the sky are two, long, indistinct constellations that can be located and traced on a clear moonless evening. These are Eridanus the Po River and Cetus the Whale. Eridanus meanders in the region between Cetus, Orion, and Lepus.

Spring Constellations
The prominent constellations in the spring (see Figure A3.7) are the three zodiacal constellations Cancer the Crab, Leo the Lion, and Virgo the Virgin. Dominating the entire southern sky is the long, faint constellation Hydra the Sea Serpent. The brightest and most beautiful constellation is Leo, and its asterism is a sickle and a triangle. The sickle represents the head of the lion; the triangle, its hind quarters. To the west of the sickle is a hazy spot of light, which is the Praesepe open cluster also called the "Beehive", located between two faint stars. This star group marks the constellation of Cancer, whose stars are very dim. To the east of the triangle in Leo is the constellation of Virgo. The location of this constellation is also marked by the continuation of the curved line through the handle of the Big Dipper, through the bright star Arcturus in Boötes, and to the bright star Spica in Virgo. Spica represents the sheaf of wheat in Virgo's left hand.

Summer Constellations
There are several prominent summer constellations shown in Figure A3.8: Boötes the Bear Driver; Corona Borealis the Northern Crown; Hercules the Kneeler; Serpens the Serpent; Ophiuchus the Serpent Carrier; and the zodiacal constellations of Libra the Scales, Scorpius the Scorpion, and Sagittarius the Archer.

The continuation of the curved line through the handle of the Big Dipper locates the star Arcturus, which marks the bottom of the kite, the asterism of Boötes. To the east of Boötes is a semicircle of stars which represent the constellation of Corona Borealis. To the east of Corona Borealis is Hercules, who appears in a kneeling position. Its asterism is the letter H. The asterism of Ophiuchus is a triangle on top of a large vertical rectangle which represents the body of Ophiuchus. Around his body is the constellation Serpens the Serpent. The most prominent of the three zodiacal constellations is Scorpius. Its arrangement of stars clearly outlines the head, body, and tail of the scorpion. Its brightest star, Antares, marks the head of the scorpion.

FIGURE A3.5
AUTUMN SKIES

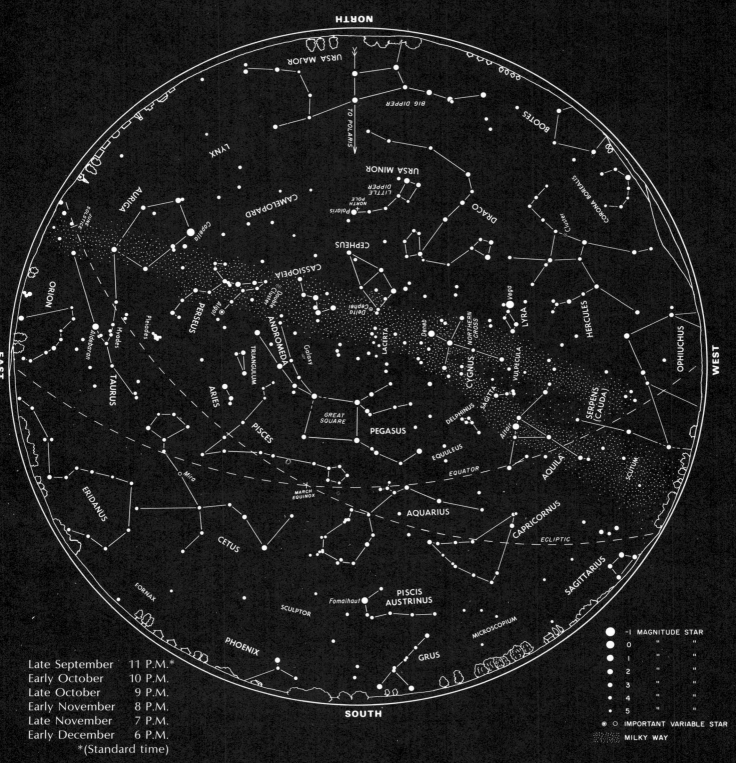

Late September   11 P.M.*
Early October   10 P.M.
Late October   9 P.M.
Early November   8 P.M.
Late November   7 P.M.
Early December   6 P.M.
        *(Standard time)

These charts show the constellations as seen from approximately 40 degrees north latitude (the average latitude of the continental United States, around which most of the population is grouped). To best use

CHARTS BY GEORGE LOVI

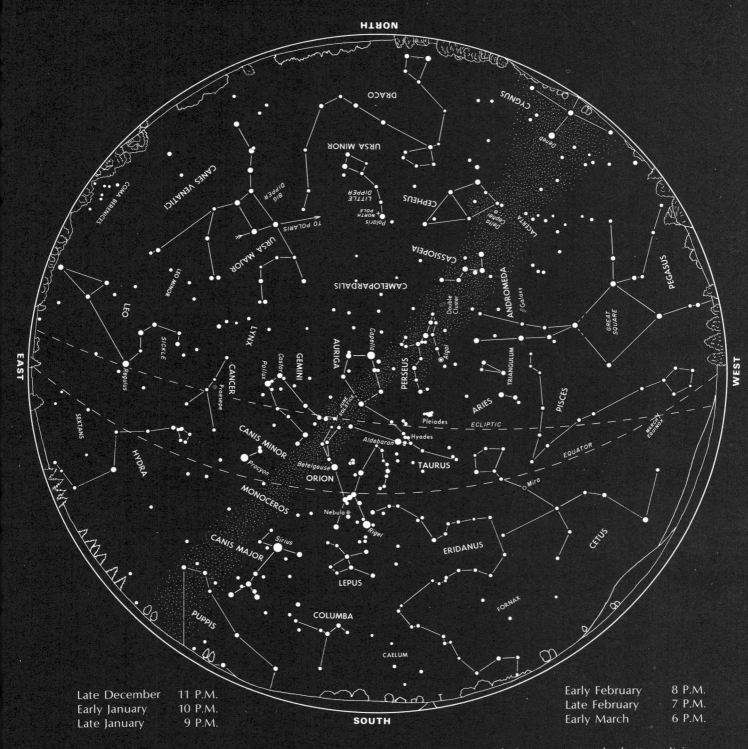

FIGURE A3.6
WINTER SKIES

Late December 11 P.M.
Early January 10 P.M.
Late January 9 P.M.

Early February 8 P.M.
Late February 7 P.M.
Early March 6 P.M.

these charts, rotate the chart so that the label for the direction you are facing appears at the bottom. It is not necessary to follow an exact time schedule. Although all constellations that are visible above the horizon at the stated times are indicated, only the more prominent and/or most important constellations have lines connecting their principal stars.

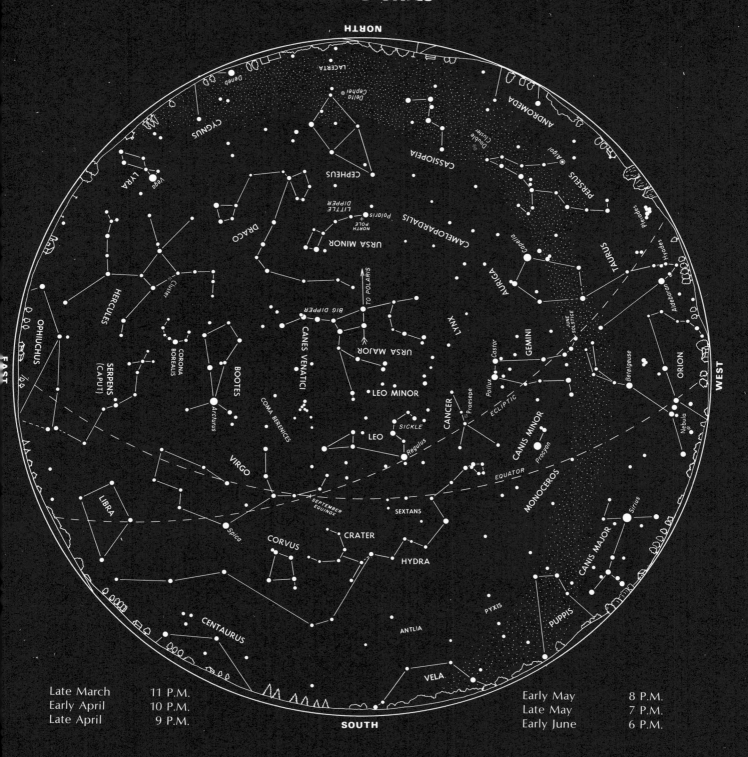

FIGURE A3.7
SPRING SKIES

Late March   11 P.M.
Early April   10 P.M.
Late April   9 P.M.

Early May   8 P.M.
Late May   7 P.M.
Early June   6 P.M.

Two coordinate lines are shown: the celestial equator and the ecliptic. It is important to remember that the moon and the planets stay close to the ecliptic—so that they often appear as strange "stars" within the

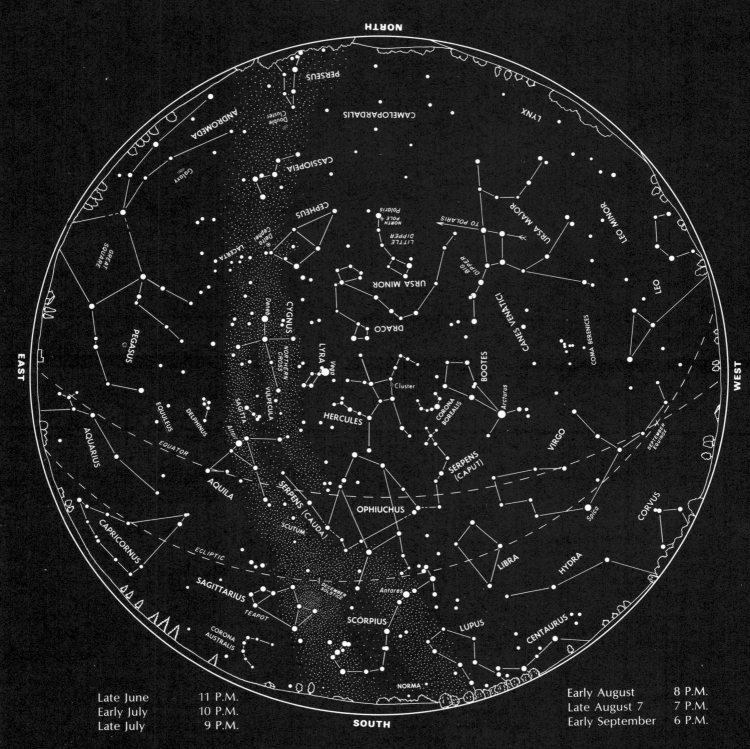

FIGURE A3.8
SUMMER SKIES

Late June	11 P.M.
Early July	10 P.M.
Late July	9 P.M.

Early August	8 P.M.
Late August 7	7 P.M.
Early September	6 P.M.

constellations that lie along this circle. Notice also the four season points along the ecliptic—the equinoxes and solstices—which indicate the sun's position at the beginning of each season.

Appendix 4
Time Systems

INTRODUCTION

Time basically is measured by the earth's rotation. Today atomic clocks are able to measure time so accurately that they monitor the rotation that is not exactly uniform (see text, page 106). But we still must go back to the basic fundamentals of reckoning time built on the earth's rotation, as described in the following paragraphs.

A convenient reference position from which to count time is the observer's celestial meridian. Either the vernal equinox (Υ) or the sun serve as "hour hands" to denote the passage of time, since both markers are carried westward by the rotation of the sky.

SIDEREAL TIME

This kind of star time is used to set the telescope on an object whose right ascension and declination (see Appendix 3) are known. The telescope is positioned according to the object's declination and hour angle. The latter is obtained by subtracting the right ascension from the known sidereal clock time. One *sidereal day* is equal to the interval between two successive crossings of the vernal equinox over the observer's meridian. It corresponds to one complete rotation of the earth on its axis. The local *sidereal time* (LST) is given by the local hour angle of the vernal equinox (LHA Υ), which is always equal to the sum of an object's right ascension and hour angle (See Figure A3.4). It progresses uniformly at the rate of 15° to the hour and can be read from a sidereal clock whose rate is adjusted to that of the earth's rotation. The clock reads $0^h0^m0^s$ at the instant the vernal equinox crosses the meridian; $1^h0^m0^s$ when the vernal equinox has moved 15° west of the meridian, corresponding to a local hour angle of 1 hour; $2^h0^m0^s$ when it is 30° west of the meridian, and so on around the sky through 24^h, when a new sidereal day begins. For example, when the sidereal clock reads $20^h36^m42^s$ (LST) the LHA Υ is 309°10.5' west of the meridian.

SOLAR TIME

We naturally use the sun as a time marker since it governs our daily lives. *Apparent solar time* is given by the local hour of the sun (LHA$\odot$) $+ 12^h$ to start the solar day at midnight. One apparent solar day is equal to the interval between two consecutive passages of the sun over the observer's meridian. Time measured by the apparent or real sun is slightly variable for two reasons: (1) its apparent annual motion along the ecliptic varies a little because of the earth's orbital eccentricity (recall Kepler's law of areas, p. 29); (2) the movement along the ecliptic when projected onto the celestial equator where hour angle is measured varies slightly throughout the year. A more suitable clock time is one that does not change in an irregular manner from day to day even though the amount is very small.

This is accomplished by introducing an imaginary sun called the *mean sun*. It moves uniformly eastward along the celestial equator at a daily rate that is equal to the average daily rate of the real sun so that both suns arrive at the vernal equinox simultaneously one year later if that is where they start together. *Mean solar time* is equal to the hour angle of the mean sun $+12^h$. Since mean solar time progresses uniformly, clocks can run on this kind of time, which is kept by us in everyday life. The difference in time between the apparent and the mean suns is called the *equation of time* and is tabulated for every day in the year in the American Ephemeris and Nautical Almanac. The greatest difference between the two times, which varies during the year, is about ± 15 minutes.

The *tropical year,* the year of the seasons, is equal to 365.2422 mean solar days. It represents the time it takes the sun to complete one revolution around the ecliptic with respect to the vernal equinox. The *sidereal year* is the period of the sun's revolution with respect to the stars, equivalent to the period of the earth's orbital revolution. The mean solar day is longer than the sidereal day by 3^m56^s of solar time because of the earth's annual motion around the sun; see Figure A4.1. For purposes of illustration the diagram is not drawn to scale.

STANDARD TIME

Obviously the local mean time is different at places separated by differences in longitude. To get around this inconvenience, the *standard time* system was introduced in 1884. There are 24 time zones, each theoretically 15° wide with the zero time zone centered on Greenwich and proceeding east or west of Greenwich. Through the approximate center of each zone runs the standard meridian, whose local mean time is the standard time within the zone. In the United States and Canada the standard meridians are at 60°, 75°, 90°, 105°, 120°, 135°, and 150°. In the continental United States the time zones are eastern standard time (EST), central standard time (CST), mountain standard time (MST), and Pacific standard time (PST), respectively 5^h, 6^h, 7^h, and 8^h behind Greenwich mean time (GMT) or universal time (UT). The time changes by

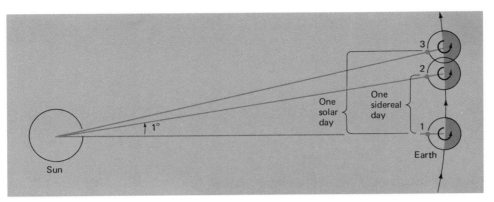

FIGURE A4.1
Why the solar day is longer than the sidereal day. With the observer at position 1, the sun is on the celestial meridian. After a 360° revolution relative to the stars, the earth has moved about 1° along its orbit and the ob-

server will be at position 2, where the sun is 1° east of the celestial meridian. The earth must rotate one additional degree, which is done in approximately 4 minutes, in order to bring the sun to the observer's meridian in position 3.

one hour as the traveler crosses a boundary zone. The zones frequently have irregular boundaries to suit the local conditions.

The successive time zones run either to the west or to the east of Greenwich until they meet halfway around the world at 180° longitude at the international date line, which passes through the center of the 12^h zone. The half portion of the zone west of the line is one day ahead of the other half east of the line, even though the standard time is the same in each half. A traveler crossing the date line from Tokyo to San Francisco, for example, gains a day; one traveling in the opposite direction loses a day. The former could celebrate two birthdays on successive days while the latter would have no birthday. What is the time right now in London? In Tokyo? If you have a shortwave radio, you can pick up the time signals broadcast around the clock transmitted by the Bureau of Standards station at Fort Collins, Colorado, WWV, on 2.5, 5, 10, 15, 20, and 25 MHz.

Daylight saving time keeps the time of the zone the same as that of the zone immediately to the east. For example, if the pacific daylight saving time is 11 A.M., add 7 hours to obtain the time in London instead of the usual 8 hours.

EARLY CALENDARS
The calendar of ancient peoples was based on the lunar cycle of 29.5 days because changes in the phases of the moon were readily apparent. Since 12 lunar months cannot be contained in a tropical year a whole number of times, people later added an extra month from time to time to bring the seasons back on schedule. Attempts to synchronize the lunar month with the year never proved satisfactory. The Egyptians were the first to base their calendar on a tropical year of 365¼ days. Their year consisted of 12 months of 30 days each with 5 extra days at the end of the year

set aside for celebrations. The early Romans employed a complicated lunar calendar of 10 months. Later the months of January and February were added, but it was not until 153 B.C. that the beginning of the year was changed from March to January. The Roman priests in whom the calendar was entrusted manipulated the calendar to serve their own ends, usually political in nature.

JULIAN CALENDAR
The hitherto imprecise Roman calendar was replaced by one that became known as the *Julian calendar* by order of Julius Caesar with the advice of the astronomer Sosigenes. The year before its adoption, on January 1, 45 B.C., had to be made 445 days long in order to bring the seasons back on schedule. The Julian calendar consisted of 365 days with every fourth year a leap year of 366 days. Emperor Augustus declared that his month, August, should also be 31 days long, the same time his predecessor had for the month of July. It is said he "stole" a day from February.

In A.D. 321 Emperor Constantine introduced the seven-day week and set aside Sunday as the first day of the week to be considered as a Christian day of worship. At the Council of Nice, convened in A.D. 325, the religious authorities established the rule for Easter Sunday to avoid conflict with the Hebrew Passover as follows: Easter Sunday is the first Sunday after the fourteenth day of the moon (around the time of full moon) on or after the date of the vernal equinox. Since 45 B.C. this date had fallen back three days in four centuries from 25 March to 21 March.

GREGORIAN CALENDAR
Now the true length of the tropical year is 365^{d}5^{h}48^{m}46^s, or about 11^{m}14^s shorter than 365¼ days. Hence the longer calendar year results in a discrepancy of 3 days in 400 years.

By A.D. 1582 the accumulated difference amounted to 10 days, so that the date of the vernal equinox had retreated from 21 March at the time of the Council of Nice to 11 March. Accordingly, Pope Gregory XIII called upon the astronomer Clavius to revise the Julian calendar. It was decreed that 10 days be dropped from the calendar. The day following 4 October, Thursday, 1582, thereby became 15 October, Friday, 1582. To avoid a future calendar discrepancy, only the century year divisible by 400 was to be a leap year (1600 and 2000). The new style *Gregorian calendar* was readily adopted in the Catholic countries. The Lutherans and Protestants finally came around in 1700 when 11 days had to be dropped. When Great Britain and the American colonies changed to the Gregorian calendar in 1752, 2 September was followed by 14 September. Early in this century other countries in Europe fell in line. Lastly, in 1923 those countries that adhered to the Greek Orthodox Church decided to adopt the Gregorian calendar. By then 13 days had to be omitted from the calendar. The present calendar is sufficiently accurate so that the accumulated error between it and the tropical year amounts to only 1 day in 3300 years.

Appendix 5
Owning Your Own Telescope

Observing through a telescope the craters and mountains on the moon, the changing phases of Venus, Jupiter's belts and the motion of its satellites, Saturn's rings, the sunspots, the Orion Nebula, the Pleiades open star cluster, the Hercules globular cluster, the double star Castor, the Andromeda galaxy, and as many other heavenly sights as your telescope can resolve is a pleasure that eclipses the television tube. Many thousands have made their own telescopes or bought them.

Your investment in a telescope needs careful fitting to your own requirements: how much to spend; the size, type, and quality of the instrument; and how often and where you will use it. Before you buy, seek advice from an amateur astronomy club, staff of a planetarium, or an astronomy teacher. Glean useful information on telescopes, kits, and parts from the telescope advertisements and articles in *Sky and Telescope* and in *Astronomy,* two widely read magazines available by subscription and in most public libraries. From these sources, you can also find people willing to sell telescopes privately.

Most amateurs work with reflectors not refractors. A 4-inch refractor may cost several times more than a 4-inch reflector. But a refractor needs less maintenance and has more stable optics. A 6-inch reflector is a satisfactory choice for good overall optical viewing. If you are mechanically inclined and want to build your own telescope, you may save about half the cost of the least expensive 6-inch reflector on the market (about $240).

The standard 6-inch Newtonian reflector has a tube several feet long and a fairly-heavy equatorial mounting tripod for stability. Several very lightweight Schmidt-Cassegrain telescopes of excellent quality can be bought at a much higher cost. They are light, have a short tube, and are easily transportable to viewing sites outside population centers with their annoying sky pollution. The following sources should help you find and use a telescope.

TELESCOPE INFORMATION

Brown, Sam. *All About Telescopes.* Barrington, N.J.: Edmund Scientific Co. You can obtain additional information in inexpensive brochures from the same company. Consult the Edmund catalogue.

Consumer Reports. p. 707. November, 1973.

Edmund Catalogue. Barrington, N.J.: Edmund Scientific Co.

Howard, N. E. *Standard Handbook for Telescope Making* New York: Crowell, 1959.

Keene, G. T. *Stargazing with Telescope and Camera.* New York: Amphoto, 1969.

Page, T., and Page, L. W. *Telescopes: How to Make and Use Them.* New York: Macmillan, 1966.

Paul, H. E. *Telescopes for Skygazing.* New York: American Photographic Book Publishing Co., 1965.

Thompson, A. J. *Making Your Own Telescope.* Cambridge, Mass.: Sky Publishing Corp., 1947.

LOCATING CELESTIAL BODIES

Amateur Astronomer's Handbook. New York: Macmillan, 1971.

Howard, N. E. *The Telescope Handbook and Star Atlas.* New York: Crowell, 1967.

Mayall, R. N., and Mayall, M. W. *A Beginner's Guide to the Skies.* New York: G. P. Putnam's Sons, 1960.

Menzel, D. H. *A Field Guide to the Stars and Planets.* Boston: Houghton Mifflin, 1975.

Miurden, J. *The Amateur Astronomer's Handbook.* New York: Crowell, 1968.

Norton, A. P. *Norton's Star Atlas and Reference Handbook.* 16th ed. Cambridge, Mass: Sky Publishing Corp., 1973.

Sidgwick, J. B. *Observational Astronomy for Amateurs.* New York: Macmillan, 1971.

Webb, T. W. *Celestial Objects for Common Telescopes.* New York: Dover Publications, 1962.

Whitney, Charles. *Whitney's Star Finder.* New York: Knopf, 1977.

Glossary

absolute brightness: the intrinsic brightness or a measure of the rate at which radiant energy is emitted.

absolute magnitude: the apparent magnitude a celestial body would have if placed at a distance of 10 parsecs (32.6 light years) from the sun.

absolute zero: theoretical temperature at which molecules and atoms of a gas have as an average zero kinetic energy. It corresponds to 0 K on the Kelvin (absolute) scale, −273.16°C on the Celsius (centigrade) scale, and −459.69°F on the Fahrenheit scale.

absorption (of energy): the conversion of electromagnetic energy to some other form by matter when electromagnetic radiation is incident upon it.

absorption line: the absence of radiation at one wavelength or several adjacent wavelengths in a continuous spectrum.

absorption spectrum: continuous spectrum crossed by dark lines or with a number of discrete wavelengths missing.

acceleration: moving body's rate of change in speed or direction, or both.

accretion theory: the formation of a planet, star, or other body by the addition of small pieces of matter to it.

achromatic: optical system relatively free from any color defect.

active galaxy: galaxies with active nuclei that are very luminous and emit both thermal and nonthermal radiation.

airglow: faint, steady, visible emission in earth's upper atmosphere.

albedo: the fraction of incident light reflected by a body.

algae: one-cell or multicell aquatic plants found in damp places.

alpha particle: $^4He_2$ nucleus of the helium atom possessing two protons and two neutrons.

altazimuth: mounting arrangement of a telescope permitting it to be rotated horizontally and vertically.

altitude: the height above the surface of a planet. The astronomical coordinate measuring the angular distance above the horizon.

amino acids: constituents of proteins containing the amino (NH_2) and carboxyl (COOH) groups of compounds.

amplitude: maximum displacement from the equilibrium position in wave motion.

angstrom: unit of length in spectroscopic wavelength measurements equal to 10^{-8} centimeter. Its symbol is Å.

angular momentum: measure of the quantity of rotation possessed by a spinning body about an axis or external point.

angular velocity: the rate at which a body rotates as expressed by an angular change of position per unit time.

annular eclipse: solar eclipse in which the moon's disk does not quite cover the sun's disk, leaving a bright ring of sunlight around the sun.

antapex: point on the sky from which the sun appears to be receding relative to its stellar neighbors. It is exactly opposite to the apex.

antimatter: matter identical in behavior to ordinary matter but containing the oppositely charged or neutral counterparts of ordinary matter. Example: the normal proton carries a positive charge; the antiproton carries an equal but negative charge.

aperture: diameter of a telescope objective or radio dish or antenna structure.

aperture synthesis: radio telescope arrangement that employs a combination of fixed and movable smaller parabolic antennas at one location to achieve a high degree of resolution.

apex: point on the sky in the constellation Hercules toward which the sun appears headed relative to its closest stellar neighbors.

aphelion: point in the orbit of a body orbiting the sun farthest from the sun.

apogee: point in orbit at which any body circling earth is farthest from earth.

apparent brightness: the brightness as perceived by an observer at some distance from the radiating body (see also *inverse-square law of light*).

apparent magnitude: apparent brightness of a celestial body based on a logarithmic scale of brightness.

association: a physical grouping of primarliy young stars, such as the O and B associations of T associations.

A star: stars of spectral type A with surface temperatures of about 7,500 to 10,000 K in whose spectrum the Balmer lines of hydrogen attain their greatest strength.

asterism: the configuration of stars or "catch figure" used to identify a constellation. Example: the Big Dipper is the asterism for Ursa Major.

asteroid: one of thousands of small solid bodies revolving in orbits, chiefly between Mars and Jupiter in the asteroid belt.

astrology: pseudo-science that claims human events can be predicted from the positions that the sun, moon, and planets occupy in the zodiac at different times.

astrometric binary: double star having an invisible companion whose presence is inferred from the wobbly motion of the visible component.

astrometry: division of astronomy dealing with precise determinations of the positions, motions, and parallaxes of heavenly bodies.

astronomical unit: the mean distance between earth and the sun (about 150 million kilometers).

astronomy: science that deals with the celestial bodies— their descriptions, radiations, movements, structures, physical and chemical characteristics, origin, evolution, and arrangement.

astrophysics: field of astronomy concerned with the physical, chemical, and thermal properties of celestial bodies accessible to direct observation and also inferred from theory.

asymptotic branch: slightly curved path in the upper right of the H-R diagram, contiguous with the upper part of the red giant branch, along which many stars near the end of their life cycle evolve.

atmosphere: the outermost gaseous medium or layers surrounding a planet or a star.

atom: the smallest unit of a chemical element possessing the properties of the element (see also *subatomic particles*).

atomic clock: electronically actuated mechanism whose clock rate is precisely governed by the naturally occurring vibration frequencies of certain atoms (cesium, rubidium, hydrogen) or molecules (ammonia).

atomic mass unit: standard unit that corresponds to one twelfth of the mass of the carbon isotope $^{12}C_6$ ($= 1.6604 \times 10^{-24}$ g).

atomic number: number assigned to an atom corresponding to the number of protons in its nucleus.

atomic weight: mass of an element in atomic mass units.

aurora: sporadic visible emission in the upper atmosphere most prominent at high latitudes in both hemispheres. Called *aurora borealis,* "northern lights," in the northern hemisphere; *aurora australis* in the southern hemisphere.

autumnal equinox: point of intersection between the ecliptic and the celestial equator that the sun crosses as it moves from north to south of the equator.

azimuth: horizontal arc from the north point of the horizon measured clockwise to the object's position projected on the horizon.

Balmer series: series of lines in the visible and ultraviolet spectral regions arising from transitions between the second energy level of the hydrogen atom and its higher levels.

barred spiral galaxy: a spiral-type galaxy whose spiral arms extend from a barlike feature containing the nucleus.

barycenter: center of mass around which two bodies orbit.

basalt: a type of igneous rock resulting from the cooling of lava, common on the terrestrial planets.

base: in biochemistry, one of the four chemical structures made up of carbon, nitrogen, and oxygen molecules in the DNA molecule (adenine, guanine, cytosine, thymine). In the RNA molecule, thymine is replaced by uracil.

basin: large depressed plain known as a mare on the moon, Mercury, or Mars.

beta particle: electron.

big bang: a cosmological theory in which a fireball explosion of dense, superhot matter initiated the expansion of the universe.

binary star: pair of stars in orbit around each other.

biosphere: portion of earth's atmosphere, ground, and water where life can flourish.

bipolar group: a large sunspot group divided into two portions with opposite magnetic polarities.

blackbody: ideal body capable of absorbing all radiation falling on it and reemitting it without loss.

black dwarf: small dense compact star, no longer radiating, which is the final state of a white dwarf or results when a star is too small to initiate nuclear fusion.

black hole: superdense configuration that a body assumes when it collapses gravitationally in such a way that its powerful gravitational field prevents radiation from emerging into external space.

blue stellar object: non-radio-emitting quasar identified by its bright ultraviolet radiation and pronounced red shift.

Bode's law: empirical rule whose progression of numbers represents the approximate distances of the planets in astronomical units from the sun.

Bohr atom: model of the atom developed by Niels Bohr in which the electrons orbit the nucleus at specfied distances.

bolide: unusually bright meteor that sometimes explodes with a loud sound into fiery fragments.

bolometer: thermal detector employed to measure infrared radiation.

bolometric magnitude: the magnitude of a celestial body for all wavelengths. It is a measure of the energy emitted in all wavelengths per unit time.

bow shock wave: conical interface formed between the free flow of a gas in a medium and an obstacle it encounters.

breccia: impacted mixture of lunar soil and rock fragments.

bright-line spectrum: emission spectrum of bright lines observed in the spectrum of an incandescent gaseous body at low pressure.

B star: a blue-white star of spectral type B with a surface temperature of about 11,000 to 28,000 K, whose spectrum is characterized by absorption lines of neutral helium.

Cambrian: paleontological/geological 100 million-year period that began about 600 million years ago. It was characterized by an explosive growth of marine life, principally the invertebrates.

canals (of Mars): the straight-line markings on Mars thought to be actual water canals, but found not to exist after spacecraft visits to Mars.

capture theory: the formation of a planet-satellite or binary system in which one body captures the other by gravity.

carbonaceous: refers to carbon-bearing material.

carbon cycle: see *CNO cycle*.

Cassegrain focus: optical arrangement that permits light from the primary mirror of the reflecting telescope to be reflected back through the central hole of the mirror by a small convex mirror placed in front of the prime focus.

Cassini division: 2000-kilometer gap between the outer and middle rings of Saturn.

celestial equator: great circle that represents the extension of earth's equator projected onto the sky.

celestial mechanics: field of astronomy that deals with the gravitational motions of celestial bodies, primarily in the solar system.

celestial meridian: the great circle on the celestial sphere containing the north and south points of the horizon and the observers zenith and nadir.

celestial pole(s): extension of earth's axis of rotation on the sky.

celestial sphere: the imaginary sphere centered on the earth to which the stars are affixed.

center of mass: the location in a body at which all its mass can be concentrated without affecting its response to gravity.

central force: the name for a force located at the center of a system to which it gives definition. Example: the gravitational attraction of the sun is a central force for the orbital motion of the planets.

centrifugal force: apparent force causing an object to move away from the center of rotation.

cepheid: pulsating variable star of the giant or supergiant class with a period of pulsation between a fraction of a day and 50 days.

cgs system: a metric system of measurement using centimeters, grams, and seconds.

Chandler wobble: minute bodily shifting of earth with respect to its axis of rotation in a roughly circular, nearly annual motion up to 15 meters in diameter.

Chandrasekhar limit: theoretical limit below which a star can evolve presumably without mishap into a white dwarf. Its value is 1.4 solar masses.

chemical differentiation: the process in the formation of a planet in which, if the planet is molten, the heavy elements sink to the center and the light elements rise to the surface.

chondrite: stony type of meteorite in which are found spherules composed of glassy material believed to be the most primitive condensed matter in the early solar nebula.

chondrule: small, glasslike, round body found in stony meteorites. It is believed to have crystallized from molten droplets present during the initial stages of condensation of the solar nebula.

chromatic aberration: defect in an optical system due to the failure to bring different wavelengths to a common focus.

chromosome: threadlike string of genes in DNA, which is part of every nucleus in a living cell.

chromosphere: narrow pinkish portion of the sun's atmosphere lying immediately above the photosphere.

circle: one member of a family of mathematical curves called conic sections for which every point on the curve is equidistant from its center.

circumpolar stars: for a given latitude, those stars close enough to the observable celestial pole that never set, or those stars close enough to the unobservable celestial pole that never rise.

clock paradox: also known as the twin paradox, in which one of the twins stays home while the other makes a high-speed, roundtrip journey to a distant star and returns younger in age than the twin who stayed on earth.

closed universe: a cosmological model of the universe with a finite volume in which space is curved.

cluster: a physical grouping of stars bound either temporarily or permanently by gravity.

CNO cycle: carbon-nitrogen-oxygen thermonuclear fusion process, which results in hydrogen nuclei being converted into helium nuclei plus energy with the aid of the carbon isotope $^{12}C_6$ catalyst.

color excess: difference between the observed color index of a star, whose light has been reddened in passing through interstellar material, and the normal unreddened color index.

color index: difference in magnitude between two spectral colors of the celestial object (most often between the blue and yellow colors).

color-luminosity diagram: Hertzsprung-Russell plot of the stars in which the horizontal axis is the color index and vertical axis is the apparent or absolute magnitude.

coma: gaseous envelope immediately surrounding the nucleus of a comet.

comet: small body consisting of frozen gases and dust in orbit around the sun. As it nears the sun, its surface material vaporizes to form a large head, or coma, at whose center lies a bright nucleus where its mass is concentrated.

comparison spectrum: the emission spectrum of a known element placed above and below a stellar spectrum as a reference for wavelength determination.

conduction: movement of heat from one point to another by collisions of more energetic particles with adjacent, less energetic particles.

configuration of the planets: the positions of the planets relative to the earth-sun line (see also *elongation*).

conic section: curve formed by the intersection of a plane and a circular cone. The angle at which the plane cuts the cone determines whether the curve is a circle, ellipse, parabola, or hyperbola.

conjunction: the planetary configuration in which the planet's angular distance (elongation) from the sun is either zero or a minimum.

conservation of energy: principle stating that energy can be converted from one form into another but it cannot be destroyed in the process; the total amount of energy remains the same.

constellation: arbitrary grouping of stars within a bounded area of the sky named after mythological heroes, animals, or other objects.

continental drift: gradual separation of the continents due to sea-floor spreading during the last 200 million years at the rate of a few centimeters per year.

continuous spectrum: uninterrupted band of emission produced by a body radiating energy over a continuous range of wavelengths.

continuum: (1) atomic: the continuous spectral region toward the violet, adjacent to the head of the series limit of the atom's spectral lines. (2) Space: the space-time environment in four-dimensional space.

convection: process in which energy is transferred from point to point by flowing currents in a gas or liquid.

convex lens: lens thicker in the middle than at its edges.

core: the central part of a planet, star or other astronomical body which is surrounded by some outer layers.

corona: outermost portion of the sun's atmosphere best observed during a total eclipse of the sun.

coronagraph: special telescope carefully designed to photograph the chromosphere and inner corona of the sun without the intervention of an eclipse of the sun.

cosmic rays: highly energetic particles impacting earth's atmosphere. Originating in outer space, they consist mostly of protons with a sprinkling of heavier atomic nuclei.

cosmological constant: mathematical term involving a repulsive force, which Einstein introduced into his field equations in the general theory of relativity to counteract the self-gravitation of the universe.

cosmological principle: assumption that the universe-at-large looks the same to all observers.

cosmology: branch of astronomy concerned with the origin, evolution, and structure of the universe.

coudé focus: optical system that permits the beam of light from the primary mirror of the reflecting telescope to be directed down the hollow polar axis of the instrument to a remote focal position that remains fixed regardless of the position of the telescope.

crust: outermost layer of a terrestrial planet.

cytoplasm: slightly viscous fluid surrounding a cell's nucleus.

dark-line spectrum: see *absorption spectrum*.

dark nebula: opaque interstellar cloud of gas and dust that dims or obscures the light of stars behind it.

deceleration parameter (*q*): negatively varying quantity that is a function of the rate of slowdown of the expanding universe and the Hubble constant *H*.

declination: celestial coordinate that corresponds to the angular distance of a body north or south of the celestial equator.

declination axis: axis of the telescope, perpendicular to the polar axis, that permits the telescope to be swung in a north–south direction.

de-excite: the process in which the electron spontaneously gives up energy creating a photon and makes the transition to a lower energy orbit.

deferent: in the Ptolemaic system, the large circle centered on the earth upon whose circumference the center of a smaller circle (epicycle) revolves, carrying the planet on its rim.

degenerate gas: very high-density, high-temperature condition of matter different from the perfect-gas condition in which the electrons or nucleons are restricted in the number of energy states they may occupy.

density: the mass per unit volume of a material body.

density wave: wave crest of matter that spirals outward from the center of a spiral galaxy through its disk at a constant rotational rate, generated by gravitational variations within the disk.

descending node: intersection between the orbit of a body and some reference plane, such as the ecliptic plane in the solar system, through which the body passes from north to south of the reference plane.

differentiation: physical and chemical separation of an inhomogeneous medium into various layers.

diffraction: slight bending of light waves around a sharp edge.

diffraction pattern: of a star, bright central disk of a stellar image surrounded by alternating, concentric, less bright and dark rings.

dipole: magnetized object that possesses north and south magnetic poles. A bar magnet and earth are two examples.

dispersion: spreading of light into its various wavelengths as, for example, by a prism or grating.

distance modulus: difference between the apparent magnitude (*m*) and the absolute magnitude (*M*) of a celestial body; that is, $m - M$, from which a celestial body's distance may be derived by application of the inverse square law of light.

DNA: deoxyribonucleic acid—the inheritance material consisting of long chains of nucleotides present in the chromosomes of all living matter.

Doppler effect: apparent shift in wavelength or frequency as a result of the relative line-of-sight motion between the observer and the source of radiation.

dust tail: dusty component of a comet's tail, consisting of dust grains evaporated from the head of the comet.

dwarf star: star of moderate or lesser mass and luminosity on the main sequence. (The sun is a yellow dwarf star.)

dynamo: the process in which the motion of electrical charges generates a magnetic field.

dyne: unit of force; it represents the force needed to accelerate a mass of one gram one centimeter per second each second.

eccentricity: in an ellipse, the numerical ratio of the distance of the focus from the center of the ellipse to the length of the semimajor axis.

eclipsing binary: two generally close stars orbiting each other in a plane viewed edgewise, or nearly so, from earth, resulting in the mutual eclipse of one star by the other.

ecliptic: great circle extension of earth's orbit on the celestial sphere; the apparent yearly path of the sun in the sky.

ecology: study of the interactions of living things with their environment.

ecoshell: thermally habitable region around a star where life can flourish.

electric field: the field of force created by electrical charges, such as that about the electron.

electromagnetic energy: the energy in electric and magnetic fields, such as that transported by electromagnetic waves.

electromagnetic force: the force associated with electrical charges and their motion.

electromagnetic radiation: energy transmitted with the velocity of light (299,800 km/s) in the form of gamma, X-ray, ultraviolet, visible, infrared, microwave, and radio waves, or portions thereof.

electromagnetic spectrum: the name given to the continuous variation of wavelength for electromagnetic waves, such as gamma rays, X rays, ultraviolet, visible, infrared, microwaves and radio waves.

electron: negatively charged particle that is the basic outer component of the atom.

electron shells: the grouping of electron orbits of similar, but distinct, energies in the atom.

electron volt: kinetic energy gained by an electron moving across an electric potential of one volt.

ellipse: closed path traced by a moving point whose distance from two fixed points (foci) remains constant.

elliptical galaxy: spherical or oval-shaped galaxy lacking a spiral structure and containing little or no gas and dust.

elongation: the angular separation between the sun and a planet measured in angular units.

emission: the release of energy by a physical system, such as the emission of light.

emission spectrum: spectrum containing bright lines.

energy: the ability of a physical system to do work when it changes from one describable state to another.

epicycle: smaller circle in the Ptolemaic system along which the planet moves while the center of the circle revolves about its deferent.

equatorial mounting: telescope arrangement that permits the instrument to be rotated about an axis parallel to earth's axis of rotation in order to follow the rotation of the sky.

equinoxes: two points of intersection between the celestial equator and the ecliptic. These are the vernal equinox, through which the sun passes on or about March 21, and the autumnal equinox, through which the sun passes on about September 22.

erg: unit of energy, defined as the work expended by a force of one dyne moving through a distance of one centimeter.

eruptive variable: variable star characterized by sudden explosive or erratic light outbursts.

escape velocity: minimum speed necessary to escape permanently from the gravitational field of a body.

Euclidean: refers to the flat space geometry as perceived by our senses.

event: occurrence specified by *both* time and place in the space-time world of four dimensions in the theory of relativity.

event horizon: surface in relativistic space around a black hole inside of which no photons of light can escape and will ever reach us.

evolutionary cosmological model: time-dependent model of an expanding universe that evolves from its superhot state of highly condensed matter originally contained in a small volume of space.

excite: to energize an atom or molecule from a lower to a higher energy state by absorption of a quantum of energy.

exobiology: study of extraterrestrial life in space.

exosphere: outermost fringe of earth's atmosphere.

exploding galaxy: galaxy in which powerful explosions are occurring in its central region and which is emitting non-thermal radiation.

faculae: enhanced bright regions best observed near the sun's edge.

field of force: region of space at each point of which a given physical quantity (a force) has some definite value; for example, a gravitational or magnetic field.

filament: slender wisp or drawn-out form of gaseous material moving outward from an object.

filtergram: photograph of the solar disk and/or its atmosphere obtained with a narrow bandpass color filter.

fireball: exceptionally brilliant meteor.

fission: the spontaneous disintegration of the nucleus of the atom.

flare: sudden energetic eruption of radiation in the sun's chromosphere.

flare star: orange or red dwarf star that exhibits sudden, brief, unpredictable outbursts of radiation.

flash spectrum: bright-line spectrum of the sun's chromospheric layer momentarily observed after the start and before the finish of the total phase of the eclipse.

flux: rate of flow of energy.

flyby: spacecraft that samples the environment of a planet or satellite as it curves around it without being captured.

focal length: distance from a lens or mirror to the focus.

focal ratio: focal length of a lens or mirror divided by its aperture. The smaller the focal ratio, the greater the speed of the optical system.

focus: place where the light rays from an object in an optical system converge to form the image of the object.

forbidden lines: spectral lines originating in a gaseous medium of exceedingly low density, where the probability of their occurrence is high compared with that under ordinary laboratory conditions.

force: a push or pull that causes a body to change its state of motion.

Fraunhofer lines: most prominent dark lines in the solar spectrum first mapped by J. Fraunhofer in 1814.

free fall: motion of an unpowered particle or object moving toward a body's center of gravitational attraction.

free-free radiation: radiation emitted by an electron as it approaches an atomic nucleus and is deflected by it without capture.

frequency: number of electromagnetic waves that pass by a given point each second.

F star: a star of spectral type F with a surface temperature of about 6000 to 7500 K in whose spectrum absorption lines of Ca II K of hydrogen are prominent.

fusion: thermonuclear synthesis of heavier elements from lighter ones.

g: acceleration of 980 cm/s^2 at earth's surface arising from earth's gravitational force.

G: constant of gravitation in Newton's law of gravitation, $F = Gm_1m_2/d^2$. Its numerical value is 6.67×10^{-8} $cm^3/g \cdot s^2$.

Galactic cluster: open star cluster in the Galaxy.

Galactic equator: great circle in the sky passing through the central plane of the Milky Way.

galactic halo: outer, nonflattened stellar portions of a galaxy. In our Galaxy, an approximately spheroidal distribution of stars.

Galactic latitude: number of degrees of an object above or below the Galactic equator.

Galactic longitude: number of degrees of an object measured along the Galactic equator northward from the direction of the Galactic center.

galaxy: a large system of stars, dust, and gas held together by the mutual gravitational attraction of its members.

Galilean satellites: four natural satellites of Jupiter discovered by Galileo: Io, Europa, Ganymede, Callisto.

gamma rays: highest-energy photons of radiation possessing the shortest wavelengths (less than about 0.1 angstrom) and energies greater than 100,000 electron volts.

gas: that state of matter in which the constituent particles maintain no permanent relationship to each other.

gaseous nebula: a diffuse collection of gaseous material emitting radiation and lying in between the stars of a galaxy.

gauss: scientific unit of magnetic field strength.

gegenschein: faint counterglow patch of light observed in the night sky opposite to the sun's direction.

gene: unit in the chromosome that determines a unique set of inheritable characteristics.

general theory of relativity: Einstein's theory of the nature of the gravitational force in the space-time geometry of four dimensions.

geosphere: solid portion of earth.

giant star: large star of higher than average luminosity.

gibbous: phase of the moon between first quarter and full, or between full and last quarter.

globular cluster: compact spheroidal assemblage of tens of thousands of stars found in the halo portion of a galaxy.

globule: small, roundish patch of dark nebulosity that may be the precursor of a protostar.

grains: small, almost microscopic, solid particles found in large interstellar clouds.

granule: one of many thousands of small dark/bright short-lived patches on the sun's surface, which gives the sun a grainy appearance.

grating: optical surface (transmissive or reflective) upon which is ruled a large number of finely spaced grooves. A beam of light impinging upon it is broken into several spectral orders on each side of the central image.

gravitational collapse: rapid contraction of a body whose gravitational force greatly exceeds its normal outward-balancing, gas-pressure force.

gravitational deflection of light: slight bending in the light path experienced by a light ray skimming past the limb of a massive body.

gravitational force: the field of force exerted by all particles of matter.

gravitational potential: stored energy that is convertible to heat energy when a gaseous body such as a star contracts gravitationally.

gravitational radiation: weak oscillations, traveling with the speed of light, emitted by a highly accelerated body, as predicted by the theory of relativity. Also known as *gravity waves*.

gravitational red shift: red shift in wavelength experienced by a photon leaving the surface of a massive object, as predicted by the theory of relativity.

gravity: force due to the mutual attraction between two masses directed along the line of their centers and having a magnitude inversely proportional to the square of the distance between their centers.

greenhouse effect: the heating of an atmosphere by absorption of infrared radiation trying to escape out through the atmosphere.

ground state: the lowest energy orbit of the electron in the atom.

G star: stars of spectral type G are yellowish stars with surface temperature of about 5000 to 6000 K in whose spectra the H and K lines of Ca II are dominant.

H I region: volume of interstellar space where the hydrogen remains neutral and optically dark.

H II region: volume of interstellar space occupied by ionized hydrogen.

hadrons: heavy class of elementary particles ranging from the mesons to the protons, neutrons, and still heavier particles up to nearly 3000 electron masses.

half-life: the length of time for half the amount of a radioactive nucleus to spontaneously disintegrate.

halo: see *galactic halo*.

HEAO: *high energy astronomical observatory* spacecraft designed to study X-ray and gamma ray radiations from celestial sources.

helium burning: fusion of helium nuclei into carbon nuclei occurring in a star's core.

helium flash: somewhat sudden and rapid ignition of helium in the electron-degenerate core of a star, after which the star settles down to fusing helium quietly into carbon.

Helmholtz contraction: the slow contraction of a diffuse gas region in which gravitational potential energy is converted to some other form of energy, generally thermal energy.

hertz: unit of frequency of electromagnetic radiation, equivalent to one cycle per second.

Hertzsprung gap: nearly vacant region existing between the upper portion of the main sequence and the left end of the giant branch in the H-R diagram.

Hertzsprung-Russell (H-R) diagram: plotted positions of the stars according to their absolute magnitudes or luminosities on a vertical scale against their spectral class or temperature or color index on a horizontal scale.

horizon: the great circle on the celestial sphere marked by a plane perpendicular to the zenith-nadir axis and tangent to the earth at the point of the observer.

Hubble constant: constant of proportionality (*H*) in the Hubble law of recession of the galaxies. Its presently determined value is about 17 km/s/megalight year.

Hubble law of recession: proportional relationship between the velocities (red shift) of the galaxies and their distances.

Hubble sequence of galaxies: classification of galaxies into an ordered arrangement depending on their appearance and structure.

hydrocarbon: any chemical containing a mixture of carbon and hydrogen.

hydrogen burning: fusion of hydrogen nuclei into helium nuclei in the core of stars.

hydrosphere: water portions of earth.

hydrostatic equilibrium: balance between pressure forces and gravitational forces in a star's layers.

hyperbola: open-ended curve of a conic section formed by the intersection of a plane with a right-circular cone at any angle between the axis of the cone and its slant edge.

igneous: adjectival description of molten lava or basalt rock that has solidified.

image tube: electronic device in which the optical image falls on a photocathode surface to emit electrons whose flow can be intensified and magnetically guided to fall on a fluorescent screen where it is recorded by a camera.

inclination: angle between the orbital planes of two bodies; or the angle of the equator of a body with respect to its orbital plane.

inertia: reluctance of a body (proportional to its mass) to change its state of motion or rest when a force acts on it.

inertial force: force needed to overcome the inertia of a body in order to change its motion.

inertial frame of reference: coordinate system of an observer moving uniformly in space to which all his measurements are referred.

inferior conjunction: position of an inferior planet, as viewed from earth, when it is between the sun and earth.

inferior planet: a planet whose orbit lies inside earth's orbit (Mercury or Venus).

infrared: portion of the electromagnetic spectrum extending from the visible red end toward the longer wavelengths up to about one millimeter.

intensity: the rate at which radiant energy propagates along the ray of electromagnetic radiation.

interference: phenomenon produced when waves of the same wavelength from different parts of a radiating source reinforce or interfere with each other to form a fringe pattern.

interferometer: apparatus employed to detect and measure interference from two or more coherent wave trains from the same source and, in astronomy, to measure the angular width of minute celestial sources and to determine their position on the sky with great accuracy.

interplanetary medium: the matter composed of gas and some dust lying between the planets.

interstellar lines: dark lines superimposed on the spectra of stars resulting from absorption of starlight passing through the interstellar clouds.

interstellar medium: region of gas and dust particles in the Milky Way system between the stars.

inverse-square law of light: relationship between the apparent brightness of a light source and the distance from the source: the apparent brightness varies inversely as the square of the distance.

ion: atom that has lost or gained one or more electrons than its normal complement.

ionization: stripping an atom of one or more of its electrons.

ionosphere: upper atmospheric region from about 50 to 400 kilometers where the air is ionized into discrete layers (D, E, F_1, and F_2 layers) by the sun's ultraviolet and X-ray radiation.

ion tail: filamentary tail of a comet (separate from the dust tail), resulting from the interaction of the solar wind with ions in the comet's head.

irregular galaxy: galaxy without symmetrical form.

isotope: atom similar to another atom with the same number of protons but with a different number of neutrons in the nucleus.

isotropic: the same in every direction.

Jovian planet: any one of the four outer large planets (Jupiter, Saturn, Uranus, or Neptune).

Kepler's laws of planetary motion: three empirical laws that J. Kepler discovered. See *law of elliptic orbits, law of areas,* and *law of periods.*

kinetic energy: energy of motion possessed by a moving body. It is equal to one-half the product of its mass and the square of its velocity.

kinetic pressure: pressure (force per unit area) arising from random motions of the particles in a gas as a consequence of its temperature.

kinetic temperature: temperature that a gas assumes as the result of the distribution of velocities of the gas particles.

K star: stars of spectral type K are cool, orange to red stars with surface temperatures of about 3600 to 5000 K in whose spectra the H and K lines of Ca II reach their greatest strength.

law of areas: Kepler's second law of planetary motion. The straight line connecting the sun and an orbiting planet sweeps out equal areas in equal intervals of time.

law of elliptic orbits: Kepler's first law of planetary motion. Each planet moves around the sun in an ellipse with the sun at one focus of the ellipse.

law of periods: Kepler's third law of planetary motion. The square of the orbital period of a planet around the sun is proportional to the cube of its mean distance from the sun.

leptons: class of light elementary particles that constitute the electrons, muons, and their associated neutrinos.

liberation: apparent oscillatory motion of the moon in an east-west direction as viewed from earth.

light curve: plot that shows the change in the magnitude of a variable star (plotted vertically) versus the time (plotted horizontally).

light year: distance that light travels in one year. Its value is 9.46×10^{12} kilometers.

limb: edge of the apparent disk of a celestial body.

liquid: that state of matter in which the constituent particles maintain only a temporary relation to each other.

lithosphere: stony crust and upper mantle of earth down to an approximate depth of 50 kilometers.

Local Group: small group of bunched galaxies, including our Galaxy, consisting of about 20 known members spread over a diameter of about three million lightyears.

local supercluster: apparent clumping of a supersystem of galaxies scattered over a volume of space about 130 million lightyears in diameter. The Local Group is one subunit of the supercluster.

long-period comet: comet whose orbital period is greater than about 200 years.

long-period variable: red variable star with an amplitude variation of several magnitudes over a period between approximately 200 to 400 days.

Lorentz contraction factor: the term $\sqrt{1 - (v^2/c^2)}$, first introduced by the Dutch physicist H. A. Lorentz. It appears in the formulas for relativistic length, mass, and time intervals.

luminosity class: one of the divisions (Ia, Ib, II, III, IV, V) into which the stars are arranged according to their luminosity and spectral class.

luminosity function: relative number of stars of a specific absolute magnitude in a given volume of space.

Lyman alpha line: first spectral line of the ultraviolet Lyman series at 1216 angstroms, produced by electron transitions between the ground level and the second level of the hydrogen atom.

Lyman series: ultraviolet hydrogen series of spectral lines arising from transitions to or from the ground level of the atom.

Magellanic Clouds: pair of irregular galaxies visible to the naked eye in the southern skies. They are our closest optical extragalactic objects.

magnetic field: region surrounding a magnetized body that acts on electrical particles or currents within its range.

magnetic pole: one of the two diametrically opposite points of a spherical body, or the end points of a bar magnet, where the flux of the magnetic lines of force is maximum.

magnetosphere: complete magnetic field that surrounds earth or any other magnetized planet.

magnitude: brightness of a celestial body based on a logarithmic scale of intensity to which the eye naturally responds.

main sequence: major distributional segment of the stars running diagonally across the H-R diagram from the upper left to the lower right.

main sequence fitting: superposition of the main sequence of a star cluster (plot of apparent magnitude versus color index) over the calibrated main sequence of a star cluster such as the Hyades (plot of absolute magnitude versus color index) to derive the distance modulus of the first cluster and hence its distance.

major axis: longest diameter of an ellipse.

mantle: intermediate layer between the core of a solid astronomical body and its outer crust.

mare: dark marking on the moon, Mercury, or Mars. (Latin name for *sea*; plural, *maria*.)

mascon: concentrated mass below the crust of the moon or a solid planet; it gravitationally disturbs a spacecraft flying over it.

mass: amount of matter contained in a body. It is a measure of the inertia possessed by a body when acted on by a force.

mass-luminosity relation: graph of the absolute magnitudes of the main sequence stars against their masses.

mass unit: standard reference against which atomic masses are compared. It is equal to one twelfth of the mass of the carbon-12 isotope ($= 1.6604 \times 10^{-24}$ gram).

mesosphere: intermediate atmospheric layer above the stratosphere extending approximately between 30 and 100 kilometers above earth's surface.

meteor: luminous trail left behind by the passage of a tiny cosmic particle (meteoroid) through earth's atmosphere.

meteorite: extraterrestrial metallic or stony object that survives flight through earth's atmosphere and then lands.

meteoroid: solid particle or body of small dimensions in extraterrestrial space.

meteor shower: bright streaks appearing to radiate from a common point in the sky; caused by a swarm of meteoroids entering earth's atmosphere.

micron: unit of measure equal to one-millionth of a meter.

microwave: radio spectral region in the millimeter to centimeter wavelengths.

microwave radiation: low-temperature radiation that pervades all of space, believed to be a relic of the primeval fireball (the big bang) that initiated the expansion of the universe.

Milky Way: the name we give to our Galaxy or the band of stars visible in our summer skies.

minimum energy transfer orbit: trajectory of a space vehicle launched from earth toward a planet using the least amount of rocket propellant by coasting most of the way.

minor planet: see *asteroid*.

molecule: aggregate of two or more atoms of the same or different elements forming a compound.

momentum: product of a body's mass and its velocity (*mv*).

monochromatic: light of a single wavelength or color.

M star: stars of spectral type M are cool red stars with surface temperatures of less than about 3600 K whose spectra are dominated by molecular bands, especially TiO.

muon: elementary charged particle of about 207 electron masses with a half-life of about 1.5-millionths of a second. It decays into an electron and a neutrino. Also known as *mu-meson*.

mutation: change in hereditary material. It is produced at random by environmental or other factors.

nadir: the point on the celestial sphere opposite the zenith or directly below the observer.

natural selection: chance adaptations that involve a continual adjustment of organisms to their environment, permitting them to survive.

neap tide: lowest tide, which occurs twice each month near the first or last quarter of the moon.

nebula: bright or dark cloud of gas and dust, which may contain stars.

nebular hypothesis: theory that supposes the solar system originated from a condensed cloud of gas and dust; first put into mathematical form by Laplace (1796) and modified by a number of modern physicists and astronomers.

negative space curvature: infinite space continuum whose curvature is hyperbolic and open-ended. In the three-dimensional version, the surface is saddle-shaped.

neutrino: elementary particle without mass or charge emitted during a nuclear reaction.

neutron: neutral elementary particle of about the same mass as the proton. Protons and neutrons are the basic constituents of atomic nuclei.

neutron star: gravitationally collapsed star of very small dimensions and enormously high density, composed mainly of neutrons that may be the core remnant of a supernova.

Newtonian laws of motion: three laws formulated by Isaac Newton: (1) a body remains at rest or moves at constant speed along a straight line so long as no external force acts upon it; (2) the acceleration of a body is proportional to the force acting on it and inversely proportional to its mass; (3) to every force there is an equal and opposite counterforce.

node: one of the two points of intersection, 180° apart, between the orbit of a celestial body and a plane of reference such as the ecliptic.

nonthermal radiation: radiation produced by a body that is not related to its thermal energy.

nova: star that suddenly erupts into an object of great brilliance, surpassing the sun's luminosity by a factor of hundreds of thousands to millions of times and then fading more slowly.

nuclear force: that force exerted by subatomic particles that is responsible for form, shape, and motion in the subatomic world of the nucleus.

nucleic acid: substance of the DNA molecule.

nucleon: either the proton or neutron inside the atomic nucleus.

nucleotide: repeating section in the double-twisted DNA chain, consisting of three units: a base, a phosphate, and a sugar, constituting half of the cross-rung of the DNA helix.

nucleus: central portion of an atom, a comet, a galaxy, or a cell.

nutation: small 18.6-year period of oscillation superimposed on earth's 25,800-year period of precession.

OAO: orbiting astronomical observatory. Two were placed in orbit, launched in 1968 and 1972.

objective: main lens or mirror of the telescope.

objective prism: thin, large prism placed in front of the telescope objective. It produces a spectrum of each star in the field of view of the telescope.

oblate spheroid: solid formed by rotating an ellipse about its short axis.

observatory satellite: earth-orbiting satellite, including orbiting astronomical observatory (OAO); orbiting geophysical observatory (OGO); orbiting solar observatory (OSO).

Occam's Razor: rule of medieval philosopher William of Occam that one should not burden the explanation of a phenomenon with involved interpretations.

occultation: transitory blocking of an object's light by the passage of a larger intervening body. Examples include the moon passing over a star or planet and the shadow of Jupiter passing over one of its satellites.

opacity: reduction in the intensity of light as it passes through the layers of a medium (in stars through the gaseous layers).

open star cluster: somewhat loose assemblage of stars, numbering dozens to hundreds, with various degrees of central condensation. In our Galaxy it is also known as a *galactic cluster.*

open universe: a cosmological model of the universe with an infinite volume and no boundaries.

opposition: position of a superior planet when it is closest

to earth. At this time it is 180° from the sun's direction.

orbit: path of a body subjected to the gravitational force of another body.

organic compound: compound that contains carbon.

OSO: orbiting solar observatory, of which nine were placed in orbit (1962–1975).

O star: stars of spectral type O are very hot blue stars with surface temperatures of about 35,000 K, whose spectra are dominated by the lines of singly ionized helium.

outgassing: venting of volatile gases from the heated interior of a solid body.

oxides: chemical compounds containing oxygen.

ozone: ultraviolet-absorbing layer of O_3 molecules between 20 and 35 kilometers high.

P (seismic) waves: longitudinal waves of an earthquake that cause earth's inner material to expand and contract alternately. Also called *primary,* or *pressure,* or *compression waves.*

Pangaea: supercontinent into which all the present continents were merged over 200 million years ago.

panspermia: theory that microorganisms floating in space or attached to interstellar dust particles can germinate and start the evolutionary chain of life when they encounter a hospitable sterile planet.

parabola: conic section formed by a plane passing parallel to one side of a cone. The eccentricity of a parabola equals 1.

parallax: apparent shift in the position of an object when observed from two different places.

parsec: unit of distance corresponding to the distance of a body whose parallax equals one arc second.

passband filter: filter that is transparent to electromagnetic radiation in a very narrow spectral range.

payload: useful instrumented cargo of a rocket system.

peculiar galaxy: abnormally shaped galaxy that emits nonthermal radiation.

penumbra: portion of the shadow from which part of the light source is excluded when a body passes over the light source.

perfect cosmological principle: proposition in the steady state theory that the universe looks the same everywhere at all times.

perfect gas: ideal gas whose pressure increases directly with the temperature and the density.

perihelion: point in the path of a body orbiting the sun where it is closest to the sun.

perturbation: disturbance in the normal movement of an orbiting body arising from an external force, usually gravitational.

phase: the repeating portions of a cyclic phenomena, such as the varying shape of the sunlit portion of the moon during its monthly orbit or the relationship between the crests and troughs of a wave.

photoconductor: any light sensor that converts light into a flow of electrons.

photoelectric: pertaining to the phenomenon whereby photons impinging on a light-sensitive metal generate a flow of electrons or electric current.

photometer: instrument used to measure the intensity of a light source.

photomultiplier: small evacuated tube within which a photoelectric cell and its multiplying stages provide an amplified flow of electric current when the cell is exposed to light.

photon: unit carrier of electromagnetic radiation.

photosphere: light-emitting, visible surface of the sun.

photosynthesis: buildup of organic compounds within plants by absorption of water, carbon dioxide, and solar energy.

pion: subatomic particle with a mass equal to 270 electron masses. It may be neutral or charged positively or negatively. Also known as *pi-meson.*

plage: bright, disturbed area of the solar surface.

Planck's constant (*h*): universal constant that connects the energy of the photon (*E*) to its frequency (*f*) through the equation $E = hf$.

Planck's law of radiation: formula giving the relation between the temperature and the energy emitted by a blackbody at any wavelength.

planet: one of the principal nonluminous bodies in orbit around the sun or another star. There are nine in the solar system.

planetary nebula: slowly expanding envelope of gas surrounding a small, hot, central star.

planetesimals: small solid bodies believed to have formed during the condensing stage of the solar nebula.

plasma: hot ionized gas that is electrically conductive.

polar axis: axis of an equatorially mounted telescope about which the telescope can be swung to follow the diurnal motions of the stars. It is always parallel to earth's axis of rotation.

polarized radiation: electromagnetic radiation whose transverse vibration is confined to a fixed plane (plane-polarized light) or to one that rotates (circularly polarized light).

polymer: giant organic molecule formed from thousands of smaller organic molecules linked together.

Population I stars: younger stars found in greatest numbers in the outer portions of the Galactic disk.

Population II stars: older stars inhabiting mainly the central and halo portions of the Galactic system.

population types: classification of the stars into two main groups, Population I and Population II, and intermediate types based on differences in age, chemical composition, spectral properties, velocities, and location in a galaxy.

positive space curvature: space continuum whose curvature is spherical or ellipsoidal, resulting in a closed universe.

positron: positivley charged electron or antielectron.

potential energy: energy acquired as a result of position in a gravitational field.

Precambrian: geologic era between the time when earth's crust formed about 4.5 billion years ago to about 1 billion years ago.

precession of equinoxes: conical movement of earth's axis of rotation about the vertical to the plane of the ecliptic in a period of 25,800 years. The phenomenon causes the equinoxes to slide westward along the ecliptic about 50 seconds of arc per year.

precession of Mercury's orbit: the slow eastward rotation of Mercury's major axis in its own plane.

pressure: the force per unit area exerted by a mechanic force or a field of force.

primeval fireball: high-powered explosion from a superhot, superdense state of condensed matter (the big bang) that supposedly initiated the expansion of the universe.

principle of equivalence: Einstein's declaration that a gravitational force cannot be distinguished from an inertial force; hence a gravitational field can be replaced by an accelerated system.

proper motion: angular change of a star's direction from the sun in one year.

protein: large molecule composed of hundreds to thousands of amino acids joined together by peptide links making up the DNA molecule.

proton: positively charged particle that is part of the nucleus of every atom. It is 1836 times heavier than the electron.

proton-proton chain: sequence of thermonuclear reactions that builds up helium from hydrogen with the release of energy inside the cores of main sequence stars. Also known as the *p-p chain.*

pulsar: very small, highly condensed, rapidly spinning star emitting a narrow beam of electromagnetic radiation, observed as a fast pulse from earth. Believed to be the neutron core remnant of a supernova explosion.

pyrheliometer: thermometer-type of instrument used to measure the intensity of solar radiation.

quadrature: configuration of the moon or a planet corresponding to its position when it is 90° from the earth-sun line.

quantum: finite or discrete amount of a quantity. The term is frequently applied to the discrete bundle of energy possessed by the photon.

quasar: popular designation for a quasi-stellar object.

quasi-stellar object (QSO): general class of stellar-appearing objects believed by most astronomers to be very distant, because of their large red shifts, and also highly luminous. One group, quasi-stellar sources, emits strong radio energy; the other group, which is often called blue stellar objects, emits no detectable radio energy.

radar: technique of bouncing pulsed radio waves from a body and observing their echoes. Derived from *radio detection and ranging.*

radar astronomy: field in which radio signals, pulsed at high power, are transmitted from earth and reflected from the object of interest.

radial velocity: component of an object's motion that lies in the line of sight, producing the Doppler shift in the spectral lines of the body.

radiant: place in the sky from which meteors diverge during a meteor shower. The shower is named after the constellations in which the radiant appears.

radiation belt: zone of high intensity radiation within the magnetosphere of a planet.

radiation pressure: pressure exerted on a body by electromagnetic radiation.

radioactive: describing the spontaneous breakdown of atomic nuclei (normally the heaviest atoms) into lighter nuclei with the ejection of alpha particles, electons, and gamma photons.

radioactive half-life: see *half-life.*

radio astronomy: branch of astronomy in which radio telescopes are used to observe electromagnetic radiations longer than about 1 millimeter in wavelength from celestial sources.

radio galaxy: galaxy that emits strong radio radiation.

radio telescope: highly sensitive receiving system employing a parabolic reflector or other highly directional antenna to locate and study radio sources.

ray: bright elongated streaks radiating from certain craters on the moon and Mercury.

red giant: large cool star of high luminosity.

red giant branch: nearly vertical string of red giants extending from the middle to upper part of the right-hand side of the Hertzsprung-Russell diagram.

red shift: displacement of spectral lines toward the longer wavelengths arising from the Doppler shift.

reflection nebula: dusty gas cloud that reflects the light of nearby stars.

refraction: change in the direction of light rays passing from one medium into another of different density.

regolith: pulverized debris on the surface of a planet produced by meteoritic bombardment of the surface material.

regression of the nodes: westward (backward) slippage of the nodes of the orbit of an easterly revolving body with respect to a fundamental plane of reference.

relativistic: pertains to an object moving at an appreciable fraction of the speed of light.

relativity: principle that postulates the equivalence of the description of the universe, in terms of physical laws, by observers in their respective frames of reference, whether in uniform or nonuniform motion relative to each other.

resolving power: ability of a telescope to separate fine details in an image.

retrograde motion: apparent westerly motion of a plane as viewed from earth, contrary to its usual easterly movement among the stars.

revolution: motion of a body around an axis external to the body.

right ascension: coordinate in the equatorial system of measurement. It measures the arc along the celestial equator from the vernal equinox eastward to the position of the celestial body. It is similar to the measurement of longitude on earth.

rille: canyon or gorge found in the surface of the moon or Mars.

RNA: ribonucleic acid molecule, containing a single string of adenine, cytosine, guanine, and uracil in place of thymine. It exists in three forms: messenger RNA, transfer RNA, and ribosome RNA and is involved in the manufacture of amino acids.

Roche limit: the distance within which tidal forces can disrupt particles trying to adhere to each other.

rotation: turning of a body around an axis inside the body.

RR Lyrae star: pulsating variable of the Population II cepheid group with a period less than one day.

saros: period of 18 years and 11 days in which eclipses repeat themselves.

Schwarzschild radius: critical radius of the event horizon reached by a gravitationally collapsing body between the point of visibility as a highly compressed body and nonvisibility as a black hole.

scintillation: twinkling effect observed when light from a very small radiating source passes through a turbulent medium.

scintillation counter: device consisting of a fluorescent substance that emits a tiny flash of light when struck by a fast-moving particle. The light is amplified by a photocell that records its intensity.

seeing: the astronomical term used to denote our atmosphere's influence on image quality.

seismic: pertaining to earthquakes.

Seyfert galaxy: class of spiral galaxy that exhibits intense, irregular, electromagnetic radiations within a small active nucleus.

shock wave: conical pattern in space produced when an object moves at supersonic speed in a gaseous medium and creates a disturbed wake.

short-period comet: comet that orbits the sun in a period less than about 200 years.

sidereal period: one complete revolution of a celestial body with respect to a fixed point in the heavens, such as a star.

signs of the zodiac: twelve equally spaced constellation divisions centered on the ecliptic, through which the sun passes monthly in succession.

silicates: mineral compositions largely containing silicon and oxygen.

solar constant: rate at which solar radiation is received on a unit surface perpendicular to the incident radiation per unit of time at earth's mean distance, just outside earth's atmosphere.

solar nebula: fragmented portion of an interstellar cloud that has begun to contract under its own gravitation, eventually leading to the formation of a planetary system.

solar wind: continuous stream of charged particles (mostly protons and electrons) ejected radially from the sun at high velocities.

solid: that state of matter in which the constituent particles maintain a permanent relation to each other.

sounding rocket: instrumented rocket that ascends up to a maximum altitude of about 150 kilometers before falling back to earth's surface.

space reddening: selective scattering of light from a distant object by dust particles in interstellar space.

It is about twice as great in the blue spectral region as in the red spectral region.

Space Telescope: 2.4-meter reflecting telescope to be placed in orbit around earth in the 1980s.

space-time: four-dimensional world of space and time as visualized in the theory of relativity. An *event* is located in the space-time continuum analogous to a *point* in three-dimensional space.

space velocity: true motion of the star in space relative to the sun.

special theory of relativity: part of Einstein's theory relating to observers moving uniformly with respect to each other.

speckle photography: technique of taking very short exposures of a star, which freezes the image in a set position on each photograph, and computer-analyzing the many photographs to obtain the star's unsmeared, better-resolved image.

spectrogram: photographic plate on which the spectrum of an object is recorded by the telescope.

spectrograph: basically the same as the spectroscope except that the eyepiece is replaced by a photographic plate for recording the spectrum.

spectroheliogram: photograph of the sun taken with a spectroheliograph.

spectroheliograph: spectrograph modified to photograph the solar disk or the chromosphere in the light of a single spectral line, either in the red hydrogen alpha line or the violet H or K line of calcium.

spectroscope: optical instrument containing a prism or grating with appropriate lenses to permit direct viewing of the spectrum of a radiating source.

spectroscopic binary: double star whose components are not observed separately in a telescope but whose binary character is revealed by the periodic Doppler shift of spectral lines.

spectroscopic parallax: derivation of the star's distance (or parallax) from its apparent magnitude and its absolute magnitude on the basis of its luminosity and spectral characteristics.

spectroscopy: that branch of physics and astronomy dealing with the color or wavelength composition of composite or white light.

spectrum: spreading out of the energy of a radiating source into its component wavelengths by means of a prism, grating, or other dispersing device.

spectrum-luminosity diagram: Hertzsprung-Russell diagram that exhibits the relationship between the absolute magnitudes of stars and their spectral classes.

spherical aberration: failure of light rays striking all parts of a lens or a mirror with a spherical surface to converge at the same focal setting.

spicule: small spikelike protrusion arising within the chromosphere.

spiral galaxy: large system of stars and interstellar matter having the shape of a flattened disk with outlying spiral arms.

state of motion: the motion of a body is some frame of reference as denoted by its velocity.

static universe: nonexpanding or noncontracting universe in equilibrium.

steady state cosmology: model in which it is assumed that the density of matter within the universe remains constant as the universe expands.

Stefan-Boltzmann law: formula that relates the emission (E) by a blackbody to the fourth power of its temperature (T): $E \propto T^4$.

stellar association: sparse aggregation of young population I stars having a common origin and found in the outer gas-dust regions of the Galaxy.

stratosphere: narrow atmospheric zone that lies above the lowest level of earth's atmosphere, the troposphere. It extends from about 11 to 25 kilometers above sea level and has a constant temperature of $-55°C$.

strong nuclear force: nuclear "glue" or binding force that holds the nucleons together against the disruptive repulsive force of the positively charged protons. It operates within the nuclear domain ($\sim 10^{-13}$ centimeter).

subatomic particles: the constituent parts of the atom, such as the electron, proton, neutron, etc.

sunspot: dark marking visible on the sun's surface. Although the sunspot temperature is about 4500 K, it appears dark by comparison with the brighter and hotter photospheric background.

supercluster: a cluster of clusters of galaxies.

supergiant: large massive star of the greatest luminosity.

superior conjunction: position of a superior planet, as viewed from earth, when it is in the same direction as the sun and farthest from earth.

superior planet: planet whose orbit lies outside earth's orbit.

supernova: exploding star that suddenly attains a luminosity up to 100 million times the sun's brightness.

surface gravity: acceleration at the surface of a body arising from its gravitational force.

S waves: transverse waves of an earthquake that cause earth's material to vibrate perpendicular to the direction of the waves' travel. Also called *secondary* or *shear* waves.

synchronous: locking of the rotation of a spinning body with its orbital revolution in which the rotation is an integer or simple fraction of its revolution.

synchrotron radiation: continuous polarized radiation emitted by fast-moving electrons spiraling around the magnetic lines of force.

synodic month: period of the moon's phases, 29.53 days.

synodic period: time between consecutive similar configurations of a planet; for example, between successive inferior conjunctions or oppositions.

tangential velocity: component of the star's motion that is at right angles to the line of sight. Also known as *transverse velocity*.

telemetry: technique of remote measurement.

terrestrial planet: Mercury, Venus, Earth, or Mars.

thermal equilibrium: balancing between incoming and outgoing radiation that keeps the internal temperature of the star constant at any point.

thermal radiation: radiation produced by a body because it is hotter than its surroundings. In producing radiation, the body reduces its thermal energy.

thermocouple: heat-measuring apparatus consisting of two strips of metals joined together at one end. A meter connected to the free ends registers the current generated when the juncture is exposed to infrared radiation.

thermonuclear: refers to a nuclear reaction that is triggered by particles of high thermal energy.

thermonuclear fusion: nuclear reaction in which light atoms are coalesced into heavier atoms with the release of energy.

thermosphere: atmospheric layer extending from about 100 to 400 kilometers. It is characterized by a constantly rising gas-kinetic temperature with height.

thrust: pushing force developed by an aircraft engine or rocket engine as a result of the discharge of high-velocity gases through a nozzle.

tidal force: unequal gravitational pull exerted on parts of a body, tending to deform its shape.

time-dependent cosmological model: any nonstatic model in which the universe evolves (contracts or expands) with time.

time dilation: stretching or slowing down of time seen by an observer with respect to an object moving at high speed in a frame of reference different from that of the observer.

trajectory: path described by any moving object.

transfer orbit: in interplanetary travel, that portion of the ellipse in which the spacecraft moves, tangent to the orbits of both the departure planet (earth) and the target body.

transit: apparent passage of a celestial body across the face of another larger celestial body, such as the passage of Mercury across the face of the sun.

trench: deep rift or fracture produced at the edge of a continental plate when it collides with an oceanic plate.

Trojan asteroid: one of a group of stable, orbiting, minor planets located at or near the vertex of an equilateral triangle with Jupiter and the sun occupying the other vertices.

troposphere: bottom layer of earth's atmosphere, in which our weather takes place. Its height averages about 11 kilometers.

T Tauri variable: low-temperature dwarf star with bright emission lines subject to erratic changes in light. It is found in the vicinity of dark nebulosity and is believed to be the forerunner of a main sequence star.

turnoff point: critical departure position where an evolving star turns off the main sequence after exhausting its core supply of hydrogen and is on the way to becoming a red giant.

21-centimeter radiation: radiation at 1420 megahertz emitted by neutral hydrogen when the bound electron reverses its spin from that which was in the same direction as the proton.

twinkling: random light fluctuations observed when radiation from an apparently very small light source passes through a disturbed gaseous medium.

UBV system: standardized magnitude and color system employed by astronomers for comparing intensities in the ultraviolet (U), blue (B), and yellow or visual (V) spectral regions.

ultraviolet: portion of the electromagnetic spectrum that extends from the shortest visible waves at about 3900 angstroms to about 100 angstroms.

umbra: dark central region of a sunspot. In an eclipse it is the dark central part of the shadow cast by an illuminated body.

universe: everything that we know exists and all that will be found in the future make up the universe.

Van Allen radiation belts: two principal zones of high intensity radiation surrounding earth within the magnetosphere; named after their discoverer.

variable star: any star that exhibits intrinsic changes in light.

velocity curve: plot of the change in radial velocity (Doppler shift) against time.

vernal equinox: point of intersection on the sky between the sun's path (ecliptic) and the celestial equator, reached by the sun on about March 21.

very long baseline interferometry: technique of employing two or more widely separated radio telescopes to observe simultaneously a minute angular source in order to obtain a high degree of resolution.

visual binary: physically related double star whose components can be resolved in a telescope.

visual magnitude: magnitude corresponding to the measurement of the visual light in the approximate spectral range of 4000–7000 angstroms.

water hole: narrow radio spectral region between 1420 and 1668 megahertz in which radiation from neutral hydrogen (H) and hydroxyl (OH) is observed in the interstellar clouds; so called because the dissociated products of water (H_2O) are H and OH.

watt: unit of power equivalent to the expenditure of ten million ergs per second.

wave: the propagation of a disturbance.

wavelength: distance between successive crests or troughs of a wave.

wave mechanics: mathematical theory of quantum mechanics that forms the basis of the modern concept of atomic phenomena in terms of the interactions of radiation with matter.

weak nuclear force: basic nuclear force involved in radioactive decay. It is characterized by a slow nuclear reaction rate (for example, 17 minutes on the average for neutron decay into a proton, electron, and antineutrino) in comparison to the strong nuclear force, which reacts in a very short time (for example, a neutron is ejected from a nucleus in 10^{-21} second).

weight: force of attraction experienced by an object resting on the surface of an attracting body.

white dwarf: star that has collapsed gravitationally into a small, very dense and faint object after expending its nuclear fuel.

white hole: hypothetical opposite of the black hole. Out of it matter and radiation emerge from another universe into our own universe.

Wien's displacement law: simple formula stating that the wavelength at which the peak energy of a radiating blackbody occurs varies inversely with the temperature of the body. As the temperature rises, the maximum of the blackbody's energy curve is displaced toward the shorter wavelengths.

worm hole: restricted neck joining two separate universes in the curved space geometry around a black hole.

W Virginis star: prototype of a special class of cepheid variables belonging to Population II.

X-ray binary: double star in which one component emits X-ray radiation.

X rays: electromagnetic radiation whose wavelengths are approximately between 1 and 100 angstroms. Their spectral region lies between the gamma and ultraviolet wavelengths.

Zeeman effect: splitting or widening of the spectral lines of a radiating soruce in the presence of a magnetic field. The phenomenon was discovered in the laboratory by Dutch physicist P. Zeeman in 1896.

zenith: point in the sky that is immediately overhead.

zero-age main sequence (ZAMS): principal branch in the H-R diagram reached by stars that have evolved to stability as the result of hydrogen burning.

zero space curvature: flat space continuum characterized by Euclidean geometry; there is no warping of space as there is in positive or negative curvature.

zodiac: see *signs of the zodiac.*

zodiacal light: faint band of light that tapers upward from the horizon, following the course of the ecliptic. It apparently results from the reflection of sunlight by interplanetary dust in the plane of the solar system.

zone of avoidance: irregular band along the Milky Way where few or no galaxies appear as a consequence of severe obscuration of light by interstellar dust in or near the Galactic plane.

Selected Readings

Chapter 1 Introduction to the Cosmos

Bok, B. J. and Jerome, L. E. *Objections to Astrology.* Prometheus Books, 1975.

Bronowski, J. *A Sense of the Future.* M.I.T Press, 1977.

Bronowski, J. *Science and Human Values.* Harper & Row, 1972.

Collin, R. *The Theory of Celestial Influence.* Samuel Weiser, 1973.

Davidson, M. *The Stars and the Mind: A Study of the Impact of Astronomical Development on Human Thought.* Gordon Press, 1976.

Dessaur, C. I. *Science Between Culture and Counter Culture.* Humanities Press, 1975.

Friedman, H. *The Amazing Universe,* National Geographic Society, 1975.

Gingerich, O., ed. *The Nature of Scientific Discovery.* Smithsonian Institution Press, 1975.

Goran, M. *Science and Anti-Science.* Ann Arbor Science Publishers, 1974.

Hawkins, G., and White, J. B. *Stonehenge Decoded.* Doubleday, 1965.

Hodson, F. R., ed. *The Place of Astronomy in the Ancient World.* Oxford University Press, 1974.

Holton, G., ed. *Science and Culture.* Beacon Press, 1965.

Holton, G. *Thematic Origins of Scientific Thought: Kepler to Einstein.* Harvard University Press, 1973.

Holton, G. *The Scientific Imagination.* Cambridge University Press, 1978.

Hoyle, F. *From Stonehenge to Modern Cosmology.* W. H. Freeman, 1972.

Lindsay, J. *Origins of Astrology.* Barnes & Noble, 1971.

Merton, R. K. *The Sociology of Science.* University of Chicago Press, 1973.

Roszak, T. *Where the Wasteland Ends.* Doubleday, 1973.

White, L. T. *Machina ex Deo: Essays in the Dynamism of Western Culture.* M.I.T Press, 1971.

Wolke, R. L., ed. *Impact: Science on Society.* Saunders, 1975.

Chapter 2 Earlier Descriptions of the Cosmos

Abetti, G. *The History of Astronomy.* Schuman, 1952.

Adamczewski, J. *Nicolaus Copernicus and His Epoch.* Scribners, 1974.

Berry, A. A. *Short History of Astronomy.* Dover, 1961.

Blacker, C., and Loewe, M., eds. *Ancient Cosmologies.* Rowman and Littlefield, 1975.

Friedemann, C., et al. *Astronomy: A Popular History.* Van Nostrand Reinhold, 1975.

Gingerich, O. "Copernicus and Tycho." *Scientific American,* December 1973.

Koestler, A. *Watershed: A Biography of Johannes Kepler.* Doubleday, 1960.

Kuhn, T. S. *The Copernican Revolution.* Vintage Books, 1959.

Lerner, L. S., and Gosselin, E. A. "Giordano Bruno." *Scientific American,* April 1973.

Ley, W. *Watchers of the Skies: An Informal History of Astronomy from Babylon to the Space Age.* Viking, 1963.

Lindsay, J. *Origins of Astrology.* Barnes & Noble, 1971.

Lockyer, J. N. *Dawn of Astronomy.* M.I.T. Press, 1964.

Pannekoek, A. *A History of Astronomy.* Rowman and Littlefield, 1961.

Pedersen, O., and Pihl, M. *Early Physics and Astronomy.* Neale Watson Academic Publications, 1974.

Shapere, D. *Galileo.* University of Chicago Press, 1974.

Van Der Waerden, B. L., et al. *Science Awakening: The Birth of Astronomy.* Oxford University Press, 1974.

Wilson, C. "How Did Kepler Discover His First Two Laws?" *Scientific American,* March 1972.

Chapter 3 Widening Cosmological Horizons

Barnett, L. *The Universe and Dr. Einstein.* Morrow, 1957.

Bronowski, J. *Ascent of Man.* Little, Brown and Company, 1973.

Coleman, J. A. *Relativity for the Layman.* Macmillan, 1959.

Gamow, G. *Mr. Tompkins in Paperback.* Cambridge University Press, 1965.

Gardner, M. *Relativity for the Millions.* Macmillan, 1966.

Hoffman, B., and Dukes, H. *Albert Einstein: Creator and Rebel.* Viking, 1973.

Hoskin, M. *History of Science Library: William Herschel and the Construction of the Heavens.* American Elsevier, 1970.

Kaufmann, W. J. *The Cosmic Frontiers of General Relativity.* Little, Brown and Company, 1977.

Manuel, F. E. *A Portrait of Newton.* Harvard University Press, 1968.

Ronan, C. *Newton and Gravitation,* Grossman, 1968.

Shankland, R. S. "The Michelson-Morley Experiment." *Scientific American,* November 1964.

Struble, M. *The Web of Space-Time: A Step by Step Exploration of Relativity.* Westminster Press, 1973.

Struve, O., and Zebergs, V. *Astronomy of the 20th Century.* Macmillan, 1962.

Toben, B., and Sarfatti, J. *Space-Time and Beyond: Toward an Explanation of the Unexplainable.* Dutton, 1975.

Toulmin, S., and Goodfield, J. *The Fabric of the Heavens.* Harper & Row, 1961.

Whitrow, G. J., ed. *Einstein: The Man and His Achievement.* Dover, 1973.

Chapter 4 Celestial Radiation and the Atom

Adler, I. *The Story of Light.* Harvey House, 1971.

Asimov, I. *Inside the Atom.* Abelard-Schuman, 1966.

Cohen, B. L. *The Heart of the Atom: The Structure of the Atomic Nucleus.* Doubleday, 1967.

Ellis, R. H., Jr. *Knowing the Atomic Nucleus.* Lothrop, Lee, and Shepard, 1973.

Hoffman, B. *Strange Story of the Atom.* Dover, 1959.

Malville, K. *A Feather for Daedalus.* Cummings, 1975.

Chapter 5 Telescopes and Their Accessories

American Institute of Aeronautics and Astronautics. *Large Space Telescope, A New Tool for Science.* 1290 Avenue of the Americas, New York, 1974.

Asimov, I. *Eyes on the Universe: A History of the Telescope.* Houghton Mifflin, 1975.

Becker, T. W. *Exploring Tomorrow in Space.* Sterling, 1972.

Hey, J. S. *The Evolution of Radio Astronomy.* Neale Watson Academic Publications, 1975.

Kopal, Z. *Telescopes in Space.* Hart, 1970.

NASA. *Future Space Programs.* U.S. Government Printing Office, 1975.

———. *The Space Telescope.* U.S. Government Printing Office, 1976.

———. *Telescopes and Space Exploration.* U.S. Government Printing Office, 1976.

Page, T., and Page L. W. *Space Science and Astronomy.* Macmillan, 1976.

Page, T., and Page L. W. *Telescopes.* Macmillan, 1966.

Smith, F. G. *Radio Astronomy.* Penguin, 1974.

Strong, J. *Search the Solar System: The Future of Unmanned Spaceflight.* Crane-Russak, 1973.

Verschuur, G. L. *Starscapes.* Little, Brown and Company, 1977.

von Braun, W. *Space Frontier.* Holt, Rinehart, and Winston, 1971.

von Braun, W., and Ordway, F. I. *History of Rocketry and Space Travel.* Crowell, 1975.

Woodbury, D. O. *The Giant Glass of Palomar.* Dodd, Mead, 1970.

Chapter 6 The Earth-Moon: A Double Planet

Alter, D. *Pictorial Guide to the Moon.* Crowell, 1973.

Carrigan, C. R., and Gubbins, D. "The Source of the Earth's Magnetic Field." *Scientific American,* February 1979.

Cherrington, E. H., Jr. *Exploring the Moon through Binoculars.* McGraw-Hill, 1969.

Cook, A. H. *The Physics of the Earth and the Planets.* Halsted Press, 1973.

Cooper, H. S. F. *Moon Rocks.* Dial, 1970.

Darden, L. *The Earth in the Looking Glass.* Doubleday, 1974.

Gamow, G. *A Planet Called Earth.* Viking, 1970.

Gamow, G., and Stubbs, H. G. *Moon.* Abelard-Schuman, 1971

Gass, I. G., Smith, P. J., and Wilson, R. C. L., eds. *Understanding the Earth: A Reader in the Earth Sciences.* M.I.T. Press, 1971.

Glen, W. *Continental Drift and Plate Tectonics.* Merrill, 1975.

Jordan, T. H. "The Deep Structure of the Continents." *Scientific American,* January 1979.

Kopal, Z. *New Photographic Atlas of the Moon.* Taplinger, 1970.

Marsden, B. G., and Cameron, A. G. W. *The Earth-Moon System.* Plenum Press, 1966.

Mason, B., and Melson, W. G. *Lunar Rocks.* Wiley, 1970.

Matthews, W. H., III. *Invitation to Geology: The Earth through Time and Space.* Natural History Press, 1971.

Meadows, D. L., et al. *Dynamics of Growth in a Limited World.* Wright-Allen, 1974.

Moorbath, S. "Oldest Rocks and the Growth of Continents." *Scientific American,* March 1977.

Mutch, T. A. *Geology of the Moon.* Princeton University Press, 1973.

NASA. *Mission to Earth: Landsat Views the World.* U.S. Government Printing Office, 1975.

Press, F., and Siever, R., eds. *Planet Earth.* W. H. Freeman, 1974.

Taylor, S. R. *Lunar Science: A Post-Apollo View.* Pergamon Press, 1975.

Wilson, J. T., et al. *Introduction to Continents Adrift.* W. H. Freeman, 1972.

Chapter 7 The Solar System

Callatay, V. de, and Dollfus, A. *Planets: A History of Man's Inquiry through the Ages.* University of Toronto Press, 1976.

Cameron, A. G. W. "The Origin and Evolution of the Solar System." *Scientific American,* September 1975.

Cornell, J., and Hayes, E. N., eds. *Man and Cosmos.* Norton, 1975.

Nieto, M. N. *The Titius-Bode Law of Planetary Distances: Its History and Theory.* Pergamon Press, 1972.

Page, T., and Page, L. W. *Origin of the Solar System.* Macmillan, 1966.

———. *Wanderers in the Sky.* Macmillan, 1965.

Schramm, D. N., and Clayton, R. N. "Did a Supernova Trigger the Formation of the Solar System?" *Scientific American,* October, 1978.

Scientific American. *The Solar System.* W. H. Freeman, 1975.

Whipple, F. L. *Earth, Moon, and Planets.* Harvard University Press, 1968.

Williams, I. P. *The Origin of the Planets.* Hager, 1975.

Chapter 8 The Inner Solar System: The Terrestrial Planets

Arvidson, R. E., Binder, A. B., and Jones, K. L. "The Surface of Mars." *Scientific American,* March 1978.

Bonestell, C., and Clarke, A. C. *Beyond Jupiter: The Worlds of Tomorrow.* Little, Brown and Company, 1973.

Bradbury, R., et al. *Mars and the Mind of Man.* Harper & Row, 1973.

Grossman, L. "The Most Primitive Objects in the Solar System." *Scientific American,* February 1975.

Hartmann, W. K. "Cratering in the Solar System." *Scientific American,* January 1977.

Hartmann, W. K., and Raper, O., eds. *The New Mars.* NASA SP-337, U.S. Government Printing Office, 1974.

Hoyt, W. G. *Lowell and Mars.* University of Arizona Press, 1976.

Knight, D. C. *Meteors and Meteorites.* Watts, 1969.

Leovy, C. B. "The Atmosphere of Mars." *Scientific American,* July 1977.

McCall, G. J. *Meteorites and Their Origins.* Halsted, 1973.

Nourse, A. E. *Asteroids.* Watts, 1975.

Richardson, R. S. *Getting Acquainted with Comets.* McGraw-Hill, 1967.

Veuerka, J. "Phobos and Deimos." *Scientific American,* February 1977.

Wetherill, G. W. "Apollo Objects." *Scientific American,* March 1979.

Chapter 9 Outer Solar System: The Jovian Planets

Asimov, I. *How Did We Find Out about Comets?* Walker, 1975.

Bonestell, C., and Clarke, A. C. *Beyond Jupiter: The Worlds of Tomorrow.* Little, Brown and Company, 1973.

Gary, G. A., ed. *Comet Kohoutek.* NASA, Washington, D.C. 1975.

Ley, W. *Comets: Visitors from Afar.* McGraw-Hill, 1969.

NASA. *The Planets Uranus, Neptune, and Pluto.* NASA SP-8103, National Technical Information Service, Springfield, Va., 1971

Richardson, R. S. *Getting Acquainted with Comets.* McGraw-Hill, 1967.

Soderblom, L. A. "The Galilean Moons of Jupiter." *Scientific American,* January 1980.

Chapter 10 Life in the Solar System

Ayala, F. J. "The Mechanisms of Evolution." *Scientific American,* September 1978.

Barber, V. "Theories of the Chemical Origin of Life on the Earth." *Mercury,* September–October 1974.

Dickerson, R. E. "Chemical Evolution and the Origin of Life." *Scientific American,* September 1978.

Ehrlich, P. R., and Ehrlich, A. H. *Population, Resources, Environment: Issues in Human Ecology.* W. H. Freeman, 1972.

Horowitz, N. H. "The Search for Life on Mars." *Scientific American,* November 1977.

Kiefer, I. "Chemical Evolution." *Smithsonian,* May 1972.

Lawless, J., et al. "Organic Matter in Meteorites." *Scientific American,* June 1972.

Mayr, E. "Evolution." *Scientific American,* September 1978.

Oparin, A. I. *Life: Its Nature, Origin, and Development.* Academic Press, 1962.

Ponnamperuma, C. *The Origins of Life.* Dutton, 1972.

Scientific American. *The Biosphere.* W. H. Freeman, 1970.

Spinar, Z. *Life Before Man.* American Heritage, 1972.

Washburn, S. L. "The Evolution of Man." *Scientific American,* September 1978.

Chapter 11 The Sun: Our Bridge to the Stars

Abetti, G. *The Sun.* Macmillan, 1957.

Bahcall, J. H. "Neutrinos from the Sun." *Scientific American,* July 1969.

Bray, R. J., and Loughhead, R. E. *Sunspots.* Halsted, 1964.

Eddy, J. A. "The Case of the Missing Sunspots." *Scientific American,* May 1977.

Ellison, M. A. *The Sun and its Influence.* American Elsevier, 1968.

Gamow, G. *A Star Called the Sun.* Viking, 1964.

Kiepenheuer, K. *The Sun.* University of Michigan Press, 1959.

Menzel, D. H. *Our Sun.* Harvard University Press, 1959.

Moore, P. *Sun.* Norton, 1968.

Parker, E. N. "The Sun." *Scientific American,* September 1975.

Zirin, H. *The Solar Atmosphere.* Blaisdell, 1966.

Chapter 12 The Nature of Stars

Abt, H. A. "The Companions of Sunlike Stars." *Scientific American,* April 1972.

Aller, L. H. *Atoms, Stars, and Nebulae.* Rev. ed. Harvard University Press, 1971.

Bok, B. J., and Bok, P. F. *The Milky Way.* Harvard University Press, 1974.

Bova, B. *The Milky Way Galaxy: Man's Exploration of the Stars.* Holt, Rinehart, and Winston, 1961.

Branley, F. M. *Milky Way: Galaxy Number One.* Crowell, 1969.

Burbidge, M., and Burbidge, G. "Sub-dwarf Stars." *Scientific American,* June 1961.

Eggen, O. "Stars in Contact." *Scientific American,* June 1968.

Glasby, J. S. *Variable Stars.* Harvard University Press, 1968.

Huang, S. "The Origin of Binary Stars." *Sky and Telescope,* December 1967.

Jaki, S. L. *The Milky Way: An Elusive Road for Science.* Neale Watson Academic Publications, 1975.

Merrill, P. W. *Space Chemistry.* University of Michigan Press, 1963.

Page, T., and Page, L. W. eds. *Starlight—What It Tells About the Stars.* Macmillan, 1967.

Percy, J. "Pulsating Stars." *Scientific American,* June 1975.

Smart, W. M. *Some Famous Stars.* McKay, 1950.

Whitney, C. A. *Discovery of Our Galaxy.* Knopf, 1971.

Chapter 13 The H-R Diagram and The Internal Structure of Stars

Jastrow, R. *Red Giants and White Dwarfs.* Harper & Row, 1971.

King, I. "The Dynamics of Star Clusters." *Sky and Telescope,* March 1971.

Limber, D. "The Pleiades." *Scientific American,* November 1962.

Lynds, B. T., ed. *Dark Nebulae, Globules, and Protostars.* University of Arizona Press, 1971.

Meadows, A. J. *Stellar Evolution*. Pergamon, 1967.

Mihalas, D. "Interpreting Early-type Stellar Spectra." *Sky and Telescope*, August 1973.

Page, T., and Page, L. W., eds. *Evolution of Stars*. Macmillian, 1967.

Schwartzchild, M. *Evolution of Stars*. Dover, 1976.

Struve, O. *Stellar Evolution*. Princeton University Press, 1950.

Chapter 14 Interstellar Matter-The Birthplace of Stars

Aller, L. H. *Atoms, Stars and Nebulae*. Harvard University Press, 1971.

Bok, B. J. "The Birth of Stars." *Scientific American*, August 1972.

Bok, B. J., and Bok, P. F. *The Milky Way*. Harvard University Press, 1974.

Chaisson, E. J. "Gaseous Nebulas." *Scientific American*, December 1978.

Dickinson, D. F. "Cosmic Masers." *Scientific American*, June 1978.

Dickman, R. L. "Bok Globules." *Scientific American*, June 1977.

Dufay, J. *Galactic Nebulae and Interstellar Matter*. Dover, 1968.

Glasby, J. S. *Dwarf Novae*. American Elsevier, 1970.

Gordon, M. A., and Burton, W. B. "Carbon Monoxide in the Galaxy." *Scientific American*, May 1979.

Gorenstein, P., and Tucker, W. "Supernova Remnants." *Scientific American*, July 1971.

Heiles, C. "The Structure of the Interstellar Medium." *Scientific American*, January 1978.

Herbst, W., and Assousa, G. E. "Supernovas and Star Formation." *Scientific American*, August 1979.

Maran, S. "The Gum Nebula." *Scientific American*, December 1971.

Miller, J. S. "Structure of Emission Nebulas." *Scientific American*, October 1974.

Page, T., and Page, L. W., eds. *Stars and Clouds of the Milky Way*. Macmillan, 1968.

Smith, G. F. *Radio Astronomy*. Penguin, 1974.

Verschuur, G. L. *The Invisible Universe*. Springer-Verlag, 1974.

_____. *Starscapes*. Little, Brown and Company, 1977.

Weymann, R. J. "Stellar Winds." *Scientific American*, August 1978.

Zeilik, M. "The Birth of Massive Stars." *Scientific American*, April 1978.

Chapter 15 The Death of Stars

Asimov, I. *The Collapsing Universe: The Story of Black Holes*. Walker, 1975.

Bova, B. *In Quest of Quasars*. Macmillan, 1970.

Burbidge, G. R., and Burbidge, M. *Quasi-stellar Objects*. W. H. Freeman, 1967.

Calder, N. *Violent Universe*. Viking, 1970.

Golden, F. *Quasars, Pulsars, and Black Holes: A Scientific Detective Story*. Scribners, 1976.

Gursky, H., and van den Heuvel, E. "X-ray Emitting Double Stars." *Scientific American*, March 1975.

Hoyle, F. *Galaxies, Nuclei, and Quasars*. Harper & Row. 1965.

Iben, I. "Globular Cluster Stars." *Scientific American*, July 1976.

Kahn, F. D., and Palmer, H. P. *Quasars, Their Importance in Astronomy and Physics*. Harvard University Press, 1967.

Kaufmann, W. J. *The Cosmic Frontiers of General Relativity*. Little, Brown and Company, 1977.

Levitt, I. M. *Beyond the Known Universe*. Viking, 1974.

Nicolson, I., , and Moore, P. *Black Holes in Space*. Orbach and Chambers, 1974.

Ostriker, J. "The Nature of Pulsars," *Scientific American*, January 1971.

Penrose, R. "Black Holes." *Scientific American*, May 1972.

Schramm, D., and Arnett, W. "Supernovae: The Origin of the Chemical Elements, Cosmic Rays, Neutron Stars and Maybe Even Black Holes." *Mercury*, May–June 1975.

Shipman, H. L. *Black Holes, Quasars, and the Universe*. Houghton Mifflin, 1976.

Stephenson, F. R., and Clark, D. H. "Historical Supernovas." *Scientific American*, June 1976.

Thorne, K. A. "The Search for Black Holes." *Scientific American*, December 1974.

Verschuur, G. L. *Starscapes*. Little, Brown and Company 1977.

Wheeler, J. O. "After the Supernova, What?" *American Scientist*, January–February 1973.

Wilson, S. G. *Cosmic Rays*. Springer-Verlag, 1976.

Chapter 16 The Milky Way Galaxy

Bok, B. "The Spiral Structure of Our Galaxy." *Sky and Telescope*, December 1969 and January 1970.

Bok, B. J., and Bok, P. *The Milky Way*. 4th ed. Harvard University Press, 1974.

Clark, G. W. "X-ray Stars in Globular Clusters." *Scientific American*, October 1977.

Geballe, T. R. "The Central Parsec of the Galaxy." *Scientific American*, May 1979.

Gordon, M. A., and Burton, W. B. "Carbon Monoxide in the Galaxy." *Scientific American*, May 1979.

Jaki, S. L. *The Milky Way, an Elusive Road for Science*. Science History Publications, 1972.

Linsley, J. "The Highest-Energy Cosmic Rays." *Scientific American*, July 1978.

Sanders, R., and Wrixon, G. "The Center of the Galaxy." *Scientific American*, April 1974.

Struve, O., and Zebergs, V. *Astronomy in the 20th Century*. Macmillan, 1962.

Weaver, H. "Steps Toward Understanding the Large Scale Structure of the Milky Way." *Mercury*, September–October 1975 and January–February 1976.

Whitney, C. A. *The Discovery of Our Galaxy*. Knopf, 1971.

Chapter 17 The Galaxies

Berendzen, R., Hart, R., and Seeley, D. *Man Discovers the Galaxies*. Neale Watson Academic Publications, 1976.

Disney, M. J., and Veron, P. "BL Lacertae Objects." *Scientific American,* August 1977.

Gorenstein, P., and Tucker, W. "Rich Clusters of Galaxies." *Scientific American,* November 1978.

Gribben, J. *Galaxy Formation: A Personal View.* Halsted, 1976.

Groth, E. J., Peebles, J. E., Seldner, M., and Soneira, R. M. "The Clustering of Galaxies." *Scientific American,* November 1977.

Hey, J. S. *The Radio Universe.* Pergamon, 1975.

Hodge, P. W. *Galaxies and Cosmology.* McGraw-Hill, 1966.

Jones, K. G. *Messier's Nebulae and Star Clusters.* American Elsevier, 1969.

————. *The Search for the Nebulae.* Neale Watson Academic Publications, 1975.

Meier, D. L., and Sunyaev, R. L. "Primeval Galaxies." *Scientific American,* November 1979.

Mitton, S. *Exploring the Galaxies.* Scribner, 1976.

Page, T., and Page, L. W., eds. *Beyond the Milky Way.* Macmillan, 1969.

Rees, M., and Silk, J. "The Origin of Galaxies." *Scientific American,* June 1970.

Ronan, C. *Discovery of the Galaxies.* Grossman, 1969.

Rubin, V. "The Dynamics of the Andromeda Nebula." *Scientific American,* June 1973.

Sandage, A. *The Hubble Atlas of Galaxies.* Carnegie Institution, 1961.

Schmidt, M., and Bello, F. "The Evolution of Quasars." *Scientific American,* May 1971.

Shapley, H. *Galaxies.* Harvard University Press, 1972.

Strom, R., et al. "Giant Radio Galaxies." *Scientific American,* August 1975.

Strom, S. E., and Strom, K. M. "The Evolution of Disk Galaxies." *Scientific American,* April 1979.

Toomre, A., and Toomre, J. "Violent Tides Between Galaxies." *Scientific American,* December 1973.

Verschuur, G. L. *The Invisible Universe.* Springer-Verlag. 1974.

————. *Starscapes.* Little, Brown and Company, 1977.

Woltjer, L., ed. *Galaxies and the Universe.* Columbia University Press, 1968.

Chapter 18 Cosmology

Dickson, F. P. *The Bowl of Night.* M.I.T. Press, 1969.

Eiseley, L. *The Unexpected Universe.* Harcourt, Brace, 1972.

Gamow, G. *The Creation of the Universe.* Viking, 1961.

Gott, J. R., Gunn, J. E., and Tinsley, B. M. "Will the Universe Expand Forever?" *Scientific American,* March 1976.

Hawking, S. W. "The Quantum Mechanics of Black Holes." *Scientific American,* January 1977.

Hinkelbein, A. *Origins of the Universe.* Watts, 1973.

Hodge, P. W. *Concepts of the Universe.* McGraw-Hill, 1969.

Kaufmann, W. J. *The Cosmic Frontiers of General Relativity.* Little, Brown and Company, 1977.

————. *Relativity and Cosmology.* 2nd ed. Harper & Row, 1977.

Lang, K. R., and Mumford, G. S. "A New Look at the Hubble Diagram." *Sky and Telescope,* February 1976.

Motz, L. *The Universe: Its Beginning and End.* Scribners, 1975.

Muller, R. A. "The Cosmic Background Radiation and the New Aether Drift." *Scientific American,* May 1978.

Pasachoff, J., and Fowler, W. A. "Deuterium in the Universe." *Scientific American,* May 1974.

Schatzman, E. *The Origin and Evolution of the Universe.* Basic Books, 1966.

Schramm, D. N. "The Age of the Elements." *Scientific American,* January 1974.

Scientific American. *Cosmology + 1.* W. H. Freeman, 1975.

Shipman, H. *Black Holes, Quasars, and the Universe.* Houghton Mifflin, 1976.

Silk, J. *The Big Bang: The Creation and Evolution of the Universe.* W. H. Freeman, 1980.

Singh, J. *Great Ideas and Theories of Modern Cosmology.* Dover, 1961.

Webster, A. "The Cosmic Background Radiation." *Scientific American,* August 1974.

Chapter 19 Exobiology: Life on Other Worlds

Berendzen, R., ed. *Life Beyond Earth and the Mind of Man.* NASA, Scientific and Technical Information Office, 1973.

Bidpath, I. *World Beyond: A Report on the Search for Life in Space.* Harper & Row, 1976.

Bracewell, R. N. *The Galactic Club.* W. H. Freeman, 1975.

Christian, J. L., ed. *Extraterrestrial Intelligence: The First Encounter.* Prometheus Books, 1976.

Dole, S. H. *Habitable Planets for Man.* American Elsevier, 1970.

Drake, W. R. *Gods and Spacemen throughout History.* Regnery, 1975.

Hynek, J. A., and Vallee, J. *The Edge of Reality: A Progress Report on Unidentified Flying Objects.* Regnery, 1975.

Klass, P. J. *UFOs Explained.* Random House, 1975.

MacGowan, R. A., and Ordway, F. I. *Intelligence in the Universe.* Prentice-Hall, 1966.

Maruyama, M., and Harkins, A., eds. *Cultures Beyond the Earth.* Vintage, 1975.

Ponnamperuma, C., and Cameron, A. G. W., eds. *Interstellar Communication.* Benjamin, 1963.

Sagan, C., ed. *Communication with Extraterrestrial Intelligence.* M.I.T. Press, 1973.

Sagan, C. *Other Worlds.* Bantam, 1975.

————. *The Cosmic Connection: An Extraterrestrial Perspective.* Doubleday, 1973.

Sagan, C., and Drake, F. "Search for Extraterrestrial Intelligence." *Scientific American,* May 1975.

Sagan, C., and Page, T. *UFOs: A Scientific Debate.* Norton, 1974.

Shklovskii, I. S., and Sagan, C. *Intelligent Life in the Universe.* Holden-Day, 1966.

Sneath, P. H. *Planets and Life.* Funk and Wagnalls, 1974.

Sullivan, W. *We Are Not Alone.* McGraw-Hill, 1973.

Young, R. K., and Veldman, D. J. *Extraterrestrial Biology.* Holt, Rinehart, and Winston, 1965.

General Astronomical Information

Allen, C. W. *Astrophysical Quantities, 3rd Ed.* Athlone Press, 1973.

Brandt, J. C., and Muran, S. P. *The New Astronomy and Space Science Reader.* W. H. Freeman, 1977.

Dorschner, J., et al. *Astronomy: A Popular History.* Van Nostrand Reinhold, 1975.

Flammarion Book of Astronomy. Simon and Schuster, 1964.

Gingerich, O., ed. *New Frontiers in Astronomy.* W. H. Freeman, 1975.

Hopkins, J. *Glossary of Astronomy and Astrophysics.* University of Chicago Press, 1976.

Lang, K. R. *Astrophysical Formulae.* Springer-Verlag, 1974.

Menzel, D. H. *Astronomy.* Random House, 1970.

Moore, P. *The Picture History of Astronomy.* Jarrold and Sons, 1964.

Rohr, H. *The Beauty of the Universe.* Viking, 1972.

Rudaux, L., and deVaucouleurs, G. *Larousse Encyclopedia of Astronomy.* Prometheus Books, 1959.

Weigert, A., and Zimmermann, H. *A Concise Encyclopedia of Astronomy.* American Elsevier, 1968.

Young, L. B., ed. *Exploring the Universe.* Oxford University Press, 1971.

Periodicals

Astronomy, monthly, 757 N. Broadway, Suite 204, Milwaukee, Wisconsin 53202.

Griffith Observer, monthly, Griffith Observatory, Los Angeles, California.

Journal of the Royal Astronomical Society, bimonthly, Royal Astronomical Society of Canada, Toronto.

Mercury, bimonthly journal of the Astronomical Society of the Pacific, 1244 Noriega Street, San Francisco, California 94122.

Science News, weekly, 1719 N Street N.W., Washington, D.C.

Scientific American, monthly, 415 Madison Avenue, New York, New York 10017.

Sky and Telescope, monthly, Sky Publishing Corporation, 49 Bay State Road, Cambridge, Massachusetts.

American, March 1975. © Scientific American, Inc. All rights reserved. *Fig. 6.8:* J. R. Eyerman © Time, Inc. *Fig. 6.13:* NASA. *Fig. 6.14:* Lick Observatory. *Fig. 6.15:* top left, Lick Observatory; top right, Hale Observatories; bottom, from left to right, Lick Observatory, NASA, NASA. *Fig. 6.16:* NASA. *Fig. 6.17:* NASA. *Fig. 6.18:* NASA. *Fig. 6.19:* Donald Davis and Donald Wilhelm, U.S. Geological Survey, Menlo Park, California. *Fig. 6.20:* from *Sky and Telescope,* vol. 53, no. 3 (March 1977), p. 166. Redrawn by permission.

Chapter 7

Page 128: NASA. *Fig. 7.2:* adapted from a diagram by B. Lovell. *Page 135:* from a poem by Diane Ackerman. Copyright © 1976, Harvard Magazine, Inc. Reprinted by permission. *Fig. 7.5:* all NASA. *Fig. 7.7:* reproduced by permission of the Smithsonian Institution Press from the *Smithsonian Institution Annual Report, 1949,* "The Origin of the Earth," by Thornton Page: Plate 1. Washington, D.C.: Government Printing Office, 1950. *Figs. 7.8 and 7.9:* Redrawn by permission of the publisher from "The Evolution of the Solar System" by A. G. W. Cameron, *Scientific American,* September 1975. © Scientific American, Inc. All rights reserved.

Chapter 8

Page 148: NASA. *Figs. 8.1 and 8.2:* Jet Propulsion Laboratory, NASA. *Figs. 8.3 and 8.4:* NASA. *Fig. 8.5:* M. Ya. Marov, Institute for Applied Mathematics, USSR. *Page 157:* courtesy Yerkes Observatory. *Figs. 8.7:* left and center photographs, Lowell Observatory; right photograph, NASA. *Fig. 8.8:* NASA. *Figs. 8.9 and 8.10:* Jet Propulsion Laboratory, NASA. *Figs. 8.11–8.14:* NASA. *Figs. 8.15 and 8.16:* Jet Propulsion Laboratory, NASA. *Fig. 8.17:* redrawn by permission from William K. Hartmann, "The Smaller Bodies of the Solar System," *Scientific American,* September 1975. © Scientific American, Inc. All rights reserved. *Fig. 8.18:* courtesy of Meteor Crater Enterprises, Inc.

Chapter 9

Page 175: NASA. *Fig. 9.1:* adapted from Jet Propulsion Laboratory diagrams, courtesy *Sky and Telescope.* *Fig. 9.2–9.5:* NASA. *Figs. 9.7 and 9.8:* NASA. *Fig. 9.9:* photographs from Lick Observatory. *Fig. 9.10:* NASA. *Figs. 9.12 and 9.13:* Lick Observatory. *Fig. 9.14:* U.S. Naval Observatory. *Page 193:* Culver Pictures. *Fig. 9.16:* adapted from a diagram by NASA. *Fig. 9.18:* top, Lunar and Planetary Laboratory; bottom, Lick Observatory.

Chapter 10

Page 199: Dr. David Phillips, Rockefeller University. *Page 200:* from Arthur Guiterman, "Ode to the Amoeba," *Gaily the Troubador,* © E. P. Dutton, 1936. Renewed 1964 by Vida Linda Guiterman. Reprinted by permission of Louise Sclove. *Fig. 10.1:* Harry Wilks, Stock, Boston. *Fig. 10.2:* Stella Snead, Bruce Coleman, Inc. *Fig. 10.8:* Professor Elso Barghoorn, Biological Labs, Harvard University. *Fig. 10.9:* Institute de Paleontologie, Museum National d'Histoire Naturelle. *Fig. 10.10:* Jay H. Matternes. *Fig. 10.11:* photo by John Fields, The Trustees, The Australian Museum.

Chapter 11

Page 215: by permission, Harvard College Observatory Cambridge, Mass. *Fig. 11.1:* top row (photographs), Hale Observatories; bottom row: far left, Princeton University Observatory; center, High Altitude Observatory; upper right, Hale Observatories; lower right, Lick Observatory. *Figs. 11.2:* redrawn by permission from TIME, *The Weekly Newsmagazine.* © Time, Inc. *Fig. 11.5:* Hale Observatories. *Fig. 11.7:* Dr. Robert Leighton, California Institute of Technology.

Fig. 11.9: redrawn by permission from TIME, *The Weekly Newsmagazine.* © Time, Inc. *Fig. 11.10:* The Kitt Peak National Observatory. *Fig. 11.12:* Dr. Mitsul Kanno, Hida Observatory. *Fig. 11.13:* Hale Observatories. *Fig. 11.14:* Sacramento Peak Observatory, Association of Universities for Research in Astronomy, Inc. *Fig. 11.15:* left, NASA; center and right, Big Bear Solar Observatory. *Fig. 11.16:* High Altitude Observatory. *Fig. 11.17:* American Science and Engineering. *Fig. 11.18:* The Naval Research Laboratory. *Fig. 11.20:* Lockheed Solar Observatory. *Fig. 11.21:* from "The Case of the Missing Sunspots," by John Eddy, *Scientific American,* May 1977, © Scientific American, Inc. All rights reserved.

Chapter 12

Page 244: Hale Observatories. *Page 249:* Brown Brothers. *Fig. 12.2:* Dr. Martha Liller, Harvard College Observatory. *Fig. 12.5:* Hale Observatories. *Fig. 12.6:* Lick Observatory. *Fig. 12.8:* courtesy Yerkes Observatory. *Fig. 12.9:* after L. H. Aller. *Page 258:* Quotation reprinted with permission of Charles Scribner's Sons from "Octaves XI," *The Collected Poems of Edwin Arlington Robinson* (New York: Macmillan Co., 1935, 1937). *Fig. 12.12:* courtesy Yerkes Observatory. *Fig. 12.13:* From William K. Hartmann, *Astronomy: The Cosmic Journey,* p. 334. © 1978 by Wadsworth Publishing Company, Inc., Belmont, California 94002. Reprinted by permission of the publisher. *Fig. 12.15:* Lick Observatory. *Fig. 12.19:* Lick Observatory.

Chapter 13

Page 267: Brookhaven Laboratory. *Fig. 13.4:* European Southern Observatories. *Fig. 13.5:* The Kitt Peak National Observatory. *Fig. 13.6:* after H. L. Johnson. *Fig. 13.7:* Hale Observatories. *Fig. 13.8:* after A. Sandage and H. L. Johnson. *Fig. 13.9:* Hale Observatories. *Page 281:* courtesy Yerkes Observatory.

Chapter 14

Page 291: Hale Observatories. *Page 292:* from Robinson Jeffers, "Nova" in *Such Counsels You Gave To Me,* pp. 111–112. Copyright 1937 by Random House Inc. Reprinted by permission. *Fig. 14.1:* based on William K. Hartmann, *Astronomy: The Cosmic Journey,* p. 309. © 1978 by Wadsworth Publishing Company, Inc., Belmont, California 94002. Reprinted by permission of the publisher. *Fig. 14.2:* Hale Observatories. *Fig. 14.4:* from William K. Hartmann, *Astronomy: The Cosmic Journey,* p. 311. © 1978 by Wadsworth Publishing Company, Inc., Belmont, California 94002. Reprinted by permission of the publisher. *Fig. 14.5:* from. R. F. Knacke, "Solid Particles in Space," *Sky and Telescope* (April 1979). Redrawn by permission. *Fig. 14.7:* from "The Birth of Massive Stars," by Zeilik, *Scientific American,* April 1978. © Scientific American, Inc. All rights reserved. Photograph, Palomar Sky Survey. *Fig. 14.8:* from William K. Hartmann, *Astronomy: The Cosmic Journey,* p. 317. © 1978 by Wadsworth Publishing Company, Inc., Belmont, California 94002. Reprinted by permission of the publisher. *Fig. 14.9:* G. H. Herbig, Lick Observatory. *Page 302:* courtesy Yerkes Observatory. *Fig. 14.10:* The Kitt Peak National Observatory. *Fig. 14.11:* Hale Observatories. *Page 304:* from Bert L. Taylor, "Canopus" in *Poems* (New York: Alfred A. Knopf, Inc.). Reprinted by permission. *Fig. 14.13:* Hale Observatories. *Fig. 14.14:* courtesy of Bart J. Bok, University of Arizona. *Figs. 14.15–14.17:* Hale Observatories. *Fig. 14.18:* redrawn by permission of The University of Chicago Press from *Astrophysical Journal,* vol. 141, p. 993. All rights reserved. © 1965 by The American Astronomical Society. *Fig. 14.19:* Lick Observatory. *Page 310:* from Robert Frost, "Desert Places." Reprinted by permission of Holt, Rinehart & Winston. *Fig. 14.20:* based on diagram by Merle Walker, Lick Observatory. *Fig. 14.21:* from "The Birth of Massive Stars," by Zeilik, *Scientific American,* April 1978. © Scientific American, Inc. All rights reserved. Photograph, The Kitt Peak National Observatory.

Chapter 15

Page 313: Painting by Lois Cohen, Griffith Observatory. *Fig. 15.1:* after Icko Iben, Jr. *Page 316:* from Robinson Jeffers, "Margrave" in *Thurso's Landing,* pp. 135–137. Copyright, 1932 by Robinson Jeffers. Reprinted by permission of Random House, Inc. *Fig. 15.4:* after Icko Iben, Jr. *Fig. 15.5:* after Icko Iben, Jr. © 1971 *Astronomical Society of the Pacific,* vol. 83, p. 697. Redrawn by permission. *Fig. 15.7:* after B. Paczynski and C. R. O'Dell. *Fig. 15.9:* Lick Observatory. *Page 328:* Bell Laboratories. *Fig. 15.14:* J. Wampler and J. C. Miller, Lick Observatory. *Fig. 15.15:* redrawn by permission of the publisher from "Pulsars" by Anthony Hewish, *Scientific American,* October 1968. © Scientific American, Inc. All rights reserved. *Fig. 15.16:* from E. Groth, "Observing Properties of Pulsars," *Neutron Stars, Black Holes, and Binary X-Ray Sources,* H. Gursky and R. Ruffini, eds. (Hingham, Mass.: D. Reidel Publishing Co., 1975), pp. 119–173. Redrawn by permission. *Fig. 15.19:* left: from William Kaufmann, *The Cosmic Frontiers of General Relativity,* p. 176. Copyright © 1977 by Little, Brown and Company (Inc.). Redrawn by permission. Right: from R. Ruffini and J. A. Wheeler, "Relativistic Cosmology and Space Platforms," *Proceedings of the Conference on Space Physics,* European Space Research Organization, 1971. Redrawn by permission of ESRD (now European Space Agency). *Fig. 15.21:* adapted from NASA diagram. *Fig. 15.22:* from William Kaufmann, *The Cosmic Frontiers of General Relativity,* p. 228. Copyright © 1977 by Little, Brown and Company (Inc.). Redrawn by permission.

Chapter 16

Page 342: National Gallery, London. *Fig. 16.1:* Lund Observatory. *Fig. 16.2:* Hale Observatories. *Fig. 16.3:* Lick Observatory. *Fig. 16.4:* The Kitt Peak National Observatory. *Fig. 16.5:* Lowell Observatory. *Fig. 16.6:* redrawn from *Astrophysical Journal;* from a map prepared by Dennis Downes, Alan Maxwell, and M. L. Meeks. *Page 353:* Hale Observatories. *Fig. 16.12:* from William K. Hartmann, *Astronomy: The Cosmic Journey,* p. 362. © 1978 by Wadsworth Publishing Company, Inc., Belmont, California 94002. Reprinted by permission of the publisher. *Fig. 16.13:* courtesy G. Westerhout. *Fig. 16.15:* Huberto Gerola, Philip Seiden. *Fig. 16.16:* redrawn from *Astrophysical Journal;* from a map prepared by Dennis Downes, Alan Maxwell, and M. L. Meeks. *Fig. 16.18:* from John Archibald Wheeler, "The Universe as Home for Man," *American Scientist,* vol. 62 (1974), p. 684. Redrawn by permission. *Fig. 16.19:* from F. W. Stecker, *American Scientist* (Sept./Oct. 1978), p. 574. Redrawn by permission.

Chapter 17

Page 371: Hale Observatories. *Figs. 17.2–17.4:* Hale Observatories. *Fig. 17.5:* European Southern Observatory. *Fig. 17.6:* Hale Observatories. *Fig. 17.7:* Dr. B. A. Vorontsov-Velyaminov, Sternberg Astronomical Institute, Moscow University. *Fig. 17.8:* photograph from Hale Observatories; drawings redrawn by permission of the publisher from "Violent Tides between Galaxies" by A. Toomre and J. Toomre, *Scientific American,* December 1973. © Scientific American, Inc. All rights reserved. *Fig. 17.10:* from M. J. Geller, "Large-Scale Structure in the Universe," *American Scientist,* vol. 66 (March 1978), p. 176. Redrawn by permission. *Fig. 17.11:* Hale Observatories. *Fig. 17.12:* from William K. Hartmann, *Astronomy: The Cosmic Journey,* p. 382. © 1978 by Wadsworth Publishing Company, Inc., Belmont, California 94002. Reprinted by permission of the publisher. *Fig. 17.13:* by permission, Harvard College Observatory, Cambridge, Mass. *Fig. 17.14 and 17.15:* The Kitt Peak National Observatory. *Fig. 17.16:* Hale Observatories. *Fig. 17.18:* Hale Observatories. *Fig. 17.19:* top, from left to right, Hale Observatories, Hale Observatories, National Radio Astronomy Observatory; middle left, Hale Observatories (inset), Lick Observatory; bottom left, Hale Observatories; bottom right, by permission of the Publisher, from Wellington, Miley and van der Laan, *Nature,* vol. 244 (1974), p. 502. Redrawn by permission. Photograph by Palomar Sky Survey. *Page 395:* from George Gamow, *A Star Called the Sun.* Copyright © 1964 by George Gamow. Reprinted by permission of Viking Penguin Inc. *Fig. 17.20:* Hale Observatories. *Fig. 17.21:* Lick Observatory. *Fig. 17.22:* Dr. Alan Stockton, Institute for Astronomy, University of Hawaii.

Chapter 18

Page 401: courtesy Allen Sandage, Hale Observatories. *Page 409:* Hale Observatories. *Fig. 18.6:* redrawn by permission of The University of Chicago Press, from a diagram by Allen Sandage. © 1972 by The American Astronomical Society. All rights reserved.

Chapter 19

Page 419: courtesy of The Boston Museum of Science, and permission of Harvard College Observatory, Cambridge, Mass. *Fig. 19.1:* Hale Observatories. *Fig. 19.2:* redrawn by permission of the publisher from "Life outside the Solar System" by Su-Shu Huang, *Scientific American,* September 1960. © Scientific American, Inc. All rights reserved. *Fig. 19.3:* Sproul Observatory, Swarthmore College. *Fig. 19.4:* From *Sky and Telescope* (March 1979), p. 247. Redrawn by permission of Sky and Telescope and Sproul Observatory. *Fig. 19.5:* NASA, Ames Research Center. *Fig. 19.6:* adapted from *Project Cyclops,* p. 41, NASA, Ames Research Center. *Fig. 19.7:* reprinted by permission of the publisher from "The Search for Extraterrestrial Intelligence" by Carl Sagan and Frank Drake, *Scientific American,* May 1975. © Scientific American, Inc. All rights reserved. *Fig. 19.8 and 19.9:* NASA, Ames Research Center. *Fig. 19.11:* NASA.

Index

Numbers in italics *refer to tables and figures.*

velocity of, 43, 44, 53–54, 402
visible, 55
wave properties of, 52, 53–54, 56–58, 65
Light curve, 256, 261
Light year, 48, 245
Liquid, 68
Lithium, 414
Lithosphere, 102, 105
Local Doppler hypothesis, 397
Local Group, 380, 381–385, *382, 384,* 387, 399
Local supercluster, 385
Lockyer, Sir Norman, 52
Longitude, Galactic, 350–352, *351*
Long period variables, 258
Loop Nebula, 304, *305*
Lorentz contraction factor, 44, 44n, *44*
Lowell, Percival, 156, 157, 191
Low mass stars, 314, 315–321
Luminosity, 216, 250, 265, 270, 283
classes, 255, 288–289
function, 272
and variability, 258, 265
Lunar rocks, 114, 121, 123, 126, 127
Luther, Martin, 27
Lyman series, *71, 72,* 95, 95n

M11, 317
M31. *See* Andromeda galaxy
M33, 382, *383*
M41, 317
M67, 317–318
M81, *392,* 394
M82, *375, 392,* 394
M84, *386*
M86, *386*
M87 (Virgo A), 391, *392–393,* 394
M92, *319*
M101, 357, *357*
Mach, Ernst, 35
McMath Solar Telescope, *221*
Macrospicules, 233
Magellan, Ferdinand, 382
Magellanic Clouds, 302, 347, 373, 382–383, *385*
Magnetic field, 53, *53,* 137
of earth, 107–108, *107,* 126
of Mercury, 151
of moon, 123
of nebulae, 306
of neutron stars, 331, *332*
and radiation, 181, *181*
of stars, 255
of sun, 227–228, 233, 237, 239
Magnetometers, 137
Magnetopause, *107*
Magnetosphere

of earth, 107–108, *107,* 126, 233–234
of Jupiter, 180–182
Magnitude
absolute, 248, 268
apparent, 247–248, *250,* 412
bolometric, *250,* 412
color, 249
median, 258
visual, *248,* 249
Main sequence, 255, 264, 276, 314, 422
definition of, 268–270, 272
and stellar evolution, 285–288, *287,* 310–311
Major axis of ellipse, *30,* 31
Malthus, Thomas, 208
Mantle
of earth, 99, 101, *101,* 102, *102,* 104, 126
of moon, 124, *125*
Marconi, Guglielmo, 428
Mare Imbrium, *118,* 120, 123
Mare Orientale, *122*
Maria
gravitational anomaly of, 125
of Mercury, 137
of moon, 116, *116, 117,* 120, 121, 123, 124, 137
Mariner 2, 153
Mariner 5, 153
Mariner 7, 165
Mariner 9, 158, 160, 210
Mariner 10, 149, *150,* 151, *151,* 156, 157
Mars, 2, 26, 95, 102, 146, 156–168, 173
atmosphere of, 141, 157–161, *160,* 174
canals of, 157, 163
composition of, 129, 163
crust of, *164,* 166–167
dust storms on, 160
internal structure of, *140*
life on, 142, 146, 209, 210–212, 213
maps of, 136, *158–159*
mascons on, 167
polar caps of, 156, 160, *160,* 165
properties of, 129, *133,* 156, 160
quakes on, 166
satellites of, *134,* 167–168
seasons on, 156–157
space missions to, 136, 137, *138,* 156–157
terrain of, 95, 137, 140, *158–159,* 162–167, *163*
water on, 160–161, 166
Mars and Its Canals (Lowell), 157
Mars as the Abode of Life (Lowell), 157
Mascons, 125, 167
Masers, 298–299
Mass, 73
of atomic particles, 65
definition of, 34

equivalence of energy and, 281–282, 285, 287
of galaxies, 272–273
inertial, vs. gravitational, 42, 44–45
Lorentz's formula for, 44, *44*
luminous vs. dynamic, 386
and space curvature, 43, 46
of stars, 262, 263–264, 265, *326*
Mass-luminosity relation, 263–264, *264,* 281, 285, 287, 381, 394
Mass transfer in binaries, 336–337
Mathematical models
of massive stars, 325
of stars, 283, 289
of sun, *284,* 287
of white dwarfs, 321
Matter
distribution of, 417
dual nature of, 73
in evolution of universe, 414
states of, 68
Matter era, 415
Mauna Kea Observatory, 93, *93*
Maunder, E. Walter, 242
Maunder minimum, 242
Maxwell, James Clerk, 53, 73
Mayall reflector telescope, *82*
Mechanical forces, 34
Mercury, 2, 24–26, 130, 146, 149–152, 173, 174
atmosphere of, 140, 141, 151
composition and properties of, 129, *133*
internal structure of, *2, 140,* 151–152
life on, 212
maps of, 136
orbit of, 21–22, 130, 132, *133,* 142, 149
period of rotation of, 136, 149
terrain of, 102, 137, 140, 149–151, *150, 151*
Mesopotamians, 15
Mesosphere, 108
Messier, Charles, 346, 346n
Meteorites, 121, 135, 170, 171–173, 174
dating of, 100, 172
isotropic anomalies in, 146
organic molecules in, 209, *210,* 420
Meteoroids, 2, 121, 135, 170
Meteors, 135, 170–171, 174
showers, 170–171, *171*
telescopic, 170
visual, 170
Michelson, Albert, 43, 252
Michelson-Morley experiment, 43, 44, 411
Microdensitometer tracing, *256*
Micron, 56
Microwaves, 56, 409, 410–411, *411,* 418
window, 429, *430*
Mid-Atlantic Ridge, 102